RNA AND DNA EDITING

RNA AND DNA EDITING

Molecular Mechanisms and Their Integration into Biological Systems

Edited by

HAROLD C. SMITH

Department of Biochemistry
and Biophysics University of Rochester

WILEY-INTERSCIENCE

A JOHN WILEY & SONS, INC., PUBLICATION

Published by John Wiley & Sons, Inc., Hoboken, New Jersey
Published simultaneously in Canada

Library of Congress Cataloging-in-Publication Data:

RNA and DNA editing : molecular mechanisms and their integration into biological systems /
[edited by] Harold C. Smith.
p. ; cm.
Includes bibliographical references and index.
ISBN 978-0-470-10991-5 (cloth)
1. RNA editing. 2. DNA. 3. Nucleotide sequence. 4. Genetic transcription. I. Smith, Harold C.
[DNLM: 1. RNA Editing. 2. Base Sequence. 3. Genomics. 4. Transcription, Genetic.
QU 475 R6265 2008]
QH450.25.R57 2008
572.8'8–dc22

2007039289

Printed in the United States of America
10 9 8 7 6 5 4 3 2 1

CONTENTS

PREFACE

IN 1986, Rob Benne's research group published their finding of a posttranscriptional process in which mitochondrial messenger RNAs were altered by uridine insertions and deletions, a process he referred to as RNA editing. The finding explained a paradox that the mitochondrial genome of protozoa such as Trypanosomes encoded a scarcity of proteins and many of the genes appeared to have disrupted open reading frames or lacked a translational start codon. Benne's publication took the scientific community by surprise. The known mechanisms for nucleotide modification in RNA and alternative mRNA splicing could simply not accommodate the finding that Trypanosoma mitochondrial mRNAs contained multiple insertions of one or more non-genomically encoded uridines with no apparent consensus flanking sequence at the sites of insertion.

By the early 1990s, several forms of insertion/deletion and base modification editing had been described in amoeba, flagellates, Physarum, mammalian viruses, plants, and the kidney, intestine, liver, and neuronal tissues of mammals. However, many in the scientific community remained unaware of this emerging frontier and the sporadic nature of the identification of editing in different organisms, tissues, and organelles, and the diversity of editing mechanisms led others to question the significance that editing mechanisms would have in understanding cellular systems. For these early years the field collectively had an orphan status, finding outlets for its new discoveries largely in "catch-all" sessions at diverse scientific society meetings.

Beginning in 1994, RNA editing realized solidarity through three international conferences on RNA editing and modification organized independently by Harold Smith and Steve Hajduk (1994, Albany Conference, Rensselaerville, NY, USA), Glenn Bjork, Ted Maden, and Henri Grosjean (1994, EMBO Workshop, Aussois, France), and Paul Sloof and Rob Benne (1996, EMBO Workshop, Maastricht, The Netherlands). The first text dedicated to the topic of RNA editing was edited by Rob Benne in 1993.* The inaugural Gordon Research Conference dedicated to RNA editing and modification was led by Smith and co-chaired by Jonatha Gott and Maureen Hanson in 1997. By 1998, many of the RNA editing systems that are known today had been identified, and it was at this time that Grosjean and Benne co-edited a comprehensive text on RNA modification and editing.† The field has grown rapidly and gathered momentum as we learn how RNA and DNA editing mechanisms influence, and are influenced by, other biochemical pathways in the cell.

***RNA Editing: The Alteration of Protein Coding Sequences of RNA*, Benne, R. (ed.), Ellis Horwood Series in Molecular Biology, Prentice Hall, Englewood Cliffs, NJ, 1993.

†*Modification and Editing of RNA,* Grosjean, H., and Benne, R. (eds.), *American Society of Microbiology Press, Washington, D.C., 1997.*

The topic of this book is RNA and DNA editing. The chapters were written from the perspective of the *next generation* of investigators who were formerly trainees in the field or have been newly drawn to it. The authors suggest open questions to pursue while evaluating the context of discoveries and methodologies that have led researchers to this threshold. The vitality of this text lies in its cutting-edge perspective and in its fresh introspective treatment of the progress to date. The target audience of the book are not only the aficionados of the field, but also academics and members of the private sector who are seeking to learn about the field and explore its new applications.

RNA editing is a process in which the nucleotide sequence of RNA is altered from the genomic code. Editing is accomplished through nucleotide insertion, nucleotide deletion, or base modification. It is distinguished from other forms of RNA modification in that the consequence of RNA editing is a change in the diversity and/or abundance of proteins expressed in the proteomes of organisms, their tissues, or organelles. RNA modifications that diversify RNA function or produce a gain or loss of RNA function are also considered editing. Within this rubric, numerous alterations to nucleotides have been documented affecting coding and noncoding sequences of messenger RNAs (mRNAs) as well as transfer RNAs (tRNAs), ribosomal RNAs (rRNAs), and spliceosomal RNAs (UsnRNAs).

As might be anticipated, coordination of editing activity is essential relative to other cellular pathways involving RNA such as transcription, RNA processing, and translation. Our appreciation of this regulation has grown through the characterization of the biological occurrence of RNA editing and the macromolecules that contribute to editing mechanisms. In this regard, the factors involved in RNA substrate recognition and catalysis are diverse, ranging from lone enzymes with both substrate recognition and catalytic activity to macromolecular complexes containing both protein and small RNAs as guides for substrate recognition and multiple proteins to carry out and coordinate editing activity. In A-to-I and C-to-U base modification editing, one editing factor or editosome serves multiple sites. In other systems, such as plant organellar C-to-U editing and organellar guide RNA-dependent mRNA, UsnRNA, and rRNA editing and modification, there is more complexity and a large number of site-specific editing factors.

A recent development in the field is the identification of select members in the family of cytidine deaminase editing enzymes that use single-stranded DNA as a substrate. DNA editing is mutagenic and is responsible for diversification of the genomic coding capacity for immunoglobulins and also serves in antiviral host defense. Another exciting discovery is that A-to-I RNA editing can regulate the production of interference RNA (RNAi) and thereby may constitute an important cellular mechanism for modulating the abundance of individual sequences within the transcriptome. A-to-I RNA editing also can modulate gene silencing through RNAi-dependent regulation of the specificity and activity of the machinery involved in DNA and histone modification, leading to chromatin remodeling.

Given these considerations, RNA and DNA editing will be discussed in four thematic areas to provide a contextual map for this field. Part I, "Diversification of the Proteome through RNA and DNA Editing," highlights how editing regulates protein expression through A-to-I base modification of mRNA, dC-to-dU modification of immunoglobulin genes for somatic hypermutation and immunoglobulin class switch recombination, guide RNA-dependent uridine insertion and deletion editing of mitochondrial RNA, and C-to-U and U-to-C mRNA editing in plant chloroplast and mitochondrial transcripts. This chapter explores the question "Why are nucleic acid sequences edited instead of encoded genomically?" through discussion of the occurrence of editing sites within transcriptomes and their distribution within individual RNAs. Depending on the biological system, editing can be seen through the lens of diversification, repair, or mutagenesis. Paramount in these discussions are mechanisms that govern RNA editing site selectivity and specificity and restrict the chromosomal domains targeted for DNA editing. Regulation depends on the temporal control of site-specific editing factor expression, subcellular localization, their interaction with nucleic acids, and the composition of individual editosomes. The reader will appreciate how diversity in *cis*- and *trans*-acting factors in different species, or in different organelles within the same species, contributes to different patterns of editing activity and thereby enables plasticity in each biological system.

Part II, "Functional Coordination of RNA Editing with Other Cellular Mechanisms," brings to the forefront why RNA and DNA editing is essential for cell survival and adaptation. This section profiles base modification of RNA and DNA and guided RNA editing as examples where cells require editing to produce functional tRNAs, process rRNA, splice pre-mRNA, regulate the stability of mRNAs, and control RNAi and viral infectivity. In some instances, editing at different sites within the same RNA is interdependent and requires coordination of the activity of different editosomes or transport of editing enzymes or their substrates within the cell and its organelles. In other examples, RNA editing site selectivity is coordinated through the interaction of A-to-I editing enzymes with the C-terminus or RNA polymerase II. In this way, editing factors have immediate access to nascent transcripts and can carry out editing before pre-mRNA splicing deletes introns that participate in RNA secondary structure necessary for editing site recognition. Transcription also makes available single-stranded DNA within chromosomes that can be targeted for mutational DNA editing leading to diversification of the genomic sequences encoding the variable regions of antibodies (as described in the prior section). Reverse transcription, coupled to RNase H activity, also regulates editing activity by exposing single-stranded viral DNA during replication for mutagenic DNA editing as a form of host defense.

The global role of RNA editing in cellular regulation is emphasized in this section of the book through several examples. Modification editing of U2 spliceosomal RNA is essential for U2-snRNP splice site binding specificity and spliceosome activity. The stability of select mRNAs is affected by binding of the factors responsible for C-to-U mRNA editing in mammalian cells to AU-rich elements in mRNA. And, modulation of RNAi production by A-to-I RNA editing is described as a mechanism for regulating gene silencing by affecting the specificity and activity of the enzymes that carry out DNA and histone modifications. The exquisite level of integration of editing with other biochemical pathways and cellular functions described in all of the chapters will lead the reader to the inescapable conclusion that RNA and DNA editing have significant roles in biology that includes, and goes well beyond, codon sequence changes and reading frame alterations.

A long-sought goal in the field has been to use our understanding of editing sites and editing factors to discover novel editing substrates and new biological roles for editing. Part III, "Predictive Studies," underscores the power of computational approaches in identifying novel editing sites and predicting the biological consequences of editing at these sites. Historically, computational analyses have been used sporadically to validate sequences as having been edited; however, computational methods have developed to the point where comparative sequence analyses enable genome wide predictions of edited mRNA sequences. Computational approaches have also advanced comparative phylogenic analyses of edited sequences. These studies have provided unique insights into the origins of editing systems, their evolution, and an understanding of the conserved, minimally essential functional domains within editing factors.

Highly related to these discussions is Part IV, "Structural Approaches," which is the final section of the text. Structural biology is an enabling technology for basic science, biomedical research, and drug development. The structural basis for function is more conserved in many instances than is primary nucleotide or amino acid sequence. Comparative structural analyses have been vital in predicting RNA secondary structure of the substrates for A-to-I editing and guide RNAs as well as the functional folds within enzymes in both A-to-I and C-to-U families of deaminases. Comparative structural analysis suggests conserved protein folds and implicate, in some instances, ancient phylogenic origins for components of editing machinery. Importantly, computational and structural studies suggest the reaction chemistry that enzymes catalyze, and they aim to predict the physical constraints in macromolecules that determine substrate and editing site specificity.

The selection of chapters and organization of the book was conceived with multiple purposes in mind. The text serves as a reference for background information in the field. It provides an opportunity for the newest contributors to the editing field to express their vision for the future. The perspectives voiced by these authors are anticipated to be provocative and are intended to motivate discussion, lead to new experiments, and promote collaboration. Finally, this book is intended to promote new hypotheses and models to springboard the next generation of discovery in the field.

Harold C. Smith

ACKNOWLEDGMENTS

THE EDITOR is grateful to the authors whose expertise and perspectives are reflected in the sixteen chapters of this text. The editor is also grateful to colleagues and government and private funding agencies for their contributions to the vitality and growth of the editing field. Special thanks are due to Drs. Juan Alfonzo, Ryan Bennett, Andrea Bottaro, Bernie Brown, Gordon Carmichael, Nic Davidson, Donna Driscoll, Rubin Harris, Celeste MacElrevey, David Mathews, Stu Maxwell, Tom Meier, Mike Mulligan, Marie Öhman, Eric Phizicky, Laurie Read, Larry Simpson, Joseph E. Wedekind and Yi-Tao Yu for critical reading of portions of the text and helpful suggestions.

CONTRIBUTORS

Juan D. Alfonzo, Department of Microbiology, the Ohio State University, Columbus, OH 43210. Email: alfonzo.1@osu.edu

M. Wadud Bhuiya, Department of Chemistry, Wake Forest University, Winston-Salem, NC 27109. *Present address*: Department of Biology, Brookhaven National Laboratory, Upton, NY 11973. Email: bhuiya@bnl.gov

Valerie Blanc, Department of Medicine, Washington University School of Medicine, St. Louis, MO 63110. Email: vblance@IM.WUSTL.edu

Bernard A. Brown II, Department of Chemistry, Wake Forest University, Winston-Salem, NC, 27109. *Present address*: Department of Molecular and Structural Biochemistry, North Carolina State University, Raleigh, NC 27695. Email: brownba2@gmail.com

Gordon G. Carmichael, Department of Genetics and Developmental Biology, University of Connecticut Health Center, Farmington, CT 06030. Email: carmichael@nso2.uchc.edu

Ling-Ling Chen, Department of Genetics and Developmental Biology, University of Connecticut Health Center, Farmington, CT 06030. Email: LLchen@student.uchc.edu

Jorge Cruz-Reyes, Department of Biochemistry and Biophysics, Texas A&M University, College Station, TX 77843. Email: cruzrey@tamu.edu

Nicholas O. Davidson, Department of Medicine, Washington University School of Medicine, St. Louis, MO 63110. Email: nod@im.wustl.edu

Zachary L. Demorest, Department of Biochemistry, Molecular Biology and Biophysics, Minneapolis, MN 55455. Email: demo0034@umn.edu

Dylan E. Dupuis, Department of Biological Sciences, Lehigh University, Bethlehem, PA 18015.

Eli Eisenberg, School of Physics and Astronomy, Tel Aviv University, Tel Aviv 69978, Israel. Email: elieis@post.tau.ac.il

Keith Gagnon, Department of Structural and Molecular Biochemistry, North Carolina State University, Raleigh, NC 27695. Email: keith_gagnon@ncsu.edu

Willemijn M. Gommans, Department of Biological Sciences, Lehigh University, Bethlehem, PA 18015.

Michael W. Gray, Program in Evolutionary Biology, Canadian Institute for Advanced Research, Department of Biochemistry and Molecular Biology, Dalhousie University, Halifax, NS B3H 1X5, Canada. Email: m.w.gray@dal.ca

Reuben S. Harris, Department of Biochemistry, Molecular Biology and Biophysics, University of Minnesota, Minneapolis, MN 55455.
Email: rsh@umn.edu

Alfredo Hernandez, Department of Biochemistry and Biophysics, Texas A&M University, 2128 TAMU College Station, TX 77843. Email: ahernandez@tamu.edu

John Karijolich, Department of Biochemistry and Biophysics, University of Rochester School of Medicine and Dentistry, Rochester, NY 14642.
Email: john_karijolich@urmc.rochester.edu

Erez Y. Levanon, Department of Genetics, Harvard Medical School, Boston, MA 02115. Email: elevanon@genetics.med.harvard.edu

Senjie Lin, Department of Marine Sciences, University of Connecticut, Groton, CT 06340. Email: senjie.lin@uconn.edu

Stefan Maas, Department of Biological Sciences, Lehigh University, Bethlehem, PA 18015. Email: swm3@lehigh.edu

Celeste MacElrevey, Department of Biochemistry and Biophysics, University of Rochester School of Medicine and Dentistry, Rochester, NY 14642.
Email: Celeste_MacElrevey@urmc.rochester.edu

Donna A. MacDuff, Department of Biochemistry, Molecular Biology and Biophysics, Minneapolis, MN 55455. Email: macd0110@umn.edu

Jill E. McCane, Department of Biological Sciences, Lehigh University, Bethlehem, PA 18015.

E. Stuart Maxwell, Department of Structural and Molecular Biochemistry, North Carolina State University, Raleigh, NC 27695. Email: tu_maxwell@ncsu.edu

U. Thomas Meier, Department of Anatomy and Structural Biology, Albert Einstein College of Medicine, Bronx, NY 10461. Email: meier@aecom.yu.edu

Jun-ichi Obokata, Center for Gene Research, Nagoya University, Nagoya, 464-8602, Japan. Email: obokata@gene.nagoya-u.ac.jp

Steven M. Offer, Department Biochemistry, Molecular Biology and Biophysics, Minneapolis, MN 55455. Email: offer014@umn.edu

Marie Öhman, Department of Molecular Biology and Functional Genomics, Stockholm University, S-106 91 Stockholm, Sweden.
Email: marie.ohman@molbio.su.se

F. Nina Papavasiliou, Laboratory of Lyphocyte Biology, the Rockefeller University, New York, NY 10021. Email: papavasilou@mail.rockefeller.edu

Evan M. Ritter, Department of Chemistry, Wake Forest University, Winston-Salem, NC 27109. *Present address*: 504 Murdocksville Road, West End, NC, 27376. Email: ritterem@gmail.com

Ann M. Sheehy, Department of Biology, College of the Holy Cross, Worcester, MA 01610. Email: asheehy@holycross.edu

Toshiharu Shikanai, Graduate School of Agriculture, Kyushu University, Fukuoka 812-8581, Japan. Email: shikanai@agr.kyushu-u.ac.jp

David Stephenson, Department of Biochemistry and Biophysics, University of Rochester School of Medicine and Dentistry, Rochester, NY 14642. Email: david_stephenson@urmc.rochester.edu

Nicholas E. Tatalias, Department of Biological Sciences, Lehigh University, Bethlehem, PA 18015.

Joseph E. Wedekind, Department of Biochemistry and Biophysics, University of Rochester School of Medicine and Dentistry, Rochester, NY 14642. Email: joseph_wedekind@urmc.rochester.edu

Yi-Tao Yu, Department of Biochemistry and Biophysics, University of Rochester School of Medicine and Dentistry, Rochester, NY 14642. Email: yitao_yu@urmc.rochester.edu

Huan Zhang, Department of Marine Sciences, University of Connecticut, Groton, CT 06340. Email: huan.zhang@uconn.edu

Xinxin Zhang, Department of Molecular and Structural Biochemistry, North Carolina State University, Raleigh, NC 27695. Email: xinxin_Zhang@ncsu.edu

Jing Zhou, Department of Genetics and Developmental Biology, University of Connecticut Health Center, Farmington, CT 06030. Email: zhou@student.uchc.edu

PART I

DIVERSIFICATION OF THE PROTEOME THROUGH RNA AND DNA EDITING

CHAPTER 1

DIVERSIFYING EXON CODE THROUGH A-TO-I RNA EDITING

Willemijn M. Gommans
Dylan E. Dupuis
Jill E. McCane
Nicholas E. Tatalias
Stefan Maas

AN INCREASING number of gene transcripts are found to be subject to recoding by RNA editing. RNA-targeted recoding leads to the substitution of single amino acids in the resulting proteins with subtle or sometimes drastic impact on protein function. New strategies to search for edited genes in mammals have accelerated the discovery of new targets and promise to reveal the many roles of RNA editing in gene regulation.

1.1 INTRODUCTION AND BACKGROUND

According to the central dogma, protein-coding sequences in eukaryotic genomes directly predict the primary structure of the encoded protein. However, processes such as alternative splicing of exons result in the inclusion or omission of protein domains and subdomains and thereby substantially extend the repertoire of expressible protein variants (1). Often, the occurrence and extent of alternative splicing is not predictable from analysis of genomic DNA sequences. Other posttranscriptional RNA modifications also contribute to the complexity of the proteome. One such important mechanism is RNA editing by adenosine modification (2–4), where single adenosine bases are converted into inosine. Since inosine is interpreted by the translation machinery as guanosine (5), A-to-I modification often results in nonsynonymous codon changes leading to protein variants with single amino acid substitutions. To date, it is impossible to predict with reasonable confidence a recoding event in mRNA from analyzing genomic sequence data. In this chapter we are reviewing the current knowledge regarding the prevalence and consequences of A-to-I recoding events in eukaryotic transcripts and discuss recent strategies for identifying and characterizing

RNA and DNA Editing: Molecular Mechanisms and Their Integration into Biological Systems,
Edited by Harold C. Smith

recoding editing sites in translated sequences as well as A-to-I editing events in micro RNA transcripts.

1.1.1 Initial Discovery and Context of A-to-I RNA Editing and ADARs

It came as a surprise when in 1991 an A/G discrepancy between genomic and cDNA sequences of the mammalian glutamate-gated ion channel subunit GluR-2 (6) turned out to be due to an adenosine base modification on the RNA level. Editing of this adenosine nucleotide results in the conversion of a glutamine codon into a codon specifying arginine. In fact, this single nucleotide substitution turned out to dominantly regulate ion-permeability in heteromeric receptor molecules and up to today represents maybe the most significant, intriguing, and puzzling case of adenosine modification editing in mammals (see Section 1.2.3).

The initial discovery of adenosine-modification editing quickly led to the identification of several other cases of recoding in nervous-system-specific transcripts, such as additional GluRs (7, 8) and 5HT2C-R (9). In each case a single nucleotide change resulting in an amino acid substitution could be linked to a change in protein function. Since unedited and edited protein variants often are co-expressed in the same cells RNA editing was soon recognized as a potentially important mechanism to diversify genetic information with the ability to enhance the complexity of the eukaryotic transcriptome and proteome.

At the time that the editing event in GluR-2 mRNA was discovered, neither the cellular machinery responsible for this adenosine base substitution nor the molecular mechanism at play was known. The observed A-to-G change in the cloned cDNAs was thought to be a result of either an adenosine modification process that alters this purine into another purine base functionally equivalent to a guanosine, such as hypoxanthine or due to a mechanism that involves removal of the base or of the whole nucleotide followed by introduction of the guanosine.

Interestingly, there was an enzyme known for a long time that converts adenosine mononucleotides to hypoxanthin nucleotides (also termed inosine). This evolutionary conserved adenosine deaminase (ADA) mediates an important step in eukaryotic and prokaryotic nucleotide metabolism. The ADA enzyme is well-studied and has become an important therapeutic target as ADA deficiency leads to various types of immune disorders (10). ADA modifies adenosine mononucleotides employing a hydrolytic deamination mechanism. However, the enzyme is not active on adenosines present in the context of DNA or RNA molecules. In addition to the modification of mononucleotides by ADA, the modification of genomically encoded adenosines to inosines in transfer RNAs (tRNA) has long been known (for reference see 11) and represents a critical feature for the degeneracy of the genetic code (wobble base in the anticodon of several tRNAs). The reaction mechanism and enzyme responsible for generating the wobble base was only recently revealed (12) (see below).

More importantly, a few years before the discovery of adenosine modification editing in pre-mRNAs, a novel enzymatic activity had been discovered that specifically targets adenosines embedded in dsRNA molecules (13, 14). Initially, it was described as dsRNA unwinding activity, but the actual nature of the molecular process was soon identified as adenosine-to-inosine modifications through an analysis of reaction products (15).

With the establishment of *in vitro* systems for RNA editing based on glutamate receptor transcript minigenes and cellular extracts, the chemical mechanism of the observed A-to-G changes in mRNAs was also soon shown to be the result of A-to-I deamination, catalyzed by a zinc-dependent protein factor (16–18). Furthermore, the cis-acting features in editing targets were characterized, identifying the requirement for partial double-stranded (ds)RNA secondary structures but with no obvious primary sequence signatures (8, 19, 20). This clearly distinguished the A-to-I editing mechanism from the mammalian C-to-U deamination process that involves secondary structure elements in addition to a primary sequence motif (mooring sequence) that guides the RNA modification machinery (see Chapter 11). *In vitro* editing systems also accelerated the isolation and cloning of the first A-to-I RNA editing enzyme from mammals (21–23). It turned out that the responsible protein (initially termed dsRAD, or DRADA, which later was renamed ADAR1) had in fact been investigated by several laboratories as either an interferon-induced protein with potential antiviral functions (24) or as the dsRNA-specific A-to-I editing activity in mammalian cells (see Section 1.1.3) (25, 26). Cloning of the first mammalian ADAR (ADAR1) was followed by the identification of ADAR2 (27) and ADAR3 (28), as well as ADAR homologs in other vertebrates (29, 30), flies (31) and worms (32) (see Section 1.1.3). Also, related enzymes responsible for tRNA-specific A-to-I editing were cloned and characterized in several species (33–35).

The C-to-U editing enzyme (APOBEC1) is remotely sequence-related to the first adenosine-targeting editing enzyme ADAR1, and it is believed that APOBEC1 cytidine deaminase and the deaminase domain of ADARs may share a common ancestor gene (36, 37). Interestingly, neither ADAR1 nor APOBEC1 shows primary sequence homology to adenosine deaminase (ADA), and their predicted (ADAR1) or known (APOBEC1) three-dimensional structures also differ substantially from that of ADA even though the reaction mechanisms catalyzed by ADA, ADAR, and APOBEC1 are highly similar.

1.1.2 Important Cases of Recoding by A-to-I Modification in Pre-mRNA

The first mammalian editing events that were characterized affect several subunits of glutamate-gated ion channels (7, 8) and a prominent serotonin receptor subunit (9). These proteins were all found to be modulated in function by single or multiple site-selective adenosine modifications within their primary transcripts. Serendipity played a central role in the identification of these targets. Only recently, systematic screening methods designed to identify recoding events caused by A-to-I editing have been developed (38–41) leading to the identification of a few additional targets (see Table 1.1).

Overall, the notion that A-to-I RNA recoding editing may be particularly significant for the nervous system is supported by the preponderance of brain tissue-specific editing events. Particularly in the fly (*Drosophila melanogaster*), the

TABLE 1.1 A-to-I Editing in the Coding Regions of Mammalian, Invertebrate and Viral Genes

A. Mammalian Genes					
Function	Gene (Accession Number)	aa Substitution	ADAR[a]	Functional Impact	Ref.
Serotonin receptor	5-HT_{2c}R (NM_000868)	I156V, I156M, N158S, N158D, N158G, I160V	ADAR1 & ADAR2 (B, C and E-site), ADAR1 (A-site), ADAR2 (D-site)	Reduced efficacy G-protein coupling	9
Glutamate receptor	GluR-2 (NM_000826)	R763G Q606R	ADAR1 & ADAR2 ADAR2	Decreased Ca^{2+} permeability; alteration maturation and cellular trafficking; faster recovery from desensitization	8 6
Glutamate receptor	GluR-5 (NM_175611)	Q621R	?	Variation in ion permeability	6
Glutamate receptor	GluR-6 (NM_175768)	I567V, Y571C, Q621R	? (Y571C: ADAR2)	Variation in ion permeability	6
Glutamate receptor	GluR-3 (NM_000828)	R775G	ADAR1 & ADAR2	Faster recovery from desensitization	8
Glutamate receptor	GluR-4 (NM_000829)	R765G	ADAR1 & ADAR2	Faster recovery from desensitization	8
Potassium channel	$hK_v1.1$ (NM_000217)	I400V	ADAR2	Faster recovery from desensitization	41, 59
Unknown	BC10 (NM_006698)	Y2C, Q5R, K15R	?	?	39, 40
Cross-linking actin filaments	FLNA (NM_001456)	Q2333R	?	?	40
FMR1 interacting protein	CYFIP2 (NM_001037333)	K320E	?	?	40
Chloride channel	Gabra-3 (NM_000808)	I342M	ADAR1 & ADAR2	?	103
A-to-I editing enzyme	ADAR2[b] (NM_001033049)	Intronic editing leads to frameshift	ADAR2	Alternative splicing	86

B. Invertebrate and Viral Genes

Function	Gene (Accession number)	Organism	aa Substitution	Functional Impact	Ref.
Potassium channel	$SqK_v1.1$ (U50543)	Squid	12 recoding sites	Altered channel kinetics; reduced ability to form tetramers	114
Potassium channel	sq K_v2 (Y14390)	Squid	12 recoding sites	Altered channel kinetics (channel closure rate & altered slowest time constant)	115
Basic fibroblast growth factor	bFGF (X16627)	*Xenopus*	Hypermutation	Unknown	116
Editing enzyme	dADAR (AF208535)	*Drosophila*	S437G	Editing activity	31, 117
Sodium channel	Para (NM_001042816)	*Drosophila*	Q473R K1455R N1587S	Unknown	118
Glutamate-gated chloride channel	GluRIIE (CG31201)	*Drosophila*	I27V, K241R, N345S	Unknown	119
Voltage gated calcium channel	cac (NM_206693)	*Drosophila*	S514G, I815M, N839S, N906S, S937G, M1016V, N1185S, N1368G, N1580D, R1602G	Unknown	120
Amine receptor	DopEcR (CG18314)	*Drosophila*	I316V Stop323W	Unknown	104
Hydrogen-transporting two-sector ATPase	CG13167	*Drosophila*	I9V	Unknown	104
Protein phosphatase type 1, regulator	CG9619	*Drosophila*	S160G	Unknown	104
Transporter activity	Spinster (CG8428)	*Drosophila*	N67G	Unknown	104

(*Coninued*)

TABLE 1.1 (***Continued***)

Function	Gene (Accession number)	Organism	aa Substitution	Functional Impact	Ref.
YT521-B	CG12076	*Drosophila*	Q636R	Unknown	104
Unknown	Tetraspanin 33B (CG14936)	*Drosophila*	Two silent sites	Unknown	104
Unknown	4f-rnp	*Drosophila*	Hypermutation	Unknown	121
Sodium channel	DSCI (CG9071)	*Drosophila*	M1174V, I1199V only *D. pseudoobscura*)	Unknown	41
Potassium channel	Sh (CG12348)	*Drosophila*	K178E, K178G, K178R, I360M, I464V, T489A, Q491R	Unknown	41
Potassium channel	Eag (CG10952)	*Drosophila*	K467R, Y548C, N567D, K699R	Unknown	41
Potassium channel	Slo (CG10693)	*Drosophila*	N264D, S977G	Unknown	41
Calcium sensor	Syt (CG3139)	*Drosophila*	I365V, K377R, I381V, I403M	Unknown	41
SNARE binding	unc-13 (CG2999)	*Drosophila*	S2371G	Unknown	41
SNARE protein	cpx (CG32490)	*Drosophila*	I124M, N129D, N129G, N129S	Unknown	41
Unknown	stnB (CG40306)	*Drosophila*	T1186A	Unknown	41
Adaptor protein	lap (CG2520)	*Drosophila*	T372A	Unknown	41
nAChRα subunit	Dα5 (CG32975)	*Drosophila*	I504V, T553A, I554V, I558M	Unknown	41

nAChR subunit	Dα6	*Drosophila*	N133S, I134V, H138R, N139G, N139S, N139D, I156M, N187S	Unknown	122
nAChR subunit	Amelα6 (ortholog of Dα6)	Apis mellifera	N164S, K176R, I181M, T184A	Unknown	123
nAChRβ subunit	ARD (CG11348)	*Drosophila*	R56G, I73M	Unknown	41
nAChRβ subunit	SBD (CG6798)	*Drosophila*	T278A	Unknown	41
GABA receptor	Rdl (CG10537)	*Drosophila*	R122G, I283V, N295G, M360V	Unknown	41
GTPase	Rab26 (GH21984)	*Drosophila*	K365R	Unknown	38
Ral GTPase activator	Rlip (GH01995)	*Drosophila*	I229V, E230G, K233E, E254G, K265R	Unknown	38
Rab3 guanyl-nucleotide exchange factor	Rab3-GEF (HL01222)	*Drosophila*	Q2022R, S2054G	Unknown	
Promotes synaptic vesicle budding	endoA (GH12907)	*Drosophila*	K129R, K137E	Unknown	38
Synapsin	Syn (CG3985)	*Drosophila*	R20G	Reduced PKA phosphorylation *in vitro*	124
AP-2 subunit	α-Adaptin (RH30202)	*Drosophila*	T207A	Unknown	38
Kinesin-dependent axonal transport	Syd (GH19969)	*Drosophila*	S983G	Unknown	38

(*Coninued*)

TABLE 1.1 (*Continued*)

Function	Gene (Accession number)	Organism	aa Substitution	Functional Impact	Ref.
Ca^{2+} binding protein	Cpn (GH08002)	*Drosophila*	S402G	Unknown	38
K^+ dependent Na^+, Ca^{2+} antiporter	Nckx30C (HL01989)	*Drosophila*	K365R	Unknown	38
K^+ dependent Na^+, Ca^{2+} antiporter	CG1090 (GH23040)	*Drosophila*	S358G	Unknown	38
Na^+, K^+ exchanging ATPase	Atpα (GH23483)	*Drosophila*	Y390C	Unknown	38
Ca^{2+} binding, acyltransferase activity	CG32699 (HL01250)	*Drosophila*	I175M	Unknown	38
Trc kinase activator	Mob1 (RH70633)	*Drosophila*	N91D	Unknown	38
G-protein coupled receptor	Boss (GH10049)	*Drosophila*	T529A, T533A	Unknown	38
Potassium channel	SK (GH16664)	*Drosophila*	Y377C	Unknown	38
Chloride channel	CG31116 (GH23529)	*Drosophila*	K232R, T259A, K268R, E269G	Unknown	38
Actin nucleation factor	Spir (GH13327)	*Drosophila*	K339R	Unknown	38
Regulator of actin filament formation	Atx2 (GH01409)	*Drosophila*	K320R, K337R	Unknown	38
Structural constituent of cytoskeleton	CG32245 (GH04632)	*Drosophila*	N297D	Unknown	38
ATPase	CG32809 (GH23439)	*Drosophila*	K179R	Unknown	38
Phosphatidylinositol transporter	Retm (GH05975)	*Drosophila*	Q245R	Unknown	38
Unknown	CG1552 (GH14443)	*Drosophila*	K121R	Unknown	38
Unknown	CG31531 (GH25780)	*Drosophila*	K679E	Unknown	38

Unknown	CG3556 (GH17087)	*Drosophila*	I572V	Unknown	38
Unknown	CG9801 (GH23026)	*Drosophila*	S345G	Unknown	38
Unknown	I(1)G0196 (GH02989)	*Drosophila*	Q1148R, S1172G, Q1176R	Unknown	38
Unknown	CG12001 (HL01040)	*Drosophila*	I325V	Unknown	38
Unknown	CG30079 (HL05615)	*Drosophila*	I127M, T303A, Q343R, Q358R, S360G	Unknown	38
nAChR subunit	α7-2 (homolog to Dα6 sites)	*H. virescens*	N133S, N139G, N139S, N139D, I156M	Unknown	122
nAChR subunit	Mdalpha6 (ortholog of Dα6)	*Musca domestica*	N129S, I130V, H134R, N135S, N135D, N135G, N137S, I152M, N183S	Unknown	125
Viral replication	HDAg-p24 (AJ307077)	Hepatitis delta virus	Stop196W	Switch from viral replication to packaging	62

[a]Preferential editing enzyme for that target site

[b]Causes indirect change in amino acid sequence

large number of neuronal editing targets and the fact that the complete elimination of the A-to-I editing machinery results in a specific neurological phenotype (42) demonstrates a critical role of editing for neural function. However, in mammals, as well as in the fly, the list of non-neuronal recoding targets is steadily growing, though knowledge regarding the physiological significance of recoding editing affecting non-neuronal transcripts is largely lacking.

1.1.2.1 Mammalian Glutamate Receptor Subunits Ionotropic glutamate receptors (iGluRs) constitute an important class of neurotransmitter receptors in the central nervous system that mediate fast excitatory neurotransmission and have been implicated in mechanisms of plasticity, such as learning and memory (43). A total of five glutamate gated ion channel subunits have been shown to be recoded at single positions within their mRNAs affecting a total of eight codons (for recent detailed reviews on the sites and regulatory roles of ion-channel receptor editing, see reference 44 and references therein). Most significantly, the GluR-2 subunit is A-to-I edited at a critical position in the ion channel molecule, which constitutes the molecular determinant for Ca-permeability (6) and in addition regulates channel trafficking (45) and receptor assembly (46). Editing at this position will therefore influence all these processes. The editing event alters a glutamine (Q) codon (CAG) to a codon (CIG) specifying arginine (R). This Q/R site of editing is further remarkable in that virtually 100% of GluR-2 pre-mRNA molecules are edited and therefore almost no GluR-2(Q) protein is present in the brain. The physiological significance of this recoding event became evident when transgenic mice with impaired RNA editing function were engineered. This resulted in mice with a severe epileptic phenotype and premature death (47, 48). It was shown that the reduction in editing at the Q/R site from 100% to ~60% results in a drastically increased conductance and Ca^{2+}-influx into principal neurons. These alterations were directly responsible for the observed phenotype, because mice that carry a genomic mutation fixing the RNA editing event in the genome showed a wild-type phenotype despite the editing deficiency (48). Why is the critical arginine codon generated by almost complete editing of the Q/R site and not genomically specified? Currently, it cannot be ruled out that nonedited versions of the GluR-2 subunit have a function during early development or within specific neuronal cell types. However, that function is dispensable for survival and normal development as judged by the lack of a discernable phenotype in transgenic mice that cannot produce nonedited GluR-2 (49). A selective deficiency in GluR-2 Q/R site editing has been implicated in a number of pathological phenotypes in humans (see reference 50 for review), such as amyotrophic lateral sclerosis (ALS). In ALS, the ensuing increase in glutamate receptor Ca^{2+}-permeability of affected neurons due to a decreased editing activity may be responsible or a contributing factor to neuronal death (51).

Another editing site changing an arginine (R) codon into a glycine (G) codon is shared between GluR subunits GluR-2, -3, and -4. Here the single amino acid alteration regulates kinetic properties of the heteromeric receptor channel (8) and also modulates receptor biogenesis (52). The extent of editing at the R/G position varies between the different GluR subunits and between neuronal cell types. It also undergoes significant regulation during embryogenesis changing from low level editing extents during early embryonic stages to high levels in adult individuals (8). The glutamate receptor subunits GluR-5 and -6 are also edited at one or three sites (7), respectively. Here the recoding events modulate the ion-permeability and kinetic properties of the corresponding ion channels (44).

1.1.2.2 Serotonin Receptor Another prominent and well-studied example of A-to-I RNA editing is the serotonin receptor subtype 5-HT_{2C}, which is important for neuronal pathways influencing sensory and motor processes, as well as behaviors. The 5-HT_{2C} receptor is part of a G-protein-coupled transmembrane receptor that couples serotonin neurotransmitter action to intracellular signaling pathways. This mainly leads to the activation of phospholipase C, which, in turn, results in a rise in intracellular inositol phosphates and diacylglycerol. The latter elicits protein kinase activation and induces an increase in intracellular Ca^{2+} concentration. A-to-I RNA editing in 5-HT_{2C} transcripts affects five major sites, which are all located within the same second intracellular loop of the receptor protein (9). Overall, the higher the extent of modification by editing at these sites, the less sensitive the receptor becomes to serotonin activation, which is the result of a decreased G-protein coupling efficiency (53, 54). The 5-HT receptors have been implicated in the etiology of several neurological and behavioral disorders, such as depression, anxiety and schizophrenia. Intriguingly, changes in the RNA editing patterns of 5-HT_{2C} transcripts have been observed in brains of people that suffered from suicidal depression (55). Mice treated with fluoxetine (a serotonin uptake inhibitor) show the converse type of change in the RNA editing pattern of 5-HT_{2C} sequences. These data indicate that the editing extent at these modification sites may be able to change in response to external signals, such as different levels of synaptic serotonin (55). In agreement with these observations, treatment of cells with the cytokine IFN-alpha resulted in alterations in the editing pattern of 5-HT_{2C} mRNA, which may link the observed depression in patients undergoing cytokine therapy to fluctuations in editing activity (56). Recently, increasing evidence has linked changes in mood and behavior to alterations of serotonin receptor editing (for review see reference 57).

1.1.2.3 Kv1.1 Potassium Channel The mammalian Kv1 subfamily of potassium channels plays an essential role in membrane hyperpolarization during an action potential and in the propagation of action potentials along the plasma membrane (58). The tetrameric receptors form a diverse group of ion channels due to the existence of several subunits and also due to A-to-I RNA modification. Editing of the human Kv1.1 transcripts modulates the kinetics of channel inactivation (59). The editing event in the human Kv1.1 mRNA is related to the site of editing in *Drosophila melanogaster* shaker potassium channels and has independently evolved at the equivalent (analog) site in the *D. melanogaster* Shab potassium channel (59).

This more recently reported editing site in humans stands out because the pre-mRNA that undergoes modification does not contain introns, which means that the partially base-paired RNA fold-back structure is comprised entirely of exonic sequences (41). For each of the cases described above, the molecular determinants for site-selective and efficient editing involve a partially double-stranded RNA fold-back structure in the substrates that is formed between exonic sequences that surround the to-be-edited adenosine and partially complementary sequence elements in a neighboring intron (see also Section 1.1.3 below).

1.1.2.4 Additional Recoding Targets in Vertebrates, Invertebrates, and Viruses A number of additional recoding events have been reported in mammals (see Table 1.1 and Section 1.1.2), for which there are currently no experimental data available regarding the physiological impact of editing (38–41). Interestingly, they also include non-neuronal transcripts.

In addition, a few examples exist where recoding events are predicted from A/G discrepancies that are only detected in transcripts derived from pathological tissues or cells, such as cancer (prox1, PCNP) and lupus erythematosus (60).

Intriguingly, the hepatitis delta virus (HDV) utilizes the A-to-I editing machinery to regulate viral replication. Within the antigenome of this virus, a site-specific adenosine to inosine modification converts a stop codon into a tryptophan codon (61, 62). This leads to the expression of an HDV antigen variant that suppresses replication and enhances the late stages of the viral life cycle (61). It seems that the viral genome has evolved in a way to utilize the host cell's RNA editing machinery for productive replication.

To date, a total of 77 targets for A-to-I editing have been identified in the fruit fly *D. melanogaster*. Few of these targets, of which the majority are expressed specifically in neurons, have been directly investigated with respect to the consequences of editing for protein function.

1.1.3 Cis-Acting Features for A-to-I Editing

The requirement for a partially double-stranded RNA fold-back structure for editing was first established for the GluR-2 Q/R editing site through meticulous analysis of editing extents in minigene substrates that tested the validity of computer-predicted RNA secondary structures of GluR-2 pre-mRNA transcripts (19). The partially base-paired region in the RNA is formed between sequences flanking the to-be-edited adenosine and a partially complementary sequence [termed the editing site complementary sequence (ECS)], which is often located within a downstream or upstream intron (19, 63, 64). Mutations that weaken the predicted RNA fold surrounding the editing site strongly impair or abolish editing at the Q/R site, whereas other mutations that restore the structure boost the levels of site-selective modification (19). Also, the modified adenosine in the GluR-2 Q/R site structure is in a base-paired configuration and changing the base pair into a mismatch decreases editing efficiency (19). Although this seems to be a rather simple set of parameters determining what constitutes an editing substrate, the analogous analysis of RNA fold-back structures governing editing at other sites revealed that the process is much more complex as in some cases the to-be-edited adenosine may be mismatched (8) or may be part of a loop structure embedded in base-paired regions (64). Based on the available data to date, it is not possible to define structural or sequence requirements that would allow straightforward screening for edited genes in sequence or structure databases (see Section 1.2.2).

Apart from the requirement for a partially base-paired structure, both ADAR1 and ADAR2 show certain nearest-neighbor preferences. The ADAR1 enzyme preferentially targets adenosines that are preceded by U = A > C > G, and ADAR2 displays a 5′-preference of U = A > C = G as well as a 3′-neighbor preference of U = G > C = A (4, 65). These properties may be related to the reaction mechanism for deamination that is believed to involve a flipping of the adenosine into the enzyme's active site, similar to the mechanism of action employed by DNA methyltransferases (4). Certain sequence environments will make access for the enzyme to the to-be-edited adenosine easier. In summary, there is still an unclear picture with respect to the molecular parameters that determine substrate specificity and editing efficiency in natural recoding targets for A-to-I editing.

1.1.4 Properties of the A-to-I Editing Machinery

The family of mammalian ADAR proteins (ADAR1, ADAR2, and ADAR3) share a common general domain architecture (see Figure 1.1A) mainly comprised of two or three double-stranded RNA binding domains (dsRBDs) and a catalytic adenosine deaminase domain toward the C-terminus (for recent reviews on ADARs see references 66 and 67). Only ADAR1 and ADAR2 have been characterized functionally, whereas ADAR3, although closely sequence related to ADAR2, has not yet been assigned a function and does not display adenosine deaminase activity in established assay systems (28, 68). *In vitro* and in cellular assays, ADAR1 and ADAR2 exhibit site-selective and efficient RNA editing activity apparently without any proteinacious co-factor requirements (27, 69).

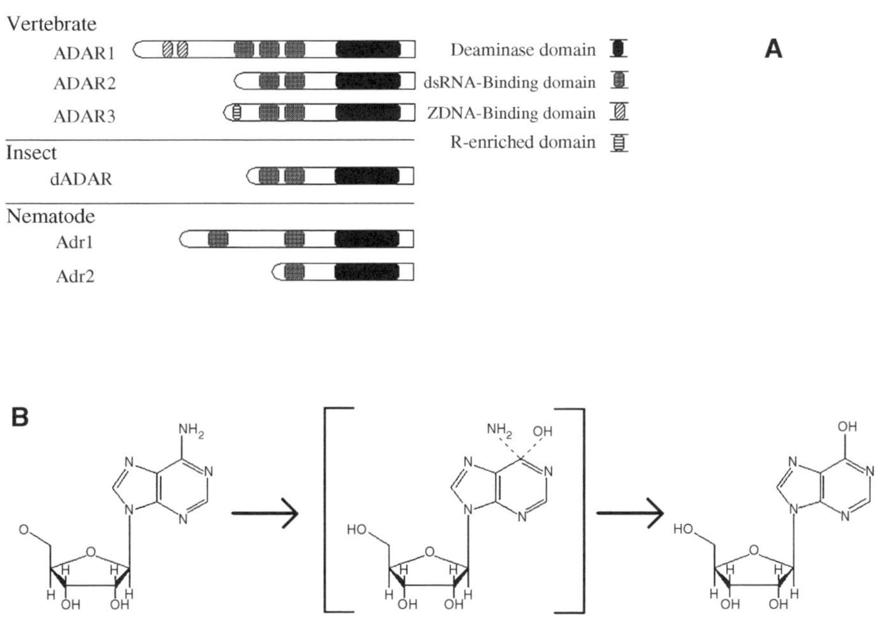

Figure 1.1 Molecular players and mechanism of A-to-I editing. (A) Schematic representation of ADAR domain structures from vertebrates, insects, and nematodes. (B) Depiction of adenosine hydrolytic deamination mechanism with transition state.

Zinc is known to be involved in the catalytic mechanism, where it activates a water molecule that initiates the nucleophilic attack at the C-6 of the targeted adenosine. However, recently it was shown that ADARs are active as dimers (70–73) and also that a small molecule inositol-6-phosphate (IP6) is an essential co-factor for function (74). A crystal structure of the catalytic domain of ADAR2 in conjunction with functional experiments demonstrated that IP6 is critical for protein folding and formation of the catalytic site (74). The mechanism of substrate recognition and site-selectivity of ADARs is not well understood, and ADAR1 and ADAR2 are known to display distinct, but overlapping specificity on known, physiological editing targets. For example, both enzymes seem to be highly active on the glutamate receptor R/G editing sites in GluR-2, -3, and -4. However, the Q/R site of GluR-B, the I/V site of the Kv1.1 potassium channel as well as the D site of the 5-HT_{2C} serotonin receptor are preferentially edited by ADAR2. The enzyme ADAR1 displays preference for the B site of 5-HT_{2C} as well as the amber/W site in the HDV antigenome (27, 56, 75, 76).

One intriguing property of ADARs is that when encountering an extended, perfectly double-stranded RNA molecule, ADARs will promiscuously deaminate up to $\sim$ 50% of all adenosines (77). Further deaminations are probably prevented due to the progressive loss of the substrate's double-strandedness such that the dsRBDs are unable to bind. This highlights that, intrinsically, the ADAR enzymes lack a particular site-selectivity but that most likely the overall three-dimensional shape and sequence environment of edited adenosines in the substrate RNAs provide the specificity seen in physiological recoding targets (4, 78). It has also been suggested that dsRBDs of ADARs may in some cases specifically interact with other structural RNA motifs such as a loop region, thereby mediating site-selective editing (79, 80).

Other organisms, such as insects (31) and nematodes (81), have A-to-I editing machineries in the form of a single ADAR enzyme (i.e., dADAR in *Drosophila melanogaster*) or a single heterodimeric adenosine deaminase targeting mRNAs (*C. elegans* adr-1 and adr-2). They harbor single dsRBDs, and their catalytic domain sequence is closely related to those of vertebrate ADARs. Plants, fungi, and yeast lack ADAR enzymes and RNA-directed adenosine deaminase activity.

Overall, the ADAR enzymes are ubiquitously expressed in most, if not all, cell types (21, 23, 27). An exception is ADAR3, which is detected only in the central nervous system (28). The diversity of ADARs is further enhanced through the expression of alternative splice variants (82–84) and the utilization of alternative promoters (85) and even by A-to-I RNA editing. Mammalian ADAR2 as well as *Drosophila* dADAR edit their own pre-mRNAs (31, 86). In the case of ADAR2, intronic editing creates a novel splice site that leads to expression of a truncated, nonfunctional enzyme (86). In dADAR, site-selective editing alters an amino acid (S-to-G substitution) that was shown to modulate enzymatic activity (31). Alternative promoter usage creates two main ADAR1 transcript variants that differ substantially with respect to subcellular localization and function (87). ADAR1-p110 is expressed from a constitutive promoter and is active primarily within the nucleus (85). Expression of ADAR1-p150 is driven by an interferon inducible promoter and generates an enzyme that harbors a unique N-terminal domain, which conveys specific binding affinity to DNA and RNA in Z-conformation (88) and is actively shuttled between the nucleus and cytoplasm (89). In contrast, both ADAR1-p110 and ADAR2 are dynamically associated with the nucleolus and might relocalize to the nucleoplasma when substrates for editing are expressed (90, 91). Moreover, ADAR2 exists in alternative splice forms that might differ in their RNA editing efficiencies or specificities (83, 84).

1.2 MAIN QUESTIONS IN THE FIELD AND APPROACHES

A critical aspect of understanding the overall impact of A-to-I RNA editing on the regulation of gene expression and for transcriptome and proteome diversity is to delineate all A-to-I editing targets followed by characterization of the functional consequences of editing. Until recently, only a relatively small number of ADAR substrates were known (4) and the A-to-I RNA editing targets seemed to be largely brain-specific. However, the editing machinery was shown to be functionally expressed in many cell types (21, 23, 27), indicating potential additional targets for these enzymes in other tissues. In fact, increasing evidence from biochemical studies further indicated that A-to-I editing occurs in other tissues as well (92) and that overall many more edited genes should exist (see Section 1.2.1).

Previously, ADAR substrates were discovered mainly by coincidence when comparing the cDNA sequence of a cloned gene to their genomic counterparts, in which an adenosine in the genomic sequence appeared as guanosine in the cDNA molecule. In particular, cases with high editing extents, such as the glutamate receptor GluR-2 Q/R site (>99.9% editing), have a higher chance of being detected, also enhanced by the fact that the ensuing amino acid substitution involves a functionally important residue. However, in recent years, several distinct and partially complementary screening methods have been developed to detect novel ADAR substrates.

1.2.1 Biochemical Versus Computational Approaches

The first technique that was developed to systematically screen for RNA molecules that have been modified by the A-to-I editing machinery involves a biochemical procedure to isolate inosine-containing mRNA molecules (93). In this approach, the inosine-containing RNA molecules are chemically modified and then preferentially cleaved at the phosphodiester bond 3′ to the inosine nucleotide. The reaction products are subsequently cloned and sequenced followed by analysis of full lengths cDNAs spanning the identified cleavage sites. This method was successful in that it led to the identification of initially five new cases of A-to-I editing in *C. elegans* mRNAs (94). Interestingly, all detected editing sites were present in the noncoding region of the RNA molecules, and this study therefore gave a first hint that editing might have additional functions besides the alteration of specific codons. In a follow-up study, five additional substrates in *C. elegans* were discovered, as well as 19 novel editing sites in transcripts from human brain (95). Similar to the previous study, all detected editing events were within noncoding regions and the editing of 15 out of the 19 discovered human RNA substrates occurred within transposon derived repeat elements. This finding led to the speculation that repetitive sequences may generally be frequent targets of the editing machinery.

The isolation and cloning of inosine-containing mRNAs using this technique is relatively laborious, and a high background of false positives makes it impractical to use for the comprehensive identification of all editing targets (94, 95).

The recent availability of complete genome sequences and annotations, including the human genome, has made it possible to conduct specific database searches to identify edited genes (see Figure 1.2 for schematic overview of screening methods to identify editing targets). The "smoking gun" of A-to-I RNA editing is an A/G discrepancy between a gene's cDNA sequence and its genomic counterpart. However, this feature is not sufficient to distinguish genuine RNA editing sites from A/G discrepancies that are the result of a single-nucleotide polymorphism (SNP), a sequencing error in the database, or a mutation introduced during cDNA cloning. Several laboratories embarked on the genome-wide computational screening for A/G discrepancies and through application of statistical methods were able to show that human Alu-type repeat elements present in mRNA sequences are a major target for A-to-I RNA editing (65, 96–98). Alu repeats are approximately 300 bp in length and consist of two monomers linked by an A-rich region. These repeats are highly abundant within the human genome and can occur both in forward and reverse orientation (for review see reference 99). If present within the same mRNA molecule, two oppositely oriented Alu elements can form a partially double-stranded structure generating a substrate for ADAR enzymes. In fact, when validating candidate Alu elements for editing *in vivo*, it turned out that if the two interacting repeat elements are within a few kilobases of one another, the occurrence of editing could always be

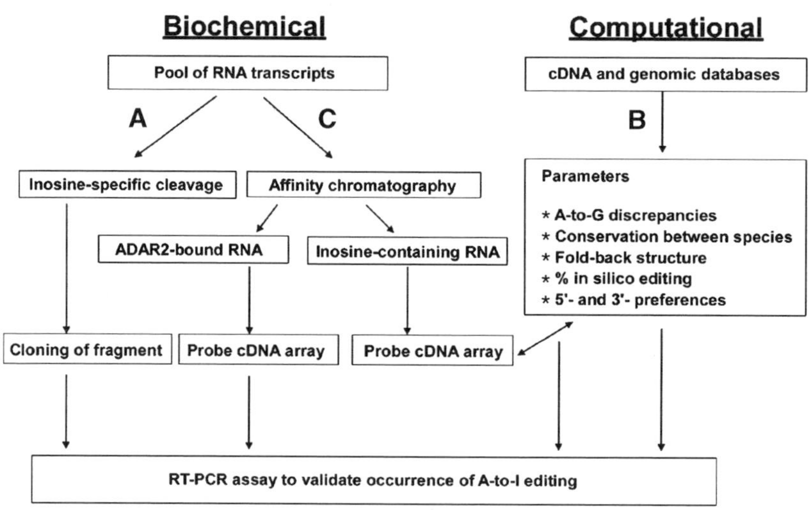

Figure 1.2 Screening strategies for identification of recoding targets. Flowchart outlining experimental strategies for delineation of editing events that lead to recoding. Approaches that use biochemical methods for the initial selection of candidate sequences (either through inosine-specific cleavage protocol or affinity chromatography using either ADAR or inosine-specific antibodies) are shown on one side. Approaches that start out by computationally filtering likely editing candidates are indicated on the right. See text for discussion of individual discussions.

confirmed experimentally (65). Between the different studies, several thousand mRNAs with a total of more than 20,000 editing sites were annotated. The main reasons for the only moderate level of overlap between the identified target sets are the different databases used for analysis (for example, either including or excluding EST sequences) and different stringencies when filtering the datasets. Since exonic Alu elements are almost always located within the noncoding regions of an mRNA molecule, it will be intriguing to see if Alu repeat editing may influence the stability, processing, or transport of mRNA molecules. In a few cases it was shown that Alu-mediated editing can destroy or create predicted splice sites (65), which represents another example of how A-to-I editing may regulate gene expression. Because Alu repeat elements are only present in primate genomes, repeat element editing in rodents was shown to occur at a much lower overall rate. This raises the intriguing question of whether editing has a specific role in primate evolution (see Chapter 13 for in-depth review on RNA editing in Alu-type repeats).

Nevertheless, rodent genomes contain other types of repetitive elements, which differ in the sequence composition (100) but which also can give rise to RNA foldback structures and subsequent A-to-I editing. In one case this has been shown to regulate the expression of a mRNA in mouse (101) (see also Chapter 13).

Alu-repeat-mediated editing targets belong to a class of editing substrates that are characterized by low site-selectivity and multiple site modification and are almost all localized in nontranslated sequences of mRNAs or introns. Figure 1.3 depicts the spectrum of currently known targets of A-to-I editing according to site-selectivity of modification, total rate of editing (inosine content), double-strandedness of the RNA fold-back structure, and the prevalence among identified substrates. Recoding targets for editing are located at the very end of the spectrum with the highest site-selectivity, the lowest relative content of inosine per transcript molecule, and the lowest double-strandedness of the substrate RNA.

The many editing events identified in noncoding mRNA sequences and within introns could explain the high amount of inosine that has been detected in mammalian

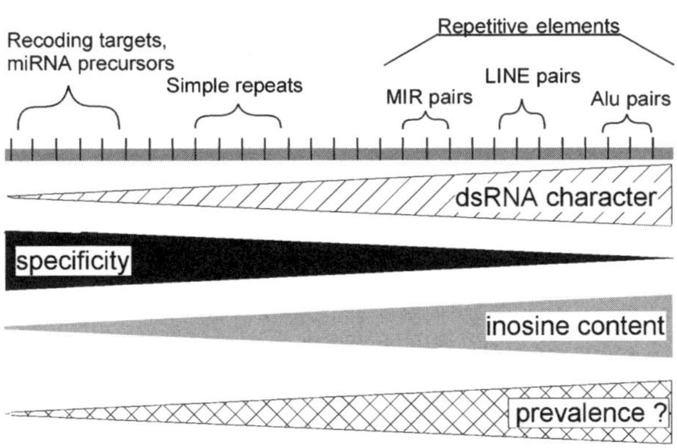

Figure 1.3 Spectrum of known A-to-I RNA editing targets. The known types of RNA editing targets are shown according to double-strandedness of the RNA fold directing editing, the site-selectivity of the editing event, total inosine content in the message, and the presumed prevalence relative to all known editing targets.

mRNA preparations (92). However, in light of the fact that rodents lack Alu elements, which constitute the main target of the RNA editing machinery in primates, a significant portion of the existing A-to-I editing events may still await discovery. If so, then site-selective targets of recoding may constitute a major fraction of these missing substrates.

Because of the predominance of inosine-containing RNAs that are edited in Alu repeats, experimental approaches to identify other sites of editing are challenging in primates. Ohlson et al. elegantly avoided this problem by using mouse brain tissue samples to selectively screen for ADAR2 specific substrates using affinity chromatography (102). For the detection of novel targets, ADAR2-specific antibodies were used to immunoprecipitate RNA molecules in complex with ADAR2. Reverse transcription and micro-array analysis of these co-precipitated RNA molecules resulted in the detection of up to 200 potential substrates (102), one of which, GABA(A) receptor subunit alpha3, has recently been verified *in vivo* (103).

In another approach to directly detect and isolate inosine-containing RNAs, an antibody was developed that selectively binds to inosine (104). Immunoprecipitated RNA molecules from wild-type and mutant (ADAR$^{-/-}$) flies were reverse transcribed and hybridized to a cDNA array. Comparison between wild-type and ADAR$^{-/-}$ cDNA array data led to the identification of 500 putative ADAR target genes (104). In addition, a database search was performed in which *Drosophila* cDNA sequences were compared to their genomic counterparts, as a genomically encoded adenosine will appear as guanosine in the expressed sequence after editing. This resulted in the detection of 800 genes that show an A/G discrepancy within the coding region. Ultimately, by comparing these two groups of putative editing targets, 62 genes were present in both groups. However, editing still has to be proven experimentally for most of these genes (104).

A prerequisite for editing is the presence of a double-stranded RNA structure. This suggests that the sequence surrounding the editing site may be conserved between species to preserve this RNA fold. Indeed, comparative genomics between 18 *Drosophila* species has demonstrated an almost complete absence of mutations in the close vicinity of the exonic edited sites (41). Using this knowledge, Hoopengardner et al. were able to discover 16 novel edited genes in *Drosophila* and one novel site in humans. Several different groups subsequently conducted database searches for finding novel edited human genes, taking into account this conservation between species as well as the established cis-sequence preferences of ADARs (39, 40). This led to the discovery of four novel human genes edited within the coding region.

Moreover, the total of known edited targets in *Drosophila* recently doubled by using *D. melanogaster* genomic and cDNA databases in a computational screen for A-to-G discrepancies, leading to the identification of 27 edited genes (38).

All of the above-described methods have proven to be effective for the discovery of novel edited genes. However, often already known substrates have been missed and there are a high number of false positives detected in each study. The methods for finding novel ADAR substrates still have to be optimized and developed further to allow for the identification of all recoding sites of editing.

1.2.2 Editing of miRNA Sequences

Since double-stranded RNA molecules represent targets not only for ADARs, but also for other dsRNA-binding proteins, such as the components of the RNA silencing pathways, it is not surprising that the natural pathways of RNA editing and RNA silencing are interrelated (reviewed in reference 105). Based on several studies, it has been suggested that there may often be a competition between the editing and silencing machineries for dsRNA molecules. The outcome with respect to the fate of the RNA may largely depend on which set of enzymes engages the RNA first. According to another model, RNA editing may be a nuclear event that can induce silencing on the chromatin level (106). See Chapter 10 for a detailed discussion on the interplay between RNA editing and RNA silencing.

Another related finding has been the discovery that some members of the large family of micro RNA genes are subject to A-to-I RNA editing (see Table 1.2 for list of known edited miRNAs) (107–110). Micro RNAs (miRNAs) constitute a very large and still growing class of RNA molecules, each about 20–22 nt long and may be involved in the regulation of up to one-third of all protein-coding genes through a mechanism of translational repression (111). The analysis of cis-acting preferences of ADAR enzymes and the characterization of RNA secondary structures of known recoding targets raised the question of whether miRNA precursor molecules might also represent ADAR targets. A hallmark feature of all miRNAs is that they are expressed

TABLE 1.2 Validated Edited miRNAs

Origin	miRNA	Edited Positions[a]	Consequence	Reference
Mammalian	miRNA-142	−12, 4, 5, 6, 9, 12, 19, 40, 50, 55, 62	No processing by Drosha-DGCR8; degradation by Tudor-SN	110[b]
	miRNA-376 cluster	4, 44	Altered target gene silencing	112
	miRNA-22	Human: 58, 62, 71, 80 Mouse: 12, 63, 64	Unknown	107
	miRNA-151	49	Unknown	109[c]
	miRNA-197	14	Unknown	109[c]
	miRNA-223	20	Unknown	109[c]
	miRNA-376a	9, 49	Unknown	109[c]
	miRNA-379	10	Unknown	109[c]
	miRNA-99a	13	Unknown	109[c]
Kaposi sarcoma-associated virus (KSHV)	miRNA-K12-10	47	Unknown	108

[a]Relative to the 5′ end of the mature miRNA.

[b]Editing *in vitro* also detected in three other miRNAs.

[c]Reference reports possible editing of six other miRNA that may be annotated on the wrong genomic strand.

as precursor molecules that form a hairpin structure from which subsequently the mature miRNA is excised (see Figure 1.4). These hairpin structures are reminiscent of known to be RNA fold-back structures that are subject to editing. The pre-mRNA transcripts of miRNAs (termed pri-miRNAs) are usually several hundred nucleotides long and are first processed into an approximately 70 to 90-nt pre-miRNA in the nucleus and then, after export into the cytoplasm, a second processing step generates the mature approximately 20 to 22-nt functional miRNA.

The occurrence of A-to-I RNA editing in a miRNA molecule was first described for human and mouse miRNA22 (107). Although the editing levels in native human and mouse tissues are low, it was shown in cellular assays that ADARs specifically target the same nucleotide positions that are also modified *in vivo*. The observed editing events were located both within and outside of the embedded mature miRNA sequence and predicted that either miRNA maturation or the targeting specificity of the mature miRNA could be affected. The next miRNA that was found to undergo

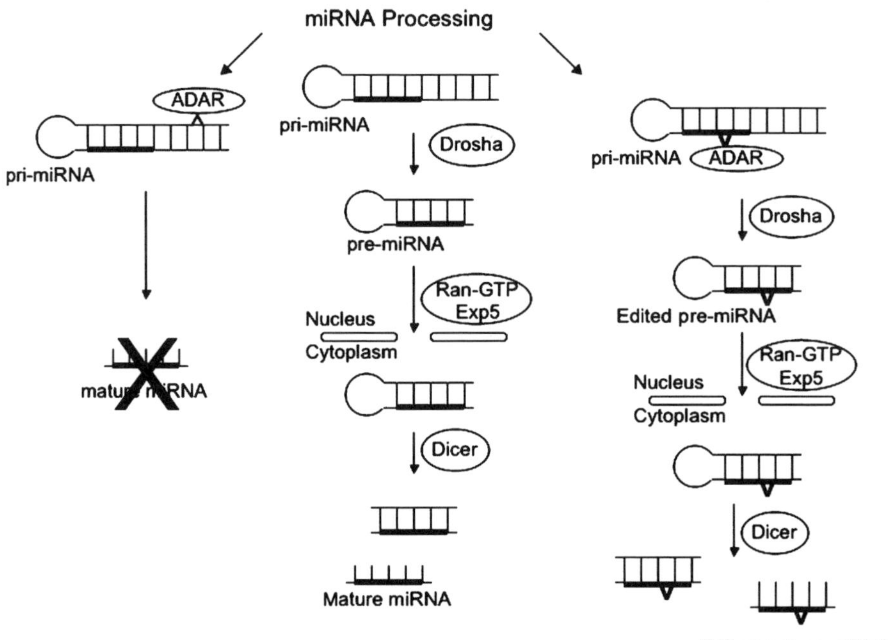

Figure 1.4 miRNA editing. The central panel outlines the biogenesis pathway for miRNAs with the main steps of pri-miRNA synthesis, pre-miRNA generation through Drosha cleavage, nuclear export (involving the Ran-GTP/Exp5 pathway), and final processing by Dicer in the cytosol. To the left the potential consequence of RNA editing in miRNA precursor sequences outside of the mature miRNA region is indicated. Due to editing, the RNA structure is altered in such a way that Drosho is not able to process the RNA further. This scenario may be involved in the regulation of miRNA 142 (110). To the right, the possible consequence of editing within the mature miRNA sequence is depicted. If ADARs modify the miRNA sequence and processing proceeds, altered, mature miRNA is formed, which may exhibit changed target specificity. An example for such an event is miRNA 376, where the edited miRNA has a very different target spectrum than the nonedited version (112). See text for details.

A-to-I editing was the Karposi-sarcoma-associated virus miRNA-K12-10b (108). High levels of edited, mature miRNA-K12-10b were observed in host cells; however, also here the functional implications of editing have not been addressed. A first insight into the consequences of A-to-I editing of miRNA precursors was provided by the finding that editing of two sites in miRNA-142 leads to suppression of further maturation and potentially targets edited pri-miRNA-142 transcripts for degradation by Tudor-SN (110). Consequently, in editing-deficient mice, expression of mature miRNA-142 was increased (110).

Another intriguing example of miRNA editing is miRNA-376. In this case, mature edited miRNA-376 could be detected and the editing event has a strong impact on miRNA target predictions (112). It could be shown that a target specific for the edited miRNA-376 sequence was upregulated in ADAR knockout mice (112).

Currently, it is estimated that between 6% and 10% of all miRNA genes are subject to A-to-I modification (109). The identification of editing events in miRNAs demonstrate that miRNA transcripts are subject to post-transcriptional modification and that the functions of miRNAs may not be completely deducible when simply analyzing their genomic sequences. RNA editing increases the diversity of the “miRNome” and should be considered when analyzing miRNA function or when contemplating the use of miRNAs as laboratory tools or for developing therapeutics.

1.3 FUTURE DIRECTIONS: EVOLUTION OF EDITING SITES AND MACHINERY

An unresolved, fundamental question regarding RNA editing is: How do novel editing sites arise in nature? This question is related to the questions of how does an editing site evolve to become more or less prevalent, and how may RNA editing shape the evolution of phenotypes and species?

First of all it is important to consider in what way RNA editing may be beneficial for an organism and how it compares to evolving additional protein subunits or splice variants as a mechanism to generate complexity and diversity. An important insight is the fact that RNA editing often seems to target codons of invariant or highly conserved amino acid residues. This is the result of an at least two-step process, first enabling editing at the novel site and then the process of establishing it permanently and optimizing the level of editing at the site.

Recoding messages on the RNA level to a certain extent enables the organism to produce protein variants that would otherwise not be allowed because DNA-based changes are much more drastic either–or decisions that will immediately fix the change in one allele of the gene. For highly conserved residues, such drastic changes are not permissive and will likely be selected against.

One current model of how generally novel A-to-I RNA editing sites arise involves “continuous probing” of the sequence space for novel sites by nature (36). It is based on the finding that the essential feature for editing is the substrate structure—an RNA fold that directs the ubiquitously expressed editing machinery to a specific adenosine for modification. RNA folding is inherently complex and dynamic, meaning that various alternative three-dimensional structures exist for a given pre-mRNA sequence, and, importantly, small changes within the primary sequence can lead to drastic changes in outcome. The "continuous probing" model postulates that there is ongoing, low-level background editing at many sites in pre-mRNA, whereas the recoding editing events that have been characterized to date represent the few modification sites whose associated RNA folds have been selected for and evolved to support significant editing levels.

What is currently the evidence that would support this model? And can it be tested experimentally? A critical prerequisite for the continuous probing of RNA structures is that the secondary and tertiary structures of pre-mRNAs be less constraining than, for example, the protein-coding sequences themselves. Since eukaryotic pre-mRNAs usually contain long noncoding introns along with relatively shorter exons, mutations within intronic sequences are often tolerable with no impact on gene expression. Novel RNA structures involving highly conserved exonic sequences and rapidly evolving introns may therefore arise at a high rate.

Intriguingly, it has been experimentally demonstrated that species-specific editing sites within the same exon of the synaptotagmin I gene can be interchanged simply by introducing a few point mutations within distant intronic sequences (113). It is also well known that small changes introduced into an intronic sequence involved in a RNA fold that mediates editing can drastically enhance or impair editing rates and site specificity (19).

Could there be an additional advantage provided to an organism maintaining a low-level background editing throughout the transcriptome? It is possible that the latent ability to edit certain sites could represent a survival advantage for the organism if environmental conditions change. It could then adapt much faster by increasing the production of the edited variant. One way to test the "continuous background editing model" is to try to detect the ongoing low-level of background editing. This will be challenging since the error rate of any detection method to be used will likely be as high as or exceed the background level of editing. Maybe proteomics approaches can give the answer if it becomes possible to specifically detect recoded protein variants. Alternatively, model systems with editable reporter genes may be useful in detecting low-level A-to-I modifications.

If continuous editing occurs, the plasticity of eukaryotic transcripts may be much more pronounced than previously anticipated and could have important consequences on how to approach evolutionary questions of adaptability, variation, and speciation.

REFERENCES

1. Baltimore, D. (2001) Our genome unveiled. *Nature* **409**, 814–816.
2. Maas, S., Rich, A., and Nishikura, K. (2003) A-to-I RNA editing: Recent news and residual mysteries. *J Biol Chem* **278**, 1391–1394.
3. Schaub, M., and Keller, W. (2002) RNA editing by adenosine deaminases generates RNA and protein diversity. *Biochimie* **84**, 791–803.
4. Bass, B. L. (2002) RNA editing by adenosine deaminases that act on RNA. *Annu Rev Biochem* **71**, 817–846.

5. Basilio, C., Wahba, A. J., Lengyel, P., Speyer, J. F., and Ochoa, S. (1962) Synthetic polynucleotides and the amino acid code. *Proc Natl Acad Sci USA* **48**, 613–616.
6. Sommer, B., Kohler, M., Sprengel, R., and Seeburg, P. H. (1991) RNA editing in brain controls a determinant of ion flow in glutamate-gated channels. *Cell* **67**, 11–19.
7. Kohler, M., Burnashev, N., Sakmann, B., and Seeburg, P. H. (1993) Determinants of Ca^{2+} permeability in both TM1 and TM2 of high affinity kainate receptor channels: Diversity by RNA editing. *Neuron* **10**, 491–500.
8. Lomeli, H., Mosbacher, J., Melcher, T., Hoger, T., Geiger, J. R., Kuner, T., Monyer, H., Higuchi, M., Bach, A., and Seeburg, P. H. (1994) Control of kinetic properties of AMPA receptor channels by nuclear RNA editing. *Science* **266**, 1709–1713.
9. Burns, C. M., Chu, H., Rueter, S. M., Hutchinson, L. K., Canton, H., Sanders-Bush, E., and Emeson, R. B. (1997) Regulation of serotonin-2C receptor G-protein coupling by RNA editing. *Nature* **387**, 303–308.
10. Hershfield, M. S. (1998) Adenosine deaminase deficiency: Clinical expression, molecular basis, and therapy. *Semin Hematol* **35**, 291–298.
11. Grosjean, H., Auxilien, S., Constantinesco, F., Simon, C., Corda, Y., Becker, H. F., Foiret, D., Morin, A., Jin, Y. X., Fournier, M., *et al.* (1996) Enzymatic conversion of adenosine to inosine and to N1-methylinosine in transfer RNAs: a review. *Biochimie* **78**, 488–501.
12. Gerber, A. P., and Keller, W. (1999) An adenosine deaminase that generates inosine at the wobble position of tRNAs. *Science* **286**, 1146–1149.
13. Bass, B. L., and Weintraub, H. (1987) A developmentally regulated activity that unwinds RNA duplexes. *Cell* **48**, 607–613.
14. Wagner, R. W., and Nishikura, K. (1988) Cell cycle expression of RNA duplex unwindase activity in mammalian cells. *Mol Cell Biol* **8**, 770–777.
15. Bass, B. L., and Weintraub, H. (1988) An unwinding activity that covalently modifies its double-stranded RNA substrate. *Cell* **55**, 1089–1098.
16. Melcher, T., Maas, S., Higuchi, M., Keller, W., and Seeburg, P. H. (1995) Editing of alpha-amino-3-hydroxy-5-methylisoxazole-4-propionic acid receptor GluR-B pre-mRNA *in vitro* reveals site-selective adenosine to inosine conversion. *J Biol Chem* **270**, 8566–8570.
17. Rueter, S. M., Burns, C. M., Coode, S. A., Mookherjee, P., and Emeson, R. B. (1995) Glutamate receptor RNA editing *in vitro* by enzymatic conversion of adenosine to inosine. *Science* **267**, 1491–1494.
18. Yang, J. H., Sklar, P., Axel, R., and Maniatis, T. (1995) Editing of glutamate receptor subunit B pre-mRNA *in vitro* by site-specific deamination of adenosine. *Nature* **374**, 77–81.
19. Higuchi, M., Single, F. N., Kohler, M., Sommer, B., Sprengel, R., and Seeburg, P. H. (1993) RNA editing of AMPA receptor subunit GluR-B: A base-paired intron-exon structure determines position and efficiency. *Cell* **75**, 1361–1370.
20. Casey, J. L., and Gerin, J. L. (1995) Hepatitis D virus RNA editing: specific modification of adenosine in the antigenomic RNA. *J Virol* **69**, 7593–7600.
21. Kim, U., Wang, Y., Sanford, T., Zeng, Y., and Nishikura, K. (1994) Molecular cloning of cDNA for double-stranded RNA adenosine deaminase, a candidate enzyme for nuclear RNA editing. *Proc Natl Acad Sci USA* **91**, 11457–11461.
22. Hurst, S. R., Hough, R. F., Aruscavage, P. J., and Bass, B. L. (1995) Deamination of mammalian glutamate receptor RNA by *Xenopus* dsRNA adenosine deaminase: Similarities to *in vivo* RNA editing. *RNA* **1**, 1051–1060.
23. O'Connell, M. A., Krause, S., Higuchi, M., Hsuan, J. J., Totty, N. F., Jenny, A., and Keller, W. (1995) Cloning of cDNAs encoding mammalian double-stranded RNA-specific adenosine deaminase. *Mol Cell Biol* **15**, 1389–1397.
24. Weier, H. U., George, C. X., Greulich, K. M., and Samuel, C. E. (1995) The interferon-inducible, double-stranded RNA-specific adenosine deaminase gene (DSRAD) maps to human chromosome 1q21.1–21.2. *Genomics* **30**, 372–375.
25. O'Connell, M. A., and Keller, W. (1994) Purification and properties of double-stranded RNA-specific adenosine deaminase from calf thymus. *Proc Natl Acad Sci USA* **91**, 10596–10600.
26. Kim, U., Garner, T. L., Sanford, T., Speicher, D., Murray, J. M., and Nishikura, K. (1994) Purification and characterization of double-stranded RNA adenosine deaminase from bovine nuclear extracts. *J Biol Chem* **269**, 13480–13489.

27. Melcher, T., Maas, S., Herb, A., Sprengel, R., Seeburg, P. H., and Higuchi, M. (1996) A mammalian RNA editing enzyme. *Nature* **379**, 460–464.
28. Melcher, T., Maas, S., Herb, A., Sprengel, R., Higuchi, M., and Seeburg, P. H. (1996) RED2, a brain-specific member of the RNA-specific adenosine deaminase family. *J Biol Chem* **271**, 31795–31798.
29. Slavov, D., Crnogorac-Jurcevic, T., Clark, M., and Gardiner, K. (2000) Comparative analysis of the DRADA A-to-I RNA editing gene from mammals, pufferfish and zebrafish. *Gene* **250**, 53–60.
30. Slavov, D., Clark, M., and Gardiner, K. (2000) Comparative analysis of the RED1 and RED2 A-to-I RNA editing genes from mammals, pufferfish and zebrafish. *Gene* **250**, 41–51.
31. Palladino, M. J., Keegan, L. P., O'Connell, M. A., and Reenan, R. A. (2000) dADAR, a *Drosophila* double-stranded RNA-specific adenosine deaminase is highly developmentally regulated and is itself a target for RNA editing. *RNA* **6**, 1004–1018.
32. Hough, R. F., Lingam, A. T., and Bass, B. L. (1999) *Caenorhabditis elegans* mRNAs that encode a protein similar to ADARs derive from an operon containing six genes. *Nucleic Acids Res* **27**, 3424–3432.
33. Gerber, A., Grosjean, H., Melcher, T., and Keller, W. (1998) Tad1p, a yeast tRNA-specific adenosine deaminase, is related to the mammalian pre-mRNA editing enzymes ADAR1 and ADAR2. *EMBO J* **17**, 4780–4789.
34. Maas, S., Gerber, A. P., and Rich, A. (1999) Identification and characterization of a human tRNA-specific adenosine deaminase related to the ADAR family of pre-mRNA editing enzymes. *Proc Natl Acad Sci USA* **96**, 8895–8900.
35. Maas, S., Kim, Y. G., and Rich, A. (2000) Sequence, genomic organization and functional expression of the murine tRNA-specific adenosine deaminase ADAT1. *Gene* **243**, 59–66.
36. Maas, S., and Rich, A. (2000) Changing genetic information through RNA editing. *Bioessays* **22**, 790–802.
37. Gerber, A. P., and Keller, W. (2001) RNA editing by base deamination: More enzymes, more targets, new mysteries. *Trends Biochem Sci* **26**, 376–384.
38. Stapleton, M., Carlson, J. W., and Celniker, S. E. (2006) RNA editing in *Drosophila melanogaster*: New targets and functional consequences. *RNA* **12**, 1922–1932.
39. Clutterbuck, D. R., Leroy, A., O'Connell, M. A., and Semple, C. A. (2005) A bioinformatic screen for novel A-I RNA editing sites reveals re-coding editing in BC10. *Bioinformatics* **21** (11), 2590–2595.
40. Levanon, E. Y., Hallegger, M., Kinar, Y., Shemesh, R., Djinovic-Carugo, K., Rechavi, G., Jantsch, M. F., and Eisenberg, E. (2005) Evolutionarily conserved human targets of adenosine to inosine RNA editing. *Nucleic Acids Res* **33**, 1162–1168.
41. Hoopengardner, B., Bhalla, T., Staber, C., and Reenan, R. (2003) Nervous system targets of RNA editing identified by comparative genomics. *Science* **301**, 832–836.
42. Palladino, M. J., Keegan, L. P., O'Connell, M. A., and Reenan, R. A. (2000) A-to-I pre-mRNA editing in *Drosophila* is primarily involved in adult nervous system function and integrity. *Cell* **102**, 437–449.
43. Mayer, M. L., and Armstrong, N. (2004) Structure and function of glutamate receptor ion channels. *Annu Rev Physiol* **66**, 161–181.
44. Seeburg, P. H., and Hartner, J. (2003) Regulation of ion channel/neurotransmitter receptor function by RNA editing. *Curr Opin Neurobiol* **13**, 279–283.
45. Greger, I. H., Khatri, L., and Ziff, E. B. (2002) RNA editing at arg607 controls AMPA receptor exit from the endoplasmic reticulum. *Neuron* **34**, 759–772.
46. Greger, I. H., Khatri, L., Kong, X., and Ziff, E. B. (2003) AMPA receptor tetramerization is mediated by Q/R editing. *Neuron* **40**, 763–774.
47. Brusa, R., Zimmermann, F., Koh, D. S., Feldmeyer, D., Gass, P., Seeburg, P. H., and Sprengel, R. (1995) Early-onset epilepsy and postnatal lethality associated with an editing-deficient GluR-B allele in mice. *Science* **270**, 1677–1680.
48. Higuchi, M., Maas, S., Single, F. N., Hartner, J., Rozov, A., Burnashev, N., Feldmeyer, D., Sprengel, R., and Seeburg, P. H. (2000) Point mutation in an AMPA receptor gene rescues lethality in mice deficient in the RNA-editing enzyme ADAR2. *Nature* **406**, 78–81.
49. Kask, K., Zamanillo, D., Rozov, A., Burnashev, N., Sprengel, R., and Seeburg, P. H. (1998) The AMPA receptor subunit GluR-B in its Q/R site-unedited form is not essential for brain development and function. *Proc Natl Acad Sci USA* **95**, 13777–13782.

50. Maas, S., Kawahara, Y., Tamburro, K. M., and Nishikura, K. (2006) A-to-I RNA Editing and Human Disease. *RNA Biol* **3** (1), 1–9.
51. Kawahara, Y., Ito, K., Sun, H., Aizawa, H., Kanazawa, I., and Kwak, S. (2004) Glutamate receptors: RNA editing and death of motor neurons. *Nature* **427**, 801.
52. Greger, I. H., Akamine, P., Khatri, L., and Ziff, E. B. (2006) Developmentally regulated, combinatorial RNA processing modulates AMPA receptor biogenesis. *Neuron* **51**, 85–97.
53. Herrick-Davis, K., Grinde, E., and Niswender, C. M. (1999) Serotonin 5-HT2C receptor RNA editing alters receptor basal activity: Implications for serotonergic signal transduction. *J Neurochem* **73**, 1711–1717.
54. Price, R. D., and Sanders-Bush, E. (2000) RNA editing of the human serotonin 5-HT(2C) receptor delays agonist-stimulated calcium release. *Mol Pharmacol* **58**, 859–862.
55. Gurevich, I., Tamir, H., Arango, V., Dwork, A. J., Mann, J. J., and Schmauss, C. (2002) Altered editing of serotonin 2C receptor pre-mRNA in the prefrontal cortex of depressed suicide victims. *Neuron* **34**, 349–356.
56. Yang, W., Wang, Q., Kanes, S. J., Murray, J. M., and Nishikura, K. (2004) Altered RNA editing of serotonin 5-HT(2C) receptor induced by interferon: Implications for depression associated with cytokine therapy. *Brain Res Mol Brain Res* **124**, 70–78.
57. Gardiner, K., and Du, Y. (2006) A-to-I editing of the 5HT2C receptor and behaviour. *Brief Funct Genomic Proteomic* **5**, 37–42.
58. Dolly, J. O., and Parcej, D. N. (1996) Molecular properties of voltage-gated K+ channels. *J Bioenerg Biomembr* **28**, 231–253.
59. Bhalla, T., Rosenthal, J. J., Holmgren, M., and Reenan, R. (2004) Control of human potassium channel inactivation by editing of a small mRNA hairpin. *Nat Struct Mol Biol* **11**, 950–956.
60. Laxminarayana, D., Khan, I. U., and Kammer, G. (2002) Transcript mutations of the alpha regulatory subunit of protein kinase A and up-regulation of the RNA-editing gene transcript in lupus T lymphocytes. *Lancet* **360**, 842–849.
61. Polson, A. G., Ley, H. L., 3rd Bass, B. L., and Casey, J. L. (1998) Hepatitis delta virus RNA editing is highly specific for the amber/W site and is suppressed by hepatitis delta antigen. *Mol Cell Biol* **18**, 1919–1926.
62. Polson, A. G., Bass, B. L., and Casey, J. L. (1996) RNA editing of hepatitis delta virus antigenome by dsRNA-adenosine deaminase. *Nature* **380**, 454–456.
63. Egebjerg, J., Kukekov, V., and Heinemann, S. F. (1994) Intron sequence directs RNA editing of the glutamate receptor subunit GluR2 coding sequence. *Proc Natl Acad Sci USA* **91**, 10270–10274.
64. Herb, A., Higuchi, M., Sprengel, R., and Seeburg, P. H. (1996) Q/R site editing in kainate receptor GluR5 and GluR6 pre-mRNAs requires distant intronic sequences. *Proc Natl Acad Sci USA* **93**, 1875–1880.
65. Athanasiadis, A., Rich, A., and Maas, S. (2004) Widespread A-to-I RNA editing of Alu-containing mRNAs in the human transcriptome. *PLoS Biol* **2**, e391.
66. Keegan, L. P., Leroy, A., Sproul, D., and O'Connell, M. A. (2004) Adenosine deaminases acting on RNA (ADARs): RNA-editing enzymes. *Genome Biol* **5**, 209.
67. Valente, L., and Nishikura, K. (2005) ADAR gene family and A-to-I RNA editing: diverse roles in posttranscriptional gene regulation. *Prog Nucleic Acid Res Mol Biol* **79**, 299–338.
68. Chen, C. X., Cho, D. S., Wang, Q., Lai, F., Carter, K. C., and Nishikura, K. (2000) A third member of the RNA-specific adenosine deaminase gene family, ADAR3, contains both single- and double-stranded RNA binding domains. *RNA* **6**, 755–767.
69. Maas, S., Melcher, T., Herb, A., Seeburg, P. H., Keller, W., Krause, S., Higuchi, M., and O'Connell, M. A. (1996) Structural requirements for RNA editing in glutamate receptor pre-mRNAs by recombinant double-stranded RNA adenosine deaminase. *J Biol Chem* **271**, 12221–12226.
70. Cho, D. S., Yang, W., Lee, J. T., Shiekhattar, R., Murray, J. M., and Nishikura, K. (2003) Requirement of dimerization for RNA editing activity of adenosine deaminases acting on RNA. *J Biol Chem* **278**, 17093–17102.
71. Chilibeck, K. A., Wu, T., Liang, C., Schellenberg, M. J., Gesner, E. M., Lynch, J. M., and MacMillan, A. M. (2006) FRET analysis of in vivo dimerization by RNA-editing enzymes. *J Biol Chem* **281**, 16530–16535.

72. Gallo, A., Keegan, L. P., Ring, G. M., and O'Connell, M. A. (2003) An ADAR that edits transcripts encoding ion channel subunits functions as a dimmer. *EMBO J* **22**, 3421–3430.
73. Valente, L., and Nishikura, K. (2007) RNA binding-independent dimerization of adenosine deaminases acting on RNA and dominant negative effects of nonfunctional subunits on dimer functions. *J Biol Chem* **282**, 16054–16061.
74. Macbeth, M. R., Schubert, H. L., Vandemark, A. P., Lingam, A. T., Hill, C. P., and Bass, B. L. (2005) Inositol hexakisphosphate is bound in the ADAR2 core and required for RNA editing. *Science* **309**, 1534–1539.
75. Sato, S., Wong, S. K., and Lazinski, D. W. (2001) Hepatitis delta virus minimal substrates competent for editing by ADAR1 and ADAR2. *J Virol* **75**, 8547–8555.
76. O'Connell, M. A., Gerber, A., and Keller, W. (1997) Purification of human double-stranded RNA-specific editase 1 (hRED1) involved in editing of brain glutamate receptor B pre-mRNA. *J Biol Chem* **272**, 473–478.
77. Polson, A. G., and Bass, B. L. (1994) Preferential selection of adenosines for modification by double-stranded RNA adenosine deaminase. *EMBO J* **13**, 5701–5711.
78. Lehmann, K. A., and Bass, B. L. (1999) The importance of internal loops within RNA substrates of ADAR1. *J Mol Biol* **291**, 1–13.
79. Stefl, R., and Allain, F. H. (2005) A novel RNA pentaloop fold involved in targeting ADAR2. *RNA* **11**, 592–597.
80. Stefl, R., Xu, M., Skrisovska, L., Emeson, R. B., and Allain, F. H. (2006) Structure and specific RNA binding of ADAR2 double-stranded RNA binding motifs. *Structure* **14**, 345–355.
81. Tonkin, L. A., Saccomanno, L., Morse, D. P., Brodigan, T., Krause, M., and Bass, B. L. (2002) RNA editing by ADARs is important for normal behavior in *Caenorhabditis elegans*. *EMBO J* **21**, 6025–6035.
82. Liu, Y., George, C. X., Patterson, J. B., and Samuel, C. E. (1997) Functionally distinct double-stranded RNA-binding domains associated with alternative splice site variants of the interferon-inducible double-stranded RNA-specific adenosine deaminase. *J Biol Chem* **272**, 4419–4428.
83. Lai, F., Chen, C. X., Carter, K. C., and Nishikura, K. (1997) Editing of glutamate receptor B subunit ion channel RNAs by four alternatively spliced DRADA2 double-stranded RNA adenosine deaminases. *Mol Cell Biol* **17**, 2413–2424.
84. Gerber, A., O'Connell, M. A., and Keller, W. (1997) Two forms of human double-stranded RNA-specific editase 1 (hRED1) generated by the insertion of an Alu cassette. *RNA* **3**, 453–463.
85. Kawakubo, K., and Samuel, C. E. (2000) Human RNA-specific adenosine deaminase (ADAR1) gene specifies transcripts that initiate from a constitutively active alternative promoter. *Gene* **258**, 165–172.
86. Rueter, S. M., Dawson, T. R., and Emeson, R. B. (1999) Regulation of alternative splicing by RNA editing. *Nature* **399**, 75–80.
87. Patterson, J. B., and Samuel, C. E. (1995) Expression and regulation by interferon of a double-stranded-RNA-specific adenosine deaminase from human cells: Evidence for two forms of the deaminase. *Mol Cell Biol* **15**, 5376–5388.
88. Herbert, A., Alfken, J., Kim, Y. G., Mian, I. S., Nishikura, K., and Rich, A. (1997) A Z-DNA binding domain present in the human editing enzyme, double-stranded RNA adenosine deaminase. *Proc Natl Acad Sci USA* **94**, 8421–8426.
89. Poulsen, H., Nilsson, J., Damgaard, C. K., Egebjerg, J., and Kjems, J. (2001) CRM1 mediates the export of ADAR1 through a nuclear export signal within the Z-DNA binding domain. *Mol Cell Biol* **21**, 7862–7871.
90. Desterro, J. M., Keegan, L. P., Lafarga, M., Berciano, M. T., O'Connell, M., and Carmo-Fonseca, M. (2003) Dynamic association of RNA-editing enzymes with the nucleolus. *J Cell Sci* **116**, 1805–1818.
91. Sansam, C. L., Wells, K. S., and Emeson, R. B. (2003) Modulation of RNA editing by functional nucleolar sequestration of ADAR2. *Proc Natl Acad Sci USA* **100**, 14018–14023.
92. Paul, M. S., and Bass, B. L. (1998) Inosine exists in mRNA at tissue-specific levels and is most abundant in brain mRNA. *EMBO J* **17**, 1120–1127.
93. Morse, D. P., and Bass, B. L. (1997) Detection of inosine in messenger RNA by inosine-specific cleavage. *Biochemistry* **36**, 8429–8434.
94. Morse, D. P., and Bass, B. L. (1999) Long RNA hairpins that contain inosine are present in *Caenorhabditis elegans* poly(A)+ RNA. *Proc Natl Acad Sci USA* **96**, 6048–6053.

95. Morse, D. P., Aruscavage, P. J., and Bass, B. L. (2002) RNA hairpins in noncoding regions of human brain and *Caenorhabditis elegans* mRNA are edited by adenosine deaminases that act on RNA. *Proc Natl Acad Sci USA* **99**, 7906–7911.
96. Blow, M., Futreal, P. A., Wooster, R., and Stratton, M. R. (2004) A survey of RNA editing in human brain. *Genome Res* **14**, 2379–2387.
97. Kim, D. D., Kim, T. T., Walsh, T., Kobayashi, Y., Matise, T. C., Buyske, S., and Gabriel, A. (2004) Widespread RNA editing of embedded alu elements in the human transcriptome. *Genome Res* **14**, 1719–1725.
98. Levanon, E. Y., Eisenberg, E., Yelin, R., Nemzer, S., Hallegger, M., Shemesh, R., Fligelman, Z. Y., Shoshan, A., Pollock, S. R., Sztybel, D. *et al.* (2004) Systematic identification of abundant A-to-I editing sites in the human transcriptome. *Nat Biotechnol* **22**, 1001–1005.
99. Batzer, M. A., and Deininger, P. L. (2002) Alu repeats and human genomic diversity. *Nat Rev Genet* **3**, 370–379.
100. Neeman, Y., Levanon, E. Y., Jantsch, M. F., and Eisenberg, E. (2006) RNA editing level in the mouse is determined by the genomic repeat repertoire. *RNA* **12**, 1802–1809.
101. Prasanth, K. V., Prasanth, S. G., Xuan, Z., Hearn, S., Freier, S. M., Bennett, C. F., Zhang, M. Q., and Spector, D. L. (2005) Regulating gene expression through RNA nuclear retention. *Cell* **123**, 249–263.
102. Ohlson, J., Enstero, M., Sjoberg, B. M., and Ohman, M. (2005) A method to find tissue-specific novel sites of selective adenosine deamination. *Nucleic Acids Res* **33**, e167.
103. Ohlson, J., Pedersen, J. S., Haussler, D., and Ohman, M. (2007) Editing modifies the GABA(A) receptor subunit alpha3. *RNA* **13**, 698–703.
104. Xia, S., Yang, J., Su, Y., Qian, J., Ma, E., and Haddad, G. G. (2005) Identification of new targets of Drosophila pre-mRNA adenosine deaminase. *Physiol Genomics* **20**, 195–202.
105. Nishikura, K. (2006) Editor meets silencer: Crosstalk between RNA editing and RNA interference. *Nat Rev Mol Cell Biol* **7**, 919–931.
106. Wang, Q., Zhang, Z., Blackwell, K., and Carmichael, G. G. (2005) Vigilins bind to promiscuously A-to-I-edited RNAs and are involved in the formation of heterochromatin. *Curr Biol* **15**, 384–391.
107. Luciano, D. J., Mirsky, H., Vendetti, N. J., and Maas, S. (2004) RNA editing of a miRNA precursor. *RNA* **10**, 1174–1177.
108. Pfeffer, S., Sewer, A., Lagos-Quintana, M., Sheridan, R., Sander, C., Grasser, F. A., van Dyk, L. F., Ho, C. K., Shuman, S., Chien, M., *et al.* (2005) Identification of microRNAs of the herpesvirus family. *Nat Methods* **2**, 269–276.
109. Blow, M. J., Grocock, R. J., van Dongen, S., Enright, A. J., Dicks, E., Futreal, P. A., Wooster, R., and Stratton, M. R. (2006) RNA editing of human microRNAs. *Genome Biol* **7**, R27.
110. Yang, W., Chendrimada, T. P., Wang, Q., Higuchi, M., Seeburg, P. H., Shiekhattar, R., and Nishikura, K. (2006) Modulation of microRNA processing and expression through RNA editing by ADAR deaminases. *Nat Struct Mol Biol* **13**, 13–21.
111. Bushati, N., and Cohen, S. M. (2007) microRNA Functions. *Annu Rev Cell Dev Biol* **23**,, 175–205.
112. Kawahara, Y., Zinshteyn, B., Sethupathy, P., Iizasa, H., Hatzigeorgiou, A. G., and Nishikura, K. (2007) Redirection of silencing targets by adenosine-to-inosine editing of miRNAs. *Science* **315**, 1137–1140.
113. Reenan, R. A. (2005) Molecular determinants and guided evolution of species-specific RNA editing. *Nature* **434**, 409–413.
114. Rosenthal, J. J., and Bezanilla, F. (2002) Extensive editing of mRNAs for the squid delayed rectifier K^+ channel regulates subunit tetramerization. *Neuron* **34**, 743–757.
115. Patton, D. E., Silva, T., and Bezanilla, F. (1997) RNA editing generates a diverse array of transcripts encoding squid Kv2 K^+ channels with altered functional properties. *Neuron* **19**, 711–722.
116. Saccomanno, L., and Bass, B. L. (1999) A minor fraction of basic fibroblast growth factor mRNA is deaminated in *Xenopus* stage VI and matured oocytes. *RNA* **5**, 39–48.
117. Keegan, L. P., Brindle, J., Gallo, A., Leroy, A., Reenan, R. A., and O'Connell, M. A. (2005) Tuning of RNA editing by ADAR is required in *Drosophila*. *EMBO J* **24**, 2183–2193.
118. Hanrahan, C. J., Palladino, M. J., Ganetzky, B., and Reenan, R. A. (2000) RNA editing of the *Drosophila* para $Na(^+)$ channel transcript. Evolutionary conservation and developmental regulation. *Genetics* **155**, 1149–1160.
119. Semenov, E. P., and Pak, W. L. (1999) Diversification of *Drosophila* chloride channel gene by multiple posttranscriptional mRNA modifications. *J Neurochem* **72**, 66–72.

120. Peixoto, A. A., Smith, L. A., and Hall, J. C. (1997) Genomic organization and evolution of alternative exons in a *Drosophila* calcium channel gene. *Genetics* **145**, 1003–1013.
121. Petschek, J. P., Scheckelhoff, M. R., Mermer, M. J., and Vaughn, J. C. (1997) RNA editing and alternative splicing generate mRNA transcript diversity from the *Drosophila* 4f-rnp locus. *Gene* **204**, 267–276.
122. Grauso, M., Reenan, R. A., Culetto, E., and Sattelle, D. B. (2002) Novel putative nicotinic acetylcholine receptor subunit genes, Dalpha5, Dalpha6 and Dalpha7, in *Drosophila melanogaster* identify a new and highly conserved target of adenosine deaminase acting on RNA-mediated A-to-I pre-mRNA editing. *Genetics* **160**, 1519–1533.
123. Jones, A. K., Raymond-Delpech, V., Thany, S. H., Gauthier, M., and Sattelle, D. B. (2006) The nicotinic acetylcholine receptor gene family of the honey bee. *Apis mellifera. Genome Res* **16**, 1422–1430.
124. Diegelmann, S., Nieratschker, V., Werner, U., Hoppe, J., Zars, T., and Buchner, E. (2006) The conserved protein kinase-A target motif in synapsin of *Drosophila* is effectively modified by pre-mRNA editing. *BMC Neurosci* **7**, 76.
125. Gao, J. R., Deacutis, J. M., and Scott, J. G. (2007) The nicotinic acetylcholine receptor subunit Mdalpha6 from *Musca domestica* is diversified via post-transcriptional modification. *Insect Mol Biol* **16**, 325–334.

CHAPTER 2

ANTIBODY GENE DIVERSIFICATION BY AID-CATALYZED DNA EDITING

Donna A. MacDuff
Steven M. Offer
Zachary L. Demorest
Reuben S. Harris

THE ASTONISHINGLY diverse antibody repertoire of vertebrates is achieved by the distinct processes of V(D)J recombination, somatic hypermutation (immunoglobulin gene conversion in some animals) and class switch recombination. These processes irreversibly alter the antibody gene DNA by mechanisms that have puzzled scientists for decades. AID was identified in 1999 and, remarkably, its ability to edit DNA was soon discovered and found to be essential for somatic hypermutation, gene conversion, and class switch recombination. These discoveries provided the basis for our current understanding of the mechanisms underlying these processes, several immunodeficiency syndromes, and cancers.

2.1 INTRODUCTION

The generation of a large and highly diverse antibody repertoire is an important part of our immune response against pathogens. Antibody diversity is achieved through several distinct molecular events that drastically alter the structure and coding information of antibody genes. The first is V(D)J recombination, a site-specific recombination process that joins together the segments of the antibody genes that encode the antigen-binding domain. The second is somatic hypermutation, nucleotide substitutions that predominantly occur within antibody variable regions. The third is class switch recombination, a region-specific recombination process that changes the antibody isotype and therefore its function. Remarkably, lymphocyte-specific proteins initiate all three of these processes. V(D)J recombination is triggered by

RNA and DNA Editing: Molecular Mechanisms and Their Integration into Biological Systems,
Edited by Harold C. Smith

RAG recombinase-mediated DNA cleavage, and both somatic hypermutation and class switch recombination are triggered by AID-catalyzed deoxycytidine to deoxyuridine deamination events. Ubiquitous DNA repair factors are required subsequently to ensure that initiating lesions persist as V(D)J joins, hypermutations, or isotype switches. Overall, a very good mechanistic understanding of the role of DNA editing in antibody diversification has been achieved, and this in turn has enriched our appreciation of the causes of several human immunodeficiency syndromes and cancers.

2.2 BEFORE AID

2.2.1 Without DNA (Darkness) and with DNA (Light)

The antibody repertoire is challenged with the task of being able to recognize and effectively neutralize (directly or indirectly) any potential pathogen that can invade the body. The potentially infinite number of foreign antigens suggests that the antibody repertoire of an individual must be similarly diverse. However, with models of antibody production being limited to a finite amount of genomic DNA, discerning how such diversity is generated has been one of the most challenging problems of immunology.

Major breakthroughs in understanding the intricacies of antibody diversity are all rooted in the 1953 DNA structure reports by Watson and Crick, by Franklin and Gosling and by Wilkins (1, 2, 207). The stunningly simple antiparallel nature of the DNA double helix and the elegance of A's neatly pairing with T's, as well as G's pairing with C's, immediately revolutionized scientists' views of the gene and of heredity. Of course, these revelations also provoked speculation and experimentation toward elucidating the molecular nature of antibody genes and antibody variation. The timeline in Figure 2.1 therefore begins in 1953, and it highlights some of the major discoveries that have since led to our current understanding of antibody diversity.

2.2.2 Prominent Early Models for Antibody Diversification

One of the earliest problems with which investigators in the field grappled was whether antibodies were germline-encoded (preexisting á la Darwin and natural selection) or generated somatically (perhaps even directed by antigen á la Lamarck and directed evolution theory). Building upon much pre-DNA work in the antibody (serology) field, Jerne's natural selection theory of antibody production proposed that antigens would function by selecting preexisting, circulating antibodies. The recognition of antigen would trigger a cell-based response that would replicate and amplify the number of antigen-specific antibodies (see reference 3 and references therein). Jerne further proposed that self-reactive antibodies would be eliminated as part of a self-tolerance mechanism. The key concept carried forward from this proposal was that antigen is simply a signal that is engaged by a preexisting antibody.

In the late 1950s, Burnet published a landmark paper on antibody formation (4). Burnet's clonal selection theory of antibody formation proposed that each

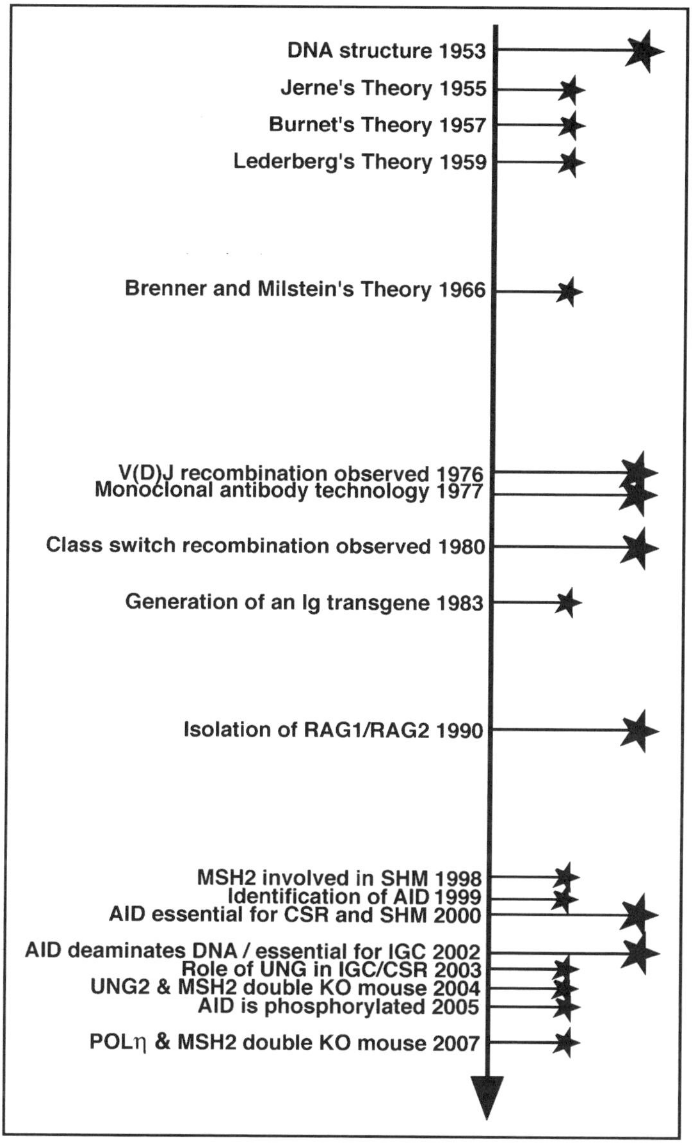

Figure 2.1 Key discoveries that contributed to understanding antibody diversity. Due to the molecule-focused nature of this review, the timeline begins with the structure of DNA and ends with some of the most recent and notable discoveries. Starred time points are separated into two tiers based roughly on the relative importance of each breakthrough.

antibody-producing lymphocyte would have the intrinsic capability to produce antibody of only one antigen-specificity. Antigen recognition by an antibody-like molecule on the cell surface was postulated to drive the expansion of the antigen-specific lymphocyte, ultimately resulting in an abundance of antigen-specific antibodies and an effective antibody response. The net result was that antigen would be neutralized and the individual would be left with a reservoir of cells capable of responding rapidly to the same antigen in the future (so-called memory B cells). Burnet further hypothesized that the individual clones would arise due to somatic mutations during embryogenesis and that self-reactive clones would be eliminated, but also acknowledged that it was difficult to envisage how clones reactive to every imaginable antigen could be produced in a few generations. Burnet's theory provided the basis for much of what we know today.

Lederberg added an important extra dimension to Burnet's theory (5). Lederberg postulated that antibody-producing B cells would be specialized and capable not only of recognizing antigen, but also of fine-tuning the response to antigen by enabling "hypermutations" to accumulate in antibody genes. Lederberg imagined interplay (signaling) between antigen and an antibody-producing B cell that would ultimately lead to a much more specific, higher-affinity antibody (coupled to selection and cell proliferation). The antigen-induced mutations that Lederberg proposed would therefore serve both to diversify the germline antibody gene repertoire and to provide the substrate for the generation of mature antibodies. Lederberg also appreciated that antigen-responsive B cells could persist for a long time in the absence of antigen (immunological memory) and that a single cell would only produce one type of antibody (6–8). Lederberg's work set the stage for other antibody diversification models, such as the nick-induced mutagenic DNA polymerization model of Brenner and Milstein (9), but concrete data in this area were sparse for many decades.

2.2.3 How Protein Sequencing Technology Enabled an Understanding of Antibody Diversity

Through the 1950s and 1960s, advances in protein chemistry helped to determine the polypeptide composition of immunoglobulin (Ig) (see Table 2.1 for a list of abbreviations) (10–12). It is now known that an antibody is composed of two heavy (H) chains and two light (L) chains linked together by disulfide bridges, as illustrated in Figure 2.2. The H chain is expressed from the IgH locus, and the L chain is expressed from either the Igλ or the Igκ locus.

The dawn of polypeptide sequencing facilitated the discovery that antibodies are remarkably variable (10, 13–19). Moreover, the variation appeared to be confined to the amino-terminal (antigen binding) regions, in contrast to the carboxy-terminal regions that appeared to be invariable. It was difficult to envisage how one part of a gene could be so variable while another part of the same gene remained constant. Several theories therefore imagined the fusion (joining) of two pieces of DNA, RNA, or protein to form a single chain (16–19). However, the notion that the genome cannot possibly have the capacity to code for as many genes as there are potential antigens stimulated Brenner and Milstein to revisit Lederberg's hypermutation theory (5). Brenner and Milstein postulated that an enzyme expressed at specific stages of differentiation of the antibody-producing cells would specifically recognize and cut the antibody gene DNA (9). They further proposed that errors could be introduced at a high frequency during repair of the DNA breaks and noted that mutant polymerases had been observed to increase mutation frequencies in other systems. Remarkably,

TABLE 2.1 Full Gene Names and Abbreviations

Abbreviation	Full Name
53BP1	p53 Binding protein 1
AID	Activation-induced deaminase
APEX1	APurinic endonuclease 1
ATM	Ataxia–telangiectasia mutated
ATLD	Ataxia–telangiectasia mutated-like disorder
BCL2	B-cell lymphoma 2
BCL6	B-cell lymphoma 6
BER	Base excision repair
BRCA2	Breast cancer associated 2
C region	Constant region
CD40L	CD40 ligand
CRM1	Chromosome region maintenance 1 (Exportin1)
CSR	Class switch recombination
CVID	Common variable immunodeficiency
dA	2′-Deoxy-adenine adenosine
dC	2′-Deoxy-cytidine
dG	2′-Deoxy-guanine guanosine
DNA-PKcs	DNA-dependent protein kinase catalytic subunit
DSB	Double-strand break
dsDNA	Double-strand DNA
dT	2′-Deoxy-thymine thymidine
EXO1	Exonuclease 1
FAS	FAS cell surface antigen
FEN1	Flap endonuclease 1
FGFR3	Fibroblast growth factor receptor 3
γH2AX	Phosphorylated histone variant H2AX
H chain	Heavy chain
HIGM	Hyper-IgM syndrome
HIGM-ED	Hyper-IgM with ectodermal dysplasia
ICOS	Inducible T-cell CO-stimulator
Ig	Immunoglobulin
IGC	Immunoglobulin gene conversion
IKKγ	Inhibitor of κ L-chain enhancer in B-cells—kinase α
KU70	KU autoantigen, 70-kDa subunit
KU80	KU autoantigen, 80-kDa subunit
L chain	Light chain
LIG4	DNA ligase IV
MLH1	Mut L homolog 1
MLH3	Mut L homolog 3
MMR	Mismatch repair
MRE11	Meiotic recombination 11a homolog
MRN	MRE11/RAD50/NBS1
MSH2	Mut S homolog 2

TABLE 2.1 (*Continued*)

Abbreviation	Full Name
MSH3	Mut S homolog 3
MSH4	Mut S homolog 4
MSH5	Mut S homolog 5
MSH6	Mut S homolog 6
MYC	MYC oncogene
NBS1	Nijmegan breakage syndrome 1
NES	Nuclear export sequence
NFκB	Nuclear factor κ B
NHEJ	Nonhomologous end joining
NLS	Nuclear localization signal
p53	Tumor suppressor protein 53
PAX5	Paired box gene 5
PCNA	Proliferating cell nuclear antigen
PIM1	Oncogene PIM1, serine/threonine kinase
PKA	Protein kinase A
PMS2	Postmeiotic segregation 2
PNK	Polynucleotide kinase
POL	Polymerase
RAD50	Radiation-sensitive 50
RAD51B	Radiation-sensitive 51B
REV1	Reversionless 1/terminal deoxycytidyl transferase
REV3	Reversionless 3/POL z catalytic subunit
RhoH/TTF	Ras homolog gene family member H/translocation three four
RPA	Replication protein A
RSS	Recombination signal sequence
SCID	Severe combined immunodeficiency
SHM	Somatic hypermutation
SMUG1	Single-strand selective monofunctional uracil DNA Glycosylase 1
S region	Switch region
SSA	Single-strand annealing
ssDNA	Single-strand DNA
TACI	Transmembrane activator and CAML interactor
UNG2	Uracil-*N*-glycosylase 2
V region	Variable region
V(D)J recombination	Variable (diversity) joining recombination
XLF	XRCC4-like factor (cernunnos)
XP-V	Xeroderma pigmentosum-variant
XRCC4	X-ray cross-complementation 4

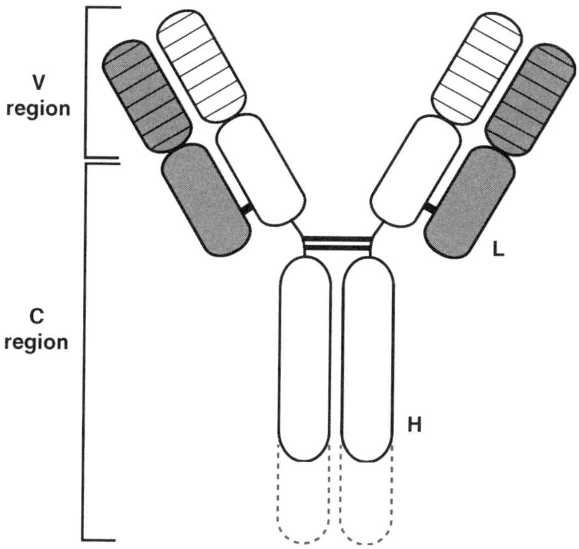

Figure 2.2 Antibody schematic. A typical antibody is composed of two heavy (H)-chain polypeptides (white) and two light (L)-chain polypeptides (gray). The amino-terminal variable (V) regions (hashed boxes) are responsible for antigen recognition. The carboxy-terminal constant (C) regions determine the effector function of the antibody. The C regions of IgM and IgE are larger (indicated by dashed lines) than those of IgD, IgG, and IgA. Lines between the H chains and between the H and L chains depict the disulfide bridging required for proper antibody structure (linkages vary for each isotype).

both the DNA joining and somatic hypermutation theories turned out to be partially correct, as discussed below.

2.2.4 Somatic DNA Rearrangements Underpin V(D)J Joining and Create the Primary Antibody Repertoire

Hozumi and Tonegawa made a breakthrough discovery while mapping the mouse Ig loci (20). They used gel fractionation and solution hybridization techniques to show that the variable and constant DNA regions were closer together in DNA isolated from somatic B cells (a mouse B-cell tumor line) than in DNA isolated from embryonic tissues (effectively representing germline DNA). They deduced that the joining of the V and C Ig gene regions, initially located some distance apart, must occur somatically. This discovery provided the first evidence for antibody gene joining at the DNA level.

We now know that during B-cell development in the bone marrow, the Ig loci must undergo a number of sequential and highly regulated somatic recombination events to produce a functional variable gene (Figure 2.3A). These site-specific V(D)J recombination reactions juxtapose a single variable (V) gene segment with a single diversity (D) gene segment (H chain only), together with a joining (J) gene segment. The various combinations of gene segments can give rise (in humans) to 320 different L chains and 10,530 different H chains. The various pairings of the κ or λ L chains with

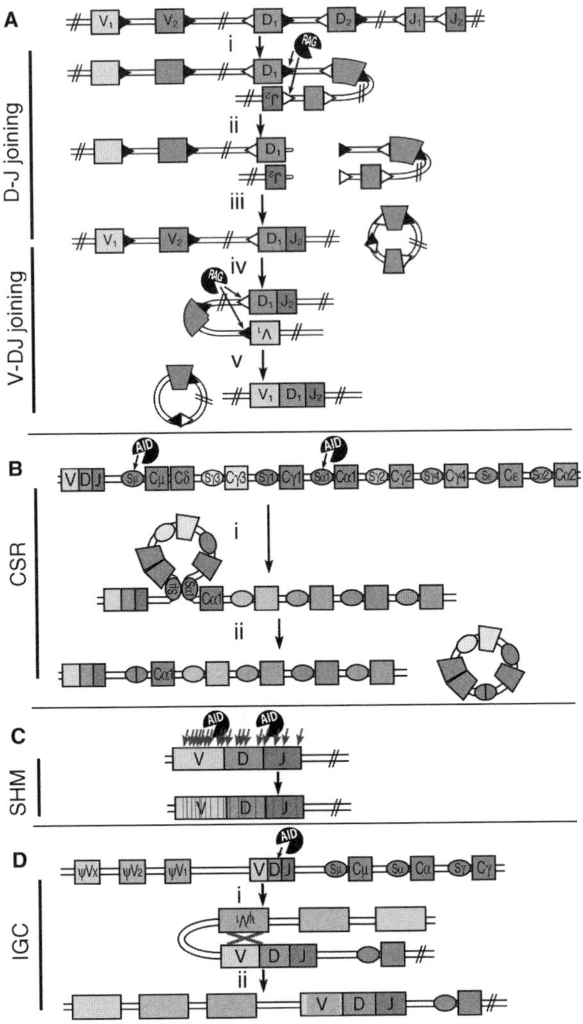

Figure 2.3 Similarities and differences between the mechanisms of V(D)J recombination, CSR, SHM, and IGC. (A) V(D)J recombination. (i, ii) RAG1 and RAG2 initiate the reaction by nicking the DNA at the D and J region RSSs (open and filled triangles, respectively). DNA hairpins are generated at the coding end breaks (left-hand fragments), and DSBs are generated at the signal ends (triangles on the right-hand fragment). (iii) DNA-PK binds to the DNA ends and activates Artemis to open the hairpins; and XRCC4/LIG4/XLF ligate the DNA ends, producing a coding joint (left) and a signal joint (right). (iv, v) The process is repeated to join a V region to the assembled DJ region. See text for details. (B) Class switch recombination. (i) CSR is initiated by AID-induced deamination events at the Sμ and a downstream S region (e.g., Sα1). (ii) Intervening DNA is excised and the prerearranged V(D)J region is joined to a new C region (e.g., Cα1). A detectable switch circle byproduct is also generated. See text for details. (C) Somatic hypermutation. Point mutations are introduced into the Ig gene variable (V) region by deamination of cytosine bases to uracils. See text for details. CSR and SHM can occur independently or simultaneously during B-cell development. (D) Immunoglobulin gene conversion. The chicken IgH locus is illustrated. (i) AID-dependent DNA cytosine deaminations in the V region lead to (ii) homologous recombination with an upstream ΨV region, resulting in the templated replacement of V gene sequence with that from the ΨV donor. See text for details. (*See color insert.*)

the H chains further multiply the potential diversity of the primary antibody repertoire sculpted by the V(D)J reaction.

Elucidating the molecular players in V(D)J recombination was not trivial. An elegant study in 1989 by Baltimore and co-workers resulted in the identification of the RAG1/RAG2 (*R*ecombination-*A*ctivating *G*ene 1 & 2) V(D)J recombinase (21, 22). Perhaps the most remarkable part of this study was the fact that both RAG1 and RAG2 were isolated on the same fragment of genomic DNA.

V(D)J recombination is initiated by the lymphoid-specific RAG1/2 recombinase and completed by the ubiquitous nonhomologous end-joining (NHEJ) machinery (Figure 2.3A). RAG1/2 initiate the V(D)J recombination reaction by generating site-specific DNA breaks. This can be imagined by likening RAG1/2 reactions to those of a cut-and-paste transposon (23–25). In essence, RAG1/2 recognizes and cuts at specific repeated DNA sites called recombination signal sequences (RSSs), much like transposases utilize short DNA repeats for transposition. RAG-mediated DNA single-stranded nicks occur adjacent to an upstream and a downstream Ig gene segment (e.g., first D to J and then V to DJ). The 3′-hydroxyl group on the nicked DNA strand then attacks a phosphodiester bond on the opposite DNA strand, generating a closed hairpin at the end of the DNA encoding the Ig gene segment (coding end). The hairpin DNA ends are then opened by the exo/endonuclease, Artemis, to form double-stranded DNA breaks (DSBs). Phosphorylation of Artemis by the DNA-dependent protein kinase catalytic subunit (DNA-PKcs) is required to activate the endonucleolytic activity. Many of the free DNA ends are subjected to end processing, whereby exonucleases remove, or polymerases add, nucleotides (e.g., terminal deoxynucleotidyl transferase). Such end processing reactions thereby further contribute to Ig gene diversity. Finally, NHEJ factors KU70/80 and DNA-PKcs bring the double-stranded ends together for ligation by a complex consisting of XRCC4 (X-ray cross-complementation factor 4), LIG4 (DNA Ligase IV), and XLF (XRCC4-like factor/Cernunnos). A byproduct of V(D)J recombination is the production of DNA circles containing the RSSs (Figure 2.3A). These molecules can be detected in cells, but they serve no known function and are ultimately degraded (26–28).

The joining of the multiple gene segments occurs sequentially, and each step must be completed successfully before the cell can develop further and rearrange the subsequent segments (a failure to produce a cell-surface-expressed Ig leads to cell death). D-to-J joining at the H-chain locus occurs first at the early pro-B-cell stage. V to DJ joining occurs subsequently at the late pro-B-cell stage. At this time, a functional H-chain molecule can be expressed on the surface of the cell with the help of a surrogate light chain, since the L chain has yet to rearrange. Expression of the pre-B-cell receptor triggers proliferation and the formation of a small pre-B-cell, which then rearranges the L-chain locus. In a process known as allelic exclusion, the recombination machinery is then shut off to prevent expression of more than one allele of the H- and L-chain genes. A functional B-cell receptor (membrane-anchored Ig) can then be expressed on the surface of the B-cell in the form of IgM (or IgD through alternative splicing). The B-cell receptor is then able to recognize antigen and the cell responds accordingly.

Lastly, as predicted by the early models, the resulting immature B cells undergo a selection process in the bone marrow to ensure that they are not self-reactive before being released into the blood. The surviving cells become mature naïve B cells, migrate into the peripheral blood and secondary lymphoid tissues, and are ready to respond to antigen (29).

2.2.5 Additional Antibody Diversity by Somatic Hypermutation (and Gene Conversion in Some Animals)

The ability to isolate antibodies from B cells before and after exposure to antigen (30, 31) and the development of powerful protein sequencing techniques (32) combined to demonstrate that much antibody variation accumulated AFTER antigen recognition (33–36). Interestingly, subsequent B-cell stimulations (immunizations) with the same antigen caused even more mutations in the antibody sequence to accumulate with clear clustering to so-called hypervariable regions (37).

In most instances, somatic hypermutation (SHM) appears in the antibody gene V region after the B cells recognize antigen (i.e., in response to antigen) (Figure 2.3C). SHM occurs primarily when B cells are located to a lymphoid micro-compartment called the germinal center (a complex, but organized, clustering of B cells, T cells, antigen-presenting cells, and other cell types). Cells expressing Ig with an increased affinity for antigen are able to successfully compete with their neighbors for antigen and positive stimuli, and they subsequently proliferate. This selection process is known as "affinity maturation." Multiple rounds of mutation and affinity maturation result in antibodies with extremely high affinity for a specific antigen. Cells with lower affinity are unable to compete for positive stimuli and undergo apoptosis (33, 38, 39). These revelations provided important steps toward understanding the full mechanism of Ig gene diversification.

Several animals, including birds, rabbits, cows, pigs, and horses, only encode a single functional V gene segment and one J gene segment and therefore gain very little diversity from V(D)J recombination (40–44). Instead, these species use a combination of SHM and a pseudogene-templated recombination process called Ig gene conversion (IGC) (Figure 2.3C). IGC is a specialized type of homologous recombination reaction that occurs between two highly similar DNA sequences (sometimes also called homeologous recombination).

Significant progress on understanding the mechanism of IGC has been made using the chicken B-cell line DT40, which is derived from a bursal B-cell lymphoma and undergoes IGC constitutively (41, 45, 46). Most models for IGC are based on homologous recombination studies in yeast. A DNA DSB within the expressed V region is thought to trigger the acquisition of a similar Ig DNA sequence from one of many nonexpressed pseudo-V gene sequences located upstream (Figure 2.3D). The net result is diversification of the expressed V region by partial replacement of its genetic information.

2.2.6 Altering Antibody Function by Class Switch Recombination (Isotype Switching)

Increasingly sophisticated gel electrophoresis and protein sequencing techniques led to the discovery of multiple distinct antibody isotypes that differ in the amino acids that make up the carboxy terminal constant (C) region (47). In humans, there are nine different constant regions or isotypes (IgA1, IgA2, IgD, IgE, IgG1, IgG2, IgG3, IgG4, and IgM), each with a different effector function. IgA is secreted across epithelia, such as the lining of the intestine, to neutralize pathogens by preventing them from binding to and entering the cells; IgE activates mast cells and the inflammatory response; IgG marks pathogens for engulfment by phagocytes, and both IgM and IgG activate the complement cascade. IgD has no known function.

A mechanistic explanation for the formation of alternative antibody isotypes came in 1980 with observations of DNA rearrangements (deletions) at the IgH locuss (48–50) (Figure 2.1). In contrast to the site-specific V(D)J recombination process, antibody isotype class switch recombination (CSR) was shown to occur between C/G-rich, repetitive, and partially palindromic DNA sequences that make up switch (S) regions, which are located immediately upstream of each C region (Figure 2.3B). A CSR event requires the excision of the C region of the expressed isotype in order to activate expression of an alternative, downstream isotype. DNA DSBs were therefore implicated in CSR, because it was difficult to imagine such large-scale deletions without DNA breakage, end-processing, and re-ligation (see Section 2.3.4.3 for a current mechanistic view). Unlike IGC, CSR was clearly not a homologous-type recombination reaction because extensively homologous S-region DNA sequences do not exist.

Four decades after the DNA structure reports and considerable model building and testing, it was clear that *both* existing germline diversity and additional somatic diversity contributed to the remarkable Ig gene repertoire in vertebrates. Unlike V(D)J recombination, which was known to be mediated by the RAG1/2 recombinase, the players required to trigger SHM, IGC, and CSR and the mechanistic details of these processes remained elusive for almost two decades.

2.3 AFTER AID

2.3.1 A Novel Deaminase Is Required for CSR, SHM, and IGC

A breakthrough discovery came with the identification and cloning of a mouse gene called *AID* by Honjo and colleagues in 1999 (51). Although a number of factors, such as DNA repair proteins, had been shown to influence CSR, an essential player had yet to be identified. Based on the observation that an inhibitor of protein synthesis (cyclohexamide) strongly inhibited CSR, they reasoned that a specific protein (or proteins) must be upregulated when the cells are induced to switch. In order to identify CSR specific factors, the Honjo laboratory used subtractive hybridization to compare the RNA populations from mouse B cells in the resting state to cells that had been induced to undergo CSR. One of seven transcripts upregulated in switch-induced cells was predicted to encode a protein of 198 amino acids, which displayed 34% amino acid identity to mouse APOBEC1 (see Chapters 10 and 16).

Interestingly, like the *APOB* mRNA editing protein, APOBEC1, AID had a cytosine/cytidine deaminase motif. This knowledge inspired the original and official GenBank name for this protein, Activation-induced cytidine deaminase (AICDA), although it is now popularly called Activation-induced deaminase (AID). However, unlike APOBEC1, AID displayed no affinity for AU-rich RNA and was unable to deaminate cytidine to uridine in *APOB* mRNA (51) (see Chapter 10). *AID* transcripts were detected specifically in germinal center B cells from immunized mice but not in T-cell populations, strongly supporting a potential role for AID in CSR. The human *AID* gene was subsequently cloned, with the DNA sequence and the mRNA expression profile proving to be highly similar to that of murine *AID* (52). The structure and position on the genomic locus was also determined; *AID* was encoded by five exons located on the small arm of chromosome 12 (12p13). The five-exon gene structure and genomic location of *AID* are similar to those of *APOBEC1*, suggesting that these two genes share a common evolutionary intermediate (discussed further in Section 2.4.4 and Figure 2.5; also see Chapter 16).

To determine the role of AID in CSR, the Honjo laboratory generated AID-deficient ($AID^{-/-}$) mice (53). At 10 weeks of age, these mice had very low or undetectable levels of switched antibody isotypes in their serum in comparison to heterozygous littermates. The presence of normal or elevated levels of IgM signified that AID is not required for V(D)J recombination. $AID^{-/-}$ splenic B cells also failed to undergo CSR when stimulated in vitro, providing evidence that AID is required for CSR. The presence of germline transcripts indicated that AID-deficiency does not affect S region accessibility, and it more likely disrupts the CSR machinery itself. Further analysis of $AID^{-/-}$ mice revealed enlarged germinal centers containing many highly activated B cells and, surprisingly, very low levels of SHM in the Ig variable (V) region. AID expression was therefore essential for both CSR of the Ig constant region and SHM in the Ig V region (53).

The elevated levels of serum IgM (hyper-IgM) observed in the $AID^{-/-}$ mice (53) helped determine the cause of an autosomal recessive human immunodeficiency called hyper-IgM syndrome type 2 (HIGM2) (54). By analyzing the segregation of microsatellite markers in a number of families with affected members, the susceptibility locus was mapped to chromosome 12p13, the same region to which human *AID* had been mapped. Sequencing of the *AID* genomic locus of 18 patients from 12 unrelated families revealed 10 different mutations in the coding regions including missense mutations, premature stop codons, and deletions leading to frameshift mutations. The phenotypes of the patients analyzed matched those of the $AID^{-/-}$ mice including defective CSR, lower levels of SHM, and enlarged lymph nodes, which could be attributed to giant germinal centers containing large numbers of B cells (54).

In light of the astonishing requirement for AID in the seemingly distinct processes of CSR and SHM, the Buerstedde and Neuberger laboratories independently determined AID's role in the third type of antibody gene diversification, IGC (55, 56). Both groups disrupted *AID* in the chicken bursal B-cell lymphoma line, DT40, and observed that IGC was completely abolished. These data demonstrated that AID is essential for all three antibody diversification processes at both the variable and constant regions. Importantly, this result also suggested that AID is responsible for an event common to the three mechanisms. The phenotype of the AID-deficient cells was complemented by both chicken and human *AID* cDNAs, indicating that AID is highly conserved. In further support of this statement, chicken AID shares greater than 85% amino acid identity with the human and mouse proteins, and shows a similar tissue distribution, being detected mostly in the B-cell compartments (bursa of Fabricius and spleen in chickens) (56).

Within four years of the identification of AID by Honjo's group, AID was established as the antibody gene diversification master protein (53, 55, 56). This brought the question of the mechanism of AID's action to the forefront: How can one protein trigger three seemingly distinct processes, and what is the target substrate for AID's predicted deaminase activity?

2.3.2 AID Is a DNA Cytosine Deaminase that Directly Triggers Antibody Diversification

Due to AID's clear homology to APOBEC1, AID was initially thought to be an RNA editing enzyme. A model was proposed by Honjo whereby AID would edit cytidines to uridines in a pre-mRNA that encodes an endonuclease that would generate the DNA breaks during CSR and SHM (57). In favor of this proposal, AID was shown to be able to bind ssRNA (58, 59). The RNA-editing model requires a new endonuclease to be synthesized after the induction and activity of AID. Consistent with this hypothesis, inhibition of protein synthesis suppressed CSR and SHM in the presence of an ectopically expressed and induced AID-estrogen receptor fusion protein (60, 61). The formation of DSBs in the S and V regions was also inhibited in these cells, as observed by chromatin immunoprecipitation analysis of γ-H2AX foci (61, 62). However, a target mRNA substrate for AID has not been identified.

Several alternative hypotheses were also proposed. Neuberger and Scott suggested that AID may function by editing the V-region and S-region sterile transcripts triggering recruitment of DNA repair machinery to the resulting RNA–DNA mismatches (63). Jacob and Bross, and subsequently Scharff and colleagues, hypothesized that AID might function even more directly in Ig gene diversification by deaminating cytosine bases in DNA (64, 65).

Neuberger and colleagues independently proposed a DNA deamination model (see Sections 2.3.3 and 2.3.4) and provided the first unambiguous evidence that AID was capable of deaminating cytosine bases in DNA (66). Expression of AID in *E. coli* triggered an increase in mutation frequency and a dC/dG to dT/dA-biased mutation pattern that were both greatly enhanced in uracil excision repair defective cells. This demonstrated that AID could trigger the accumulation of dU lesions in DNA, which in turn manifested as dT's upon DNA synthesis.

The genetic experiments of Neuberger and colleagues were soon supported by biochemical experiments demonstrating that purified AID protein can deaminate ssDNA, but not dsDNA, RNA, or DNA/RNA hybrids (58, 59, 67). Purified AID also had a preference for WRCY hotspot motifs as observed for SHM in vivo (where W = A/T, R = A/G, Y = C/T), suggesting that the targeting preference is intrinsic to the AID protein itself. Deamination of dsDNA occurred when coupled to transcription, with a preference for (but not exclusive to) the nontranscribed strand and also greater activity on G-rich strands than C-rich strands, as observed in vivo.

The demonstration of DNA deamination by AID provided an explanation for AID's ability to initiate three seemingly distinct DNA alteration processes and thereby transformed the way the field viewed Ig gene diversification events (58, 59, 66, 67). This knowledge supplied a building block on which further experiments and models were developed. The conceptual shift from RNA to DNA deamination

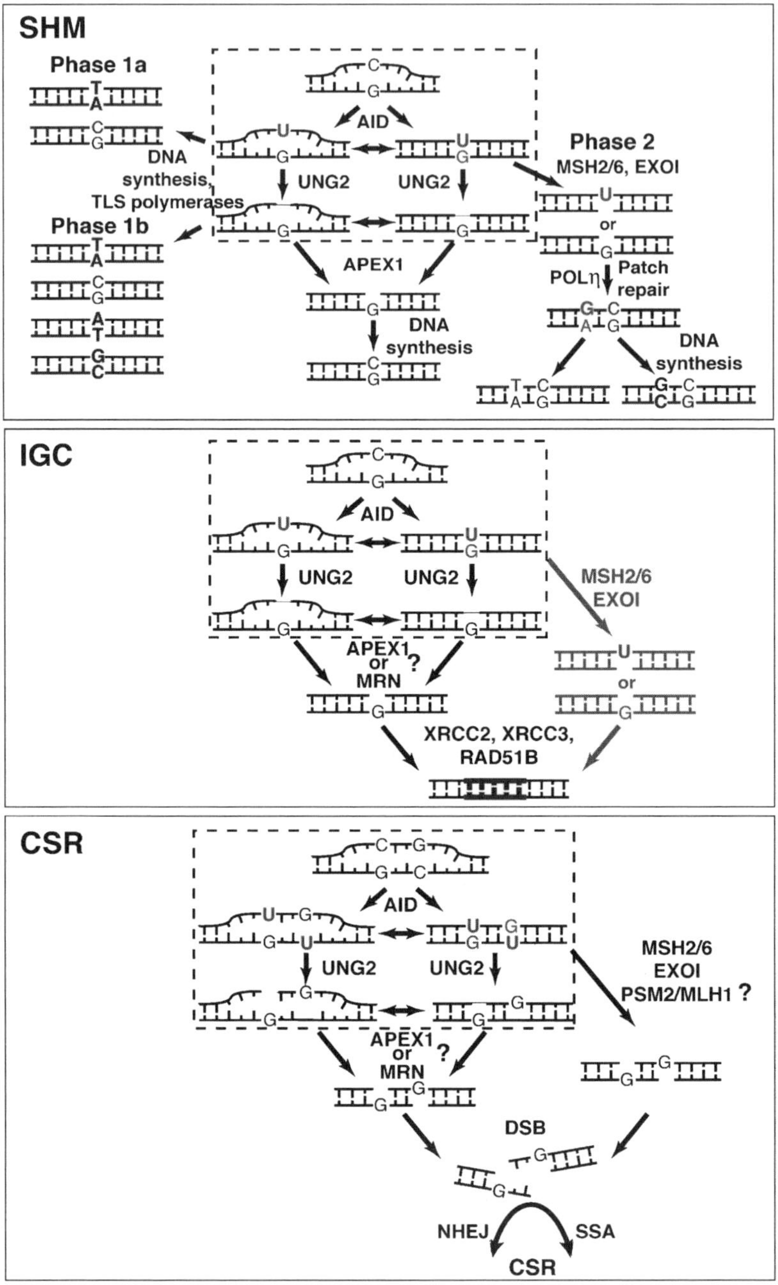

Figure 2.4 The molecular events leading to SHM, IGC and CSR. All three events are initiated by deamination of dC → dU by AID (highlighted by dashed inset boxes). Excision of U by either UNG2 and/or MSH2/6 is also a prominent feature of SHM, IGC, and CSR. Most U lesions are repaired faithfully by the UNG2-dependent base excision repair pathway, and therefore they go unnoticed (shown in the top panel but not in the others for clarity). (Top panel, SHM) Phase 1a SHM occurs by DNA synthesis over the U, generating C → T transition mutations. Virtually any

transformed the field and provided the catalyst for the study of other instances of DNA deamination (see Chapter 11). It is currently difficult to fully describe the importance of this new paradigm, since the physiological roles of the other DNA editors have not been completely revealed (see Chapters 11 and 16).

2.3.3 The Importance of Uracil Bases in DNA *In Vivo*

The DNA deamination data, along with the prior observations that several B-cell lines and cells deficient in the mismatch repair (MMR) protein, MutS Homolog 2 (MSH2), exhibited a bias toward mutations at dC/dG base pairs and that all three Ig diversification processes required AID, inspired Neuberger and colleagues to develop the DNA deamination model for AID in antibody gene diversification (66). In this model, AID deaminates cytosine bases in DNA to uracils (Figure 2.4). The noncanonical uracil could then be removed by the base-excision repair (BER) machinery. DNA synthesis over dU's would generate dC to dT transition mutations, whereas synthesis over abasic sites could generate both the transition and transversion mutations observed in SHM. Mutations at dA/dT could be generated during patch repair involving the MMR proteins, MSH2 and MSH6 (68, 69). Template-mediated repair could lead to IGC, whereas staggered DSBs generated by BER occurring on both DNA strands could lead to CSR. The DNA deamination model thereby accounted for AID's involvement in all three antibody diversification processes.

The second critical test of the DNA deamination model was to determine the role of uracil-*N*-glycosylase 2 (UNG2) in SHM, IGC, and CSR. Di Noia and Neuberger initially demonstrated that inhibition of UNG2 in DT40 by the bacteriophage protein, Ugi, caused a shift to 86% transition mutations, compared to only 38% in the controls (70). UNG2-depleted DT40 also showed higher levels of SHM, with no effect on the

DNA polymerase could promote such synthesis, since Watson–Crick base-pairing rules are obeyed, but the major replicative polymerases (α/ε) are most likely. Phase 1b SHM occurs when DNA is synthesized opposite an abasic (noninstructive) site. Such DNA synthesis requires TLS polymerases (see text). Phase 2 (A/T-biased) SHM requires MSH2/6 recognition of the mispaired bases and EXOI, which likely collaborate to create a single-stranded DNA gap that can be filled by an error-prone TLS polymerase, such as POLη. (Middle panel, IGC) Gene conversion is initiated predominantly by the excision of U by UNG2, leaving an abasic site. The endonuclease that nicks the DNA in this mechanism has not been determined. The gene conversion event appears to occur through homologous recombination because RAD51 paralogs are important (XRCC2, XRCC3, and RAD51B). MMR does not appear to be involved in IGC (indicated in gray for comparison only). (Bottom panel, CSR) Isotype switching most likely requires the excision of U and strand nicking to occur at neighboring sites on opposite strands to generate a staggered DSB. U excision by UNG2 appears to be the predominant pathway in humans and mice, with MMR playing a minor role. Alternatively, DNA synthesis could readily turn a single-strand nick into the necessary double-stranded break. The subsequent processing, joining, and ligation of the DNA ends are mediated by NHEJ or by SSA. See text for details. (*See color insert.*)

overall distribution of the mutations. In a complementary study, Neuberger and colleagues analyzed SHM and CSR in UNG2-deficient mice (71). The mice also had a bias toward transition mutations at dC/dG in the V region, but mutations at dA/dT and targeting preferences were unaffected. A CSR defect was also observed, with higher titers of serum IgM and lower titers of IgG1, IgG3, and IgA compared to the controls, indicating that the removal of uracils by UNG2 is an important part of both SHM and CSR.

Based on these data, Durandy and colleagues speculated that other hyper-IgM syndromes might be due to UNG2-deficiencies. They therefore sequenced the *UNG* genes from a group of patients with a HIGM2-like syndrome with a transition mutation bias at dC/dG base pairs but no mutations in the *AID* gene (72). Recessive mutations were observed in the *UNG2* genes, and UNG activity was undetectable in B cells from these patients. The CSR defect displayed by these patients was more severe than that seen in the UNG2-deficient mice, and DSBs were undetectable when their B cells were induced to switch in vitro. Induction of UNG2 expression was also observed when normal B cells were activated to undergo CSR, consistent with a crucial role for UNG2 in CSR.

The Honjo laboratory also observed that inhibition of UNG2 by Ugi inhibited CSR and SHM (61, 73). Surprisingly, Ugi did not appear to affect the DNA cleavage step as measured by the formation of γ-H2AX foci in the V and S regions by chromatin immunoprecipitation, suggesting that UNG2 may not be required for this step (61, 73). In support of this idea and the RNA-editing hypothesis, three UNG2 catalytic mutants were able to support CSR in UNG$^{-/-}$ B cells, although two other UNG2 catalytic site mutants did not display this activity (73). However, in a technical comment on these data, Stivers pointed out that Ugi does not completely inhibit the activity of human UNG2, even when in 10-fold excess, but it may block interaction with downstream CSR factors (74, 75). The three UNG2 catalytic mutants used also showed residual uracil excision activity, which may have been sufficient to promote DSBs and CSR (74, 76, 77).

Together, these data demonstrated that UNG2 most likely plays a key role in antibody diversification upstream of DSB formation and supported the DNA deamination model (70–72). However, UNG2-deficiency did not completely ablate CSR or affect the mutations at dA/dT base pairs, suggesting that there is a backup activity for uracil processing. This activity could be from another DNA glycosylase(s) or from MMR, since a deficiency in MSH2 or MSH6 results in a decrease in both CSR and mutations at dA/dT (68, 69). The Neuberger group therefore overexpressed another cellular DNA glycosylase, called SMUG1, in *UNG2*$^{-/-}$ mice and demonstrated that SMUG1 was unable to compensate for the UNG2 deficiency (78). They also observed that SMUG1 expression decreases during B-cell activation and that its overexpression triggers DNA repair rather than antibody diversification (79). In contrast, the *UNG2&MSH2*-double knockout mice were completely deficient for CSR and the A/T program of SHM (78). CSR is therefore triggered mainly through uracil excision by UNG2, with MMR providing a backup pathway, and the A/T program of SHM occurs mainly through uracil processing by MSH2/MSH6, with UNG2 providing a backup pathway (Table 2.2; Figure 2.4).

TABLE 2.2 Genetic Requirements for Antibody Gene Diversification[a]

Protein	V(D)J	SHM	IGC	CSR	References
AID	None	Essential	Essential	Essential	51–53
UNG2	nd[b]	Involved	Essential	Involved	70–72, 94, 101, 102
SMUG1	nd	None	None	None	78, 79
MSH2	None	Involved	Small/none	Involved	69, 78, 93, 128, 130
MSH3	nd	None	None	None	41, 56, 68, 128, 129
MSH6	nd	Involved	None	Involved	41, 56, 68, 94, 128, 129
MSH4	nd	nd	None	nd	41, 56
MSH5	nd	nd	nd	Involved	131
MLH1	nd	Small	nd	Involved	99, 112, 113, 127
MLH3	nd	Small	nd	Involved	96, 97
PMS2	nd	Small	nd	Involved	98, 100, 112, 113
EXO1	nd	Involved	nd	Involved	95
REV1	nd	Involved	None	nd	85
POL ζ	nd	Involved	nd	nd	86, 87
POL η	nd	Involved	nd	nd	91–93
POL θ	nd	Involved	nd	nd	84
PCNA	nd	Involved	nd	nd	90
MRE11	nd	nd	nd	Involved	103, 104, 121
NBS1	nd	nd	nd	Involved	103, 105, 121, 125
XRCC2	nd	None	Involved	nd	106
XRCC3	nd	None	Involved	nd	106
RAD51B	nd	None	Involved	nd	106
BRCA2	nd	None	Involved	nd	107
DNA-PKcs	Essential	Small	Antagonistic	Involved	108, 109, 139–142
KU70	Essential	None	Antagonistic	Essential	108, 109, 137
KU80	Essential	None	nd	Essential	108, 109, 137
LIG4	Essential	nd	nd	Involved	144
XRCC4	Essential	nd	nd	nd	
Artemis	Essential	None	None	None	149
XLF	Involved	nd	nd	nd	150–152
53BP1	None	nd	nd	Involved	135, 136
ATM	nd	nd	nd	Involved	133
H2AX	nd	nd	nd	Involved	134

[a]See main text for details.

[b]Not determined.

These studies demonstrated that AID almost certainly acts by deaminating cytosine bases in the Ig gene DNA to uracils, creating a dU:dG mismatch (70–72, 78). The focal question then became how differential processing of this lesion by the cellular DNA maintenance machinery leads to CSR at the switch regions and SHM (or IGC) at the variable region.

2.3.4 Processing of AID-induced Lesions: The Molecular Mechanism of Somatic-Hypermutation

Mutations have been observed in the Ig V regions of B cells at all four bases at frequencies up to a million-fold higher than other genomic loci during an immune response. This phenomenon was therefore termed "somatic hypermutation" (SHM) (37, 80, 81).

Prior to the identification of AID, Milstein, Neuberger, and colleagues observed that the MSH2-deficient mice displayed increased focusing to mutational hot spots and reduced mutations at dA/dT base pairs in the Ig V regions (69) (see Table 2.2 for a complete list of proteins involved or implicated in antibody gene diversification). Based on these observations, this group proposed two-phase model for SHM. In the first phase, mutations would occur at dG/dC base pairs and be independent of MSH2. The second phase SHM would be triggered by the first phase and result in the introduction of mutations at dA/dT base pairs by later, most evidence points toward this in fact being the case, as discussed below.

In light of the discovery that AID can mutate cytosine bases in DNA to uracils, Neuberger and colleagues further modified their previously proposed two-phase model for SHM (66). During DNA synthesis, dA would be incorporated opposite dU, yielding only dC $\rightarrow$ dT and dG $\rightarrow$ dA transition mutations (Phase 1a) (82) (Figure 2.4, top panel). Analysis of the mutational spectrum in the V regions of the UNG2&MSH2 doubly-deficient mice demonstrated that in the absence of both of these proteins, the uridine residues cannot be removed from the DNA and only Phase 1a SHM is observed (78).

Recognition and removal of dU by UNG2 would create an abasic site. The abasic site would stall the DNA replication folk and require a switch to a trans-lesion synthesis (TLS) polymerase to continue DNA synthesis over the lesion. TLS polymerases are error-prone and therefore likely to generate both transition and transversion mutations, depending on the preference of the polymerase (Phase 1b), and also to amplify SHM by creating mutations at adjacent residues. This idea stimulated a plethora of TLS polymerase knockout and mutational studies. While polymerases (POLs) θ (theta), ζ (zeta), and REV1 are all important for Phase 1b of SHM, no single polymerase is absolutely essential and each TLS POL seems to be able to partially substitute in the absence of another (83–87). However, TLS POLs ι (iota) and κ (kappa) do not appear to be involved (88, 89). Consistent with the requirement for the TLS polymerases, monoubiquitination of proliferating cell nuclear antigen (PCNA) on lysine 164, which mediates the TLS polymerase switch, also appears to be important for proper SHM (90).

Recognition and removal of dU/dG by the mismatch repair machinery was predicted to trigger short patch repair involving an error-prone TLS polymerase that makes mistakes at dA/dT (e.g., polymerase η) (Phase 2). Xeroderma pigmentosum variant (XP-V) is a genetic disorder associated with sensitivity to UV light and a predisposition to sun-induced skin cancers, due to a mutation in the *POL*η gene; the encodes the TLS POLη (eta). B cells from XP-V patients display normal frequencies of SHM, but a decrease in mutations at dA/dT (91). Mice lacking POL η displayed a bias towards mutations at dG/dC base pairs (92). This mutation pattern was noted to resemble that of the MSH2-deficient mice, suggesting a role for POL η in Phase 2 of SHM. However, whereas the MSH2-deficiency resulted in an increase in transition mutations at dG/dC, POL η-deficiency caused a transversion bias at the residual dA/dT base pairs.

MSH2&POLη-double knockout mice completely lacked mutations at dA/dT base pairs (93). Residual mutations at dA/dT were present in the *MSH2*$^{-/-}$ mice, but not in the *UNG2*$^{-/-}$*&MSH2*$^{-/-}$ mice (78), indicating that POL η can also function in patch repair during base-excision repair during Phase 1b. POL η is therefore likely to be responsible for the majority of mutations at dA/dT base pairs under normal physiological conditions but, in its absence, MSH2 may recruit another polymerase with different (mis)incorporation biases (92, 93).

MSH2 forms a heterodimer with either MSH3 or MSH6. Analysis of the rearranged V regions from mice deficient for the MSH3 and MSH6 MMR proteins revealed that MSH6, but not MSH3, is involved in SHM (68). This concept was strengthened by the observation that UNG2&MSH6 doubly deficient mice, like the *UNG2*$^{-/-}$*&MSH2*$^{-/-}$ mice, completely lack Phase 2 of SHM (94).

In keeping with the requirement of the MMR machinery for Phase 2 SHM, mice deficient for exonuclease I (EXOI) resemble the *MSH2*$^{-/-}$ and *MSH6*$^{-/-}$ mice (95). However, mice lacking the MutL Homolog 1 (MLH1), MLH3 or Post-Meiotic Segregation 2 (PMS2) MMR proteins exhibit much milder phenotypes (96–100). The reason for this discrepancy is currently unclear, but functional redundancy between these proteins is a possibility.

Overall, it is apparent that deamination of cytosine bases to uracils in DNA is mutagenic in itself, generating dC/dG → dT/dA transition mutations. The attempted repair of the uracil by the base-excision repair pathway triggers further mutations during the synthesis of DNA over abasic sites by translesion synthesis POLs θ, ζ, and REV1 (83–87). Repair of the dU:dG mismatch by the MMR machinery also promotes the amplification of mutations, possibly by recruiting POL η, which is prone to making errors at dA/dT base pairs (91–93). The tendency for error-prone rather than error-free repair is a topic that remains to be solved.

2.3.5 Processing of AID-induced Lesions: The Molecular Mechanism of Immunoglobulin Gene Conversion

Some species, such as birds, rabbits, cows, pigs, and horses, use pseudogene-templated IGC, in addition to SHM, to expand their antibody repertoire (44). Interestingly, AID was also shown to be required for IGC (55, 56). The DT40 chicken B-cell line has become the most popular model system to study IGC since it expresses AID and constitutively diversifies its Ig loci by IGC from adjacent pseudo V genes.

One would predict that a single- or double-stranded DNA break would be required to trigger the gene conversion process (Figure 2.4, middle panel). DNA breaks could arise during either base-excision repair or mismatch repair of the AID-induced dU:dG lesions. Studies in which UNG2 was inhibited by Ugi or disrupted by gene targeting in DT40 cells indicated that IGC occurs via UNG2-generated abasic site intermediates (101, 102). Accordingly, disruption of the MMR proteins MSH3, MSH4, or MSH6 does not perturb IGC in DT40 (41, 56).

DNA breaks could then be created by AP endonuclease 1 (APEX1) or another endonucleolytic activity. However, a role for APEX1 in antibody diversification has not yet been determined. Moreover, the MRE11/RAD50/NBS1 (MRN) complex, which is involved in DNA damage repair, has also been implicated in the generation of DNA breaks during IGC (103, 104). This is supported by the observation that the formation of MRE11 foci at the IgH locus was dependent on the presence of AID (105). An absolute requirement for APEX1 and/or the MRN complex for IGC has yet to be demonstrated by gene knockout or knockdown studies.

Resolution of the DNA breaks during IGC most likely involves the homologous recombination machinery. Indeed, disruption of the RAD51 paralogs, XRCC2, XRCC3, and RAD51B (106), and of BRCA2 (107) shifts the pattern of gene diversification from IGC to SHM. The involvement of RAD51 paralogs strongly indicates that IGC occurs via a homologous recombination mechanism. In contrast, disruption of the NHEJ DSB repair pathway appears to cause a redirection of lesions to the homologous recombination pathway, resulting in higher levels of IGC (108, 109). However, it is still an open question as to whether the homologous recombination that underlies IGC is entirely canonical.

2.3.6 Processing of AID-induced Lesions: The Molecular Mechanism of Class Switch Recombination

Mature B cells that have yet to encounter antigen are capable of expressing IgM and IgD (by alternative splicing of the mRNA). Once B cells have been activated by antigen and T cells, they are able to further rearrange their Ig constant region DNA by CSR (Figure 2.3B). B cells that have successfully undergone CSR express another antibody isotype, such as IgG, IgA, or IgE. Individuals who are unable to complete CSR tend to have elevated levels of IgM in their blood [Hyper IgM (HIGM) syndrome] and to be susceptible to opportunistic infections (see Section 2.4.1 for more detail on CSR-related immunodeficiency syndromes) (110).

CSR takes place by a region-specific recombination reaction between switch (S) regions that occur immediately upstream of each H-chain C region (except Cδ) (Figure 2.3B). The joining of two S regions results in the permanent excision of the intervening DNA. Switching can occur more that once, provided that there is at least one downstream C region to which the upstream sequence can be joined. As discussed above, AID is required to initiate CSR by deaminating dC to dU at a high frequency in the S-region DNA. The subsequent removal of uracil by UNG2 and single-strand nicking by APEX1 on both DNA strands has been proposed to generate a staggered DSB (Figure 2.4, bottom panel) (66). In support of this hypothesis, CSR is impaired in the absence of UNG2 or MSH2 and is completely abolished in *MSH2&UNG2*- and *MSH6&UNG2*-double knockout mice, demonstrating that excision of uracil from the S region DNA is absolutely essential to initiate CSR (69, 71, 78, 94). The more severe CSR defect observed in the absence of UNG2 compared to MSH2-deficiency indicates that uracil excision by UNG2 forms the major pathway for initiation of CSR, with backup activity provided by MSH2/MSH6 (68, 69, 71, 78).

The factors required to generate the DNA breaks required for CSR have not yet been identified, although roles of APEX1 and the MRN complex have been proposed (see Section 2.3.4.2) (66, 104). The MMR machinery also appears to be able to generate DNA breaks at dU:dG mismatches in S regions, and MLH1/PMS2 heterodimer may possess the endonuclease activity needed to nick the DNA strands in the presence of MSH3/MSH6 (111). Consistent with this idea, mice deficient for MLH1 and/or PMS2 display CSR defects similar to the MSH2-deficient mice (69, 112). However, whereas switch junctions from *MSH2*$^{-/-}$ mice show decreased microhomology compared to wild-type, PMS2, or MLH1-deficiency results in increased switch junction microhomology, suggesting that these proteins play distinct roles in CSR (98, 113). The phenotype of a *UNG2&PMS2* double knockout mouse may be informative in determining whether PMS2 is required to nick DNA in the absence of uracil excision by UNG2.

Both MMR and the MRN complex are also likely to be involved in the end-processing and final resolution of the DNA ends. The MRN complex is involved in both homologous recombination and NHEJ and has been implicated in synapsis of the DNA ends as well as in DNA damage signaling (114–117). Hypomorphic mutations cause the chromosome instability syndromes Nijmegen breakage syndrome [NBS (NBS1)] and ataxia–telangiectasia-like disorder [ATLD (MRE11)] in humans (118–120) (Table 2.3). Both NBS and ATLD patients have lower levels of switched antibody isotypes and display increased microhomology as well as more mutations and insertions at the switch region junctions (121). Null mutations are embryonically lethal in mice, making their roles in CSR difficult to define (122–124). Conditional knockout of NBS1 specifically in mouse B cells showed that NBS1 is involved in, but not required for, CSR. Whether the MRN complex functions in creating, processing, or mending the DSBs is currently unclear (105, 125).

TABLE 2.3 Antibody-Related Immunodeficiency Syndromes and Causative Factors[a]

Syndrome	Genetic Deficiency	Transmission	Up- or Downstream of AID	References
HIGM1	CD40L	XL[b]	US	155–156
HIGM2	AID	AR/AD	NA	54, 161
HIGM3	CD40	AR	US	157
HIGM4	?	?	?	166
HIGM5	UNG	AR	DS	72
HIGM-ED	IKKγ	XL	US	158
AT	ATM	AR	DS	133
ATLD	MRE11	AR	DS	121
NBS	NBS1	AR	DS	119
CVID	ICOS	AR	US	165
CVID	CD19	AR	US	165
CVID	TACI	AD	US	165
CVID	MSH5	AD	DS	131

[a]See main text for details.

[b]Abbreviations: XL, X-linked; AD, autosomal dominant; AR, autosomal recessive; US, upstream; DS, downstream; NA, not applicable; ?, unknown.

Analysis of *MSH2, MSH6, MLH1, PMS2*, and *EXO1* knock-out mice and mice expressing low levels of MSH5 also revealed diminished CSR with varying amounts of microhomology at the switch junctions (69, 95, 112, 113, 126–131). Whereas $MSH2^{-/-}$ and $EXO1^{-/-}$ mice displayed shorter microhomologies at switch junctions compared to wild-type mice, $MLH1^{-/-}$, $PMS2^{-/-}$, $MSH5^{low}$, and MSH2 ATPase mutants showed an increase in microhomology. No difference was observed for $MSH6^{-/-}$ mice. The varying phenotypes observed for the knockout mice led to suggestions that the MMR proteins may play both distinct and overlapping roles in CSR, perhaps in processing the broken DNA ends prior to ligation.

Upon formation of a DSB, histone H2AX is rapidly phosphorylated (γ-H2AX) in the chromatin for up to a megabase surrounding the break. 53BP1 is associated with chromatin containing γ-H2AX: and it recruits another kinase, ATM, that plays a central role in the signaling response to DNA damage (132). Deficiencies in H2AX, 53BP1, or ATM have all been shown to impair isotype switching, illustrating the importance of the DNA damage response in CSR (105, 125, 133–136).

The subsequent joining and ligation of the DSBs in two S regions to complete the CSR reaction was thought to be carried out primarily by the ubiquitous NHEJ machinery, due the lack of extensive homology between S regions and at switch junctions. In order to test this hypothesis, mice deficient for the Ku70 and Ku86 NHEJ proteins were analyzed for their ability to undergo CSR (137, 138). Since NHEJ is also essential for V(D)J recombination (a prerequisite for CSR), mice carrying functionally rearranged H- and L-chain transgenes knocked-in to the endogenous locus were used for these experiments. In contrast to the KU-proficient controls, B cells from these mice failed to complete CSR when stimulated in vitro, despite the presence of germline transcripts that indicated that CSR was being initiated normally, demonstrating that Ku70/86 are required for CSR.

Ku70 and Ku86 bind to DNA ends and recruit the catalytic subunit of DNA-PKcs. Together, these three proteins make up the DNA-PK holoenzyme. The exact role of DNA-PKcs in the reaction is more controversial. Whereas a null mutation abolished switching to all isotypes except IgG1 in mice (139), the mouse severe combined immunodeficiency (SCID) mutation, which truncates the protein 83 amino acids short of the C-terminus and disrupts the conserved kinase domain, did not appear to affect switching of a functionally rearranged H-chain transgene (140, 141). Targeted deletion of the kinase domain of DNA-PKcs did not appear to abrogate CSR either (142). These data have led to suggestions that the kinase activity of DNA-PKcs is dispensable for CSR, but the physical presence of the protein may be required as a scaffold for other factors and/or synapsis of the DNA ends, at least in mice.

In the NHEJ reaction, DNA-PK facilitates synapsis and processing of the broken DNA ends and their subsequent ligation by the XRCC4/LIG4 complex. XRCC4 and LIG4 are also essential for V(D)J recombination, and null mutations in each of the genes results in embryonic lethality in mice, making their roles in CSR difficult to examine (143, 144). XRCC4 is most likely required to stabilize LIG4 and to recruit it to the DNA-PK holoenzyme (145–148). The NHEJ endonuclease/exonuclease, Artemis, which is essential for V(D)J recombination, is dispensable for CSR (149). An XRCC4-like factor (XLF), also known as Cernnunos, is the most recently identified NHEJ protein (150, 151). XLF binds to the XRCC4/LIG4 complex and is required for DNA ligation (151, 152). The role of XLF in CSR is also undetermined. In the absence of the NHEJ proteins, cells must find another way to repair the breaks, such as by homologous recombination or single-strand annealing (resulting in increased homologies at the junctions), or die by apoptosis.

Overall, much progress has been made toward understanding how AID-catalyzed lesions can lead to three seemingly distinct outcomes: SHM, IGC, and CSR. The answer clearly lies in which ubiquitous DNA repair factors gain access to the uridine DNA lesion. The location (e.g., V or S regions) and frequency of lesions, as well as the cell cycle phase during which they occur, may also be crucial determinants. This promises to be a fruitful area for years to come.

2.4 HOT AREAS AND SPECULATIONS

In this final section of our review, we will endeavor to build upon the last 50 years of antibody research by helping the reader appreciate some of the ways that work toward understanding antibody diversification broadly impacts human health.

2.4.1 Immunodeficiency Syndromes Caused by Defects in AID-Mediated Ig Gene Diversification

Valuable insight about the processes involved in DNA break generation and repair during Ig diversification can be gained from understanding the genetic defects that lead to human immune deficiency syndromes, and vice versa (see Table 2.3 for a list of human immunodeficiency-related mutations). HIGM syndrome was defined originally as increased levels of IgM in the blood, but is more commonly characterized by markedly reduced serum levels of IgA, IgG, and IgE, consistent with defects in CSR (reviewed in references 153 and 154). SHM is also impaired in many, but not all, HIGM patients. Here, we discuss the HIGM syndromes in the order in which they appear in the pathway, from signaling on the cell surface to factors involved in recombining the DNA.

The first HIGM syndrome to be characterized exhibited X-linked inheritance and was determined to be due to mutations within the gene encoding CD40 ligand (CD40L), a protein expressed on the surface of T cells (155, 156). Co-stimulation of CD40, a receptor expressed on the B-cell surface, by CD40L is required for proliferation and CSR in response to BCR activation. Mutations in CD40 were later shown to be the cause of an autosomal recessive form of HIGM (HIGM3) (157).

HIGM syndrome patients with ectodermal dysplasia (HIGM-ED) also show low serum IgG and IgA with normal to elevated levels of IgM and general developmental defects in ectoderm-derived structures, resulting in abnormal or absent teeth, hair, and nails (158). HIGM-ED is caused by mutations in IKKγ (also known as NEMO), which functions in the NFκB signal transduction pathway downstream of CD40 activation (158). Signaling through this pathway ultimately leads to the nuclear translocation and activation of NFκB. NFκB can activate the transcription of many genes, including several involved in both SHM and CSR, such as AID. Thus, the phenotype of IKKγ deficiency is not limited to the humoral immune system as evidenced by the associated ectodermal defects.

Concomitant with the identification of AID as an essential component of the CSR machinery, mutations were identified in the human *AID* gene in patients with autosomal recessive HIGM2 (54) (see Section 2.3.1). Mutations resulting in complete loss of AID expression attenuated both CSR and SHM. Mapping of the missense mutations onto a structure of AID predicted from the yeast cytidine deaminase (CDD1) crystal structure revealed three classes of mutations: One class of mutations was predicted to disrupt the active site, another group was predicted to interfere with substrate binding, and the third class of mutations mapped to the protein surface, potentially disrupting interactions with regulatory cofactors (159). Interestingly, an autosomal dominant form of HIGM was ascribed to C-terminal truncations of the AID protein (160). B cells harboring this mutation were unable to undergo CSR, despite the persistence of mutations in both the S and V regions, suggesting that a factor specifically required for CSR (but not SHM) interacts with the C-terminus of AID (54, 160–163).

Mutations in UNG2 account for HIGM5, a syndrome with autosomal recessive inheritance (72). The SHM signature of HIGM5 patients is skewed toward transitions at dG/dC residues, probably due to replication over the unprocessed dU residues (see Section 2.3.3 for discussion of the role of UNG2).

Single isotype or subtype deficiencies such as IgA deficiency (IgAD) or IgG1 deficiency are the most common immunodeficiency syndromes, but they often go undiagnosed due to the relatively subtle defect in immune response (164, 165). A more severe form of antibody deficiency, common variable immunodeficiency (CVID), is characterized by near absence of both serum IgA and IgG antibodies (for reviews of CVID and IgAD, see references 164 and 165). The genetic mutations associated with CVID are listed in Table 2.3, although many more await discovery.

It is likely that the genetic defects underlying many of the HIGM [including HIGM4 (166)] and CVID cases still need to be identified. The advent of technology allowing whole-genome association studies and pyro-sequencing will greatly improve the fraction of cases for which a genetic determinant can be identified. CVID is often found intermixed with IgAD within families, supporting the idea that these diseases have common genetic components. However, both appear to have a polygenic or multifactorial mode of inheritance (164, 165, 167).

2.4.2 Regulating the DNA Mutator Activity of AID

An important hallmark of SHM and CSR is that they are B-cell-specific. These processes are particularly interesting because V and S region mutation frequencies can be a million-fold higher than those observed at other genomic loci (e.g., references 37 and 80). Therefore an important question is how the highly mutagenic activity of AID be targeted to the Ig locus and simultaneously prevented from damaging the bulk of the nuclear DNA? There will undoubtedly be many answers to this important question, and here we consider transcriptional regulation, subcellular compartmentalization, and posttranslational regulation.

It is possible that the highly restricted expression of AID to specific stages in B-lymphocyte development is sufficient to prevent the potential mistargeting of AID's activity. In situ hybridization, RT-PCR, and northern blot analyses of mouse and human tissues revealed that AID expression is mainly restricted to germinal center B cells in the lymph nodes and tonsils, although lower levels of AID transcripts were also detected in the kidney, pancreas, spleen, and fetal liver (51, 52).

AID is upregulated at the transcriptional level in B cells upon stimulation by antigen and T cells, and it is subsequently turned off after B cells have undergone affinity maturation and differentiated into plasma cells or memory B cells. The relatively short duration of AID expression thereby provides an important method of regulating AID's activity.

Another method of regulating AID appears to be by localizing the protein to the cytoplasm, away from the DNA in the nucleus. Since AID deaminates genomic DNA, it was initially presumed that AID would have a nuclear localization. Surprisingly, a study utilizing a chimeric protein of human AID fused to green fluorescent protein (GFP) led to the discovery that AID is predominately localized to the cytoplasm (168). A key study demonstrated that treatment of cells with leptomycin B (LMB), a CRM1-dependent nuclear export inhibitor, resulted in the accumulation of an AID-GFP fusion protein in the nucleus (169). This provided strong evidence that AID somehow entered the nucleus and was then exported through the CRM1 pathway.

Analysis of AID C-terminal deletion mutants led to the discovery that human AID contained a leucine-rich nuclear export sequence (NES) in its C-terminal domain (169). An AID protein that lacked the NES displayed a more nuclear localization and was more active than the wild-type protein on a nonphysiological target in fibroblasts (170). However, loss of the NES did not affect the rate of mutation of Ig genes in B cells, suggesting that the nuclear trafficking alone may not be sufficient to limit the activity of AID at the Ig loci. Sequence analysis subsequently identified a potential bipartite nuclear localization signal (NLS) near the N-terminus of human AID (169). Mutational analysis of this region confirmed that this sequence could function as an NLS (169). However, another study investigating nucleocytoplasmic trafficking of mouse AID was unable to confirm active import into the nucleus through an NLS mechanism, potentially reflecting differences between the human and mouse proteins (170). Since AID is a relatively small protein (24 kD), and nuclear pore complexes have been reported to allow passive diffusion of molecules up to 60 kD in size (e.g., reference 171), it is possible that diffusion alone is sufficient for AID to enter the nucleus, and the NES serves as a mechanism to keep it out.

Immunohistochemistry of human lymphoid tissues revealed that a small percentage of germinal center B cells have AID protein accumulated in the nucleus at any given time, suggesting that AID may transiently relocate to the nucleus at certain times during the cell cycle and at particular stages during Ig repertoire diversification (172). Whether this nuclear accumulation is due to the active import or the inhibition of nuclear export is not clear, but it is apparent that at least one of these processes can be turned on and off as necessary by an unknown mechanism. The nucleo-cytoplasmic trafficking of AID is therefore likely to provide an additional mode of regulation by keeping the mutator away from the DNA most of the time.

CSR and SHM do not always occur simultaneously in activated B cells, suggesting that SHM- and CSR-specific cofactors may exist (51, 173, 174). In support of this hypothesis, mutational analysis of AID has revealed protein domains in the

N- and C-terminal regions that may be required specifically for diversification of the V region and for rearrangement at the C region, respectively (160, 162, 163). Nussenzweig and colleagues observed that a C-terminal deletion mutant lacking the last 10 amino acids failed to support CSR in mouse B cells in vitro, but retained the ability to trigger SHM and IGC in the V region of DT40 cells (163). Surprisingly, this AID variant produced mutations in the Sμ region with a similar frequency and spectrum to the wild-type protein, suggesting that the very C-terminal domain (CTD) of AID is required to recruit a CSR-specific factor to the S region to induce DNA breakage and/or joining (163).

In a parallel study, the Honjo laboratory identified three CSR-deficient, but SHM-proficient, variants of AID from HIGM patients (160). All three mutations mapped to the CTD of AID in agreement with the results reported by Nussenzweig and colleagues (163). However, the Honjo group did not find AID-dependent mutations in the Sμ region of cells expressing the mutant proteins, indicating that interaction with an unknown factor through the CTD may be required to target AID to the S regions (162). Analysis of an AID variant that can mutate *E. coli* DNA but is unable to trigger CSR or SHM suggests that interaction with other proteins may be required for AID to deaminate DNA in mammalian cells (160). An SHM-specific interaction domain may also map to the putative NLS in the N-terminal region of AID (162). However, it is difficult to separate the effects on protein function that may be due to failure to import AID into the nucleus or failure to bind to another component.

In order to identify AID-interacting proteins, Alt and colleagues screened a phage lambda expression library with S^{35}-labeled AID purified from mouse B cells (175). They pulled out the 32-kDa subunit of the heterotetrameric RPA complex and subsequently confirmed the interaction by co-immunoprecipitation from B-cell extracts. RPA was also found to be enriched in glycerol gradient fractions of B-cell extracts that enhanced the deaminase activity of AID on a transcribed dsDNA SHM substrate in vitro. Purified recombinant RPA also augmented the activity of AID on this substrate, suggesting that RPA can modulate AID's activity on transcribed dsDNA targets. Transcription through the V and S regions is essential for SHM and CSR, respectively, and may therefore provide an opportunity for AID to transiently access the locus and mutate the ssDNA (176–180). However, it is difficult to envisage how a sequence-nonspecific ssDNA binding protein, such as RPA, might target AID specifically to Ig DNA and not the rest of the cellular DNA during replication. Further studies will be required to address these issues.

Interestingly, Alt and colleagues also observed that AID purified from B cells in the absence of phosphatase inhibitors was inactive on the transcribed SHM substrate even in the presence of RPA, suggesting that AID may be phosphorylated and that this event may be required for the interaction with RPA. The Alt laboratory therefore utilized mass spectrometry to determine the site(s) of phosphorylation in mouse AID (181). They identified two phosphorylated residues in AID purified from B cells, but not from 293 cells. One of these residues, serine 38 (S38), mapped to a PKA consensus phosphorylation site [RRX(S/T)]. Mutation of this site to alanine abrogated AID's ability to interact with RPA, to deaminate a transcribed dsDNA SHM substrate, and to support CSR in mouse B cells in vitro. These data suggest that the phosphorylation of S38 is important for the proper function of AID. The catalytic subunit α of PKA

(PKACA) co-fractionated with AID by glycerol gradient fractionation and co-immunoprecipitated from B-cell extracts. AID phosphorylation was also inhibited by a PKA-specific inhibitor, confirming that PKA can phosphorylate AID in vivo. AID isolated from nuclear extracts was better able to deamininate transcribed dsDNA than AID from cytoplasmic extracts, despite its lower abundance, indicating that AID may be preferentially phosphorylated in the nucleus (181).

In a simultaneous and independent set of experiments, Dalla-Favera and colleagues identified the regulatory subunit Iα of PKA (PKAR1A) in a complex with AID isolated from cytoplasmic extracts of human B cells (182). They also identified the phosphorylation site data reported by the Alt laboratory. Defining the conditions that determine whether or not AID is phosphorylated will most likely be a key piece in the puzzle of understanding the posttranslational regulation of AID in B lymphocytes.

AID therefore appears to be under multiple levels of regulation including both transcriptional (cell type and developmental timing) and posttranslational (subcellular localization and phosphorylation) control (51, 52, 168–170, 175, 181, 182). These observations highlight the importance of controlling a DNA mutator protein, and they imply that other modes of regulation will also exist.

2.4.3 Misregulation of AID and Cancer

The high levels of DNA mutagenesis required for antibody diversification must be tightly regulated to prevent AID from deaminating improper targets. Furthermore, the AID-dependent Ig DNA breaks that are likely to be involved in IGC (perhaps SHM) and CSR must be processed in an error-free way (processed in *cis*) to avoid the accumulation of chromosome translocations and other gross rearrangements. Thus, a consequence of inappropriate deamination and/or misprocessed AID-dependent DNA breaks will be pro-carcinogenic aberrations.

A number of B-cell cancers have been shown to contain SHM-like mutations in non-Ig genes (183). More than 50% of diffuse large cell lymphomas show SHM-like mutations in the proto-oncogenes *PIM1, MYC, RhoH/TTF*, or *PAX5* (184). Similar proto-oncogene mutations were also noted in Burkitt's lymphoma and large B-cell lymphoma (185, 186). *BCL6* and *FAS* have been shown to be mutated at low levels in mature germinal center B cells but not naive B-cell counterparts (187–189). The precise mechanism by which these non-Ig mutations accumulate is not known. However, several pieces of evidence implicate deamination by AID. Mutations are limited to actively transcribed genes and are only found downstream of the transcription start site, consistent with the activity of AID. Furthermore, these mutations are targeted to AID hot-spot motifs and exhibit a bias toward transitions (183).

Translocations between the Ig heavy chain promoter a number of other genes have been observed in B-cell cancers, such as *BCL2* in follicular lymphoma, *BCL6* in diffuse large-cell lymphoma, and *FGFR3* in multiple myeloma (183). One of the most studied B-cell malignancies, Burkitt's lymphoma, is due to reciprocal translocation between the Ig switch region and *c-MYC*. The translocation that causes Burkitt's lymphoma places the oncogene, *c-MYC,* under the control of the Ig heavy-chain promoter, causing overexpression of c-MYC. In mouse models of lymphoma, AID has been shown to be required for the accumulation of Ig-*c-MYC* translocations (190–192). Switch region translocations have been shown to occur within hours following the induction of AID expression in an UNG-dependent manner (190). Activated tumor suppressor proteins, including 53BP1, H2AX, ATM, and NBS1, likely inhibit translocation and promote lesion repair via CSR.

The carcinogenic effects of AID may not be strictly limited to B-cell cancers since AID expression has been documented in a variety of other normal human tissues, including liver, kidney, and spermatocytes (51, 193). When AID was constitutively expressed in mice, many mice developed T-cell lymphomas and tumors of the lung epithelia (194, 195). Ectopic expression of AID can be induced by a number of factors, including pathogenic infection and inflammatory cytokines, potentially triggering cancer in humans (196, 197). For example, hepatitis C infection can lead to deregulation of the NFκB signaling pathway and subsequent inappropriate expression of AID and proto-oncogene mutation in hepatic cells (197). Infection of gastric epithelial cells with *Helicobacter pylori* caused a similar aberrant expression of AID via NFκB activation leading to the accumulation of mutations in the tumor suppressor gene *p53* (196). Overall, it is formally possible that aberrant (non-developmentally controlled) transcription of AID may also contribute to carcinogenesis [e.g., AID expression has been associated with breast tumorogenesis (198)].

2.4.4 AID Is But One Member of a Much Larger Family of Polynucleotide Deaminases

AID is arguably the most ancient member of a larger family of putative and proven polynucleotide cytosine deaminases that exist in vertebrate species, ranging from fish to primates (Figure 2.5 and Chapters 10, 11 and 16).

All vertebrates also encode APOBEC2, which possesses a single, conserved zinc-binding deaminase motif (199, 200). Although this protein has not been shown to edit DNA or RNA, its profound conservation and presence in all vertebrates implies ancient and integral functions. APOBEC2 is expressed predominantly in cardiovascular tissues. APOBEC4 appears in mammals, birds, and frogs, but not fish, and is possibly testis-specific (201). However, it is already clear that the functions of these proteins will not be particularly obvious, because APOBEC2-deficient mice appeared phenotypically normal (202).

Mammals have two additional AID-related proteins, APOBEC1 and APOBEC3. In fact, a reasonable argument can be made that a pre-mammalian *AID* gene duplicated (at least twice) and diverged to yield the *APOBEC1* and *APOBEC3* genes [Figure 2.5 (199, 203)]. The mRNA editing protein APOBEC1, alluded to earlier in this chapter (Sections 2.3.1 and 2.3.2), is the main subject of Chapter 10. However, for purposes of the present discussion, it is notable that although APOBEC1 can also deaminate cytosine bases in DNA, enforced expression of this protein in AID-deficient splenic B cells could not promote CSR (i.e., it does not appear to have the capacity to substitute for AID) (203, 204). Future studies are likely to shed light on the determinants of the RNA and DNA substrate specificities of APOBEC1 and AID, as well as on the intriguing possibility that APOBEC1 has physiological function(s) in addition to mRNA editing.

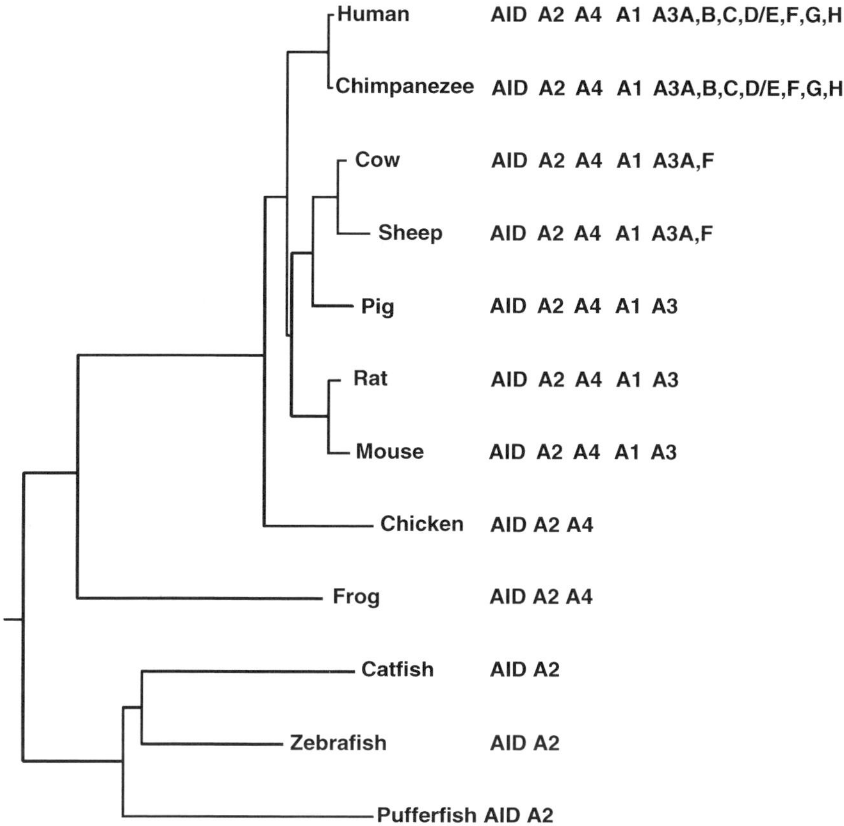

Figure 2.5 The distribution of AID and its family members in vertebrates. Rooted neighbor-joining tree with branch lengths created in ClustalW using full-length AID protein sequences. AID and APOBEC2 (A2) are present in vertebrate species from fish to primates. APOBEC4 (A4) appears in frogs, birds, and mammals. APOBEC3 (A3) and APOBEC1 (A1) are only found in mammals. The APOBEC3 locus has undergone an expansion from one gene in mice, rats, and pigs, to two genes in cows and sheep, and to seven genes in primates.

APOBEC3 proteins have had considerable fanfare recently because several of these proteins have demonstrated potent anti-retrovirus and anti-retrotransposon activities. A defining feature of the mammalian APOBEC3 proteins is that nearly all have been shown to deaminate cytosine bases within single-strand DNA. None have shown RNA editing activity, even in head-to-head biochemical experiments testing recombinant protein on DNA versus RNA substrates (205). Although deamination-independent restriction activities have been documented, many viruses (such as the AIDS virus, HIV-1) and transposons are clearly attacked by the APOBEC3 proteins during the vulnerable ssDNA stage of reverse transcription. The prevailing model indicates that APOBEC3 proteins gain access to assembling retrovirus and retro-transposon ribonucleoprotein particles, travel with the particles until reverse transcription initiates, and, finally, deaminate first cDNA strand cytidines during reverse transcription. The resulting dU residues either (a) trigger the degradation of the potentially pathogenic DNA or (b) template the insertion of second strand cDNA adenosines, events that can ultimately manifest as strand-specific dC/dG → dT/dA transition mutations (dC-to-dT on the first cDNA strand and dG-to-dA on the second cDNA strand, which is also the coding strand). Overall, the APOBEC3 proteins constitute an important facet of the innate immune response to both endogenous and exogenous retroelements.

The remarkable anti-retroelement activities of the APOBEC3 proteins and their obvious biochemical and evolutionary linkages to AID have led to the hypothesis that AID might have a more direct role in restricting pathogens (206). Specifically, AID might also work as a retrovirus and retrotransposon restriction factor. This is supported by observations indicating that AID can be expressed in a variety of different tissues outside the B-cell compartment, notably germ cells (see Section 2.4.3). We have further speculated that such a retroelement restriction activity of AID may have preexisted its antibody diversification functions and, moreover, that present-day AID proteins may still utilize this activity (206). Overall, a clearer understanding of the role of AID in both adaptive and innate immunity is anticipated.

2.5 CONCLUSIONS

The most important conclusion from recent studies on antibody diversity is that the critical step—the initiating lesion—is provided by an enzyme (AID) that catalyzes the conversion of Ig gene DNA cytosine bases to uracils. AID-catalyzed DNA lesions occur in a purposeful, directed, and developmentally regulated manner. Prior to these insights, it was believed that the conversion of DNA cytosines to uracils was exclusively a spontaneous event. Therefore, studies on the DNA deaminase, AID, provided a crucial conceptual advance that has and will continue to influence mechanistic studies in the broad area of DNA editing. Already, evolutionary descendents of an ancestral AID protein, the present-day APOBEC3 proteins, have demonstrated DNA deamination-dependent anti-retrovirus and anti-retrotransposon activities. The next half-century will undoubtedly be filled with additional exciting advances toward understanding the broader physiological roles of this important family of enzymes.

ACKNOWLEDGMENTS

The authors thank Drs. W. Brown and E. Hendrickson for helpful comments on this manuscript and also thanks R. LaRue for advice on Figure 2.5. The authors' work on antibody gene diversification has been supported by a Burroughs Wellcome Fund Hitching–Elion Fellowship, a Searle Scholarship, a University of Minnesota McKnight Land Grant Assistant Professorship, a University of Minnesota Biomedical Genomics Seed Grant, a University of Minnesota Biomedical Leukemia Research Fund Grant, and an NIH grant GM080437. D.M. was supported in part by a University of Minnesota Graduate School Doctoral Dissertation Fellowship, and S.O. was also supported in part by NIH grant AI067152.

REFERENCES

1. Watson, J. D., and Crick, F. H. (1953) Molecular structure of nucleic acids; a structure for deoxyribose nucleic acid. *Nature* **171**, 737–738.
2. Franklin, R. E., and Gosling, R. G. (1953) Evidence for 2-chain helix in crystalline structure of sodium deoxyribonucleate. *Nature* **172**, 156–157.
3. Jerne, N. K. (1955) The Natural-selection theory of antibody formation. *Proc Natl Acad Sci USA* **41**, 849–857.
4. Burnet, F. (1957) A modification of Jerne's theory on antibody production using the concept of clonal selection. *Aust J Sci* **20**, 67–69.
5. Lederberg, J. (1959) Genes and antibodies. *Science* **129**, 1649–1653.
6. White, R. G. (1958) Antibody production by single cells. *Nature* **182**, 1383–1384.
7. Nossal, G. J., and Lederberg, J. (1958) Antibody production by single cells. *Nature* **181**, 1419–1420.
8. Coons, A. H. (1957) The application of fluorescent antibodies to the study of naturally occurring antibodies. *Ann NY Acad Sci* **69**, 658–662.
9. Brenner, S., and Milstein, C. (1966) Origin of antibody variation. *Nature* **211**, 242–243.
10. Titani, K., and Putnam, F. W. (1965) Immunoglobulin structure: Amino- and carboxyl-terminal peptides of type I Bence-Jones proteins. *Science* **147**, 1304–1305.
11. Putnam, F. W., and Easley, C. W. (1965) Structural studies of the immunoglobulins. I. The tryptic peptides of Bence-Jones proteins. *J Biol Chem* **240**, 1626–1638.
12. Porter, R. R. (1958) Separation and isolation of fractions of rabbit gamma-globulin containing the antibody and antigenic combining sites. *Nature* **182**, 670–671.
13. Milstein, C. (1966) Comparative peptide sequences of kappa and lambda chains of human immunoglobins. *J Mol Biol* **21**, 203–205.
14. Milstein, C. (1966) Variations in amino-acid sequence near the disulphide bridges of Bence-Jones proteins. *Nature* **209**, 370–373.
15. Hilschmann, N., and Craig, L. C. (1965) Amino acid sequence studies with Bence-Jones proteins. *Proc Natl Acad Sci USA* **53**, 1403–1409.
16. Cioli, D., and Baglioni, C. (1966) Origin of structural variation in Bence Jones proteins. *J Mol Biol* **15**, 385–388.
17. Smithies, O. (1965) Disulfide-bond cleavage and formation in proteins. *Science* **150**, 1595–1598.
18. Dreyer, W. J., and Bennett, J. C. (1965) The molecular basis of antibody formation: A paradox. *Proc Natl Acad Sci USA* **54**, 864–869.
19. Burch, P. R., Burwell, R. G., and Rowell, N. R. (1965) Aetiology of multiple sclerosis. *Br Med J* **1**, 723.
20. Hozumi, N., and Tonegawa, S. (1976) Evidence for somatic rearrangement of immunoglobulin genes coding for variable and constant regions. *Proc Natl Acad Sci USA* **73**, 3628–3632.
21. Oettinger, M. A., Schatz, D. G., Gorka, C., and Baltimore, D. (1990) RAG-1 and RAG-2, adjacent genes that synergistically activate V(D)J recombination. *Science* **248**, 1517–1523.
22. Schatz, D. G., Oettinger, M. A., and Baltimore, D. (1989) The V(D)J recombination activating gene. *RAG-1. Cell* **59**, 1035–1048.
23. Jones, J. M., and Gellert, M. (2004) The taming of a transposon: V(D)J recombination and the immune system. *Immunol Rev* **200**, 233–248.
24. Schatz, D. G. (1999) Transposition mediated by RAG1 and RAG2 and the evolution of the adaptive immune system. *Immunol Res* **19**, 169–182.
25. Brandt, V. L., and Roth, D. B. (2004) V(D)J recombination: How to tame a transposase. *Immunol Rev* **200**, 249–260.
26. Spicuglia, S., Franchini, D. M., and Ferrier, P. (2006) Regulation of V(D)J recombination. *Curr Opin Immunol* **18**, 158–163.
27. Jung, D., Giallourakis, C., Mostoslavsky, R., and Alt, F. W. (2006) Mechanism and control of V(D)J recombination at the immunoglobulin heavy chain locus. *Annu Rev Immunol* **24**, 541–570.
28. Lieber, M. R., Ma, Y., Pannicke, U., and Schwarz, K. (2004) The mechanism of vertebrate nonhomologous DNA end joining and its role in V(D)J recombination. *DNA Repair (Amst)* **3**, 817–826.

29. Janeway, C. A., Travers, P., Walport, M., and Schlomchik, M. J. (2005) Immunobiology, 6th ed, Garland Science, New York.
30. Kohler, G., and Milstein, C. (1975) Continuous cultures of fused cells secreting antibody of predefined specificity. *Nature* **256**, 495–497.
31. Cotton, R. G., and Milstein, C. (1973) Letter: Fusion of two immunoglobulin-producing myeloma cells. *Nature* **244**, 42–43.
32. Sanger, F., and Tuppy, H. (1951) The amino-acid sequence in the phenylalanyl chain of insulin. *2. The investigation of peptides from enzymic hydrolysates., Biochem J* **49**, 481–490.
33. Wagner, S. D., and Neuberger, M. S. (1996) Somatic hypermutation of immunoglobulin genes. *Annu Rev Immunol* **14**, 441–457.
34. Betz, A. G., Neuberger, M. S., and Milstein, C. (1993) Discriminating intrinsic and antigen-selected mutational hotspots in immunoglobulin V genes. *Immunol Today* **14**, 405–411.
35. Weiss, U., and Rajewsky, K. (1990) The repertoire of somatic antibody mutants accumulating in the memory compartment after primary immunization is restricted through affinity maturation and mirrors that expressed in the secondary response. *J Exp Med* **172**, 1681–1689.
36. Weigert, M. G., Cesari, I. M., Yonkovich, S. J., and Cohn, M. (1970) Variability in the lambda light chain sequences of mouse antibody. *Nature* **228**, 1045–1047.
37. Betz, A. G., Rada, C., Pannell, R., Milstein, C., and Neuberger, M. S. (1993) Passenger transgenes reveal intrinsic specificity of the antibody hypermutation mechanism: Clustering, polarity, and specific hot spots. *Proc Natl Acad Sci USA* **90**, 2385–2388.
38. Berek, C., and Milstein, C. (1987) Mutation drift and repertoire shift in the maturation of the immune response. *Immunol Rev* **96**, 23–41.
39. Sarvas, H., and Makela, O. (1970) Haptenated bacteriophage in the assay of antibody quantity and affinity: Maturation of an immune response. *Immunochemistry* **7**, 933–943.
40. Maizels, N. (2005) Immunoglobulin gene diversification. *Annu Rev Genet* **39**, 23–46.
41. Arakawa, H., and Buerstedde, J. M. (2004) Immunoglobulin gene conversion: Insights from bursal B cells and the DT40 cell line. *Dev Dyn* **229**, 458–464.
42. Thompson, C. B., and Neiman, P. E. (1987) Somatic diversification of the chicken immunoglobulin light chain gene is limited to the rearranged variable gene segment. *Cell* **48**, 369–378.
43. Reynaud, C. A., Anquez, V., Dahan, A., and Weill, J. C. (1985) A single rearrangement event generates most of the chicken immunoglobulin light chain diversity. *Cell* **40**, 283–291.
44. Reynaud, C. A., Anquez, V., Grimal, H., and Weill, J. C. (1987) A hyperconversion mechanism generates the chicken light chain preimmune repertoire. *Cell* **48**, 379–388.
45. Sale, J. E. (2004) Immunoglobulin diversification in DT40: A model for vertebrate DNA damage tolerance. *DNA Repair (Amst)* **3**, 693–702.
46. Sonoda, E., Morrison, C., Yamashita, Y. M., Takata, M., and Takeda, S. (2001) Reverse genetic studies of homologous DNA recombination using the chicken B-lymphocyte line, DT40. *Philos Trans R Soc Lond B Biol Sci* **356**, 111–117.
47. Agarwal, S. C. (1964) Nomenclature for human immunoglobulins. *Bull World Health Organ* **30**, 447–450.
48. Maki, R., Traunecker, A., Sakano, H., Roeder, W., and Tonegawa, S. (1980) Exon shuffling generates an immunoglobulin heavy chain gene. *Proc Natl Acad Sci USA* **77**, 2138–2142.
49. Kataoka, T., Kawakami, T., Takahashi, N., and Honjo, T. (1980) Rearrangement of immunoglobulin gamma 1-chain gene and mechanism for heavy-chain class switch. *Proc Natl Acad Sci USA* **77**, 919–923.
50. Davis, M. M., Calame, K., Early, P. W., Livant, D. L., Joho, R., Weissman, I. L., and Hood, L. (1980) An immunoglobulin heavy-chain gene is formed by at least two recombinational events. *Nature* **283**, 733–739.
51. Muramatsu, M., Sankaranand, V. S., Anant, S., Sugai, M., Kinoshita, K., Davidson, N. O., and Honjo, T. (1999) Specific expression of activation-induced cytidine deaminase (AID), a novel member of the RNA-editing deaminase family in germinal center B cells. *J Biol Chem* **274**, 18470–18476.
52. Muto, T., Muramatsu, M., Taniwaki, M., Kinoshita, K., and Honjo, T. (2000) Isolation, tissue distribution, and chromosomal localization of the human activation-induced cytidine deaminase (AID) gene. *Genomics* **68**, 85–88.

53. Muramatsu, M., Kinoshita, K., Fagarasan, S., Yamada, S., Shinkai, Y., and Honjo, T. (2000) Class switch recombination and hypermutation require activation-induced cytidine deaminase (AID), a potential RNA editing enzyme. *Cell* **102**, 553–563.
54. Revy, P., Muto, T., Levy, Y., Geissmann, F., Plebani, A., Sanal, O., Catalan, N., Forveille, M., Dufourcq-Labelouse, R., Gennery, A., Tezcan, I., Ersoy, F., Kayserili, H., Ugazio, A. G., Brousse, N., Muramatsu, M., Notarangelo, L. D., Kinoshita, K., Honjo, T., Fischer, A., and Durandy, A. 2000 Activation-induced cytidine deaminase (AID) deficiency causes the autosomal recessive form of the Hyper-IgM syndrome (HIGM2). *Cell* **102** 565–575.
55. Harris, R. S., Sale, J. E., Petersen-Mahrt, S. K., and Neuberger, M. S. (2002) AID is essential for immunoglobulin V gene conversion in a cultured B cell line. *Curr Biol* **12**, 435–438.
56. Arakawa, H., Hauschild, J., and Buerstedde, J. M. (2002) Requirement of the activation-induced deaminase (AID) gene for immunoglobulin gene conversion. *Science* **295**, 1301–1306.
57. Kinoshita, K., and Honjo, T. (2001) Linking class-switch recombination with somatic hypermutation. *Nat Rev Mol Cell Biol* **2**, 493–503.
58. Dickerson, S. K., Market, E., Besmer, E., and Papavasiliou, F. N. (2003) AID mediates hypermutation by deaminating single stranded DNA. *J Exp Med* **197**, 1291–1296.
59. Bransteitter, R., Pham, P., Scharff, M. D., and Goodman, M. F. (2003) Activation-induced cytidine deaminase deaminates deoxycytidine on single-stranded DNA but requires the action of RNase. *Proc Natl Acad Sci USA* **100**, 4102–4107.
60. Doi, T., Kinoshita, K., Ikegawa, M., Muramatsu, M., and Honjo, T. (2003) De novo protein synthesis is required for the activation-induced cytidine deaminase function in class-switch recombination. *Proc Natl Acad Sci USA* **100**, 2634–2638.
61. Nagaoka, H., Ito, S., Muramatsu, M., Nakata, M., and Honjo, T. (2005) DNA cleavage in immunoglobulin somatic hypermutation depends on de novo protein synthesis but not on uracil DNA glycosylase. *Proc Natl Acad Sci USA* **102**, 2022–2027.
62. Begum, N. A., Kinoshita, K., Muramatsu, M., Nagaoka, H., Shinkura, R., and Honjo, T. (2004) De novo protein synthesis is required for activation-induced cytidine deaminase-dependent DNA cleavage in immunoglobulin class switch recombination. *Proc Natl Acad Sci USA* **101**, 13003–13007.
63. Neuberger, M. S., and Scott, J. (2000) Immunology. RNA editing AIDs antibody diversification? *Science* **289**, 1705–1706.
64. Jacobs, H., and Bross, L. (2001) Towards an understanding of somatic hypermutation. *Curr Opin Immunol* **13**, 208–218.
65. Martin, A., Bardwell, P. D., Woo, C. J., Fan, M., Shulman, M. J., and Scharff, M. D. (2002) Activation-induced cytidine deaminase turns on somatic hypermutation in hybridomas. *Nature* **415**, 802–806.
66. Petersen-Mahrt, S. K., Harris, R. S., and Neuberger, M. S. (2002) AID mutates *E. coli* suggesting a DNA deamination mechanism for antibody diversification. *Nature* **418**, 99–103.
67. Chaudhuri, J., Tian, M., Khuong, C., Chua, K., Pinaud, E., and Alt, F. W. (2003) Transcription-targeted DNA deamination by the AID antibody diversification enzyme. *Nature* **422**, 726–730.
68. Wiesendanger, M., Kneitz, B., Edelmann, W., and Scharff, M. D. (2000) Somatic hypermutation in MutS homologue (MSH)3-, MSH6-, and MSH3/MSH6-deficient mice reveals a role for the MSH2-MSH6 heterodimer in modulating the base substitution pattern. *J Exp Med* **191**, 579–584.
69. Rada, C., Ehrenstein, M. R., Neuberger, M. S., and Milstein, C. (1998) Hot spot focusing of somatic hypermutation in MSH2-deficient mice suggests two stages of mutational targeting. *Immunity* **9**, 135–141.
70. Di Noia, J., and Neuberger, M. S. (2002) Altering the pathway of immunoglobulin hypermutation by inhibiting uracil-DNA glycosylase. *Nature* **419**, 43–48.
71. Rada, C., Williams, G. T., Nilsen, H., Barnes, D. E., Lindahl, T., and Neuberger, M. S. (2002) Immunoglobulin isotype switching is inhibited and somatic hypermutation perturbed in UNG-deficient mice. *Curr Biol* **12**, 1748–1755.
72. Imai, K., Slupphaug, G., Lee, W. I., Revy, P., Nonoyama, S., Catalan, N., Yel, L., Forveille, M., Kavli, B., Krokan, H. E., et al. (2003) Human uracil-DNA glycosylase deficiency associated with profoundly impaired immunoglobulin class-switch recombination. *Nat Immunol* **4**, 1023–1028.

73. Begum, N. A., Kinoshita, K., Kakazu, N., Muramatsu, M., Nagaoka, H., Shinkura, R., Biniszkiewicz, D., Boyer, L. A., Jaenisch, R., and Honjo, T. (2004) Uracil DNA glycosylase activity is dispensable for immunoglobulin class switch. *Science* **305**, 1160–1163.
74. Stivers, J. T. (2004) Comment on "Uracil DNA glycosylase activity is dispensable for immunoglobulin class switch." *Science* **306**, *2042; author reply* 2042.
75. Scaramozzino, N., Sanz, G., Crance, J. M., Saparbaev, M., Drillien, R., Laval, J., Kavli, B., and Garin, D. (2003) Characterisation of the substrate specificity of homogeneous vaccinia virus uracil-DNA glycosylase. *Nucleic Acids Res* **31**, 4950–4957.
76. Drohat, A. C., Jagadeesh, J., Ferguson, E., and Stivers, J. T. (1999) Role of electrophilic and general base catalysis in the mechanism of *Escherichia coli* uracil DNA glycosylase. *Biochemistry* **38**, 11866– 11875.
77. Mol, C. D., Arvai, A. S., Slupphaug, G., Kavli, B., Alseth, I., Krokan, H. E., and Tainer, J. A. (1995) Crystal structure and mutational analysis of human uracil-DNA glycosylase: Structural basis for specificity and catalysis. *Cell* **80**, 869–878.
78. Rada, C., Di Noia, J. M., and Neuberger, M. S. (2004) Mismatch recognition and uracil excision provide complementary paths to both Ig switching and the A/T-focused phase of somatic mutation. *Mol Cell* **16**, 163–171.
79. Di Noia, J. M., Rada, C., and Neuberger, M. S. (2006) SMUG1 is able to excise uracil from immunoglobulin genes: Insight into mutation versus repair. *EMBO J* **25**, 585–595.
80. Rajewsky, K., Forster, I., and Cumano, A. (1987) Evolutionary and somatic selection of the antibody repertoire in the mouse. *Science* **238**, 1088–1094.
81. Kim, S., Davis, M., Sinn, E., Patten, P., and Hood, L. (1981) Antibody diversity: Somatic hypermutation of rearranged VH genes. *Cell* **27**, 573–581.
82. Xue, K., Rada, C., and Neuberger, M. S. (2006) The in vivo pattern of AID targeting to immunoglobulin switch regions deduced from mutation spectra in msh2$^{-/-}$ ung$^{-/-}$ mice. *J Exp Med* **203**, 2085–2094.
83. Ross, A. L., and Sale, J. E. (2006) The catalytic activity of REV1 is employed during immunoglobulin gene diversification in DT40. *Mol Immunol* **43**, 1587–1594.
84. Zan, H., Shima, N., Xu, Z., Al-Qahtani, A., Evinger, A. J., III Zhong, Y., Schimenti, J. C., and Casali, P. (2005) The translesion DNA polymerase theta plays a dominant role in immunoglobulin gene somatic hypermutation. *EMBO J* **24**, 3757–3769.
85. Simpson, L. J., and Sale, J. E. (2003) Rev1 is essential for DNA damage tolerance and non-templated immunoglobulin gene mutation in a vertebrate cell line. *EMBO J* **22**, 1654–1664.
86. Zan, H., Komori, A., Li, Z., Cerutti, A., Schaffer, A., Flajnik, M. F., Diaz, M., and Casali, P. (2001) The translesion DNA polymerase zeta plays a major role in Ig and Bcl-6 somatic hypermutation. *Immunity* **14**, 643–653.
87. Diaz, M., Verkoczy, L. K., Flajnik, M. F., and Klinman, N. R. (2001) Decreased frequency of somatic hypermutation and impaired affinity maturation but intact germinal center formation in mice expressing antisense RNA to DNA polymerase zeta. *J Immunol* **167**, 327–335.
88. Martomo, S. A., Yang, W. W., Vaisman, A., Maas, A., Yokoi, M., Hoeijmakers, J. H., Hanaoka, F., Woodgate, R., and Gearhart, P. J. (2006) Normal hypermutation in antibody genes from congenic mice defective for DNA polymerase iota. *DNA Repair (Amst)* **5**, 392–398.
89. Shimizu, T., Azuma, T., Ishiguro, M., Kanjo, N., Yamada, S., and Ohmori, H. (2005) Normal immunoglobulin gene somatic hypermutation in Pol kappa-Pol iota double-deficient mice. *Immunol Lett* **98**, 259–264.
90. Arakawa, H., Moldovan, G. L., Saribasak, H., Saribasak, N. N., Jentsch, S., and Buerstedde, J. M. (2006) A role for PCNA ubiquitination in immunoglobulin hypermutation. *PLoS Biol* **4**, e366.
91. Zeng, X., Winter, D. B., Kasmer, C., Kraemer, K. H., Lehmann, A. R., and Gearhart, P. J. (2001) DNA polymerase eta is an A-T mutator in somatic hypermutation of immunoglobulin variable genes. *Nat Immunol* **2**, 537–541.
92. Delbos, F., De Smet, A., Faili, A., Aoufouchi, S., Weill, J. C., and Reynaud, C. A. (2005) Contribution of DNA polymerase eta to immunoglobulin gene hypermutation in the mouse. *J Exp Med* **201**, 1191–1196.
93. Delbos, F., Aoufouchi, S., Faili, A., Weill, J. C., and Reynaud, C. A. (2007) DNA polymerase eta is the sole contributor of A/T modifications during immunoglobulin gene hypermutation in the mouse. *J Exp Med* **204**, 17–23.

94. Shen, H. M., Tanaka, A., Bozek, G., Nicolae, D., and Storb, U. (2006) Somatic hypermutation and class switch recombination in Msh6$^{(-/-)}$Ung$^{(-/-)}$ double-knockout mice. *J Immunol* **177**, 5386–5392.
95. Bardwell, P. D., Woo, C. J., Wei, K., Li, Z., Martin, A., Sack, S. Z., Parris, T., Edelmann, W., and Scharff, M. D. (2004) Altered somatic hypermutation and reduced class-switch recombination in exonuclease 1-mutant mice. *Nat Immunol* **5**, 224–229.
96. Wu, X., Tsai, C. Y., Patam, M. B., Zan, H., Chen, J. P., Lipkin, S. M., and Casali, P. (2006) A role for the MutL mismatch repair Mlh3 protein in immunoglobulin class switch DNA recombination and somatic hypermutation. *J Immunol* **176**, 5426–5437.
97. Li, Z., Peled, J. U., Zhao, C., Svetlanov, A., Ronai, D., Cohen, P. E., and Scharff, M. D. (2006) A role for Mlh3 in somatic hypermutation. *DNA Repair (Amst)* **5**, 675–682.
98. Ehrenstein, M. R., Rada, C., Jones, A. M., Milstein, C., and Neuberger, M. S. (2001) Switch junction sequences in PMS2-deficient mice reveal a microhomology-mediated mechanism of Ig class switch recombination. *Proc Natl Acad Sci USA* **98**, 14553–14558.
99. Phung, Q. H., Winter, D. B., Alrefai, R., and Gearhart, P. J. (1999) Hypermutation in Ig V genes from mice deficient in the MLH1 mismatch repair protein. *J Immunol* **162**, 3121–3124.
100. Kong, Q., and Maizels, N. (1999) PMS2-deficiency diminishes hypermutation of a lambda1 transgene in young but not older mice. *Mol Immunol* **36**, 83–91.
101. Saribasak, H., Saribasak, N. N., Ipek, F. M., Ellwart, J. W., Arakawa, H., and Buerstedde, J. M. (2006) Uracil DNA glycosylase disruption blocks Ig gene conversion and induces transition mutations. *J Immunol* **176**, 365–371.
102. Di Noia, J. M., and Neuberger, M. S. (2004) Immunoglobulin gene conversion in chicken DT40 cells largely proceeds through an abasic site intermediate generated by excision of the uracil produced by AID-mediated deoxycytidine deamination. *Eur J Immunol* **34**, 504–508.
103. Yabuki, M., Fujii, M. M., and Maizels, N. (2005) The MRE11-RAD50-NBS1 complex accelerates somatic hypermutation and gene conversion of immunoglobulin variable regions. *Nat Immunol* **6**, 730–736.
104. Larson, E. D., Cummings, W. J., Bednarski, D. W., and Maizels, N. (2005) MRE11/RAD50 cleaves DNA in the AID/UNG-dependent pathway of immunoglobulin gene diversification. *Mol Cell* **20**, 367–375.
105. Reina-San-Martin, B., Nussenzweig, M. C., Nussenzweig, A., and Difilippantonio, S. (2005) Genomic instability, endoreduplication, and diminished Ig class-switch recombination in B cells lacking Nbs1. *Proc Natl Acad Sci USA* **102**, 1590–1595.
106. Sale, J. E., Calandrini, D. M., Takata, M., Takeda, S., and Neuberger, M. S. (2001) Ablation of XRCC2/3 transforms immunoglobulin V gene conversion into somatic hypermutation. *Nature* **412**, 921–926.
107. Hatanaka, A., Yamazoe, M., Sale, J. E., Takata, M., Yamamoto, K., Kitao, H., Sonoda, E., Kikuchi, K., Yonetani, Y., and Takeda, S. (2005) Similar effects of Brca2 truncation and Rad51 paralog deficiency on immunoglobulin V gene diversification in DT40 cells support an early role for Rad51 paralogs in homologous recombination. *Mol Cell Biol* **25**, 1124–1134.
108. Cook, A. J., Raftery, J. M., Lau, K. K., Jessup, A., Harris, R. S., Takeda, S., and Jolly, C. J. (2007) DNA-dependent protein kinase inhibits AID-induced antibody gene conversion. *PLoS Biol* **5**, e80.
109. Tang, E. S., and Martin, A. (2006) NHEJ-deficient DT40 cells have increased levels of immunoglobulin gene conversion: Evidence for a double strand break intermediate. *Nucleic Acids Res* **34**, 6345–6351.
110. Etzioni, A., and Ochs, H. D. (2004) The hyper IgM syndrome—an evolving story. *Pediatr Res* **56**, 519–525.
111. Kadyrov, F. A., Dzantiev, L., Constantin, N., and Modrich, P. (2006) Endonucleolytic function of MutLalpha in human mismatch repair. *Cell* **126**, 297–308.
112. Schrader, C. E., Edelmann, W., Kucherlapati, R., and Stavnezer, J. (1999) Reduced isotype switching in splenic B cells from mice deficient in mismatch repair enzymes. *J Exp Med* **190**, 323–330.
113. Schrader, C. E., Vardo, J., and Stavnezer, J. (2002) Role for mismatch repair proteins Msh2, Mlh1, and Pms2 in immunoglobulin class switching shown by sequence analysis of recombination junctions. *J Exp Med* **195**, 367–373.

114. Tauchi, H., Kobayashi, J., Morishima, K., van Gent, D. C., Shiraishi, T., Verkaik, N. S., vanHeems, D., Ito, E., Nakamura, A., Sonoda, E., Takata, M., Takada, S., Matsuura, S., and Komatsu, K. (2002) Nbs1 is essential for DNA repair by homologous recombination in higher vertebrate cells. *Nature* **420**, 93–98.
115. Grenon, M., Gilbert, C., and Lowndes, N. F. (2001) Checkpoint activation in response to double-strand breaks requires the Mre11/Rad50/Xrs2 complex. *Nat Cell Biol* **3**, 844–847.
116. de Jager, M., Dronkert, M. L., Modesti, M., Beerens, C. E., Kanaar, R., and van Gent, D. C. (2001) DNA-binding and strand-annealing activities of human Mre11: Implications for its roles in DNA double-strand break repair pathways. *Nucleic Acids Res* **29**, 1317–1325.
117. Buscemi, G., Savio, C., Zannini, L., Micciche, F., Masnada, D., Nakanishi, M., Tauchi, H., Komatsu, K., Mizutani, S., Khanna, K., Chen, P., Concannon, P., Chessa, L., and Delia, D. (2001) Chk2 activation dependence on Nbs1 after DNA damage. *Mol Cell Biol* **21**, 5214–5222.
118. Stewart, G. S., Maser, R. S., Stankovic, T., Bressan, D. A., Kaplan, M. I., Jaspers, N. G., Raams, A., Byrd, P. J., Petrini, J. H., and Taylor, A. M. (1999) The DNA double-strand break repair gene hMRE11 is mutated in individuals with an ataxia-telangiectasia-like disorder. *Cell* **99**, 577–587.
119. Varon, R., Vissinga, C., Platzer, M., Cerosaletti, K. M., Chrzanowska, K. H., Saar, K., Beckmann, G., Seemanova, E., Cooper, P. R., Nowak, N. J., Stumm, M., Weemaes, C. M., Gatti, R. A., Wilson, R. K., Digweed, M., Rosenthal, A., Sperling, K., Concannon, P., and Reis, A. 1998 Nibrin, a novel DNA double-strand break repair protein, is mutated in Nijmegen breakage syndrome. *Cell* **93**, 467–476.
120. Carney, J. P., Maser, R. S., Olivares, H., Davis, E. M., Le Beau, M., Yates, J. R., 3rd Hays, L., Morgan, W. F., and Petrini, J. H. (1998) The hMre11/hRad50 protein complex and Nijmegen breakage syndrome: Linkage of double-strand break repair to the cellular DNA damage response. *Cell* **93** 477–486.
121. Lahdesmaki, A., Taylor, A. M., Chrzanowska, K. H., and Pan-Hammarstrom, Q. (2004) Delineation of the role of the Mre11 complex in class switch recombination. *J Biol Chem* **279**, 16479–16487.
122. Zhu, J., Petersen, S., Tessarollo, L., and Nussenzweig, A. (2001) Targeted disruption of the Nijmegen breakage syndrome gene NBS1 leads to early embryonic lethality in mice. *Curr Biol* **11**, 105–109.
123. Luo, G., Yao, M. S., Bender, C. F., Mills, M., Bladl, A. R., Bradley, A., and Petrini, J. H. (1999) Disruption of mRad50 causes embryonic stem cell lethality, abnormal embryonic development, and sensitivity to ionizing radiation. *Proc Natl Acad Sci USA* **96**, 7376–7381.
124. Xiao, Y., and Weaver, D. T. (1997) Conditional gene targeted deletion by Cre recombinase demonstrates the requirement for the double-strand break repair Mre11 protein in murine embryonic stem cells. *Nucleic Acids Res* **25**, 2985–2991.
125. Kracker, S., Bergmann, Y., Demuth, I., Frappart, P. O., Hildebrand, G., Christine, R., Wang, Z. Q., Sperling, K., Digweed, M., and Radbruch, A. (2005) Nibrin functions in Ig class-switch recombination. *Proc Natl Acad Sci USA* **102**, 1584–1589.
126. Martin, A., Li, Z., Lin, D. P., Bardwell, P. D., Iglesias-Ussel, M. D., Edelmann, W., and Scharff, M. D. (2003) Msh2 ATPase activity is essential for somatic hypermutation at A-T basepairs and for efficient class switch recombination. *J Exp Med* **198**, 1171–1178.
127. Schrader, C. E., Vardo, J., and Stavnezer, J. (2003) Mlh1 can function in antibody class switch recombination independently of Msh2. *J Exp Med* **197**, 1377–1383.
128. Li, Z., Scherer, S. J., Ronai, D., Iglesias-Ussel, M. D., Peled, J. U., Bardwell, P. D., Zhuang, M., Lee, K., Martin, A., Edelmann, W., and Scharff, M. D. (2004) Examination of Msh6- and Msh3-deficient mice in class switching reveals overlapping and distinct roles of MutS homologues in antibody diversification. *J Exp Med* **200**, 47–59.
129. Martomo, S. A., Yang, W. W., and Gearhart, P. J. (2004) A role for Msh6 but not Msh3 in somatic hypermutation and class switch recombination. *J Exp Med* **200**, 61–68.
130. Ehrenstein, M. R., and Neuberger, M. S. (1999) Deficiency in Msh2 affects the efficiency and local sequence specificity of immunoglobulin class-switch recombination: Parallels with somatic hypermutation. *EMBO J* **18**, 3484–3490.
131. Sekine, H., Ferreira, R. C., Pan-Hammarstrom, Q., Graham, R. R., Ziemba, B., de Vries, S. S., Liu, J., Hippen, K., Koeuth, T., Ortmann, W., Iwahori, A., Elliott, M. K., Offer, S., Skon, C., Du, L., Novitzke, J., Lee, A. T., Zhao, N., Tompkins, J. D., Altshuler, D., Gregersen, P. K., Cunningham-Rundles, C., Harris, R. S., Her, C., Nelson, D. L., Hammarstrom, L., Gilkeson, G. S., and Behrens, T. W. 2007 Role

for Msh5 in the regulation of Ig class switch recombination. *Proc Natl Acad Sci USA* **104**, 7193–7198.

132. Anderson, L., Henderson, C., and Adachi, Y. (2001) Phosphorylation and rapid relocalization of 53BP1 to nuclear foci upon DNA damage. *Mol Cell Biol* **21**, 1719–1729.
133. Pan-Hammarstrom, Q., Dai, S., Zhao, Y., van Dijk-Hard, I. F., Gatti, R. A., Borresen-Dale, A. L., and Hammarstrom, L. (2003) ATM is not required in somatic hypermutation of VH, but is involved in the introduction of mutations in the switch mu region. *J Immunol* **170**, 3707–3716.
134. Reina-San-Martin, B., Difilippantonio, S., Hanitsch, L., Masilamani, R. F., Nussenzweig, A., and Nussenzweig, M. C. (2003) H2AX is required for recombination between immunoglobulin switch regions but not for intra-switch region recombination or somatic hypermutation. *J Exp Med* **197**, 1767–1778.
135. Manis, J. P., Morales, J. C., Xia, Z., Kutok, J. L., Alt, F. W., and Carpenter, P. B. (2004) 53BP1 links DNA damage-response pathways to immunoglobulin heavy chain class-switch recombination. *Nat Immunol* **5**, 481–487.
136. Ward, I. M., Reina-San-Martin, B., Olaru, A., Minn, K., Tamada, K., Lau, J. S., Cascalho, M., Chen, L., Nussenzweig, A., and Livak, F. et al. (2004) 53BP1 is required for class switch recombination. *J Cell Biol* **165**, 459–464.
137. Manis, J. P., Gu, Y., Lansford, R., Sonoda, E., Ferrini, R., Davidson, L., Rajewsky, K., and Alt, F. W. (1998) Ku70 is required for late B cell development and immunoglobulin heavy chain class switching. *J Exp Med* **187**, 2081–2089.
138. Casellas, R., Nussenzweig, A., Wuerffel, R., Pelanda, R., Reichlin, A., Suh, H., Qin, X. F., Besmer, E., Kenter, A., Rajewsky, K., and Nussenzweig, M. C. (1998) Ku80 is required for immunoglobulin isotype switching. *EMBO J* **17**, 2404–2411.
139. Manis, J. P., Dudley, D., Kaylor, L., and Alt, F. W. (2002) IgH class switch recombination to IgG1 in DNA-PKcs-deficient B cells. *Immunity* **16**, 607–617.
140. Cook, A. J., Oganesian, L., Harumal, P., Basten, A., Brink, R., and Jolly, C. J. (2003) Reduced switching in SCID B cells is associated with altered somatic mutation of recombined S regions. *J Immunol* **171**, 6556–6564.
141. Bosma, G. C., Kim, J., Urich, T., Fath, D. M., Cotticelli, M. G., Ruetsch, N. R., Radic, M. Z., and Bosma, M. J. (2002) DNA-dependent protein kinase activity is not required for immunoglobulin class switching. *J Exp Med* **196**, 1483–1495.
142. Kiefer, K., Oshinsky, J., Kim, J., Nakajima, P. B., Bosma, G. C., and Bosma, M. J. (2007) The catalytic subunit of DNA-protein kinase (DNA-PKcs) is not required for Ig class-switch recombination. *Proc Natl Acad Sci USA* **104**, 2843–2848.
143. Gao, Y., Sun, Y., Frank, K. M., Dikkes, P., Fujiwara, Y., Seidl, K. J., Sekiguchi, J. M., Rathbun, G. A., Swat, W., Wang, J., Bronson, R. T., Maylynn, B. A., Bryans, M., Zhu, C., Chaudhuri, J., Davidson, L., Ferrini, R., Stamato, T., Orkin, S. H., Greenberg, M. E., and Alt, F. W. (1998) A critical role for DNA end-joining proteins in both lymphogenesis and neurogenesis. *Cell* **95** 891–902.
144. Frank, K. M., Sekiguchi, J. M., Seidl, K. J., Swat, W., Rathbun, G. A., Cheng, H. L., Davidson, L., Kangaloo, L., and Alt, F. W. (1998) Late embryonic lethality and impaired V(D)J recombination in mice lacking DNA ligase IV. *Nature* **396**, 173–177.
145. Nick McElhinny, S. A., Snowden, C. M., McCarville, J., and Ramsden, D. A. (2000) Ku recruits the XRCC4-ligase IV complex to DNA ends. *Mol Cell Biol* **20**, 2996–3003.
146. Lee, K. J., Huang, J., Takeda, Y., and Dynan, W. S. (2000) DNA ligase IV and XRCC4 form a stable mixed tetramer that functions synergistically with other repair factors in a cell-free end-joining system. *J Biol Chem* **275**, 34787–34796.
147. Chen, L., Trujillo, K., Sung, P., and Tomkinson, A. E. (2000) Interactions of the DNA ligase IV-XRCC4 complex with DNA ends and the DNA-dependent protein kinase. *J Biol Chem* **275**, 26196–26205.
148. Bryans, M., Valenzano, M. C., and Stamato, T. D. (1999) Absence of DNA ligase IV protein in XR-1 cells: Evidence for stabilization by XRCC4. *Mutat Res* **433**, 53–58.
149. Rooney, S., Alt, F. W., Sekiguchi, J., and Manis, J. P. (2005) Artemis-independent functions of DNA-dependent protein kinase in Ig heavy chain class switch recombination and development. *Proc Natl Acad Sci USA* **102**, 2471–2475.

150. Buck, D., Malivert, L., de Chasseval, R., Barraud, A., Fondaneche, M. C., Sanal, O., Plebani, A., Stephan, J. L., Hufnagel, M., le Deist, F., et al. (2006) Cernunnos, a novel nonhomologous end-joining factor, is mutated in human immunodeficiency with microcephaly. *Cell* **124**, 287–299.
151. Ahnesorg, P., Smith, P., and Jackson, S. P. (2006) XLF interacts with the XRCC4-DNA ligase IV complex to promote DNA nonhomologous end-joining. *Cell* **124**, 301–313.
152. Lu, H., Pannicke, U., Schwarz, K., and Lieber, M. R. (2007) Length-dependent binding of human XLF to DNA and stimulation of XRCC4/DNA ligase IV activity. *J Biol Chem* **282**, 11155–11162.
153. Erdos, M., Durandy, A., and Marodi, L. (2005) Genetically acquired class-switch recombination defects: The multi-faced hyper-IgM syndrome. *Immunol Lett* **97**, 1–6.
154. Durandy, A., Peron, S., and Fischer, A. (2006) Hyper-IgM syndromes. *Curr Opin Rheumatol* **18**, 369–376.
155. Padayachee, M., Levinsky, R. J., Kinnon, C., Finn, A., McKeown, C., Feighery, C., Notarangelo, L. D., Hendriks, R. W., Read, A. P., and Malcolm, S. (1993) Mapping of the X linked form of hyper IgM syndrome (HIGM1). *J Med Genet* **30**, 202–205.
156. Padayachee, M., Feighery, C., Finn, A., McKeown, C., Levinsky, R. J., Kinnon, C., and Malcolm, S. (1992) Mapping of the X-linked form of hyper-IgM syndrome (HIGM1) to Xq26 by close linkage to HPRT. *Genomics* **14**, 551–553.
157. Ferrari, S., Giliani, S., Insalaco, A., Al-Ghonaium, A., Soresina, A. R., Loubser, M., Avanzini, M. A., Marconi, M., Badolato, R., Ugazio, A. G., Levy, Y., Catalan, N., Durandy, A., Tbakhi, A., Notarangelo, L. D., and Plebani, A. (2001) Mutations of CD40 gene cause an autosomal recessive form of immunodeficiency with hyper IgM. *Proc Natl Acad Sci USA* **98**, 12614–12619.
158. Jain, A., Ma, C. A., Liu, S., Brown, M., Cohen, J., and Strober, W. (2001) Specific missense mutations in NEMO result in hyper-IgM syndrome with hypohydrotic ectodermal dysplasia. *Nat Immunol* **2**, 223–228.
159. Xie, K., Sowden, M. P., Dance, G. S., Torelli, A. T., Smith, H. C., and Wedekind, J. E. (2004) The structure of a yeast RNA-editing deaminase provides insight into the fold and function of activation-induced deaminase and APOBEC-1. *Proc Natl Acad Sci USA* **101**, 8114–8119.
160. Ta, V. T., Nagaoka, H., Catalan, N., Durandy, A., Fischer, A., Imai, K., Nonoyama, S., Tashiro, J., Ikegawa, M., Ito, S., Kinoshita, K., Muramatsu, M., and Honjo, T. (2003) AID mutant analyses indicate requirement for class-switch-specific cofactors. *Nat Immunol* **4**, 843–848.
161. Imai, K., Zhu, Y., Revy, P., Morio, T., Mizutani, S., Fischer, A., Nonoyama, S., and Durandy, A. (2005) Analysis of class switch recombination and somatic hypermutation in patients affected with autosomal dominant hyper-IgM syndrome type 2. *Clin Immunol* **115**, 277–285.
162. Shinkura, R., Ito, S., Begum, N. A., Nagaoka, H., Muramatsu, M., Kinoshita, K., Sakakibara, Y., Hijikata, H., and Honjo, T. (2004) Separate domains of AID are required for somatic hypermutation and class-switch recombination. *Nat Immunol* **5**, 707–712.
163. Barreto, V., Reina-San-Martin, B., Ramiro, A. R., McBride, K. M., and Nussenzweig, M. C. (2003) C-terminal deletion of AID uncouples class switch recombination from somatic hypermutation and gene conversion. *Mol Cell* **12**, 501–508.
164. Salzer, U., and Grimbacher, B. (2006) Common variable immunodeficiency: The power of co-stimulation. *Semin Immunol* **18**, 337–346.
165. Castigli, E., and Geha, R. S. (2006) Molecular basis of common variable immunodeficiency. *J Allergy Clin Immunol* **117**, 740–746.
166. Imai, K., Catalan, N., Plebani, A., Marodi, L., Sanal, O., Kumaki, S., Nagendran, V., Wood, P., Glastre, C., Sarrot-Reynauld, F. et al. (2003) Hyper-IgM syndrome type 4 with a B lymphocyte-intrinsic selective deficiency in Ig class-switch recombination. *J Clin Invest* **112**, 136–142.
167. Vorechovsky, I., Zetterquist, H., Paganelli, R., Koskinen, S., Webster, A. D., Bjorkander, J., Smith, C. I., and Hammarstrom, L. (1995) Family and linkage study of selective IgA deficiency and common variable immunodeficiency. *Clin Immunol Immunopathol* **77**, 185–192.
168. Rada, C., Jarvis, J. M., and Milstein, C. (2002) AID-GFP chimeric protein increases hypermutation of Ig genes with no evidence of nuclear localization. *Proc Natl Acad Sci USA* **99**, 7003–7008.
169. Ito, S., Nagaoka, H., Shinkura, R., Begum, N., Muramatsu, M., Nakata, M., and Honjo, T. (2004) Activation-induced cytidine deaminase shuttles between nucleus and cytoplasm like apolipoprotein B mRNA editing catalytic polypeptide 1. *Proc Natl Acad Sci USA* **101**, 1975–1980.

170. McBride, K. M., Barreto, V., Ramiro, A. R., Stavropoulos, P., and Nussenzweig, M. C. (2004) Somatic hypermutation is limited by CRM1-dependent nuclear export of activation-induced deaminase. *J Exp Med* **199**, 1235–1244.
171. Paine, P. L., Moore, L. C., and Horowitz, S. B. (1975) Nuclear envelope permeability. *Nature* **254**, 109–114.
172. Cattoretti, G., Buttner, M., Shaknovich, R., Kremmer, E., Alobeid, B., and Niedobitek, G. (2006) Nuclear and cytoplasmic AID in extrafollicular and germinal center B cells. *Blood* **107**, 3967–3975.
173. Nagumo, H., Agematsu, K., Kobayashi, N., Shinozaki, K., Hokibara, S., Nagase, H., Takamoto, M., Yasui, K., Sugane, K., and Komiyama, A. (2002) The different process of class switching and somatic hypermutation; a novel analysis by CD27(−) naive B cells. *Blood* **99**, 567–575.
174. Liu, Y. J., Malisan, F., de Bouteiller, O., Guret, C., Lebecque, S., Banchereau, J., Mills, F. C., Max, E. E., and Martinez-Valdez, H. (1996) Within germinal centers, isotype switching of immunoglobulin genes occurs after the onset of somatic mutation. *Immunity* **4**, 241–250.
175. Chaudhuri, J., Khuong, C., and Alt, F. W. (2004) Replication protein A interacts with AID to promote deamination of somatic hypermutation targets. *Nature* **430**, 992–998.
176. Sohail, A., Klapacz, J., Samaranayake, M., Ullah, A., and Bhagwat, A. S. (2003) Human activation-induced cytidine deaminase causes transcription-dependent, strand-biased C to U deaminations. *Nucleic Acids Res* **31**, 2990–2994.
177. Bottaro, A., Lansford, R., Xu, L., Zhang, J., Rothman, P., and Alt, F. W. (1994) S region transcription per se promotes basal IgE class switch recombination but additional factors regulate the efficiency of the process. *EMBO J* **13**, 665–674.
178. Betz, A. G., Milstein, C., Gonzalez-Fernandez, A., Pannell, R., Larson, T., and Neuberger, M. S. (1994) Elements regulating somatic hypermutation of an immunoglobulin kappa gene: Critical role for the intron enhancer/matrix attachment region. *Cell* **77**, 239–248.
179. Zhang, J., Bottaro, A., Li, S., Stewart, V., and Alt, F. W. (1993) A selective defect in IgG2b switching as a result of targeted mutation of the I gamma 2b promoter and exon. *EMBO J* **12**, 3529–3537.
180. Jung, S., Rajewsky, K., and Radbruch, A. (1993) Shutdown of class switch recombination by deletion of a switch region control element. *Science* **259**, 984–987.
181. Basu, U., Chaudhuri, J., Alpert, C., Dutt, S., Ranganath, S., Li, G., Schrum, J. P., Manis, J. P., and Alt, F. W. (2005) The AID antibody diversification enzyme is regulated by protein kinase A phosphorylation. *Nature* **438**, 508–511.
182. Pasqualucci, L., Kitaura, Y., Gu, H., and Dalla-Favera, R. (2006) PKA-mediated phosphorylation regulates the function of activation-induced deaminase (AID) in B cells. *Proc Natl Acad Sci USA* **103**, 395–400.
183. Kuppers, R., and Dalla-Favera, R. (2001) Mechanisms of chromosomal translocations in B cell lymphomas. *Oncogene* **20**, 5580–5594.
184. Pasqualucci, L., Neumeister, P., Goossens, T., Nanjangud, G., Chaganti, R. S., Kuppers, R., and Dalla-Favera, R. (2001) Hypermutation of multiple proto-oncogenes in B-cell diffuse large-cell lymphomas. *Nature* **412**, 341–346.
185. Rossi, D., Cerri, M., Capello, D., Deambrogi, C., Berra, E., Franceschetti, S., Alabiso, O., Gloghini, A., Paulli, M., Carbone, A., Pileri, S. A., Pasqualucci, L., and Gaidano, G. (2005) Aberrant somatic hypermutation in primary mediastinal large B-cell lymphoma. *Leukemia* **19**, 2363–2366.
186. Gordon, M. S., Kanegai, C. M., Doerr, J. R., and Wall, R. (2003) Somatic hypermutation of the B cell receptor genes B29 (Igbeta, CD79b) and mb1 (Igalpha, CD79a). *Proc Natl Acad Sci USA* **100**, 4126–4131.
187. Muschen, M., Re, D., Jungnickel, B., Diehl, V., Rajewsky, K., and Kuppers, R. (2000) Somatic mutation of the CD95 gene in human B cells as a side-effect of the germinal center reaction. *J Exp Med* **192**, 1833–1840.
188. Shen, H. M., Peters, A., Baron, B., Zhu, X., and Storb, U. (1998) Mutation of BCL-6 gene in normal B cells by the process of somatic hypermutation of Ig genes. *Science* **280**, 1750–1752.
189. Pasqualucci, L., Migliazza, A., Fracchiolla, N., William, C., Neri, A., Baldini, L., Chaganti, R. S., Klein, U., Kuppers, R., Rajewsky, K., and Dalla-Favera, R. (1998) BCL-6 mutations in normal germinal center B cells: Evidence of somatic hypermutation acting outside Ig loci. *Proc Natl Acad Sci USA* **95**, 11816–11821.

190. Ramiro, A. R., Jankovic, M., Callen, E., Difilippantonio, S., Chen, H. T., McBride, K. M., Eisenreich, T. R., Chen, J., Dickins, R. A., Lowe, S. W., Nussenzweig, A., and Nussenzweig, M. C. (2006) Role of genomic instability and p53 in AID-induced c-myc-Igh translocations. *Nature* **440**, 105–109.
191. Duquette, M. L., Pham, P., Goodman, M. F., and Maizels, N. (2005) AID binds to transcription-induced structures in c-MYC that map to regions associated with translocation and hypermutation. *Oncogene* **24**, 5791–5798.
192. Ramiro, A. R., Jankovic, M., Eisenreich, T., Difilippantonio, S., Chen-Kiang, S., Muramatsu, M., Honjo, T., Nussenzweig, A., and Nussenzweig, M. C. (2004) AID is required for c-myc/IgH chromosome translocations in vivo. *Cell* **118**, 431–438.
193. Schreck, S., Buettner, M., Kremmer, E., Bogdan, M., Herbst, H., and Niedobitek, G. (2006) Activation-induced cytidine deaminase (AID) is expressed in normal spermatogenesis but only infrequently in testicular germ cell tumours. *J Pathol* **210**, 26–31.
194. Muto, T., Okazaki, I. M., Yamada, S., Tanaka, Y., Kinoshita, K., Muramatsu, M., Nagaoka, H., and Honjo, T. (2006) Negative regulation of activation-induced cytidine deaminase in B cells. *Proc Natl Acad Sci USA* **103**, 2752–2757.
195. Okazaki, I. M., Hiai, H., Kakazu, N., Yamada, S., Muramatsu, M., Kinoshita, K., and Honjo, T. (2003) Constitutive expression of AID leads to tumorigenesis. *J Exp Med* **197**, 1173–1181.
196. Matsumoto, Y., Marusawa, H., Kinoshita, K., Endo, Y., Kou, T., Morisawa, T., Azuma, T., Okazaki, I. M., Honjo, T., and Chiba, T. (2007) *Helicobacter pylori* infection triggers aberrant expression of activation-induced cytidine deaminase in gastric epithelium. *Nat Med* **13**, 470–476.
197. Endo, Y., Marusawa, H., Kinoshita, K., Morisawa, T., Sakurai, T., Okazaki, I. M., Watashi, K., Shimotohno, K., Honjo, T., and Chiba, T. (2007) Expression of activation-induced cytidine deaminase in human hepatocytes via NF-kappaB signaling. *Oncogene* **38**, 5587–5595 Apr 2, Epub ahead of print.
198. Babbage, G., Ottensmeier, C. H., Blaydes, J., Stevenson, F. K., and Sahota, S. S. (2006) Immunoglobulin heavy chain locus events and expression of activation-induced cytidine deaminase in epithelial breast cancer cell lines. *Cancer Res* **66**, 3996–4000.
199. Jarmuz, A., Chester, A., Bayliss, J., Gisbourne, J., Dunham, I., Scott, J., and Navaratnam, N. (2002) An anthropoid-specific locus of orphan C to U RNA-editing enzymes on chromosome 22. *Genomics* **79**, 285–296.
200. Liao, W., Hong, S. H., Chan, B. H., Rudolph, F. B., Clark, S. C., and Chan, L. (1999) APOBEC-2, a cardiac- and skeletal muscle-specific member of the cytidine deaminase supergene family. *Biochem Biophys Res Commun* **260**, 398–404.
201. Rogozin, I. B., Basu, M. K., Jordan, I. K., Pavlov, Y. I., and Koonin, E. V. (2005) APOBEC4, a new member of the AID/APOBEC family of polynucleotide (deoxy)cytidine deaminases predicted by computational analysis. *Cell Cycle* **4**, 1281–1285.
202. Mikl, M. C., Watt, I. N., Lu, M., Reik, W., Davies, S. L., Neuberger, M. S., and Rada, C. (2005) Mice deficient in APOBEC2 and APOBEC3. *Mol Cell Biol* **25**, 7270–7277.
203. Harris, R. S., Petersen-Mahrt, S. K., and Neuberger, M. S. (2002) RNA editing enzyme APOBEC1 and some of its homologs can act as DNA mutators. *Mol Cell* **10**, 1247–1253.
204. Fugmann, S. D., Rush, J. S., and Schatz, D. G. (2004) Non-redundancy of cytidine deaminases in class switch recombination. *Eur J Immunol* **34**, 844–849.
205. Iwatani, Y., Takeuchi, H., Strebel, K., and Levin, J. G. (2006) Biochemical activities of highly purified, catalytically active human APOBEC3G: Correlation with antiviral effect. *J Virol* **80**, 5992–6002.
206. MacDuff, D. A., and Harris, R. S. (2006) Directed DNA deamination by AID/APOBEC3 in immunity. *Curr Biol* **16**, R186–189.
207. Wilkins, M. H.F., Stokes, A. R., and Wilson, H. R. (1953) Molecular structure of deoxypentose nucleic acids. *Nature* **171**, 738–740.

CHAPTER 3

PROTEIN–PROTEIN AND RNA–PROTEIN INTERACTIONS IN U-INSERTION/DELETION RNA EDITING COMPLEXES

Jorge Cruz-Reyes
Alfredo Hernandez

MANY MITOCHONDRIAL primary mRNAs in early-branched kinetoplastids are plagued with frameshifts and stop codons. An array of proteins associated in a megadalton catalytic complex specifically recognizes and repairs these "faulty" sequences by inserting and deleting uridines at often hundreds of editing sites. The known macromolecular interactions suggest that U-deletion and U-insertion editing activities occur in separate and "separable" compartments. A few editing-type specific subunits appear to be dynamic components of the complexes, and several subunits are critical to their overall integrity.

3.1 A BIZARRE PHENOMENON AND ITS *RAISON D'ÊTRE*

The term RNA editing was coined after the discovery of an extraordinary form of posttranscriptional mRNA maturation in the single mitochondrion of the kinetoplastid flagellate *Crithidia faciculata* over two decades ago (1). This editing, also found in other kinetoplastids, including the widely studied *Leishmania and Trypanosoma* species, involves the specific U-insertion and U-deletion of uridylates in pre-edited (pre-)mRNA transcripts (Figure 3.1) for recent reviews see references (2–4). Additional forms of editing and various forms of base modification found in these and other organisms appear to involve very different enzymatic mechanisms (also discussed in this volume). Kinetoplastid U-insertion/U-deletion editing takes place within megadalton multi-subunit complexes known as L-complexes or ~20S

RNA and DNA Editing: Molecular Mechanisms and Their Integration into Biological Systems,
Edited by Harold C. Smith

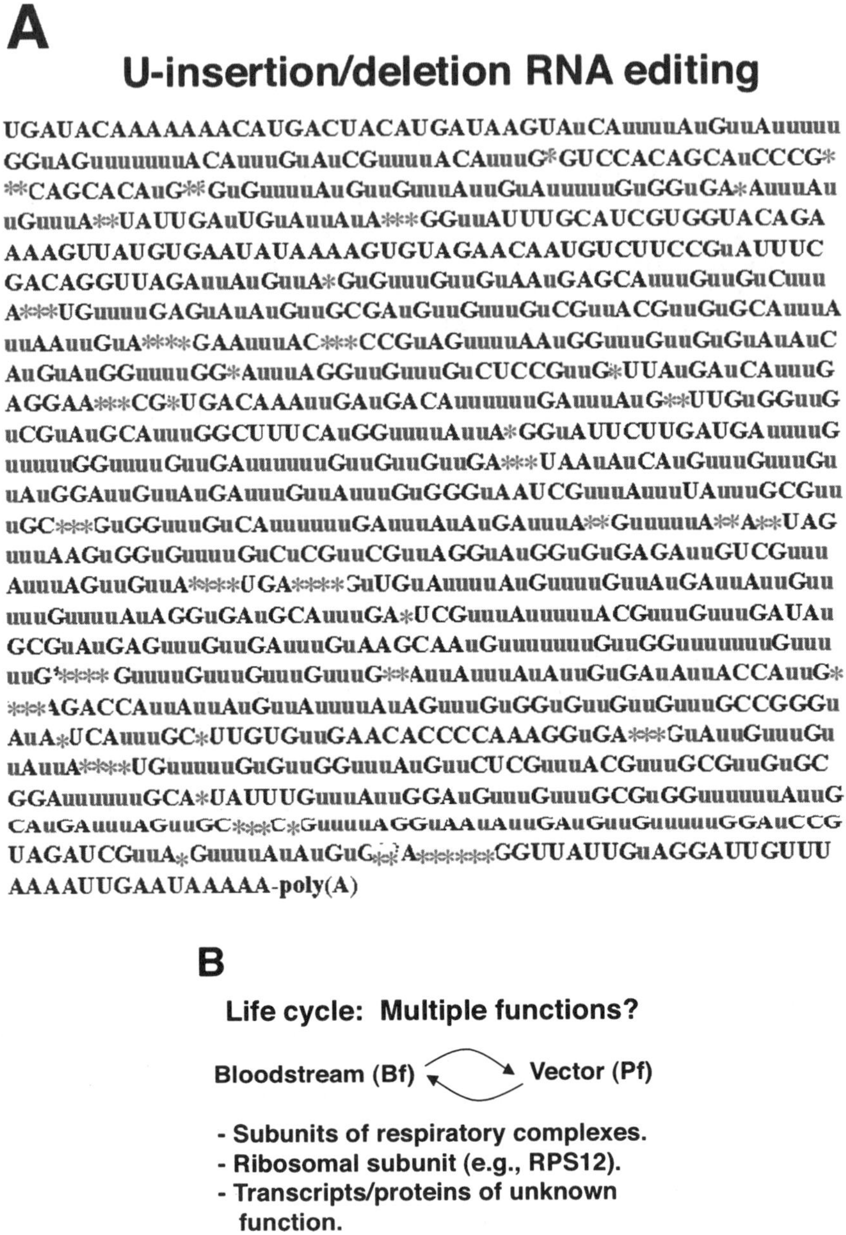

Figure 3.1 RNA editing by U-insertion and U-deletion. (**A**) A mature mRNA after extensive editing throughout its sequence. Site-specific postranscriptional U-insertions are indicated by "u," and U-deletions are indicated by asterisks. (**B**) During its life cycle, *T. brucei* procyclic (Pf) and bloodstream (Bf) forms infect the vector of transmission (tsetse fly) and mammalian hosts, respectively. RNA editing may play multiple functions because the predicted products that involve editing include subunits of respiratory complexes and ribosomes as well as transcripts or proteins of unknown functions. (*See color insert.*)

editosomes, which at least in *T. brucei* recognize thousands of pre-mRNA sites. The editing site selection follows a general 3′-to-5′ polarity in blocks of ~10–15 sites defined by trans-acting guide RNAs (gRNAs) (5). In most cases, coordinated use of multiple gRNAs is necessary, such that a newly edited block can provide a short complementary sequence to "anchor" a subsequent incoming gRNA. In this relay-type process, the first gRNA targets the 3′-most end of the pre-edited region, thus accounting for its general polarity (6). Editing generates the open-reading frame of mature mRNAs, including start and stop codons in some examples. Kineoplastids are considered among the earliest branched eukaryotes, and the origins and natural selection forces that shaped and conserved this process throughout evolution remain a mystery. However, modern U-insertion/deletion RNA editing exhibits important analogies with the process of RNA interference including the use of RNase III-type endonucleases and site-specific mRNA cleavage guided by small complementary RNAs, and is thought to play a major role in regulation of respiration in kinetoplastids (7) (see Section 3.3).

Finally, while pre-mRNA transcripts are currently considered to be cryptic, the possibility of alternative editing and generation of novel functions beyond respiration has not been eliminated. In fact, the large majority of mitochondrial pre-mRNAs contain extensive misediting (8), which is presumably "fixed" by constant proofreading. Interestingly, a recent study suggested that at least one pre-mRNA substrate in *T. brucei* undergoes differential editing and thereby generates an alternative protein product (9). However, the potential function of this protein remains to be determined.

3.2 THE CATALYTIC MECHANISM AND MACHINERY

Editing is directed by trans-acting guide RNAs (gRNA) that form canonical and G-U base pairs with pre-mRNA substrates. Editing sites (ESs) are irregularities of the gRNA/pre-edited mRNA duplexed sequences; upon editing, the complementarity between mature mRNA and gRNA is extended.

Editing of a cognate pre-mRNA is believed to initiate with the formation of a short "anchor" duplex between gRNA 5′ residues and complementary mRNA nucleotides just 3′ of the pre-edited region. Catalysis of a complete editing cycle ("full-round" editing) at each ES involves three basic steps: (a) mRNA cleavage by an endonuclease, (b) U-addition by a 3′ terminal uridylyl transferase (TUTase) or U-removal by a 3′ to 5′ U-specific exoribonuclease, and (c) RNA ligation. The scissile phosphodiester bond at each ES is at the single-/double-stranded junction indicated in Figure 3.2. Early mechanistic studies established that the U-insertion and U-deletion editing reactions use distinct enzymatic activities. Namely, the endonuclease activities in the two editing types exhibit different nucleotide requirements, the U-removal and U-addition steps are not the reverse reactions of the same enzyme, and separate ligase enzymes are preferentially used for U-deletion and U-insertion editing (10–13) (Figure 3.2). Altogether, these studies led to the hypothesis that the enzymes in each pathway, U-insertion or U-deletion, could be compartmentalized within editing complexes.

Several subsequent studies have significantly increased our understanding of the structural and functional organization, as well as molecular interactions within editing

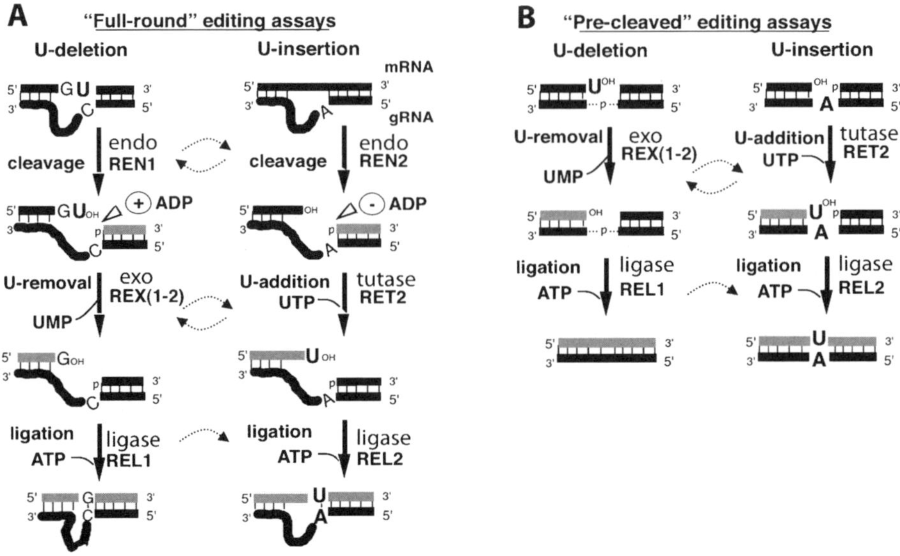

Figure 3.2 "Full-round" and "precleaved" editing reactions and assays. (**A**) Full-round editing at a specific site implies a complete reaction with three intermediate catalytic steps: cleavage, U-removal/addition, and ligation, as directed by a gRNA. These steps are usually scored by the accumulation of the (gray boxes): downstream piece, upstream piece, and full-length mRNA, respectively, using ^{32}P-end-labeled substrates. The known or proposed responsible enzymes are indicated. It is unclear whether REX1 and/or REX2U catalyze U-removal *in vivo*. An arrowhead points to the cleavage site. Nucleotide cofactors, essential (ATP), or modulatory (ADP) are indicated. There is evidence for compartmentalization of U-insertion and U-deletion activities in editing complexes but also indications of interplay between the two pathways, potentially in proofreading of misedited sites. REL1 can replace REN2 but not the converse (see text). (**B**) Precleaved editing assays score only U-removal or U-addition and ligation, using substrates that are significantly simpler than those for full-round editing.

complexes. The precise subunit composition has been debated for several years, but cumulative evidence from several groups using mass spectrometric analyses suggests that extensively purified complexes in *Trypanosoma* and *Leishmania* contain between 16 and 20 subunits (2, 3), contrary to the Simpler Protein Composition originally proposed (14). Variations in the observed protein composition may reflect distinct properties of the complexes in *Leishmania* and trypanosomes as well as differences between native complexes purified via ion-exchange chromatography or immunoprecipitation, and TAP-tagged complexes purified via affinity-chromatography. The stringency of the chosen purification protocol as well as unforeseen consequences of tag-modification or overexpression may determine the observed composition of editing complexes. Evidently, not all subunits assemble into editing complexes with the same efficiency, and further work is necessary to define a minimal core catalytic structure. It has been proposed that the mitochondrial editing holoenzyme (2) is a complicated assembly of at least three discrete particles interlinked by RNA (presumably gRNA) via low-affinity interactions. The basic editing particle containing all activities to catalyze a complete cycle of editing has been termed L-complex, 20S editosome, or 20S editing complex and two smaller proposed regulatory complexes, one of which exhibits annealing activity *in vitro* and is thought to promote base pairing of pre-mRNA and gRNA (MRP-complex), whereas the other complex bears TUTase activity proposed to mediate the 3′ maturation of gRNAs (GP complex). Several known or proposed regulatory factors have been identified, although they are not components of basic 20S editing complexes (15–25). In most studies, the activity of native or affinity-purified tagged editing complexes is determined using complete "full-round" (26, 27) or partial "precleaved" (28, 29) editing assays (Figure 3.2). In contrast to the former, precleaved assays do not score endonuclease cleavage activities, but effectively score the subsequent catalytic steps: (a) U-addition (by TUTase) or U-removal (by an exonuclease), and (b) ligation. *T. brucei* full-round *in vitro* editing was originally re-created using fractionated mitochondrial lysates from procyclics (30) and more recently from bloodstream trypanosomes (31). *In vivo* editing is usually examined using poisoned-primer extension or real-time PCR.

The architecture of multi-subunit 20S editing complexes is unknown, although the presence of both endogenous and tagged subunits in affinity-purified complexes suggests a dimeric configuration (e.g., see references 17 and 32). Also, a recent study found purified tagged complexes to differ in function and protein composition, although they share most known subunits (33). Thus, dimeric or multimeric combinations of editing complexes may be used during transcript-specific and/or developmentally regulated editing.

This chapter focuses on the basic ~20S editing complex and, in particular, discusses the current understanding and mechanistic implications of its known protein–protein and protein–RNA interactions. The identified conserved motifs of the editing subunits, and their predicted catalytic and protein- or nucleic acid-binding potential have been recently discussed (2, 3, 34), and they will not be restated in such detail here.

3.3 EXTENT OF U-INSERTION/DELETION RNA EDITING IN *TRYPANOSOMA* AND *LEISHMANIA* SPECIES

Many mitochondrial transcripts in *Trypanosoma* and *Leishmania* species, as well as in other kinetoplastids, are remodeled by U-insertion/deletion RNA editing. The extent of sequence modification is limited in some transcripts and massive in others, often creating over half the mature length. U-insertion is by far the most abundant type of editing, and a few substrates do not undergo U-deletion (e.g., CYb and COII). In *T. brucei* (an early diverging *Trypanosomatid*) and in the later-diverging *Leishmania* species, many mitochondrial transcripts are extensively edited. The main differences between *T. brucei* and *L. tarentolae* editing are that ND7 and COIII are modified throughout most of their sequences (pan-editing) in T. brucei (see Table 3.1; L. Simpson, personal communication). In the *Trypanosomatids* examined, the complexity of the gRNA population is large enough to specify all observed RNA editing. *T. brucei* exhibits significant regulation of the levels of most edited transcripts during the life cycle, in both transcript-specific and developmental-stage manners. The peak

TABLE 3.1 The Extent of RNA Editing in *Trypanosoma brucei* and *Leishmania tarentolae*.

mRNA	*Trypanosoma*	*Leishmania*
ND8	259-46 Bf	215-41
ND9	345-20 Bf	335-40
ND7	553-89 Bf (5′+3′) Pf (5′)	25-0
COIII	547-41	29-15
CYb	34-0 Pf	39-0
ATP6	447-28	106-5
G3	148-13 Bf	35-14
COII	4-0 Pf	4-0
MURF2	26-4 Bf	28-4
CR3/G4	325-40 Bf	323-4
CR4/G5	210-13 Bf	166-5
RPS12	132-28 Bf	117-32

The columns indicate the abbreviations of the known edited mRNAs, their number of U-insertion/deletions, and the *T. brucei* developmental stage exhibiting the peak of edited product (Pf and Bf stand for procyclic and bloodstream forms, respectively). "NDx" stands for NADH dehydrogenase subunit x; CYb stands for apo-cytochrome *b*; COII and COIII stands for cytochrome oxydase II and III, respectively; ATP6 stands for ATP synthase subunit 6; RPS12 stands for ribosomal protein subunit 12. The remaining predicted encoded proteins have not been identified.

of these transcripts occurs in most cases in trypanosomes that infect the mammalian host (bloodstream from "Bf"). Other transcripts either peak in trypanosomes that infect the tsetse fly vector of transmission (procyclic form "Pf") or exhibit similar levels in both stages. (Table 3.1). Since most edited sequences encode protein subunits of respiratory complexes, it has been hypothesized that RNA editing helps regulate the production of the mitochondrial respiratory system. Edited mRNAs encoding components of NADH dehydrogenase peak in Bf trypanosomes, suggesting utilization of reducing power even though they lack cytochromes and active oxidative phosphorylation. Consistent with this, edited CYb and COII mRNAs are nearly absent in Bf cells. One edited sequence in Bf trypanosomes encodes a ribosomal subunit (RPS12), implying that RNA editing may also serve a role in the control of translation. For recent reviews on the evolution and potential functions of RNA editing, see references (35–37). The studies discussed below have shown that RNA editing is essential for both procyclic and bloodstream trypanosome forms.

3.4 FUNCTIONAL STUDIES OF EDITING COMPLEX SUBUNITS

Overview. Most known subunits of editing complexes have been genetically analyzed in several labs by RNAi downregulation or gene knockouts (i.e., Stuart, Simpson, Goringer, and Sollner–Webb labs; Table 3.2). With the exception of REN2 (specific endonuclease for U-insertion), REL2 (ligase linked with U-insertion), REX2 (potential exonuclease), RET2 (TUTase), MP67 (potential endonuclease), and MP42 (most likely structural), current studies indicate that the editing subunits studied so far

TABLE 3.2 Summary of Known Editing Complex Subunits and Associated Features

Subunit	1 **Effect on Cell Growth**	2 **Effect on Stability ~20S**	3 **Protein Decay,** Fractions, or Whole (Lysate)	4 **Sub-comp.** Over-express	5 **Subunits in cbp-Subcomp** W, a, MS	6 **Protein Direct Contacts**	7 **Binds RNA**	8 **Inhibition of *In Vivo* Editing** RNAi, KO	9 **Inhibition of *In Vitro* Editing** FR, PC, D, I, cut, exo, tut, lig
REN1 (MP90 KREPB1)	Essential	Strong[39] None[38]	Moderate	Yes	MP42, MP81[W] aREL2, aREL1 (min)			Light[KO] [including CYb & COII][Note]	D-cut (s), I-cut (n)[38] PC-D (m) [lig], PC-I (n)[38] FR-D (s)[39]
REN2 (MP61 KREPB3 LC6A)	Essential	None	None	Yes	REN2 MP81[MS] MP42 RET2, MP18 aREL2			Strong [KO] [COII only by 50%]	I-cut (s), D-cut (n), PC-D (n), PC-I (n)[41]
MP67 (KREPB2)	None	None	None					Light	None
REX1 (MP100 (KREPC1 LC2)	Essential	Moderate	Moderate aREL2 (strong)	Yes				Light [gene specific][Note]	PC-D (s) [exo (s), lig (I)][42] PC-I (I) [tut (I), lig (I)]
REX2 (MP99 LC3 KREPC1 Band I)	None (see note)								
RET2 (MP57 KRET2 LC6B)	Essential	None	None (Lysate) MP81 (moderate) aREL2 (moderate)			MP81 (inc TUtase)		Strong	PC-I [tut (s)][43] PC-D (n)[43], FR-D (n)[51]
REL1 (MP52 LC7A Band IV)	Essential	None[45] Light[13]	None MP 63 (moderate)	Yes	MP63 REX2[MS] MP18 (min)[W]	MP63		Moderate [all substrates][45] [none = CYb, COII][13]	FR-D (s), FR-I (n)[13] PC-D (m) [lig], PC-I (m) [lig][45]

(continued)

TABLE 3.2 (*Continued*)

	1	2	3	4	5	6	7	8	9
Subunit	**Effect on Cell Growth**	**Effect on Stability ~20S**	**Protein Decay,** Fractions, or Whole (Lysate)	**Sub-comp.** Over-express	**Subunits in cbp-Subcomp** W, a, MS	**Protein Direct Contacts**	**Binds RNA**	**Inhibition of *In Vivo* Editing** RNAi, KO	**Inhibition of *In Vitro*** Editing FR, PC, D, I, cut, exo, tut, lig
REL2 (MP48, Band V. LC9)	None	None		Yes	MP81 RET2[w] MP18 (min)	MP63 MP81 (inc REL2 ligase)		None	None
MP81 (KREPA1 Band II, LC1)	Essential	Strong[49] None[48]	None (Lysate) aREL2 (strong)			REL2 RET2 MP18		Strong	PC-I (s) [tut, lig][48], FR-I (n)[49] I-cut (s)[48,51], D-cut (n)[51]
MP63 (KREPA2 Band III, LC4)	Essential	Strong	(Lysate) None? (note) REL1 (strong)	Yes		REL1 REL2 REX2 MP18		Strong	FR-D (s) [cut, lig],[52] I-cut (s), tut (n) exo(l) but see text
MP46 (KREPB4, LC5)	Essential	Strong	Moderate MP63					Strong	D-cut (s), I-cut (s)[53] PC-D (n), PC-I (n)
MP44 (KREPB5, LC8)	Essential	**Strong**	**Strong**					Strong	D-cut (s), I-cut (n.d.), PC-D(s)[lig][54] PC-I(s)[lig, tut?], exo(n)[see text]
MP42 (KREPA3 Band VI, LC7B)	Essential	None					Yes recom	Strong	PC-I (s) [lig seemed active][55] PC-D (n.d.)

MP24 (KREPA4, KREPA4, LC10)	Essential	Strong	Moderate	Yes	MP81 MP63[W] MP52 MP42 (TEV eluate)[Note]		Yes recom	Strong [except MURF2]	PC-I (s) [lig, tut?], PC-D (s) [lig][56] exo (n)[see text]
MP18 (KREPA6 (Band VII, LC11)	Essential	Strong	Moderate						D-cut(s), I-cut(s)[57] PC-I(m)[tut, lig]

The names of the subunits (in bold) described in this chapter are based on a nomenclature first used by the Stuart lab. Alternative names are included in brackets. Superscript numbers denote reference numbers. A complementary spreadsheet including all alternative names in *Trypanosoma* and *Leishmania* species as well as amino acid and nucleotide GeneBank sequences can be found at http://dna.kdna.ucla.edu/simpsonlab/ maintained by Dr. Larry Simpson. **Columns 1–9.** *Notes*: (i) The degree of effect is either strong (or essential), moderate, light, or none. (ii) The protein knockout or knockdown was partial or undetermined for some subunits (see text), and thus their assigned effect may be an underestimation. The specific references are indicated in the text.

(1) Effect of repression on cell growth. REX2's is a personal communication by Larry Simpson.

(2) Effect of repression on stability of ~20S complexes. Some references are indicated where there are discrepancies. MP44 repression has a particularly strong effect.

(3) Observed protein decay of several (if not all) editing subunits in sedimentation fractions and/or whole lysate (*lysate*) from repressed cells. None suggests no decay of subunits except those specified that undergo selective decay. For MP63 "?" indicates uncertainty about the fate of other subunits because only REL1 and REL2 were examined (but the aREL2 level seemed unaffected). REL2 indicates that adenylylation activity (not protein) was scored. REL1 indicates protein scored by western blotting.

(4) "Yes" indicates that overexpression of the TAP-tagged subunit causes accumulation of ~5–15S subcomplexes.

(5) Subunits identified by western (W), adenylation ("a" prefix), or mass spectrometry (MS) in calmodulin-binding protein (cbp)-tagged subcomplexes purified in affinity columns. (min) Minimal activity or protein was detected. MP24-bound subunits were detected in the TEV eluate, but not in the calmodulin eluate, suggesting weak binding.

(6) Direct protein–protein interactions observed in a yeast-two hybrid system, or *in vitro* transcribed/translated systems. MP81 increased (inc) both REL2 ligase and RET2 TUTase activities in mixtures with recombinant proteins.

(7) RNA-binding activity observed in isolated recombinant proteins.

(8) Inhibition of *in vivo* RNA editing by repression in RNAi or KO (if indicated) cell lines. For REN1*in vivo*, inhibition of both CYb and COII (U-insertion-only substrates) was noted because REN1 was only linked with U-deletion *in vitro* (see the text).

(9) Inhibition of *in vitro* RNA editing from the cell lines in column 8. The activities tested: Full-round (FR) and precleaved (PC) editing, U-Deletion (D), U-insertion (I), and specific intermediate activities: endonuclease (cut), exonuclease (exo), TUTase (tut), and ligase (l). Not determined (n.d.) activities that would be informative are indicated.

The level of inhibition is strong (s), moderate (m), light (s), or none (n).

are required to maintain the integrity of editing complexes. MP67 and REL2 are dispensable for cell growth, and the latter is also dispensable for editing *in vivo*. Thus, REL2 is the only known dispensable subunit for cell growth, complex integrity, and editing. While some subunits are critical to the integrity of entire editing complexes, others are required for the stable association of one or more subunits in the complexes. Importantly, the original hypothesis that U-insertion and U-deletion editing are separate pathways with activities potentially compartmentalized in editing complexes is generally consistent with several observations including (a) separate enzymes catalyzing each step of U-deletion and U-insertion, (b) the isolation of some catalysts and proposed structural components in separate subcomplexes that exhibit either precleaved U-insertion or precleaved U-deletion activity, and (c) the reconstitution of a full-round U-deletion cycle with recombinant subunits that prior studies specifically linked with this type of editing. However, while the higher-order structure of editing complexes are currently unknown, the attractive model of a single complex bearing two functional separable regions is most likely too simplistic. Recent observations indicate that subunits initially considered as "specific markers" of one editing type may also function in the other type. Also, functionally and compositionally diverse forms of complexes may occur *in vivo*. Such particles, if confirmed in native complexes, may bear a common complement of subunits as well as variable dynamic components.

All data discussed below derive from studies in *T. brucei* and *Leishmania* editing complexes (summarized in Table 3.2), which are considered to be very similar, and most findings should be relevant to both systems. The editing subunits in *Trypanosoma* and *Leishmania* are named based on the following prefixes: (a) MPx, *M*itochondrial *P*rotein, with "x" indicating the predicted molecular weight of the mitochondrial leader-containing precursor in *T. brucei*, or (b) based on their demonstrated or proposed enzymatic activity; REy, *R*NA *E*diting, where "y" is N for nuclease, X for exonuclease, T for TUTase, or L for ligase. (b) Alternative names in the literature are in Table 3.2.

3.4.1 REN1, REN2, and MP67. Endonuclease Homologs

Overview. These close homologs exhibit a canonical RNase III motif with strong conservation of critical residues for enzymatic activity in the prototypical bacterial and related enzymes. Recent *in vitro* studies of REN1 (RNA editing nuclease 1) and REN2 indicate that these are specialized endonucleases for U-deletion and U-insertion editing, respectively. However, downregulation of REN1 inhibited *in vivo* editing of all substrates tested including those that only undergo U-insertion (COII and CYb), indicating that REN1 may play a role in both types of editing *in vivo*. MP67 is also a potential endonuclease, but its role remains unclear. Interestingly, TAP-tagging and overexpression of each homolog led to affinity-purified editing complexes that exclude the other two related subunits. This heterogeneity in the tagged complexes, which otherwise exhibit similar composition, is very intriguing, but its physiological relevance *in vivo* remains unknown. Two other subunits, MP46 and MP44, also have significant RNase III conservation, but the absence of critical canonical residues suggests that they are not catalytic.

3.4.1.1 REN1 Conditional RNAi, regulatable double knockout (RKO) studies, and mutagenesis of RNase III-conserved amino acids initially linked this subunit with the endonucleolytic step in U-deletion editing, and showed that it is essential for *in vivo* editing and cell growth (33, 38, 39). In RKO cells, a tetracycline-regulatable ectopic REN1 allele can rescues an otherwise lethal phenotype due to the lack of both endogenous alleles. *In vitro* assays using extracts of these cell lines showed preferential inhibition of the endonuclease activity in U-deletion editing, and recombinant REN1 expressed in insect cells cleaved a U-deletion site but not an U-insertion site in model substrates (39). Furthermore, Kang et al. also showed that a mixture of *Leishmania* and *L. major* recombinant REN1, REX1, and REL1 catalyze full-round U-deletion editing (39); however, whether this reaction requires association of the proteins in a complex *in vitro* or in native complexes was not resolved. It is also important to note that the recombinant REN1 used in full-round editing *in vitro* was cytosolically expressed and purified from *Leishmania*, and potential copurification of other editing proteins was not eliminated. Trotter et al. (38) showed that RNAi of REN1 in procyclics significantly repressed this transcript and all edited mRNAs tested—including CYb and COII, which only undergo U-insertion. Furthermore, complete inhibition of REN1 by RKO performed in bloodstream form cells depleted all edited mRNAs tested and stopped cell growth by day 4 of repression.

Surprisingly, CYb and COII edited mRNAs were only reduced by 50%. This may reflect that COII edited mRNA is relatively stable and/or its editing can partially use an alternative endonuclease, presumably REN2. The REN1 requirement for CYb and COII editing may reflect interactions of this subunit with components for U-insertion editing, as the data by Panigrahi et al. (33) suggest (see below). Furthermore, REN1 could be used to begin a proofreading cycle of misedited U-insertion sites bearing extra Us; that is, upon re-cleavage (by REN1) such sites could enter U-deletion editing. In this line of thought, alternate cycles of U-insertion and U-deletion could occur until each site is accurately edited, as defined by the cognate gRNA sequence information (Figure 3.2). In fact, sharing of enzymatic activities between U-deletion and U-insertion editing has been observed before, because TUTase activity can act at U-deletion sites *in vitro* (40). RKO cell lines exhibited more robust repression of both REN1 mRNA and editing *in vivo* than did RNAi-repressed cells; however, in both cases the level of never-edited ND4 transcript remained unchanged. The level of REN1 protein was not determined due to the lack of a specific antibody. Interestingly, one study reported that neither RKO of REN1 in bloodstream form *T. brucei* nor RNAi repression in procyclics visibly affects the integrity of ~20S editing complexes (38); however, a second study of RNAi-repressed procyclics detected a partial decrease in sedimentation (to ~15S) of the complexes (39). Such contrasting outcomes may reflect either different efficiency of repression in the systems used or the length of time that cells underwent RNAi (i.e., 4 days versus 6 days, respectively), rather than differences in developmental stage.

Complexes that were affinity-purified via tagged REN1 lacked REN2 and MP67. They catalyzed full-round U-deletion editing, but only the U-addition and ligation steps (i.e., not cleavage) of U-insertion editing. Overexpression of TAP-REN1 induced accumulation of 5–10S tagged subcomplexes with partial cleavage activity for U-deletion but not for U-insertion editing (33). The subcomplexes contain MP42, MP81, some adenylatable REL2, and minimal adenylatable REL1, but MP63 was not detected. The presence of endogenous REN1 or other subunits in the subcomplexes was not examined, although the subcomplexes exhibited significant precleaved U-insertion and U-deletion editing activities. Whether such subcomplexes support full-round editing was not determined, but their composition was surprising since previous *in vivo* and *in vitro* studies linked MP81 and REL2 with U-insertion and linked MP63 and REL1 with U-deletion (32). The physiological relevance of the tagged complexes and subcomplexes is unclear, but REN1 evidently can form stable interactions with components needed for U-insertion editing. Also, whether or not REN1-containing complexes devoid of the other two close homologs occur naturally or are an artifact of overexpression is unclear. In any case, it is attractive to propose that REN1 and its homologs are dynamic components able to homodimerize or heterodimerize in native complexes; Overexpression and affinity-purification may lead to the accumulation and selection of tagged homodimers. Finally, Panigrahi et al. (33) showed that inactivating point mutations in the conserved RNase III domain prevents association of MP81 and MP42 markers in subcomplexes but not in ~20S tagged complexes. This suggests that REN1 residues expected to participate in catalysis also are relevant in structural interactions for proper folding or assembly into editing complexes.

3.4.1.2 REN2 RNAi repression in procyclics and RKO studies in bloodstream trypanosomes show that REN2 is essential for cell growth and editing (41). In these studies, after growth ceased at day 4 and day 3 of repression, respectively, cell extracts were made and analyzed. Interestingly, no changes were detected in either sedimentation of editing complexes or relative level of marker subunits examined (MP81, MP63, MP42, and REL1). However, *in vivo* editing of all substrates tested was strongly inhibited, particularly in RKO cells where conditional repression can eliminate 90–100% of both REN2 mRNA and all edited mRNAs examined, except for COII, which was only reduced by ~50%. COII pre-mRNA is the only known substrate to contain a cognate gRNA *in cis* (5), and this result suggests that its editing may involve an additional endonuclease(s), likely REN1 or MP67 (see ahead), although the level of REN2 protein was not scored due to the lack of a specific antibody. *In vivo*, all known editing substrates undergo U-insertion and U-deletion—except for COII and CYb, which only undergo U-insertion editing. Thus, neither REN1 nor MP67 can compensate for REN2. Cell lysates from REN2-repressed cells showed inhibited endonuclease activity for U-insertion but not for U-deletion, and the subsequent catalytic steps in U-insertion and U-deletion editing appeared normal. Furthermore, specific point mutations in the conserved RNase III motif specifically decreased the U-insertion endonuclease cleavage activity in sedimentation fractions from lysates (41) or in tagged-REN2 complexes (33). Altogether, the data suggest that REN2 may be a U-insertion-specific endonuclease, but enzymatic activity of isolated REN2 remains to be shown. ~20S complexes purified via tagged REN2 contained all intermediate catalytic activities for U-insertion and U-deletion editing, except for cleavage in U-deletion. This coincided with the absence of REN1 (33). Also, overexpression of TAP-REN2 accumulated 5–10S tagged subcomplexes containing MP81, MP42, and very low levels of adenylatable REL1 and REL2. Interestingly, MP63 that other studies linked to U-deletion by (see Section 3.4.5.2) was not detected in the TAP-REN2 subcomplexes, and the presence of other subunits was not examined. However, in contrast to tagged-REN1 subcomplexes described above, these particles exhibited minimal precleaved U-insertion and U-deletion editing activities, and they cleaved an U-insertion site but not a U-deletion editing site. Furthermore, point mutations in the conserved RNase III, motif, which inhibits RNA cleavage, also prevented ~20S complexes and largely accumulated 5–10S subcomplexes. Thus, these conserved residues may have a dual role in catalysis and forming stabilizing interactions in editing complexes. The latter observation appears inconsistent with a previous indication that REN2 repression does not compromise the stability of editing complexes (41). It is possible that a longer repression of REN2 is needed to destabilize the complexes, as was found with REN1 (39). Mass spectrometric (MS) analysis of the mutant REN2 subcomplexes showed MP81, RET2, and REL2 all functionally linked to U-insertion, as well as MP46 and MP18 that are critical for the stability of ~20S complexes (see Sections 3.4.5.3 and 3.4.5.7). MS of purified wild-type REN2 subcomplexes also showed these subunits except for MP46.

3.4.1.3 MP67 RNAi repression of this subunit did not affect cell growth or complex integrity, but caused a small reduction of editing *in vivo* for some of the substrates examined, particularly A6, RPS12, and COII. Both these edited substrates and MP67 mRNA were reduced by ~50% in the repressed cell line (38), but MP67 protein levels were not determined. Panigrahi et al. (33) purified tagged MP67 complexes lacking REN1 and REN2, thereby inactive in the cleavage steps in U-insertion or U-deletion editing, but active in all subsequent intermediate catalytic steps (i.e., U-removal or U-addition and ligation steps, which were scored in precleaved editing assays). MP67 appeared dispensable for cell growth, and it is unknown if it is catalytic; however, its mild effects on editing and its homology to REN1 and REN2 suggest that it may modulate their cleavage activity, possibly via formation of heterodimers. It is also conceivable that MP67 plays a specialized role in editing of specific mRNAs—for example, COII that is unique in its use of a *cis*-acting gRNA.

3.4.2 REX1 and REX2. Exonuclease Homologs

Overview. Editing complexes contain two potential exonuclease homologs but their specific roles in editing are not well-defined, and additional studies may be necessary to tie together some current observations. Cell lysates from REX1 (RNA editing exonuclease 1) RNAi-repressed procyclic *T. brucei* exhibited partial inhibition of both U-deletion and U-insertion precleaved editing *in vitro*. It was surprising that REX1 repression significantly reduced adenylatable REL2 in ~20S complexes, because other studies have linked REL2 ligase with U-insertion. Recombinant REX1 from *Leishmania* is a 3′-to-5′ exonuclease with specificity for terminal poly-U extensions. This subunit and other recombinant components were used to reconstitute full-round U-deletion *in vitro*, but it was unclear if that reaction required association of these proteins in a complex. RNAi of REX1 had little effect on editing *in vivo*, suggesting that another exonuclease, likely REX2, may complement its function. On the other hand, REX2 was found in subcomplexes containing both precleaved U-deletion activity and subunits previously linked with U-deletion such as REL2 and MP63 (that directly binds to REX2; see Section 3.4.5.2).

3.4.2.1 Functional Studies RNAi repression of REX1 decreased *T. brucei* growth but was not lethal after 10 days, although REX1 mRNA was apparently depleted after 3 days (42). Unfortunately, in this study and in other studies of editing subunits (Table 3.1), the degree of protein knockdown was not determined due to lack of a specific antibody. This complication limits our understanding of the effect that this and other subunits may have on cell growth and viability. Nevertheless, after extended repression, ~20S complexes exhibited a moderate gradual decrease and ~15S subcomplexes accumulated in mitochondrial lysates. At day 9 of repression, adenylatable REL2 was dramatically reduced in the ~20S fraction of the lysates, whereas other markers including adenylatable REL1, MP81, MP63, and MP42 were reduced to a lesser extent. REL1 accumulated freely at the top of glycerol sedimentation gradients and in subcomplexes, whereas REL2 only accumulated in subcomplexes. REL2 co-sedimented with editing markers MP81, MP42, and apparently some MP63 at 15S, but it was unclear whether tagged or endogenous REX1 was present in these fractions and whether these subunits were in the same subcomplexes. Kang et al. (42) indicated in unpublished observations that RNAi of REX1 had a small effect *in vivo* and that it was gene-specific. The binding partners of REX1 are still unknown; however, its presence evidently stabilizes editing complexes. Surprisingly, REX1 has a greater effect on the stability of REL2 than on that of REL1. This ligases have been linked with U-insertion and U-deletion editing, respectively (see 3.4.4). A mixture of *Leishmania* recombinant proteins REN1, REX1, and REL1 was reported to catalyze full-cycle U-deletion editing; however, whether association of these proteins in a subcomplex is required to reconstitute editing, or each enzyme acts individually was not resolved (39). REX1 has been proposed to be the major editing U-specific exonuclease in *T. brucei* and *Leishmania*; however, the nonlethal (slow-growth) RNAi phenotype and small associated effect on RNA editing in *T. brucei* cells suggested that its close homolog REX2 or another potential exonuclease complements REX1 *in vivo*. ~20S complexes from REX1-repressed *T. brucei* lysates exhibited both reduced U-insertion and U-deletion precleaved editing activities, possibly reflecting the observed partial breakdown of the complexes. However, precleaved assays showed a preferential inhibition of both U-removal and ligation steps in U-deletion editing. The U-addition step in U-insertion editing was slightly inhibited. Clearly, REX2 does not efficiently replace REX1 in the precleaved *in vitro* assay. Whether the editing complexes lacking REX1 retain editing endonuclease activities was not determined. Finally, REX2 RNAi did not affect cell growth in *T. brucei* (Larry Simpson, personal communication), and the *Leishmania* REX2 homolog (termed REX2*) lacks an evident exonuclease domain (42). Thus, the specific role of REX2 in editing *in vivo* or *in vitro* remains undetermined.

3.4.3 RET2. TUTase

Overview. This subunit is structurally related to a 3′ TUTase (RET1) proposed to synthesize the 3′ U-tail of gRNAs. However, RET1 is not a component of 20S editing complexes, although it can bind to them via an RNA link. RET2 (RNA editing TUTase 2) is the editing enzyme according to genetic and biochemical evidence; e.g., its genetic repression specifically blocks the U addition step in editing in vitro assays, and the isolated recombinant protein specifically adds uridines.

3.4.3.1 Functional Studies RNAi repression showed that RET2 is essential for *T. brucei* growth, which stopped after 4 days (43). RET2-protein levels were not determined, although RET2 mRNA was depleted at day 2, and the integrity of editing complexes appeared normal when examined at day 3. Editing of the mRNAs examined (ND8, COIII, and CYb), particularly ND8, was inhibited *in vivo*. Furthermore, U-addition but neither U-removal nor ligation steps was affected in precleaved editing *in vitro* assays. Recombinant RET2 from *T. brucei* (22) and *Leishmania* isolated from *Leishmania* cytosol exhibited specific 3′-terminal urydilyl transferase activity (TUTase). The latter was a derivative devoid of its putative mitochondrial leader (43). Recombinant RET2 bound to MP81 *in vitro* and the interaction stimulated TUTase activity. RET2 can be detected using specific polyclonal antibodies (22), and a fortuitous C-terminal poly-histidine in *Leishmania* RET2 can be detected using an anti-His antibody (43). RET2 repression caused a significant decrease of MP81 and adenylatable REL2 in mitochondrial extracts (43). The RET2-dependent stability of MP81 and REL2 suggested functional interactions among these subunits. This is consistent with the reported association of RET2, MP81, and REL2 in active subcomplexes for precleaved U-insertion that were produced upon REL2 overexpression in *T. brucei* (32).

3.4.4 REL1 and REL2. Ligase Homologs

Overview. These two homologs have been extensively studied, and recombinant proteins confirmed their ligase activity. Together, functional analysis of both purified editing complexes and subcomplexes containing wild-type or mutant ligases have linked REL1 (RNA editing ligase 1) with U-deletion and REL2 with U-insertion editing. However, REL1 is essential for survival whereas REL2 is dispensable, suggesting that REL1 can complement REL2 in U-insertion editing, but not the converse. Neither protein seems critical for stability of editing complexes, although one study suggested that REL1 repression had a slight effect. REL1, REX2 (potential exonuclease), and MP63 (structural) were found in active subcomplexes for precleaved U-deletion, whereas REL2, RET2 (TUTase), and MP81 (structural) were found in active subcomplexes for precleaved U-insertion editing. REL1 and REL2 directly bound MP63 and MP81, respectively, and these interactions may be regulatory. Consistent with this idea recombinant REL2 ligation was stimulated by MP81, and REL1-depletion decreased the MP63 level in editing complexes.

3.4.4.1 REL1 REL1 is essential for growth of *T. brucei* (13, 44, 45). Importantly, REL1-repression in bloodstream form *T. brucei* showed for the first time that RNA editing is essential for survival of this developmental stage in culture and in infected rats (44). In procyclics, Huang et al. showed cessation of cell growth after 1-day overexpression of an enzymatically inactive REL1 version (i.e., with a K86R mutation at the conserved KXXG motif), which replaced most endogenous REL1 in editing complexes. This was possible because any REL1 excess, not incorporated in editing complexes, is short-lived. The K86R REL1 mutation inhibited *in vivo* editing of substrates requiring U-insertion and U-deletion, but had no effect on substrates that only require U-insertion editing (i.e., COII and CYb). Furthermore, lysates of K86R mutant cells catalyzed full-round U-insertion but not U-deletion editing *in vitro*. Also, recombinant REL1 exhibited RNA ligase activity (46). Together, the data indicate a specialized role of REL1 ligase in U-deletion editing. They also imply that REL2 cannot substitute for REL1 *in vivo* and that REL2 suffices for ligation in U-insertion editing.

In a second study, REL1 RNAi in procyclics (45) depleted most REL1 protein at day 3 and inhibited cell growth at day 6. However, in contrast with Huang et al., these authors found significant reduction of both precleaved U-insertion and U-deletion *in vitro*, as well as partial inhibition of editing *in vivo* for all substrates tested including COII and CYb, which only undergo U-insertion editing. A major experimental difference in these studies that may account for the apparent discrepancies is that Huang et al. exclusively assessed the catalytic role of REL1, because the K36R mutation most likely preserved most of this protein's native structure and macromolecular interactions in editing complexes. This would explain the consistency of the data by Huang et al., with the biochemical studies that first proposed a functional link of native REL1 and REL2 with the ligation steps in full-round U-deletion and U-insertion editing, respectively (12). On the other hand, in Gao and Simpson (45), REL1 depletion eliminated all potential interactions this protein may have contributed to the overall organization and function of editing complexes. In fact, MP63 levels (but not MP81 protein or adenylatable REL2) partially decreased upon REL1 repression. Since MP63 directly binds REL1 and REL2 and was critical to the integrity of editing complexes (see Section 3.4.5.2), REL1 interactions could impact both U-deletion and U-insertion editing. Thus, the complete elimination of REL1 may have a greater impact on the function of editing complexes than its replacement with structurally native-like inactive enzyme.

Huang et al. (13) and Gao and Simpson (45) also somewhat disagree on the potential contribution of REL1 on the stability of editing complexes. Huang et al. (13) reported that KO of one of the two REL1 alleles led to a partial reduction of adenylatable REL2 in 20S complexes and corresponding buildup at lower sedimentation. This change in the distribution of adenylatable REL2 implied a partial breakdown of editing complexes, although the levels of this and other subunits were not examined. In contrast, Gao and Simpson (45) found that virtual depletion of REL1 protein after 5 days of RNAi did not affect the sedimentation of editing complexes, as judged by levels of MP81 and MP63 and adenylatable ligases. Although the reason for this discrepancy is unclear, REL1 may contribute slightly to stability of the complexes. In subsequent work, Gao et al. (46) also showed that recombinant REL1 expressed in insect cells assembled into and can functionally complement REL1-depleted editing complexes. In these add-back experiments, the wild-type protein but not a C-truncated version efficiently assembled into ~20S complexes and reconstituted their editing activity *in vitro*. Thus, the removed C-terminal region is required for binding to ligase-depleted editing complexes.

Cell lines overexpressing TAP-REL1 or TAP-REL2 accumulated subcomplexes with precleaved U-deletion or U-insertion editing activity, respectively (17, 32). This observation lent direct evidence for at least partial compartmentalization of U-deletion and U-insertion pathways in editing complexes as previously hypothesized (10–12). It is unclear, however, whether the bulky TAP-epitope or subunit overexpression caused accumulation of incompletely assembled particles. Overexpression of other tagged subunits also led to accumulation of subcomplexes (e.g., REL2, REX1, REN1, REN2, MP63, and MP24; see Table 3.2). Purified cbp-REL1 subcomplexes specifically contained REX2 (potential exonuclease) and MP63 (structural) and lacked the homolog REL2 ligase. On the other hand, at least MP18 was common to these particles and counterpart cbp-REL2 subcomplexes (32). REL1 directly bound MP63 *in vitro* and in yeast two-hybrid assays (32, 47); and downregulation of endogenous REL1 caused a decrease of MP63 in editing complexes (45), suggesting that the interaction is functional.

3.4.4.2 REL2 REL2 is dispensable for cell growth and also for editing *in vivo* and *in vitro* (45, 48, 49). In all three studies, RNAi repression eliminated all adenylatable REL2 in cell lysates, after a few days of repression. Depletion of REL2 protein was also observed (45). However, the cell growth rate, editing complex integrity and *in vivo* editing levels in induced and noninduced cells were indistinguishable during extended cultures in all studies. The presence of normal U-insertion activity in precleaved (45, 48) and full-round editing assays (49) suggests that REL2 depletion may not affect the expression or activity of RET2 and MP81 in editing complexes (contrasting the converse depletions of these subunits; see RET2 and MP81), but this was not determined. The data indicate that REL2 can be complemented by REL1 *in vivo*, but not vice versa, as was originally hypothesized based on biochemical studies of the two activities and the relaxed substrate specificity of REL1 (12, 50). Overexpression of TAP-REL2 accumulated cbp subcomplexes that specifically contained RET2 (TUTase) and MP81 (structural) but that lacked the homolog REL1 ligase. At least MP18 was shared by cbp-REL2 and cbp-REL1 subcomplexes (32). Schanufer et al. showed that MP81 directly binds REL2 *in vitro* and stimulates its ligase activity, suggesting that the interaction is regulatory, at least in the ligation step of U-insertion editing. REL2 also bound MP81 and MP63 in yeast two hybrids.

3.4.5 MP81, MP63, MP42, MP46, MP44, MP24, MP18. Structural Components

Overview. Repression of any of these subunits inhibited cell growth and was eventually lethal. All proteins, except for MP42, are evidently required for stability of ~20S editing complexes. There is evidence that at least some of these components are engaged in potential regulatory interactions with other editing subunits. MP81, MP63,

MP42, MP24, and MP18 contain potential oligonucleotide/oligosaccharide binding (OB) folds that may be involved in interactions with single-stranded RNA. This motif seems particularly conserved in the first three subunits, which also have potential C2H2 zinc-finger motifs. MP46 and MP44 have some conservation of an RNase III motif, but they may not be catalytic. Isolated recombinant MP42 and MP24 were shown to exhibit RNA-binding activity. Interestingly, repression of certain subunits such as MP44 and MP46 led to a complete and apparently specific elimination of all or some subunits of editing complexes, implying a specialized decay mechanism devoted to discard free editing components *in vivo*. The observed degree of destabilization and decay of editing complexes may be indicative of the relative extent to which each subunit helps stabilize the core scaffold of editing complexes. However, variations in the efficiency of repression achieved in each system are also expected.

3.4.5.1 MP81 This subunit is essential for growth and editing (48, 49). In the RNAi study by Drozdz et al. (48), MP81 protein was largely depleted at day 2 and was undetectable at day 3. Cell growth inhibition was evident at day 3, at which time cell lysates were analyzed. *In vivo*, the edited mRNAs tested (A6, ND7, and CYb) were significantly reduced, but the never-edited ND4 transcript remained at normal levels. In this study, the lack of MP81 expression did not affect the integrity of editing complexes, as determined by the examined MP63, REL1, and MP42 protein levels. However, adenylatable REL2 was specifically lost at days 2–3 of repression. In a precleaved U-insertion assay, both U-addition and ligation steps were clearly inhibited, but all precleaved U-deletion steps were unaffected. The U-insertion endonuclease activity was also reduced, but the U-deletion endonucleavage activity was not tested.

In O'Hearn et al. (49), cell growth inhibition was evident at day 7 of RNAi repression. Consistent with Drozdz et al. (48), adenylatable REL2 was absent and REL1 was not affected in lysates of repressed cells. However, in contrast with those authors, O'Hearn et al. (49) showed a clear reduction in editing complex sedimentation at day 5 of repression. The discrepancy between the two studies may reflect MP81-dependent stabilization of the complexes that becomes evident only after extended periods of repression. It is possible that the complex stability is maximal with multiple copies of MP81, but seriously compromised below a critical stoichiometry of this subunit.

In any case, both studies indicate a functional association between MP81 and U-insertion editing. This is in agreement with (a) the known physical association between MP81, RET2 (TUTase), and REL2 (ligase) in purified subcomplexes (32) and (b) a RET2-dependent partial stabilization of both MP81 and REL2 in ~20S editing complexes (48). Repression of either RET2 or REL2 did not compromise the integrity of editing complexes, possibly because they retained sufficient MP81. Conversely, whether MP81 was required for stable expression of RET2 and REL2 in ~20S complexes has not been determined. However, RNAi of MP81 eliminated adenylatable REL2, which presumably reflects loss of this ligase subunit(48, 49). Importantly, Schnaufer et al. (32) showed that MP81 directly interacts with REL2, RET2, and MP18 and that recombinant MP81 stimulates REL2 ligation *in vitro*. Since MP81 bears a potentially regulatory OB-fold whereas REL2 lacks one [contrary to most evolutionarily related ligases (32)], these authors speculated that MP81 could at least modulate ligation in U-insertion editing. Consistent with this scheme, Law et al. (51) reported that RNAi depletion of MP81 preferentially inhibits the catalytic U-insertion steps *in vitro*.

3.4.5.2 MP63 MP63 is required for cell growth, editing, and stability of editing complexes in *Leishmania* and *T. brucei* (47, 52). In *T. brucei*, virtually all MP63 protein was depleted at day 6 of RNAi repression, which coincides with the onset of cell growth inhibition. Cell lysates prepared at this time lacked REL1 protein, but retained normal levels of adenylatable REL2. Other subunits were not examined. The ligation activity of these lysates was not examined, nor was the presence of other editing subunits. Most likely, the loss of *in vitro* editing activity was due to the nearly complete disruption of ~20S editing complexes. TUTase activity in a precleaved editing assay was retained, but both cleavage activities for U-insertion and U-deletion editing were lost. The lysates also contained a U-exonuclease activity, assessed by removal of a terminal U-run in a synthetic transcript, but its relationship with editing remained unclear.

In *Leishmania*, Kang et al. introduced point mutations in two conserved zinc fingers (ZnF-1 and ZnF-2) of MP63 that changed critical metal-binding cysteines to glycines in each motif. The ZnF-1 mutation reduced the cells doubling time from 6 to 8 hr and caused major breakdown of ~20S editing complexes but was not lethal. ZnF-2 did not affect cell growth and caused a partial breakdown of the complexes, as determined by MP63 and adenylatable REL1 and REL2 markers. Other subunits were not examined. Both mutants described above and wild-type MP63 were overexpressed as TAP-tagged constructs. Interestingly, most MP63-TAP was found in ~10–15S subcomplexes and some in ~20S complexes, suggesting that either MP63 overexpression or its TAP epitope interferes with the assembly of normal complexes. The subcomplexes accumulated particularly with ZnF-1. These particles exhibited the same set of protein bands as ~20S complexes.

This observation and the detection of both endogenous and tagged-MP63 led the authors to propose a monomer-dimer model in which ZnF-1 is involved in stability of entire complexes and ZnF-2 may be involved in dimerization. Although assembled editing complexes contain at least two copies of MP63, the precise quaternary structure of the complexes is unknown. Thus, it appears that multiple MP63 protein interactions stabilize editing complexes, some of them mediated by its zinc fingers. Consistent with this concept, MP63 depletion (52) should remove a larger number of relevant contacts than point mutations, thereby leading to a more dramatic effect on stability (47). It is known that MP63 directly binds REL1, REL2, REX2, and MP18 (32). Importantly, the contact between MP63 (a U-deletion marker) and REL2 (a U-insertion marker) suggests that MP63 is a core component at the interface of hypothetical U-insertion and U-deletion "compartments" of editing complexes.

3.4.5.3 MP46 RNAi repression of MP46 is lethal and disrupted the integrity of editing complexes (53). About 50% of both MP46 mRNA and all edited substrates tested (A6, RPS12, MURF2, and COII) were depleted at day 3, and cell growth inhibition was evident after day 5. The sedimentation of editing complexes (assessed by MP81, MP63, MP42, and REL1 protein markers and by adelylatable ligases) was significantly and progressively reduced at days 3 and 6. These marker subunits accumulated in ~5–10S fractions (with the apparent exception of MP63) and gradually decayed to very low levels at days 3 and 6, when analyzed. This protein loss was specific for ~20S editing complexes and did not affect other mitochondrial proteins examined. Despite the reduction of several marker subunits, all editing activities scored in precleaved editing assays (i.e., TUTase, exonuclease, and ligases) were comparable in ~5–10S and ~20S fractions, but both cleavage activities in U-insertion and U-deletion editing were strongly inhibited in all fractions tested. Although the MP46-binding subunits are unknown, its strong effect on both integrity of ~20S complexes and stability of released components implied that this subunit is located in the core scaffold of the complexes. Interestingly, the disruption of the complexes due to MP46 repression triggered a relatively rapid degradation of free components.

3.4.5.4 MP44 Wang et al. (54) showed that elimination of both endogenous alleles of MP44 in bloodstream from trypanosomes is lethal, and inducible expression of ectopic MP44 suffices for normal cell growth. MP44 mRNA was dramatically reduced at day 2 of repression and cell growth stopped at day 5, but resumption of expression at this time reestablished normal growth. Functional analyses performed at day 5 of repression showed a strong decrease *in vivo* of the edited mRNAs tested (ND7, A6, and RPS12), while the never-edited transcripts (ND4 and COI) remained constant. Sedimentation analyses of cell lysates showed that all marker subunits tested (i.e., MP81, MP63, MP42, and REL1 proteins and adenylatable REL2) shifted from ~20S to ~10S fractions at day 3 of repression and were virtually loss at day 5. Consistent with the loss of components reflecting specific protein decay, the levels of the mRNA encoding one of the subunits (MP63) remained constant upon repression. As expected, all *in vitro* editing activities tested including cleavage for U-deletion, and both precleaved U-insertion and U-deletion editing were lost by day 5 of repression. Substantial U-exonuclease activity in the cell lysates was observed, but its relationship with editing was unclear. The function of MP44 is uncertain, but clearly it forms critical interactions in ~20S complexes and does not complement other subunits required for complex integrity (e.g., MP46, MP63, MP24, and MP18; see Table 3.2). Furthermore, the dramatic loss of editing complexes may reflect a specific decay mechanism that targets released editing components, rather than a general outcome of cell death.

3.4.5.5 MP42 RNAi downregulation of this subunit of editing complexes inhibited *T. brucei* growth and was ultimately lethal, but did not lead to visible loss of editing complex integrity after a few days of repression (55). Both MP42 mRNA and protein were depleted at day 2 of repression, and cell growth stopped at day 5. At this latter time, all functional analyses were performed. *In vivo* editing of CYb and ND7 substrates was significantly reduced, whereas the never-edited COI transcript remained unaltered. Also, precleaved U-insertion editing *in vitro* was dramatically inhibited in extracts from depleted cells, although ligation of pre-edited substrate seemed active. U-deletion editing activities were not tested.

The authors suggested that the structure of MP42-depleted complexes was largely retained, since adenylable ligases sedimented at ~20S and titrated recombinant MP42 reconstituted precleaved U-insertion editing activity in lysates from repressed cell. Most added back MP42 was incorporated into ~20S particles, indicating that these particles were responsible for the observed editing in mitochondrial reconstituted lysates. Interestingly, a recombinant preparation exhibited single-strand specific endo- and exo-ribonucleolytic activities, but their relationship with editing remains to be established. Consistent with their association with MP42, these activities were present in an editing-active mitochondrial fraction, but strongly reduced in an MP42-RNAi knockdown strain. Thus, it appears that the molecular interactions involving MP42 are critical for catalysis but dispensable for stability of the complexes. However, longer repression periods may have an impact on stability (as seen with REN1). A feasible complication when analyzing consequences of extended repression is the potential overlap with general cell-death effects. Overall, MP42 may play a structural role that impacts the enzymatic activity but not the integrity of editing complexes. This subunit could help control the catalysts as was proposed for its homologs MP63 and MP81, or alternatively may interact with the RNA substrate during docking or processing. In fact, recombinant MP42 was shown to exhibit RNA-binding activity (55) (see Section 3.5.1).

3.4.5.6 MP24 In Salavati et al. (56), RNAi of MP24 in procyclic trypanosomes significantly reduced this mRNA at day 2 of repression and cell growth stopped at day 6. Analyses of cell extracts at this latter time showed a dramatic reduction in all edited mRNAs tested (A6, RPS12, COIII, CYb, COII) except for MURF2. The reasons for persistence of edited MUR2 in this study are unknown, but it was proposed to reflect greater stability. This seems unlikely, however, because repression of other subunits [e.g., MP46, REN1, REN2 (38, 41, 53)] significantly reduced edited MURF2 like most other substrates tested. Thus, the effect of MP24 RNAi may be gene-specific. As expected, the never-edited transcripts COI and ND4 were not affected. A sedimentation analysis of cell lysates at day 4 of repression showed a dramatic loss of all tested subunits in ~20S complexes (MP81, MP63, REL1, MP42, and MP18), but a small amount of these proteins co-sedimented in ~5–10S fractions. Other mitochondrial proteins tested were not affected in the repressed cells. Also, all tested editing activities were strongly inhibited, although some RNA ligation and substantial U-exonuclease activity remained. However, the latter could in part reflect contaminating nucleases in the cell lysates. A TAP-MP24 construct was used to confirm that this subunit co-sediments and affinity co-purifies with marker editing proteins including MP81, MP63, MP52, and MP42. Overexpression of the TAP construct induced accumulation of subcomplexes. TEV-purified eluates contained significant REL1 and some MP63 and MP42, but calmodulin-purified eluates did not contain any of these markers, possibly reflecting weak interactions or decay of labile components.

The MP24 binding partners are unknown, although evidently some of its interactions critically stabilize editing complexes. The released editing components upon MP24 repression may be specifically targeted for degradation. Furthermore, MP24 binds RNA and exhibits preference for a substrate containing a 3′ polyU terminus. Thus, this subunit may recognize gRNAs in editing complexes (see Section 3.5.2).

3.4.5.7 MP18 RNAi repression of this subunit in procyclic trypanosomes eliminates over 90% of MP18 protein at day 3, and cells stopped growing by day 5 (57). Sedimentation analyses showed that ~20S complexes were essentially absent at day 4, as determined by 10 maker subunits (MP99, MP81, MP63, REL1, REL2, MP52, MP48, MP57, MP42, and MP18). All markers were present at ~10S, although MP18 appeared minimal. At days 3 and 6 of repression, crude lysates and sedimentation fractions showed a strong inhibition of cleavage in both U-deletion and U-insertion sites, as well as reduction of all activities in precleaved editing assays. The only exception was a U-exonuclease activity; however, there was concern that some of this activity in the lysates is not related to editing. Effects on *in vivo* editing were not examined. Upon MP18 repression, this subunit was undetectable at day 4, whereas all editing subunits tested were significantly decreased but still detectable. The relative levels of protein decay associated with repression of MP18 compared to MP44 and MP46 may reflect differences in downregulation efficiency of different systems. In any case, MP18 is critical for the general stability of editing complexes. Schnaufer et al. (32) found this subunit in subcomplexes formed upon overexpression of either REL1 or REL2 ligases. These authors hypothesized that MP18 interactions may bridge these subcomplexes in fully active ~20S particles. The subcomplexes were not reported to be prone to degradation, which may reflect the presence of MP18.

3.5 RNA–PROTEIN INTERACTIONS: ISOLATED SUBUNITS AND ASSEMBLED EDITING COMPLEXES

Overview. The previous sections described the enzymes responsible for editing activity, the protein composition, and particularly the known protein–protein interactions in 20S editing complexes. We are left with other fundamental challenging questions that remain essentially unexplored and evidently involve RNA–protein interactions, including the mechanism of substrate recognition and specificity, RNP assembly, editing site determination, and associated regulation/modulation of catalysis. Many of the subunits of the editing complex contain domains that are indicative of nucleic acid interaction (for review see references 3 and 34). The recombinant form of proteins MP42 (KREPA3) and MP24 (KREPA4) have been found to exhibit RNA binding activity in form. On the other hand, the protein–RNA contacts in assembled editing complexes are only beginning to be characterized. The first observations were made using sensitive photo-crosslinking approaches with editing substrates that carry a single photoreactive 4-thioU and ^{32}P atom at targeted editing sites. This section discusses the features of the known RNA–protein associations and potential mechanistic implications.

3.5.1 MP42

This subunit represents the first identified component of editing complexes that was shown to exhibit RNA-binding activity (55). MP42 contains two zinc fingers and a C-terminal putative OB-fold (3). Brecht et al. isolated recombinant hexahistidine-tagged

MP42 directly from *E. coli* inclusion bodies and refolded it in a Zn^{2+}-dependent manner. This protein preparation exhibited endo- and exoribonuclease as well as double-stranded (ds) nucleic acid-binding activity as determined in a surface plasmon resonance-binding assay where tagged-MP42 was covalently attached to a chip. Tagged MP42 bound a 15-base-pair (bp) dsRNA ligand with high affinity (K_d of 10 nM), and the equilibrium was rapidly reached within 2–4 min. The authors stated that it could also bind an 18-bp dsDNA and 15 to 18-nt single-stranded (ss) DNA ligands, but not to an 18-bp DNA/RNA hybrid (55). The potential role of MP42 and its protein-binding partners are still undefined. Specifically, the potential relevance of the observed associated enzymatic and RNA-binding activities in the context of editing complexes and functional editing substrates would be particularly informative.

3.5.2 MP24

Based on its high positive charge density (11% lysine), sequence analyses, and three-dimensional structure predictions, Salavati et al. hypothesized that MP24 could bind RNA as well as other editing complex subunits. MP24 was originally described as containing a C-terminal OB-fold, and computational approaches identified an S1 motif that has been observed within the OB-fold in a large number of RNA-binding proteins (34, 56). Purified recombinant hexahistidine-tagged MP24 did not show editing-associated enzymatic activities, but an electrophoretic mobility shift assay (EMSA) coupled with competition experiments suggested that MP24 could preferentially associate with the oligo (U) tail of gRNAs. Furthermore, a radiolabeled oligo (U)-containing cognate gRNA exhibited a MP24-dependent mobility shift that could be competed out with a 100-fold excess of homologous, unlabeled gRNA. A poly(U) homoribopolymer was even more effective at inhibiting the formation of the shifted product; that is, equimolar amounts of poly(U) inhibited product formation by 35%, while 10 to 100-fold excess inhibited product formation below detectable levels. Other RNA competitors, including pre-edited mA6, poly(G), poly(A), poly(C), a gRNA devoid of a U-tail, and an unrelated transcript (derived from pBlueScript), were not effective competitors (56). Additional studies will be needed to confirm the RNA-binding activity of MP24 and its apparent preferential interaction with oligo U runs in the context of ~20S editing complexes. The MP24 protein-binding partners are currently unknown, but one or more interacting subunits may enhance or modulate the MP24 recognition of editing substrates.

3.5.3 RNA–Protein Interactions in Assembled Editing Complexes

Sacharidou et al. (58) reported the first observations of direct of RNA–protein contacts between purified editing complexes and a functional model substrate for full-round editing. The editing complexes were biochemically purified using standard methods (14), which yield preparations with significant activity for full-round U-deletion and U-insertion editing (50). The model pre-mRNA contained both a sensitive photo-reactive (4-thioU) residue and a single ^{32}P at the targeted editing site (ES) (58). This methodology and some of its applications to study editing complexes have been described in detail in reference 59. The radiolabeled phosphorous corresponds to the scissile bond at this site. In the presence of purified editing complexes, UV irradiation induced direct photo-crosslinking and therefore radiolabeling of a few editing subunits. Importantly, the highly sensitive photoreactive 4-thioU requires very close proximity ($\leq$4 Å) (60) of nearby proteins for photo-crosslinking to occur. Thus, the observed interactions most likely involved editing subunits that were in intimate contact with the targeted ES. Notably, the 4-thioU and unmodified substrates were similarly active in editing. The latter is the most efficient substrate currently available to score full-round editing *in vitro* and is widely used. This model substrate consists of a downstream fragment of the native ATP synthase subunit 6 pre-mRNA ("A6" pre-mRNA) (26), and the gRNA is an enhanced version ("D33") of a cognate transcript that directs U-deletion at the first natural site (ES1) (50). Thus the robust photo-cross-linking activity observed in this study was a result of the following substrate attributes: sensitivity of the photo-reagent, ^{32}P labeling, and its high editing efficiency.

At least four cross-linked polypeptides were detected on an SDS-polyacrylamide gel with apparent M_r of 100, 60, 50, and 40 kD. Co-elution of the crosslinks with active complexes after extensive purification suggested that the interactions occurred in editing complexes. This was confirmed in immunoprecipitation assays with antibodies against four different editing subunits (58, 59). Efficient crosslinking required positioning the photoreactive 4-thioU at the single-/double-strand intersection that defines the ES. Such junction feature is a well-established determinant of functional editing sites (10), underscoring a structural preference of the cross-linkable interaction. Lastly, competition assays with homologous and heterologous transcripts indicated preferential recognition of the A6 pre-mRNA/gRNA substrate by editing complexes. Surprisingly, the purified complexes distinguished this substrate when diluted with other RNAs (e.g., tRNA) at severalfold molar excess.

A similar crosslinking assay on a minimal A6 model substrate for U-insertion editing at ES2 (61, 62) was used in combination with systematic sequence mutagenesis and 2′-hydroxyl substitutions to define features required for full-round U-insertion editing *in vitro*. Importantly, it was found that both association and catalysis are completely sequence-independent and that an internal loop (containing ES2) flanked by single RNA helical turns effectively specified the editing site. Furthermore, multiple and single 2′-hydroxyl substitutions helped assess the importance of specific RNA features for both crosslinking and catalysis. Notably, the substitutions led to comparable degrees of inhibition in mRNA cleavage and crosslinking. The incorporation of 2′-deoxy residues within both helices, along with the internal loop residues, suggested that editing complexes make relevant contacts with these regions. The downstream anchor duplex and "surprisingly" also the gRNA internal loop appeared particularly sensitive (Figure 3.3 summarizes the results of these studies). Overall, it was clear that 2′ hydroxyl substitution is an effective way to probe the importance of secondary structure in functional substrates under reaction conditions for full-round editing.

The crosslinking patterns obtained with A6 U-insertion and U-deletion editing substrates were almost identical; however, the former was a few-fold less effective, mirroring the fact that the U-deletion editing substrate is the most efficient substrate

Left		Internal Loop		Right	
Ins/Endo(*)	*X-links*	*Ins/Endo(*)*	*X-links*	*Ins/Endo(*)*	*X-links*
Moderate (L1; L2)	Moderate (L3)	None (IL1)	n.d. (IL1)	Moderate (R1)	n.d.
n.d.	n.d.	Strong (IL2)	Moderate (IL2)	Strong (R2)	Strong (R2)
		Complete (IL3)	None	n.d.	n.d.

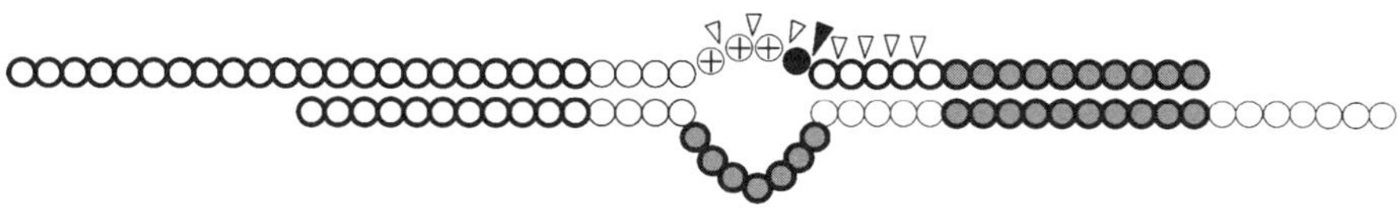

(L1) pre-mRNA protrusion or either strand of the left duplex.

(L2) 2'-Hs on both strands in duplex inhibit more than on one strand.

(L3) Tested with 2'-Hs on both strands

(IL1) Strong in combination with gRNA loop 2'-Hs.

(IL2) Even more inhibitory in combination with mRNA loop 2'-Hs.

(IL3) Complete inhibition of cleavage activity at the editing site.

(R1) Either strand of duplex. Increased with extent of 2'-H modification.

(R2) Both strands of the duplex combined. Induced cryptic cuts Δ

Figure 3.3 Summary of ribose 2′-deoxy substitutions tested on a minimal substrate for full-round U-insertion. *Upper panel*: This indicates the level of inhibition for the left and right sides of the internal loop, as well as within the loop (each region separated by a vertical dotted line). Full-round U-insertion and pre-mRNA cleavage (*Ins/Endo*) are at left, and editing complex crosslinking (*X-links*) is at right. (*) Indicates that U-insertion and cleavage are affected at comparable levels. Circle types represent the level of inhibition of Ins/Endo: ***thick line***, moderate; ***thin line***, not determined (n.d.); ***gray** (in addition to thick line)*, strong; ***with pattern***, no effect; ***black***, complete inhibition. Enzymatic and crosslinking activities not always correlate (e.g., the ES2 modification obliterates cleavage but has no effect on crosslinking. Brackets indicate clarification notes (lower panel). A filled arrowhead points to the natural ES2 for full-round insertion. dsDNA within the right duplex induced cryptic cuts at several residues (open arrowheads) flanking the editing site. *Middle panel*: Diagram of the substrate with individual residues as circles. *Lower panel*: Explanatory notes on the effect of the indicated modifications.

available for *in vitro* studies (50). Furthermore, the U-insertion editing substrate was also still preferentially recognized by editing complexes (e.g., versus tRNA and other heterologous transcripts). Thus, it is possible that a common set of proteins is involved in the recognition of relatively simple common features of both U-insertion and U-deletion substrates (58, 62). These studies suggested that editing complexes recognize the secondary structure of editing substrates, although the overall affinity of the interaction appears relatively low.

Based on the observations above, the following basic working model of substrate recognition by trypanosome RNA editing complexes *in vitro* can be proposed. First, editing complexes recognize helical regions flanking the editing site in a sequence-independent manner, most likely though interactions with 2′-hydroxyls and the phosphodiester backbone. The length of the binding region could minimally be one turn of A-form dsRNA (i.e., 11 bp). RNase III-type enzymes (such as REN1 and

REN2) that contain dsRNA binding motifs are prime candidates to mediate this interaction, because bacterial RNase III enzymes have been found to possess similar binding requirements (63). mRNA and gRNA internal loop residues may be recognized by single-stranded RNA-binding proteins primarily via their 2′-hydroxyls and base-specific contacts, the identity of editing site is established (U-insertion or U-deletion) through such interactions, and this in turn commits editing complexes to one of two catalytic pathways.

3.6 CONCLUDING REMARKS

While a map of protein–protein and protein–RNA interactions in U-insertion/deletion editing complexes is on its way in several laboratories, mechanistic studies are needed to define the role of these interactions in the order of assembly and stability of editing complexes, substrate recognition, and editing site specificity. What are the molecular interactions that commit editing complexes to U-deletion or U-insertion editing and trigger catalysis, and how are the intermediate catalytic steps and dynamic subunits (e.g., RENs) coordinated? Additional levels of complexity involve understanding the regulatory mechanisms of RNA editing during cell growth and development, the detailed 3-D architecture of editing complexes, and the kinetics of the editing holoenzyme/substrate interaction and catalysis. Is RNA editing processive or are editing complexes recycled after each ES is remodeled? Is U-insertion and U-deletion editing in mitochondria catalyzed by completely separate particles or by higher-order assemblies containing insertion and deletion sub-complexes. Overall, our increasing knowledge of the molecular basis of this process and its importance in the adaptation and survival of kinetoplastids should lead to novel molecular strategies to interfere with the life cycle and disease produced by pathogenic species.

ACKNOWLEDGMENTS

We would like to thank Dr. Juan Alfonzo, Dr. Larry Simpson, and an anonymous reviewer for their careful and insightful comments on this chapter. This work was supported by a grant from the National Institute of Health (GM67130 to J.C.-R.).

REFERENCES

1. Benne, R., Van den Burg, J., Brakenhoff, J. P., Sloof, P., Van Boom, J. H., and Tromp, M. C. (1986) *Cell* **46** (6), 819–826.
2. Simpson, L., Aphasizhev, R., Gao, G., and Kang, X. (2004) *RNA* **10** (2), 159–170.
3. Stuart, K. D., Schnaufer, A., Ernst, N. L., and Panigrahi, A. K. (2005) *Trends Biochem Sci* **30** (2), 97–105.
4. Madison-Antenucci, S., Grams, J., and Hajduk, SL. (2002) *Cell* **108**, 435–438.
5. Blum, B., Bakalara, N., and Simpson, L. (1990) *Cell* **60**, 189–198.
6. Maslov, D. A., and Simpson, L. (1992) *Cell* **70** (3), 459–467.
7. Stuart, K., Allen, T. E., Heidmann, S., and Seiwert, S. D. (1997) *Microbiol Mol Biol Rev* **61** (1), 105–120.
8. Decker, C. J., and Sollner-Webb, B. (1990) *Cell* **61**, 1001–1011.

9. Ochsenreiter, T., and Hajduk, S. L. (2006) *EMBO Rep* **7** (11), 1128–1133.
10. Cruz-Reyes, J., and Sollner-Webb, B. (1996) *Proc Natl Acad Sci USA* **93** (17), 8901–8906.
11. Cruz-Reyes, J., Rusche, L. N., Piller, K. J., and Sollner-Webb, B. (1998) *Mol Cell* **1** (3), 401–409.
12. Cruz-Reyes, J., Zhelonkina, A. G., Huang, C. E., and Sollner-Webb, B. (2002) *Mol Cell Biol* **22** (13), 4652–4660.
13. Huang, C. E., Cruz-Reyes, J., Zhelonkina, A. G., O'Hearn, S., Wirtz, E., and Sollner-Webb, B. (2001) *EMBO J* **20** (17), 4694–4703.
14. Rusche, L. N., Cruz-Reyes, J., Piller, K. J., and Sollner-Webb, B. (1997) *EMBO J* **16** (13), 4069–4081.
15. Muller, U. F., Lambert, L., and Goringer, H. U. (2001) *EMBO J* **20** (6), 1394–1404.
16. Blom, D., Burg, J., Breek, C. K., Speijer, D., Muijsers, A. O., and Benne, R. (2001) *Nucleic Acids Res* **29** (14), 2950–2962.
17. Aphasizhev, R., Aphasizheva, I., Nelson, R. E., Gao, G., Simpson, A. M., Kang, X., Falick, A. M., Sbicego, S., and Simpson, L. (2003) *EMBO J* **22** (4), 913–924.
18. Vondruskova, E., van den Burg, J., Zikova, A., Ernst, N. L., Stuart, K., Benne, R., and Lukes, J. (2005) *J Biol Chem* **280** (4), 2429–2438.
19. Pelletier, M., and Read, L. K. (2003) *RNA* **9** (4), 457–468.
20. Missel, A., Souza, A. E., Norskau, G., and Goringer, H. U. (1997) *Mol Cell Biol* **17** (9), 4895–4903.
21. Madison-Antenucci, S., Sabatini, R. S., Pollard, V. W., and Hajduk, S. L. (1998) *EMBO J* **17** (21), 6368–6376.
22. Ernst, N. L., Panicucci, B., Igo, R. P., Jr. Panigrahi, A. K., Salavati, R., and Stuart, K. (2003) *Mol Cell* **11** (6), 1525–1536.
23. Vanhamme, L., Perez-Morga, D., Marchal, C., Speijer, D., Lambert, L., Geuskens, M., Alexandre, S., Ismaili, N., Goringer, U., Benne, R., and Pays, E. (1998) *J Biol Chem* **273** (34), 21825–21833.
24. Miller, M. M., Halbig, K., Cruz-Reyes, J., and Read, L. K. (2006) *RNA* **12** (7), 1292–1303.
25. Halbig, K., Sacharidou, A., De Nova-Ocampo, M., and Cruz-Reyes, J. (2006) *Int J Parasitol* **36** (12), 1295–1304.
26. Seiwert, S. D., Heidmann, S., and Stuart, K. (1996) *Cell* **84** (6), 831–841.
27. Kable, M. L., Seiwert, S. D., Heidmann, S., and Stuart, K. (1996) *Science* **273**, 1182–1183.
28. Igo, R. P., Jr., Palazzo, S. S., Burgess, M. L., Panigrahi, A. K., and Stuart, K. (2000) *Mol Cell Biol* **20** (22), 8447–8457.
29. Igo, R. P., Jr., Weston, D. S., Ernst, N. L., Panigrahi, A. K., Salavati, R., and Stuart, K. (2002) *Eukaryot Cell* **1** (1), 112–118.
30. Seiwert, S. D., and Stuart, K. (1994) *Science* **266** (5182), 114–117.
31. Halbig, K., De Nova-Ocampo, M., and Cruz-Reyes, J. (2004) *RNA* **10** (6), 914–920.
32. Schnaufer, A., Ernst, N. L., Palazzo, S. S., O'Rear, J., Salavati, R., and Stuart, K. (2003) *Mol Cell* **12** (2), 307–319.
33. Panigrahi, A. K., Ernst, N. L., Domingo, G. J., Fleck, M., Salavati, R., and Stuart, K. D. (2006) *RNA* **12** (6), 1038–1049.
34. Worthey, E. A., Schnaufer, A., Mian, I. S., Stuart, K., and Salavati, R. (2003) *Nucleic Acids Res* **31** (22), 6392–6408.
35. Simpson, L., Thieman, O. H., Savill, N. J., Alfonzo, J. D., and Maslov, D. A. (2000) *Proc Natl Acad Sci USA* **97**, 6986–6993.
36. Lukes, J., Hashimi, H., and Zikova, A. (2005) *Curr Gene* **48**, 277–299.
37. Schnaufer, A., Domingo, G. J., and Stuart, K. (2002) *Int J Parasitol* **32**, 1071–1084.
38. Trotter, J. R., Ernst, N. L., Carnes, J., Panicucci, B., and Stuart, K. (2005) *Mol Cell* **20** (3), 403–412.
39. Kang, X., Gao, G., Rogers, K., Falick, A. M., Zhou, S., and Simpson, L. (2006) *Proc Natl Acad Sci USA* **103** (38), 13944–13949.
40. Zhelonkina, A. G., O'Hearn, S. F., Law, J. A., Cruz-Reyes, J., Huang, C. E., Alatortsev, V. S., and Sollner-Webb, B. (2006) *RNA* **12** (3), 476–487.
41. Carnes, J., Trotter, J. R., Ernst, N. L., Steinberg, A., and Stuart, K. (2005) *Proc Natl Acad Sci USA* **102** (46), 16614–16619.
42. Kang, X., Rogers, K., Gao, G., Falick, A. M., Zhou, S., and Simpson, L. (2005) *Proc Natl Acad Sci USA* **102** (4), 1017–1022.
43. Aphasizhev, R., Aphasizheva, I., and Simpson, L. (2003) *Proc Natl Acad Sci USA* **100** (19), 10617–10622.

44. Schnaufer, A., Panigrahi, A. K., Panicucci, B., Igo, R. P., Jr. Wirtz, E., Salavati, R., and Stuart, K. (2001) *Science* **291** (5511), 2159–2162.
45. Gao, G., and Simpson, L. (2003) *J Biol Chem* **278** (30), 27570–27574.
46. Gao, G., Simpson, A. M., Kang, X., Rogers, K., Nebohacova, M., Li, F., and Simpson, L. (2005) *Proc Natl Acad Sci USA* **102** (13), 4712–4717.
47. Kang, X., Falick, A. M., Nelson, R. E., Gao, G., Rogers, K., Aphasizhev, R., and Simpson, L. (2004) *J Biol Chem* **279** (6), 3893–3899.
48. Drozdz, M., Palazzo, S. S., Salavati, R., O'Rear, J., Clayton, C., and Stuart, K. (2002) *EMBO J* **21** (7), 1791–1799.
49. O'Hearn, S. F., Huang, C. E., Hemann, M., Zhelonkina, A., and Sollner-Webb, B. (2003) *Mol Cell Biol* **23** (21), 7909–7919.
50. Cruz-Reyes, J., Zhelonkina, A., Rusche, L., and Sollner-Webb, B. (2001) *Mol Cell Biol* **21** (3), 884–892.
51. Law, J. A., Huang, C. E., O'Hearn, S. F., and Sollner-Webb, B. (2005) *Mol Cell Biol* **25** (7), 2785–2794.
52. Huang, C. E., O'Hearn, S. F., and Sollner-Webb, B. (2002) *Mol Cell Biol* **22** (9), 3194–3203.
53. Babbarwal, V. K., Fleck, M., Ernst, N. L., Schnaufer, A., and Stuart, K. (2007) *RNA* **13** (5), 737–744.
54. Wang, B., Ernst, N. L., Palazzo, S. S., Panigrahi, A. K., Salavati, R., and Stuart, K. (2003) *Eukaryot Cell* **2** (3), 578–587.
55. Brecht, M., Niemann, M., Schluter, E., Muller, U. F., Stuart, K., and Goringer, H. U. (2005) *Mol Cell* **17** (5), 621–630.
56. Salavati, R., Ernst, N. L., O'Rear, J., Gilliam, T., Tarun, S., Jr., and Stuart, K. (2006) *RNA* **12** (5), 819–831.
57. Law, J. A., O'Hearn, S., and Sollner-Webb, B. (2007) *Mol Cell Biol* **27** (2), 777–787.
58. Sacharidou, A., Cifuentes-Rojas, C., Halbig, K., Hernandez, A., Dangott, L. J., De Nova-Ocampo, M., and Cruz-Reyes, J. (2006) *RNA* **12** (7), 1219–1228.
59. Cruz-Reyes, J. (2007) RNA–protein interactions in assembled editing complexes in trypanosomes, In: Gott. J. M. (ed.), RNA *Editing and Modification*, Elsevier, Amsterdams, pp. 107–125.
60. Fabre, A., (1990) Bioinorganic photochemistry. In: Morrison, H. (ed.), *Photobiochemistry and Nucleic Acids*, John Wiley & Sons, New York, pp. 379–425.
61. Cifuentes-Rojas, C., Halbig, K., Sacharidou, A., De Nova-Ocampo, M., and Cruz-Reyes, J. (2005) *Nucleic Acids Res* **33** (20), 6610–6620.
62. Cifuentes-Rojas, C., Pavia, P., Hernandez, A., Osterwisch, D., Puerta, C., and Cruz-Reyes, J. (2006) *J Biol Chem* **282**, 4265–4276.
63. Pertzev, A. V., and Nicholson, A. W. (2006) *Nucleic Acids Res* **34** (13), 3708–3721.

CHAPTER 4

MACHINERY OF RNA EDITING IN PLANT ORGANELLES

Toshiharu Shikanai
Jun-ichi Obokata

RECENTLY, *Arabidopsis thaliana* mutants specifically defective in RNA editing processes were identified, resulting in the discovery of a pentatricopeptide repeat (PPR) protein as a site-recognition factor for RNA editing in plastids. On the basis of this breakthrough, we discuss the machinery of RNA editing in plant organelles.

4.1 INTRODUCTION

RNA editing is a process of modifying genetic information posttranscriptionally on RNA molecules, and it takes place in both nuclear and organellar transcripts (1, 2). Although this process has been reported in a variety of eukaryotic organisms, the different systems involve divergent mechanisms, implying the independent origins of RNA editing. Uridine (U) insertion/deletion-type RNA editing is best understood in the kinetoplastid mitochondria of trypanosomes (3) (see Chapter 3). A guide RNA forming an incomplete double strand with the target RNA specifies several sites of RNA editing. Despite fewer examples than in plant or protozoan organelles, editing in animals is relatively well characterized. Conversion of adenosine to inosine is catalyzed by adenosines deaminase acting on RNA (ADARs); this action modifies the amino acid sequences encoded by a limited number of genes (4) (see Chapters 1, 6, 9, 12 and 15). Recently, noncoding, repetitive RNAs have been shown to be frequent targets of RNA editing, which may be involved in retrotransposons and gene silencing regulation (5). Cytidine (C)-to-U editing was first discovered in mammals in the apolipoprotein B (apoB) mRNA, in which a CAA codon (glutamine) is converted to a stop codon (6, 7) (see Chapters 10 and 16). Although C-to-U conversion is involved in both mammalian and plant editing, the characteristics of the conversion are divergent in the two systems, suggesting distinct mechanisms.

RNA and DNA Editing: Molecular Mechanisms and Their Integration into Biological Systems,
Edited by Harold C. Smith

In plants, RNA editing was discovered in mitochondria (8–10) and subsequently in plastids (11). RNA editing occurs mostly in coding regions and is essential for producing functional proteins by creating either new codons (including initiation codons) or a stop codon, resulting in the preservation of functionally important amino acids (12, 13). Although in plants, RNA editing involves C-to-U conversion, in a limited number of species, the reverse process of U-to-C can also occur. In seed plants, approximately 30 editing sites in plastids and 400–500 sites in mitochondria have been discovered. Key questions concerning RNA editing in plants are: (1) How is the target site recognized by the editing machinery with high efficiency and precision? (2) What is the identity of the editing enzyme(s) in C-to-U and/or U-to-C conversion? (3) How many independent origins does plant RNA editing have? and (4) What is the physiological function of RNA editing in plants? Although we still cannot answer these questions, here we will summarize our current knowledge on these issues.

4.2 MECHANISM OF TARGET RECOGNITION

During this decade, our understanding of how the editing machinery recognizes the target site has improved. We can draw a simple model, in which a *trans*-factor recognizes a *cis*-element that surrounds the target C residue (Figure 4.1). The model was established in the tobacco plastid editing system, for which two powerful experimental tools were available: *in vivo* analysis using plastid transformation (14) and an *in vitro* editing system (15). In tobacco, the plastid transformation technique facilitated the introduction of a chimeric gene including an RNA editing site into chloroplasts, in which the site is edited precisely (16). By introducing the deletion derivatives into chloroplasts, the sequence required for site recognition (*cis*-element) was determined. In the tobacco *petL* gene, 22 nucleotides (nt), 16 nt 5′ of the C to be edited and 5 nt 3′ of the C (−16/+5), were sufficient for editing (17). This discovery was followed by complementation

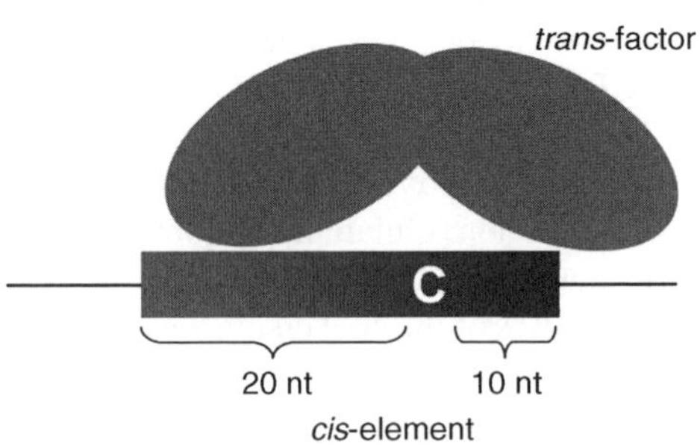

Figure 4.1 An early model of RNA editing in plant organelles. Site recognition for RNA editing requires fewer than 20 upstream nt and fewer than 10 nt downstream in some cases (*cis*-element). This *cis*-element is recognized by a site-specific factor (*trans*-factor). (*See color insert.*)

using an *in vitro* editing system in which competitor assay for the site-specific factor was quantitatively available (15). UV crosslinking using *psbL* mRNA detected a 25-kD protein that is specific to this RNA editing event, although the identity of this protein is not clear. By the electroporation of transgenes into isolated mitochondria, a similar model was shown to be true in the mitochondria of seed plants (18). In the mitochondrial *coxII* gene of wheat, 16 upstream nt are necessary for editing, and 6 nt immediately downstream of the target site are needed for precise site recognition (19). *In vitro* RNA editing systems were also established in mitochondria (20, 21) and helped to specify the sequence required for site recognition. In RNA editing of the mitochondrial *atp9* gene in pea, −15/−5 is essential for site recognition, whereas −40/−35 is required for high efficiency (22). In summary, site recognition for RNA editing requires fewer than 20 upstream nt and fewer than 10 downstream nt in some cases (Figure 4.1). This characteristic is conserved between the plastids and mitochondria of seed plants, implying that a similar system for site recognition is conserved between organelles.

Altogether, we have arrived at two major discoveries: (1) There are no strict consensus motifs in *cis*-elements among different RNA editing sites, and (2) expression of chimeric RNA containing the editing sites specifically decreases the editing extent in RNAs derived from the corresponding endogenous genes (16, 23). Although there have been conflicting results and ideas (discussed later), these results are explained by a model in which the chimeric RNA competes with endogenous mRNA for common site-specific factors present in limiting amounts (*trans*-factors). On the basis of this idea, a *trans*-acting factor is considered to recognize a single or a few *cis*-elements, implying that the genes encoding for *trans*-factors should form a large family that independently manages approximately 500 RNA editing sites present in plant organelles. Any model of RNA editing machinery has to explain this surprising feature.

4.3 PPR PROTEIN IS A *TRANS*-FACTOR IN PLASTIDS

A *trans*-factor involved in plastid RNA editing was discovered in the course of research on photosynthetic electron transport. *crr4* (*chlororespiratory reduction*) *Arabidopsis thaliana* mutants were isolated on the basis of their lack of activity of chloroplast NAD(P)H dehydrogenase (NDH) (24). NDH is involved in cyclic electron transport around photosystem I (25), and its activity is visualized by the imaging of chlorophyll fluorescence with a CCD (charge-coupled device) camera (26). The *crr4* mutants are defective in the RNA editing that creates a translational initiation codon of *ndhD* encoding a subunit of NDH (ndhD-1 site). Defects in plastid RNA editing in genes other than the 11 *ndh* genes (*ndhA-ndhK*) encoding NDH subunits probably lead to severe defects in photosynthetic activity and often in chloroplast development; the phenotypes are similar to that of mutants that are simply defective in general chloroplast function. This was the key to focusing on

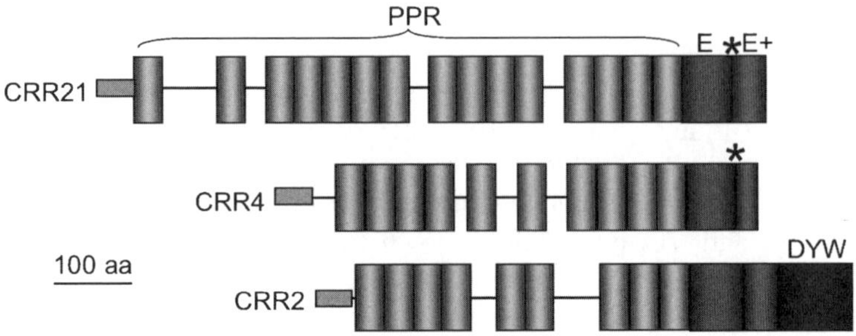

Figure 4.2 Comparison of domain structures among the PPR proteins present in *Arabidopsis* plastids. CRR4 and CRR21 (members of the E+ subgroup) are involved in RNA editing (24, 57), whereas CRR2 (a member of the DYW subgroup) is essential for intergenic RNA cleavage (26). The PPR, E, E+, and DYW motifs are indicated by boxes. Asterisks indicate the position of the 15-amino-acid motif conserved in some E+ members, including CRR4 and CRR21 (24). Horizontal boxes indicate plastid targeting signals. (*See color insert.*)

NDH activity to discover those mutants specifically defective in an RNA editing event, since the NDH complex is dispensable, at least in the absence of environmental stress (27).

Map-based cloning showed that *CRR4* encodes a member of the pentatricopeptide repeat (PPR) family (24, 28). A PPR motif is a highly degenerate unit of 35 amino acids and usually appears as tandem repeats in members of this family (Figure 4.2). In general, PPR proteins are targeted to mitochondria or plastids and, on the basis of the characteristics of the mutant phenotypes, are considered to be involved in RNA maturation processes (24, 26, 29–33) (Table 4.1). PPR proteins are also involved in the restoration of cytoplasmic male sterility (CMS), probably by modifying the expression of CMS-associated genes in mitochondria (34–42). The PPR family is specific to eukaryotes and is extraordinarily large in seed plants (28). The *Arabidopsis* genome encodes 466 members (43).

From the mutant phenotypes and their possible structures, a PPR protein is considered to be a sequence-specific RNA-binding protein (44). The idea has support in biochemical studies of some PPR proteins (45–47). Mutation in a gene encoding a PPR protein, CRR4, leads to a specific defect in RNA editing at the ndhD-1 site (24). If CRR4 is a *trans*-factor for RNA editing, then it is likely to bind the short sequence surrounding the editing site. To obtain biochemical evidence to complement the genetic findings, mature CRR4 protein was expressed in *Escherichia coli*. The purified recombinant CRR4 bound the short sequence surrounding the target site (−25/+10) in a sequence-specific manner (47). By taking this together with the genetic evidence, we concluded that CRR4 is a *trans*-factor for RNA editing at the ndhD-1 site. Although the problem remains regarding how the model can be applied generally to plant RNA editing system(s), we do know that a PPR protein is involved in site recognition for RNA editing, at least in plastids—a discovery that takes our discussion to the next step.

TABLE 4.1 List of Characterized PPR Proteins

	Localization	Group	Function	Target RNA	Refs
Maize					
CRP1	P[a]	P[c]	Translation RNA cleavage	*petA*, *psaC* *petB/petD*	29
PPR2	P	P	Ribosome accumulation	ND	91
PPR4	P	P[f]	*Trans*-splicing	*rps12*	32
GRP23	Nucleus	P[f]	Early Embryogenesis	RNA polymerase II	92
EMP4	M[b]	P	RNA stabilization?	*rps2a/rps2b*, *rps3/rpl16*, *mttb*	34
Arabidopsis					
HCF152	P	P	RNA cleavage Splicing	*psbH/petB petB*	30, 45
PGR3	P	P	RNA stabilization Translation?	*petL petL*, *ndhX*[h]	31
CRR2	P	DYW[d]	RNA cleavage	*rps7/ndhB*	26
CRR4	P	E+[e]	RNA editing	*ndhD* (ndhD-1)	24
CRR21	P	E+	RNA editing	*ndhD* (ndhD-2)	57
EMB175	P	P	Embryogenesis	ND	93
LOI1	M	DYW	Isoprenoid biosynthesis	ND	94
GUN1	P	P[g]	Plastid signal	ND	95
LOJ	M	P	Lateral organ development	ND	96
AtC401	ND	P[g]	Clock-controlled	ND	97
Rice					
OsPPR1	P	P	Chloroplast biogenesis	ND	98
Rf-1	M	P	Fertility restoration	*atp6/orf79*	38, 40, 41
Petunia					
Rf	M	P	Fertility restoration	*pcf*	35
Radish					
Rfk1	M	P	Fertility restoration	*orf125*	37
Rfo	M	P	Fertility restoration	*orf138*	36, 39
Physcomitrella					
PPR531-11	P	P	RNA cleavage	*clpP*	33

(*Coninued*)

TABLE 4.1 (*Continued*)

	Localization	Group	Function	Target RNA	Refs
Yeast					
PET309	M	P	Transcription, RNA Stabilization, translation	COX1	99
Neurospora					
cya-5	M	P	Translation	COX1	100

[a]plastids. [b]mitochondria. [c]P subfamily. [d]DYW subgroup. [e] E+ subgroup.

[f]PPR4 is fused with an RRM domain. GRP23 contains bZIP and Q-rich domains.

[g]GUN1 and AT401 contains a SMR and a protein kinase domain, respectively.

[h]One of the 11 chloroplast *ndh* genes.

Many other PPR proteins were partially characterized (28, 94).

4.4 HOW CAN THE MODEL BE GENERALIZED TO PLANT RNA EDITING?

PPR proteins that form a large family in seed plants (28) bind RNA in a sequence-specific manner (45–47). This can explain the characteristics of the *trans*-factors observed in both plastids and mitochondria. For RNA editing at the ndhD-1 site, the PPR protein CRR4 is a *trans*-factor (24, 47), suggesting that each RNA editing site might be recognized by a particular PPR protein in the plastids of seed plants. How can this model be applied generally to plant RNA editing? The answer may be found in the central question of whether editing systems in plants are of monophyletic origin. The observed divergent characteristics of RNA editing, the drastic differences in its frequency, and the presence of the reverse reaction of U-to-C imply diverse origins between plastids and mitochondria and also among different species. RNA editing has been observed in the organelles of all land plants except the liverwort, *Marchantia polymorpha* (48). In seed plants, RNA editing in two organelles is characterized by differences in the frequency of editing sites. In *Arabidopsis* mitochondria, 456 C-to-U, but no U-to-C, editing sites were identified exclusively in mRNAs (48). In contrast, only 28 C-to-U conversions were detected in *Arabidopsis* plastids (49). Despite the divergence of editing sites among species—12 sites are common to a dicot (tobacco) and a monocot (maize or rice)—the number of editing sites is conserved in seed plants, tobacco (38 sites), maize (27 sites), and rice (26 sites) (50–54). If a single PPR protein recognizes a single *cis*-element, approximately 30 genes are enough to specify RNA editing in plastids. Could the same estimate be true for the more than 400 RNA editing sites in mitochondria? As discussed later, a model in which a single *trans*-factor recognizes multiple *cis*-elements may provide a potential explanation.

Despite the large difference in frequency, editing in both organelles always coincides phylogenetically (48, 55), implying a monophyletic origin of RNA editing in the two organelles (56). If this is true, the machineries of RNA editing should target both plastids and mitochondria. CRR4 is a member of the E+ subgroup of the PPR family, and this subgroup contains 60 members. The E subgroup lacking the E+ motif contains 47 members. (The E+ motif is incomplete in CRR4.) Recently, we identified a second *Arabidopsis* mutant, *crr21*, defective in RNA editing of the ndhD-2 site in plastids (57). CRR21 is also a member of the E+ subgroup, and its C-terminal region is conserved with that of CRR4 (Figure 4.2). Between CRR4 and CRR21, the 15-amino-acid motif located at the junction between the E and E+ motifs is especially well-conserved and has been found in some members of the E+ subgroup. Although some of these members have putative targeting signals to mitochondria, there are not enough members to account for all the editing sites either by one member recognizing a single or multiple editing sites. Thus plastids and mitochondria may share the machinery of site recognition—a PPR protein—at least partly. This proposition then implies commonalities of the RNA editing machinery between the systems.

In addition to differences between organelles, plastid RNA editing in the fern *Adiantum capillus-veneris* and the hornwort *Anthoceros formosae* has characteristics that are distinct from that in plastids in seed plants. In hornwort plastids, 509 C-to-U and 433 U-to-C conversions were found (58), whereas 315 C-to-U and 35 U-to-C conversions were found in fern plastids (59). Although information from genome-wide analysis is not available, extensive editing is likely to exist also in the mitochondria of both species (55, 60, 61). The most striking feature of RNA editing in both species is a high frequency of reverse editing (U-to-C). In seed plants, no examples of U-to-C conversion were discovered in the plastids and mitochondrial editing was also nearly exclusively by C-to-U conversions (49, 62, 63), with the exception of a few examples (64). While deamination of C to U is thermodynamically very favorable, the reverse reaction is consequently highly unfavorable and would be expected to be an ATP-driven reaction (65). A prokaryotic-type C deaminase may not catalyze the reverse reaction. If the editing enzyme can manage the reverse reaction in plant RNA editing, why are frequent reverse reactions specific to *Adiantum capillus-veneris* and *Anthoceros formosae*? It is possible that the editing enzymes are not common among species, or that multiple enzymes may function even within a single system. In the near future, comparative genomics may provides a clue to clarify the evolution of RNA editing by assessing whether the genes encoding RNA editing machinery are conserved among species with the different systems.

4.5 CAN CLOSELY LOCATED EDITING SITES SHARE A *TRANS*-FACTOR?

In both mitochondria and plastids, many RNA editing sites are located close to each other. A question related to the mechanism of site recognition is whether these sites are edited independently or simultaneously by a common machinery. Clearly, resolution of this question will shed light on the related issue of how many *trans*-factors are needed to manage all the editing sites. In the *ndhB* gene of tobacco plastids, a pair of editing sites, IV and V, is separated by only 8 nt. Analysis of the deletion derivatives indicates that the −12/−2 region (corresponding to site IV) is essential for both RNA editing events *in vivo*, suggesting that the editing of adjacent sites is mechanistically coupled (23). However, both sites were independently edited in partially edited transcripts (66), suggesting that editing does not follow the exact order between two sites. This idea was supported by a result indicating that edited *ndhF* transcripts could compete with unedited transcripts (67). Consistent with these observations, CRR4 can bind both pre-edited and post-edited RNA, although CRR4 slightly prefers the pre-edited RNA, at least *in vitro* (47).

A similar example was observed in the mitochondrial *atp4* gene, in which three editing sites are clustered in 4 nt (68). The first and third sites are edited *in vitro*, but the second is not. For both events, identical *cis*-elements were required, suggesting that a single *trans*-acting factor may be involved in two editing events. Partially pre-edited transcripts could be edited *in vitro*, as could the unedited precursor molecules, indicating that initiation of the editing reaction is independent of the status of editing in the target molecules.

In both plastid *ndhB* and mitochondrial *atp4*, closely located editing sites can share the *cis*-element and a single *trans*-factor may address both editing sites. Multiple editing events take place independently, and the partially edited RNA can be a substrate for the editing of another site. These results suggest a model in which a *trans*-factor binds the *cis*-element independently of the editing status of the target RNA, and a *trans*-factor flexibly recognizes multiple target sites that are at different distances from the *trans*-factor. It is apparent that most of the *trans*-factors lack this flexible site recognition, since it may cause aberrant editing in the C residues located close to the target C site. Therefore, it is interesting to study the molecular mechanism by which the specific *cis*- and *trans*-interaction acquires this flexibility of site recognition.

In the mitochondrial *atp9* gene, two editing sites are spaced 30 nt apart; the question of whether the two editing events would influence each other was investigated (69). In contrast to the examples in plastid *ndhB* and mitochondrial *atp4* (23, 68), each editing involves a distinct *cis*-element located 20 to 30 nt upstream of each target site (69). However, the enhancer element for the first site (−40/−20 from the first site) (22) also influences the editing efficiency of the second site (50–70 nt downstream) (69). Although access by a *trans*-factor is enabled independently via the specific *cis*-element, multiple editing sites that are located relatively close to each other may be regulated via an unknown enhancer-like factor.

4.6 IS A *TRANS*-FACTOR SPECIFIC TO A SINGLE *CIS*-ELEMENT?

Even though closely located editing sites are recognized via a single *cis*-element, a huge number of *trans*-factors are still required for managing all the editing sites. Pioneer work using plastid transformation technique indicated that each individual *cis*-element corresponds to a site-specific *trans*-factor (16). This idea is based on a result demonstrating that overexpression of a transcript containing the *petL* editing site results in a decrease in the editing of transcripts originating from the *petL* endogenous gene. This competition was not observed at other endogenous editing sites. The result is in apparent conflict with the observation that overexpression of

AtndhF-2 GAUCCACUUACUUCUAUUAUGU--CAAUAUUA
AtndhB-3 GGGGCAAGCUCUUCUAUUCUGGUUCAUGGUUU
AtndhD-1 CCUGGUGUAUCUUGUCUUUACCA-CGAAUGAU
NtndhD-1 CCAAGUGUAUCUUGUCUUUACUA-CGAAUUAU
OsndhD AUCUUGUCUUUAUCACUUUAUCA-UGAGUUCU

Figure 4.3 Alignment of three putative *cis*-elements for RNA editing in *Arabidopsis*. By overexpressing ndhF-2 RNA in tobacco plastids, the extent of RNA editing was decreased in endogenous RNA in this group of genes sharing weak similarity of the *cis*-element (70). The sequences in tobacco (NtNdhD-1) and rice (OsndhD) corresponding to *Arabidopsis* NdhD-1 are also aligned. In rice, the ndhD-1 site is already encoded by T in the genome, and the similarity is weak in the upstream region. Editing sites are indicated by capital letters. Nucleotides conserved in the putative *cis*-region are in red. This figure is based on reference 70. (*See color insert.*)

transcripts carrying the ndhB-2 or ndhF-2 editing site affected the editing efficiency at several other sites (70). Interestingly, weak sequence similarities were observed in the *cis*-elements exhibiting cross-competition for the identical *trans*-factor (Figure 4.3). Although the number of PPR proteins is roughly enough for all the *cis*-elements, it is not clear whether plants conserved several hundreds of genes for RNA editing. Although the idea that a *trans*-factor recognizes a set of *cis*-elements that show sequence similarity is likely, it still lacks definitive experimental evidence to support it. Recently, we published an evidence that a single *trans*-factor is involved in ndhB-9 and ndhF-1 editing events in tobacco plastids. Kobayashi 7, Matsuo M, Sakamoto K, Wakasugi T, Yamata K, Obokata J, (2007) Two RNA editing sites with *cis*-acting elements of moderate sequence identity are recognized by an identical site-recognition protein in tobacco chloroplats. *Nucleic acids Res* in Press. Overproduction of ndhF-2 reduces the editing efficiency of ndhD-1, and the 5′-flanking sequences of both editing sites share a short conserved sequence in tobacco (70) and in *Arabidopsis* (Figure 4.3). However, *crr4* is specifically defective in ndhD-1 editing, and editing of the ndhF-2 site is not affected (24). To explain this discrepancy, it is necessary to postulate that there is a second *trans*-factor that is shared by multiple editing sites. This hypothesis may explain the developmental co-variation of the extent of RNA editing in clusters of RNA editing sites (71). It is also possible that CRR4 can recognize several *cis*-elements but that it functions essentially only at the ndhD-1 site. If the other editing sites are recognized by multiple PPR proteins alternatively, the defect in RNA editing of these sites would be not detected in *crr4*. In this case, the phenotype in such RNA editing events could be detected only in double mutants defective in the two genes encoding the PPR proteins, including CRR4. Information on PPR family members is still too limited for us to draw this conclusion, and more extensive and genome-wide analyses of the family are essential to clarify this point.

4.7 MECHANISM DETERMINING THE EFFICIENCY OF RNA EDITING

Since the majority of RNA editing events are essential for the production of functional proteins, the sites should be edited with high efficiency so as not to produce aberrant proteins. In dicot plastids, an exception is the editing that creates a translational initiation codon of *ndhD* (ndhD-1), which is recognized by CRR4 (24). In tobacco, the editing efficiency is developmentally regulated (72); this is often cited as an example suggesting the physiological function of RNA editing in the regulation of translation. However, the NDH complex consists of at least more than 14 subunits that are encoded by both nuclear and plastid genomes (73), and there is no experimental evidence suggesting that the translation of *ndhD* is a limiting point that determines the level of the NDH complex. It may be more likely that the site does not need to be edited completely, since the unedited transcript without a translational initiation codon does not encode any abnormal proteins. This initiation codon is encoded by ATG in the genome in monocots (51), suggesting that regulation via RNA editing is not physiologically essential.

Why does the efficiency of the RNA editing that creates the translational initiation codon of *ndhD* remain low even in wild-type leaves, where the NDH complex is functioning? Approximately 50% of mRNA has a translational initiation codon in tobacco and *Arabidopsis* leaves (24, 72). Because CRR4 is a site-specific *trans*-factor for this RNA editing, it is possible that the CRR4 protein level limits the efficiency of RNA editing. However, overexpression of *CRR4* under the control of the CaMV 35S promoter fails to increase the extent of editing (24). This result suggests several possibilities: (1) The CRR4 protein level is low despite the overaccumulation of mRNA via an unknown mechanism. Protein blot analysis failed to detect CRR4, although the transgene could complement the *crr4* defect in RNA editing (24). Plastids have a proteolytic system by which the stability of specific proteins is precisely regulated (74); (2) CRR4 may have to be modified to be active, a process that limits the efficiency of RNA editing; and (3) a factor other than CRR4 limits the efficiency of RNA editing. This idea is supported by the fact that the level of RNA editing at other sites of *ndh* transcripts is generally lower in the roots than in the leaves (71). We cannot eliminate the possibility that the production of some PPR proteins is precisely regulated and that the process is physiologically essential. However, we could not find evidence suggesting that the RNA editing of ndhD-1 is regulated via CRR4. Although the initiation codon of *ndhD* is not edited in the roots, where the NDH complex does not accumulate, *CRR4* is transcribed in the roots, suggesting that the *CRR4* expression is not sufficient for the RNA editing (24). It is also unclear whether the tissue-specific regulation of RNA editing efficiency is required physiologically.

In the mitochondrial *ccb206* gene in *Arabidopsis* the C24 site is partially edited (75). The efficiency of the editing depends on the ecotype. The partial editing does not affect the protein sequence, since the editing is silent. An interesting approach to determining the factor involved in editing efficiency is to map the loci that influence editing efficiency. However, it was puzzling that the lower editing efficiency observed in the Landsberg *erecta* ecotype is dominant over the higher efficiency in the Columbia ecotype (75). The lower efficiency in RNA editing cannot be simply explained by partial loss of function in the Landsberg *erecta* ecotype. Although two quantitative trait loci (QTL) affecting editing efficiency were identified, the genes responsible for the differences in efficiency were not determined. The efficiency of RNA editing may be affected by multiple factors in partial editing sites. It is possible that these factors include a general process such as transcription, translation, and RNA stabilization in mitochondria and that the process is not necessarily directly related to the RNA editing. The editing efficiency of this site may be developmentally regulated (75). Because this RNA editing is silent, it is unlikely that this regulation has any physiological function. The efficiency may be affected secondarily by the developmental status, instead of being precisely regulated.

The efficiency of some RNA editing events is low, possibly since high efficiency is not a requisite but rather a result of regulation. The partial editing may be a result of the low affinity of the *trans*-factor to a *cis*-element. It is also possible that CRR4 has low affinity for another factor required for the editing reaction, such as an editing enzyme. Since high editing efficiency is not essential in the plastid ndhD-1 and C24 of mitochondrial *ccb206*, evolution may have permitted this characteristic of *trans*-factors.

4.8 CO-EVOLUTION OF *TRANS*-FACTORS AND EDITING SITES

In contrast to the mammalian RNA editing sites that are conserved among species, the RNA editing sites in plant organelles are phylogenetically dynamic. Even closely related species exhibit different patterns of RNA editing sites (51, 62). Spinach- and maize-specific editing sites introduced into tobacco chloroplasts by plastid transformation remain unedited in heterologous systems (76, 77). These results suggest that the editing sites and *trans*-factors form a pair and are co-evolving rapidly. A similar result was also achieved with an *in vitro* editing system in plastids (78). In the tobacco *in vitro* system, a 56-kD protein is involved in the recognition of the *psbE* editing site, but the pea chloroplast extract lacks this protein and, as a possible consequence, the site is not edited in the pea system. Taken together with the *crr4* and *crr21* phenotypes specifically observed in the ndhD-1 and ndhD-2 sites, respectively (24, 57), these results imply that each *trans*-factor recognizes a single or a few *cis*-elements in plastids and that the nuclear genome does not encode *trans*-factors for noncognate editing sites.

An example of a heterologous RNA editing event was found at the *ndhA*-189 site in tobacco (79). Although tobacco (*Nicotiana tabacum*) does not contain this editing site, it is edited in tobacco plastids. At first, it was explained as a result of allopolyploidization, since a parental species, *N. tomentosiformis*, edits this site. The plastid genome of *N. tabacum* originated from another parental species, *N. sylvestris*, in which the RNA editing site is not conserved. Surprisingly, however, the site was edited even in plastids of *N. sylvestris* (80). The RNA editing site may have existed in *N. sylvestris* and been lost during evolution. *N. tabacum* may still conserve the *trans*-factor required for this RNA editing, although it is not essential. One possible explanation for this mysterious occurrence of the *trans*-factor in *N. sylvestris* is that this factor is required for other function. However, overexpression of RNA containing the *ndhA*-189 site does not alter other editing patterns, implying that the *cis*-element of *ndhA*-189 does not compete with other *cis*-elements for recruiting the *trans*-factor (79). It is still possible that the *trans*-factor has a function other than in RNA editing. Some PPR proteins have been shown to have multiple targets and functions (30, 31, 81).

Unlike in plastids, where the relationship between an editing site and its corresponding *trans*-factors is relatively simple, in mitochondria the recognition of noncognate editing sites is observed more frequently, implying differences between the two organelles in terms of the system used to manage the editing sites. The mitochondrial *cox2* gene contains 17 and 12 editing sites in wheat and potato, respectively, and eight are common to the two species. Even in heterologous mitochondria, some of the wheat-specific editing sites are recognized in potato (82). In contrast, despite the complete identity in flanking sequences with those of potato genes, the remaining wheat-specific sites were not edited in potato. A possible explanation is that plants are in the evolutionary process of losing the *trans*-acting factors for editing sites lost during evolution. The story may be plausible for the example of the plastid *ndhA*-189 site in *Nicotiana* (80), since *N. silvestris* and *N. tomentosiformis* are closely related. However, wheat and potato were divided during the early evolution of the monocot and dicot lineages. Why do plants conserve *trans*-factors that may be dispensable during a long period of evolution? We come back again to the idea that a *trans*-factor is involved in multiple RNA editing events and/or other RNA maturation processes. To advance the discussion to the next step, it is essential to identify a *trans*-factor involved in RNA editing in plant mitochondria. We hypothesize that a PPR protein has strict specificity in recognition of *cis*-elements in plastids. If PPR protein were a *trans*-factor also in mitochondria, how would the protein acquire this flexibility of site recognition? We still do not have solid evidence to exclude the involvement of RNA factor (guide RNA), as occurs in the *Trypanosome* system, in site recognition. This idea is convenient for explaining some of the characteristics of RNA editing in plant mitochondria.

4.9 WHAT IS AN EDITING ENZYME?

The most fundamental question remaining about RNA editing in plant organelles involves the enzyme catalyzing the editing reaction. Unfortunately, our limited information on this topic currently precludes any extensive discussion. In mitochondria, C-to-U conversion is not accompanied by an exchange of the sugar phosphate backbone (65) but by modification of a nucleotide (83, 84). Although there is no direct experimental evidence, the same story is believed to occur in plastid RNA editing. There are two possible enzymes catalyzing this reaction: C deaminase and transaminase (Figure 4.4). The C deaminase apobec-1 has been well characterized in the mammalian C-to-U conversion system (85), and an enzyme with similar activity may be involved in RNA editing in plant organelles. The possible involvement of a prokaryotic-type C deaminase that shares a domain with mammalian apobec-1 has been addressed by a reverse genetic approach (86). Including our extensive survey of the candidate genes in *Arabidopsis* (unpublished), the results have so far been negative for this hypothesis. The plant editing enzyme may be divergent in sequence from apobec-1, even though it catalyzes C deamination. In contrast to C deaminase, transaminase requires an amino acceptor, such as α-ketoglutarate or oxaloacetate (Figure 4.4). However, any acceptor has not been reported to be essential in the *in vitro* editing system. Furthermore, although the transamination reaction would most likely need cofactors with high molecular energy, *in vitro* RNA editing in plant mitochondria does not require added energy (87).

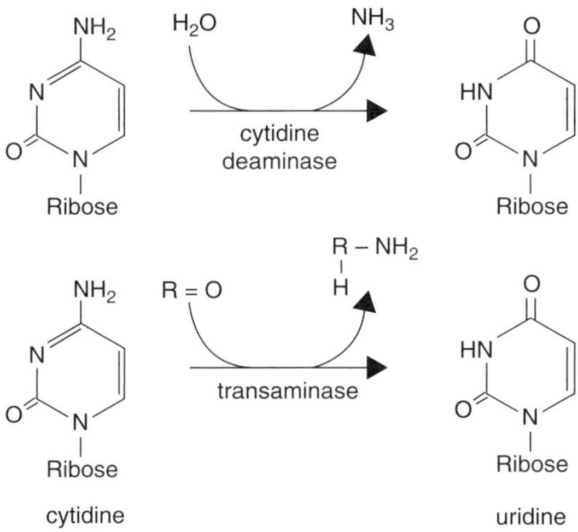

Figure 4.4 Two possible reactions occurring in C-to-U conversion in plant RNA editing.

In vitro editing reactions with plant mitochondrial extracts are not inhibited by zinc chelators, which inhibit C deaminases, including apobec-1 (21). This result does not suggest that C deaminase is the enzyme involved in RNA editing in mitochondria. However, the editing is sensitive to a zinc chelator in the *Arabidopsis* plastid extract (88). The simplest explanation is that the editing enzymes differ between plastids and mitochondria. It is also possible that multiple types of enzymes are responsible for the reactions in mitochondria. The *in vitro* editing systems have contributed to the provision of additional information on editing reactions in plants. In pea mitochondria, editing requires NTP or dNTP (21). Similar results were also reported in *Arabidopsis* chloroplasts (88). On the basis of this requirement of NTP or dNTP, RNA helicase is hypothesized to facilitate the access of editing machinery to the target site by modifying the structure of template RNA. However, it was recently shown that NTP or dNTP is required for the release of glutamate dehydrogenase (GDH) from the target RNA (87). GDH unspecifically binds RNA and inhibits the RNA editing reaction *in vitro*. Unfortunately, available information is still preliminary and it is difficult to conclude something on the editing machinery.

Another topic often referred to in discussions of the character of the editing enzyme in plants is the presence of the reverse reaction of U-to-C, especially in hornwort and a fern (58, 59). The prokaryotic-type C deaminase probably cannot mediate this reverse reaction U-to-C (65). At least in hornwort and the fern, C deaminase is unlikely to catalyze the editing reaction. It is also possible that C-to-U and U-to-C reactions are mediated by different enzymes. Although exceptions have been reported in the mitochondria of seed plants (64), it is likely that RNA editing exclusively involves C-to-U conversion in most seed plants (49). Thus, it is still possible that the editing enzyme is a C deaminase in seed plants. If this were true, then plant RNA editing would have multiple origins despite some similarity in characteristics of different RNA editing systems (56). Although no editing enzymes have been identified in plants, the editing machinery—or at least some of the editing machinery in the plastids of seed plants—contains PPR protein. Is the same story true in the system that shows the most divergence from plastid editing—that is, the system in hornwort and the fern—and especially in the case of reverse editing?

4.10 A MODEL OF EDITING MACHINERY IN PLASTIDS

In plastids a PPR protein is a *trans*-factor essential for RNA editing (24, 47). The PPR family is specific to eukaryotes and is extraordinarily large in higher plants (466 members in *Arabidopsis*) (28, 43). The members generally localize to plastids or mitochondria and are involved in RNA maturation processes (Table 4.1). The PPR family is further classified into the P subfamily and the PLS subfamily (28). The members of the P subfamily consist of a PPR motif that is conserved in length. In contrast, members of the PLS subfamily contain motifs related to a PPR motif, the PPR-like S (for short) and the PPR-like L (for long), which show variation in both length and sequence. Members of the P subfamily usually consist of a tandem array of PPR motifs. In addition to N-terminal PPR motifs, members of the PLS subfamily contain additional C-terminal motifs (28). On the basis of the presence or absence of these C-terminal motifs, the PLS subfamily is further divided into the PLS, E, E+, and DYW subgroups. The DYW subgroup with the well-conserved DYW motif is specifically conserved in land plants in which RNA editing take place, implying that this subgroup may be related to RNA editing (28). However, both CRR4 and CRR21 are members of the E+ subgroup without the DYW motif (24, 57), and a member of the DYW subgroup, CRR2, is involved in intergenic RNA cleavage in plastids rather than in RNA editing (26) (Table 4.1).

Comparison of CRR2 with CRR4 and CRR21 indicated that almost all the conserved motifs in CRR4 and CRR21 are present in CRR2 (Figure 4.2). While CRR4 and CRR21 are involved in RNA editing, CRR2 is essential for the intergenic RNA cleavage between *rps7* and *ndhB* (26). CRR2 is a member of the DYW subgroup, the members of which have long conserved C-terminal domains. Although the 15 amino acids present in CRR4 and CRR21 are highly conserved in some members of the E+ subgroup, the sequence is related to the E+ motif, which is also present in CRR2. Furthermore, the 15 amino acids are too short to explain the enzyme activity modifying C to U. Thus, the PPR proteins CRR4 and CRR21 are unlikely to have enzymatic activity in RNA editing. We propose a multi-subunit model of the RNA editing machinery in which a PPR protein functions as an RNA recognition factor (Figure 4.5). This model is similar to the RNA editing machinery of *apoB* mRNA in mammals, with the exception that the site-specific factor, a PPR protein, is part of an extraordinarily large family. This may explain the drastic difference in the frequency of RNA editing sites between two systems.

Although the domain structures are conserved between CRR4 and CRR21, the N-terminal PPR motifs are divergent in the sequence; this is consistent with their probable function in recognizing specific RNA sequences (57). In contrast,

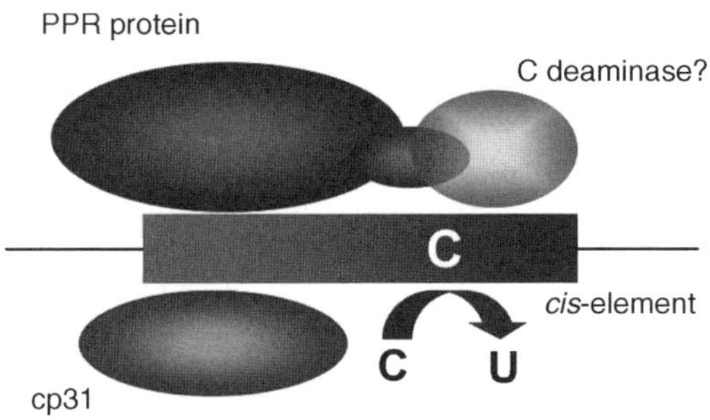

Figure 4.5 Current model of RNA editing in plastids. The PPR protein is a *trans*-factor that binds to a *cis*-element. The C-terminal region of the PPR protein has a common function in the RNA editing machinery, such as binding to an editing enzyme. The identity of the editing enzyme (C deaminase?) is still not clear. A nonspecific RNA binding protein, cp31, is probably a common factor required for RNA editing in plastids. (*See color insert.*)

the C-terminal region including the E and E+ motifs is well-conserved in the two PPR proteins. Although the C-terminal region is not required for binding to the target RNA in CRR4, it is essential for RNA editing *in vivo* (57). Furthermore, the C-terminal domains are exchangeable between CRR4 and CRR21. All these results suggest that the N-terminal region consisting of a tandem array of PPR motifs is required for the sequence-specific binding to the target RNA, whereas the well-conserved C-terminal region is essential for the common function between the two PPR proteins involved in distinct RNA editing events. As discussed above, the C-terminal region is unlikely to possess C deaminase activity. Via this region the PPR protein may interact with C deaminase or with an unknown another factor required for RNA editing.

In tobacco, the chloroplast RNA-binding protein cp31 was discovered by UVcrosslinking in the *in vitro* system, and it is a factor required for RNA editing in *ndhB* and *petL* (15). Addition of antibody against cp31 into the *in vitro* system inhibits RNA editing at both sites. cp31 is an abundant stromal protein containing two consensus-type RNA-binding domains (CS-RBD) and an N-terminal acidic domain (AD) (89). Although the exact function of cp31 in the editing reaction is unclear, it is likely one of the general components of the RNA editing machinery in plastids.

A single distinct protein, p56 or p70, is also crosslinked by UV to the *cis*-element of *psbE* or *petB*, respectively in the tobacco *in vitro* system (90). For the following reasons, these proteins also may be members of the PPR family: (1) Their molecular masses (56 kD for *psbE* and 70 kD for *petB*) are roughly similar to those of mature CRR4 (63 kD) and CRR21 (86 kD), as predicted by a computer program (24, 57). From the number of PPR motifs in the protein, the sizes of the PPR proteins are variable in this range. (2) CRR4 can bind both pre- and post-edited RNA (47). This is consistent with the results indicating that a *trans*-factor recognizes the target RNA independently of its editing status (66–68). A PPR protein may remain bound to the target RNA even after the editing reaction. (3) CRR4 has a slight preference of pre-edited RNA over post-edited RNA, at least *in vitro* (47). This result suggests that the target C residue also weakly participates in binding CRR4. This is consistent with the fact that p56 and p70 are also crosslinked by the editing sites, as well as by the *cis*-elements (90). From these results, we propose a model of RNA editing in plastids (Figure 4.5). A PPR protein interacts with the target RNA, probably independently of the editing status of the target RNA. A putative editing enzyme interacts with the PPR protein or an RNA–protein complex including the PPR protein. The C-terminal region of PPR proteins is probably required for this interaction, directly or indirectly, via unknown factors.

4.11 FUTURE DIRECTIONS

RNA editing is a process that challenges the central dogma and has been studied extensively since its discovery in plant organelles as well as in animal cells. Even though its physiological significance is obscure in plants (13), the topic is of great interest in terms of molecular evolution following the endosymbiosis of plastids and mitochondria. Components of the editing machinery have been identified (24, 47, 57), but the reaction core is still not clear. The priority should be given to clarify the editing enzyme. The next important task is to determine how the machinery is common among different systems. In the near future it may be possible to discuss the evolution of the editing machinery by comparing genome information. This may provide an answer to the fundamental question of why plants do not correct their genomic information and thus stop editing their RNA.

ACKNOWLEDGMENTS

TS was supported by a grant-in-aid for Scientific Research on Priority Areas (16085296) and for Creative Scientific Research (17GS0316) from the Ministry of Education, Culture, Sports, Science, and Technology, Japan. JO was supported by grants from the Ministry of Education, Culture, Sports, Science, and Technology, Japan (17026014, 18017014), from DAIKO Foundation (No.9106), and from NOVARTIS Foundation (Japan) for the Promotion of Science (No.18-103).

REFERENCES

1. Brennicke, A., Marchfelder, A., and Binder, S. (1999) RNA editing. *FEMS Microbiol Rev* **23**, 297–316.
2. Gott, J. M., and Emeson, R. B. (2000) Functions and mechanisms of RNA editing. *Annu Rev Genet* **34**, 499–531.
3. Stuart, K. D., Schnaufer, A., Ernst, N. L., and Panigrahi, A. K. (2005) Complex management: RNA editing in trypanosomes. *Trends Biochem Sci* **30**, 97–105.
4. Bass, B. L. (2002) RNA editing by adenosine deaminases that act on RNA. *Annu Rev Biochem* **71**, 817–846.

5. Nishikura, K. (2006) Editor meets silencer: Crosstalk between RNA editing and RNA interference. *Nature Rev Mol Cell Biol* **7**, 919–931.
6. Chen, S. H., Habib, G., Yang, C. Y., Gu, Z. W., Lee, B. R., Weng, S. A., Silberman, S. R., Cai, S. J., Deslypere, J. P., Rosseneu, M., Gotto, A. M. Jr., Li, E. H., and Chan, L. (1987) Apolipoprotein B-48 is the product of a messenger RNA with an organ-specific in-frame stop codon. *Science* **238**, 363–366.
7. Powell, L. M., Wallis, S. C., Pease, R. J., Edwards, Y. H., Knott, T. J., and Scott, J. (1987) A novel form of tissue-specific RNA processing produces apolipoprotein-B48 in intestine. *Cell* **50**, 831–840.
8. Covello, P. S., and Gray, M. W. (1989) RNA editing in plant mitochondria. *Nature* **341**, 662–666.
9. Gualberto, J. M., Lamattina, L., Bonnard, G., Weil, J. -H., and Grienenberger, J. -M. (1989) RNA editing in wheat mitochondria results in the conservation of protein sequences. *Nature* **341**, 660–662.
10. Hiesel, R., Wissinger, B., Schuster, W., and Brennicke, A. (1989) RNA editing in plant mitochondria. *Science* **246**, 1632–1634.
11. Hoch, B., Maier, R. M., Appel, K., Igloi, G. L., and Kössel, H. (1991) Editing of a chloroplast mRNA by creation of an initiation codon. *Nature* **353**, 178–180.
12. Bock, R. (2000) Sense from nonsense: How the genetic information of chloroplasts is altered by RNA editing. *Biochimie* **82**, 549–557.
13. Shikanai, T. (2006) RNA editing in plant organelles: Machinery, physiological function and evolution. *Cell Mol Life Sci* **63**, 698–708.
14. Svab, Z., and Maliga, P. (1993) High-frequency plastid transformation in tobacco by selection for a chimeric *aadA* gene. *Proc Nat Acad Sci USA* **90**, 913–917.
15. Hirose, T., and Sugiura, M. (2001) Involvement of a site-specific *trans*-acting factor and a common RNA-binding protein in the editing of chloroplast mRNAs: Development of a chloroplast *in vitro* RNA editing system. *EMBO J* **20**, 1144–1152.
16. Chaudhuri, S., Carrer, H., and Maliga, P. (1995) Site-specific factor involved in the editing of the *psbL* mRNA in tobacco plastids. *EMBO J* **14**, 2951–2957.
17. Chaudhuri, S., and Maliga, P. (1996) Sequences directing C to U editing of the plastid *psbL* mRNA are located within a 22 nucleotide segment spanning the editing site. *EMBO J* **15**, 5958–5964.
18. Farré, J. -C., and Araya, A. (2001) Gene expression in isolated plant mitochondria: high fidelity of transcription, splicing and editing of a transgene product in electroporated organelles. *Nucleic Acids Res* **29**, 2484–2491.
19. Farré, J. -C., Leon, G., Jordana, X., and Araya, A. (2001) *Cis* recognition elements in plant mitochondrion RNA editing. *Mol Cell Biol* **21**, 6731–6737.
20. Araya, A., Domec, C., Begu, D., and Litvak, S. (1992) An *in vitro* system for the editing of ATP synthase subunit 9 mRNA using wheat mitochondrial extracts. *Proc Nat Acad Sci USA* **89**, 1040–1044.
21. Takenaka, M., and Brennicke, A. (2003) *In vitro* RNA editing in pea mitochondria requires NTP or dNTP, suggesting involvement of an RNA helicase. *J Biol Chem* **278**, 47526–47533.
22. Takenaka, M., Neuwirt, J., and Brennicke, A. (2004) Complex *cis*-elements determine an RNA editing site in pea mitochondria. *Nucleic Acids Res* **32**, 4137–4144.
23. Bock, R., Hermann, M., and Kössel, H. (1996) *In vivo* dissection of *cis*-acting determinants for plastid RNA editing. *EMBO J* **15**, 5052–5059.
24. Kotera, E., Tasaka, M., and Shikanai, T. (2005) A pentatricopetide repeat protein is essential for RNA editing in chloroplasts. *Nature* **433**, 326–330.
25. Shikanai, T. (2007) Cyclic electron transport around photosystem I: genetic approaches. *Annu Rev Plant Biol* **58**, 199–217.
26. Hashimoto, M., Endo, T., Peltier, G., Tasaka, M., and Shikanai, T. (2003) A nucleus-encoded factor, CRR2, is essential for the expression of chloroplast *ndhB* in *Arabidopsis*. *Plant J* **36**, 541–549.
27. Shikanai, T., Endo, T., Hashimoto, T., Yamada, Y., Asada, K., and Yokota, A. (1998) Directed disruption of the tobacco *ndhB* gene impairs cyclic electron flow around photosystem I. *Proc Nat Acad Sci USA* **95**, 9705–9709.
28. Lurin, C., Andrés, C., Aubourg, S., Bellaoui, M., Bitton, F., Bruyere, C., Caboche, M., Debast, C., Gualberto, J., Hoffmann, B., Lecharny, A, Le Ret, M., Martin-Magniette, M. -L., Mireau, H., Peeters, N., Renou, J. -P., Szurek, B., Taconnat, L., and Small, I. (2004) Genome-wide analysis of Arabidopsis pentatricopeptide repeat proteins reveals their essential role in organelle biogenesis. *Plant Cell* **16**, 2089–2103.

29. Fisk, D. G., Walker, M. B., and Barkan, A. (1999) Molecular cloning of the maize gene *crp1* reveals similarity between regulators of mitochondrial and chloroplast gene expression. *EMBO J* **18**, 2621–2630.
30. Meierhoff, K., Felder, S., Nakamura, T., Bechtold, N., and Schuster, G. (2003) HCF152, an Arabidopsis RNA binding pentatricopeptide repeat protein involved in the processing of chloroplast *psbB-psbT-psbH-petB-petD* RNAs. *Plant Cell* **15**, 1480–1495.
31. Yamazaki, H., Tasaka, M., and Shikanai, T. (2004) PPR motifs of the nucleus-encoded factor, PGR3, function in the selective and distinct steps of chloroplast gene expression in *Arabidopsis*. *Plant J* **38**, 152–163.
32. Schmitz-Linneweber, C., Williams-Carrier, R. E., Williams-Voelker, P. M., Kroeger, T. S., Vichas, A., and Barkan, A. (2006) A pentatricopeptide repeat protein facilitates the *trans*-splicing of the maize chloroplast *rps12* pre-mRNA. *Plant Cell* **18**, 2650–2663.
33. Hattori, M., Miyake, H., and Sugita, M. (2007) A pentatricopeptide repeat protein is required for RNA processing of *clpP* pre-mRNA in moss chloroplasts. *J Biol Chem* **282**, 10773–10782.
34. Gutierréz-Marcos, J. F., Dal Pra, M., Giulini, A., Costa, L. M., Gavazzi, G., Cordelier, S., Sellam, O., Tatout, C., Paul, W., Perez, P., Dickinson, H. G., and Consonni, G. (2007) *empty pericarp4* encodes a mitochondrion-targeted pentatricopeptide repeat protein necessary for seed development and plant growth in maize. *Plant Cell* **19**, 196–210.
35. Bentolila, S., Alfonso, A. A., and Hanson, M. R. (2002) A pentatricopeptide repeat-containing gene restores fertility to cytoplasmic male-sterile plants. *Proc Nat Acad Sci USA* **99**, 10887–10892.
36. Desloire, S., Gherbi, H., Laloui, W., Marhadour, S., Clouet, V., Cattolico, L. Falentin, C., Giancola, S., Renard, M., Budar, F., Small, I., Caboche, M., Delourme, R., and Bendahmane, A. (2003) Identification of the fertility restoration locus, *Rfo*, in radish, as a member of the pentatricopeptide-repeat protein family. *EMBO Report* **4**, 588–594.
37. Koizuka, N., Imai, R., Fujimoto, H., Hayakawa, T., Kimura, Y., Kohno-Murase, J., Sakai, T., Kawasaki, S., and Imamura, J. (2003) Genetic characterization of a pentatricopeptide repeat protein gene, *orf687*, that restores fertility in the cytoplasmic male-sterile Kosena radish. *Plant J* **34**, 407–415.
38. Kazama, T., and Toriyama, K. (2003) A pentatricopeptide repeat-containing gene that promotes the processing of aberrant *atp6* RNA of cytoplasmic male-sterile rice. *FEBS Lett* **544**, 99–102.
39. Brown, G. G., Formanová, N., Jin, H., Wargachuk, R., Dendy, C., Patil, P., Laforest, M., Zhang, J., Cheung, W. Y., and Landry, B. S. (2003) The radish *Rfo* restorer gene of Ogura cytoplasmic male sterility encodes a protein with multiple pentatricopeptide repeats. *Plant J* **35**, 262–272.
40. Komori, T., Ohta, S., Murai, N., Takakura, Y., Kuraya, Y., Suzuki, S., Hiei, Y., Imaseki, H., and Nitta, N. (2004) Map-based cloning of a fertility restorer gene, *Rf-1*, in rice (*Oryza sativa* L.). *Plant J* **37**, 315–325.
41. Akagi, H., Nakamura, A., Yokozeki-Misono, Y., Inagaki, A., Takahashi, H., Mori, K., and Fujimura, T. (2004) Positional cloning of the rice *Rf-1* gene, a restorer of BT-type cytoplasmic male sterility that encodes a mitochondria-targeting PPR protein. *Theor Appl Genet* **108**, 1449–1457.
42. Wang, Z., Zou, Y., Li, X., Zhang, Q., Chen, L., Wu, H., Su, D., Chen, Y., Guo, J., Luo, D., Long, Y., Zhong, Y., and Liu, Y. G. (2006) Cytoplasmic male sterility of rice with boro II cytoplasm is caused by a cytotoxic peptide and is restored by two related PPR motif genes via distinct modes of mRNA silencing. *Plant Cell* **18**, 676–687.
43. Rivals, E., Bruyère, C., Toffano-Nioche, C., and Lecharny, A. (2006) Formation of the Arabidopsis pentatricopeptide repeat family. *Plant Physiol* **141**, 825–839.
44. Small, I. D., and Peeters, N. (2000) The PPR motif—ATPR-related motif prevalent in plant organellar proteins. *Trends Biochem Sci* **25**, 46–47.
45. Nakamura, T., Meierhoff, K., Westhoff, P., and Schuster, G. (2003) RNA-binding properties of HCF152, an *Arabidopsis* PPR protein involved in the processing of chloroplast RNA. *Eur J Biochem* **270**, 4070–4081.
46. Schmitz-Linneweber, C., Williams-Carrier, R., and Barkan, A. (2005) RNA immunoprecipitation and microarray analysis show a chloroplast pentatricopeptide repeat protein to be associated with the 5′ region of mRNAs whose translation it activates. *Plant Cell* **17**, 2791–2804.
47. Okuda, K., Nakamura, T., Sugita, M., Shimizu, T., and Shikanai, T. (2006) A pentatricopeptide repeat protein is a site-recognition factor in chloroplast RNA editing. *J Biol Chem* **281**, 37661–37667.

48. Freyer, R., Kiefer-Meyer, M. -C., and Kössel, H. (1997) Occurrence of plastid RNA editing in all major lineages of land plants. *Proc Nat Acad Sci USA* **94**, 6285–6290.
49. Giegé, P., and Brennicke, A. (1999) RNA editing in *Arabidopsis* mitochondria effects 441 C to U changes in ORFs. *Proc Nat Acad Sci USA* **96**, 15324–15329.
50. Tillich, M., Funk, H. T., Schmitz-Linneweber, C., Poltnigg, P., Sabater, B., Martin, M., and Maier, R. M. (2005) Editing of plastid RNA in *Arabidopsis thaliana* ecotypes. *Plant J* **43**, 708–715.
51. Tsudzuki, T., Wakasugi, T., and Sugiura, M. (2001) Comparative analysis of RNA editing sites in higher plant chloroplasts. *J Mol Evol* **53**, 327–332.
52. Kahlau, S., Aspinall, S., Gray, J. C., and Bock, R. (2006) Sequence of the tomato chloroplast DNA end evolutionary comparison of solanaceous plastid genome. *J Mol Evol* **63**, 194–207.
53. Sasaki, T., Yukawa, Y., Miyamoto, T., Obokata, J., and Sugiura, M. (2003) Identification of RNA editing sites in chloroplast transcripts from the maternal and paternal progenitors of tobacco (*Nicotiana tabacum*): Comparative analysis shows the involvement of distinct *trans*-factors for *ndhB* editing. *Mol Biol Evol* **20**, 1028–1035.
54. Sasaki, T., Yukawa, Y., Wakasugi, T., Yamada, K., and Sugiura, M. (2006) A simple *in vitro* RNA editing assay for chloroplast transcripts using fluorescent dideoxynucleotides: Distinct types of sequence elements required for editing of *ndh* transcripts. *Plant J* **47**, 802–810.
55. Steinhauser, S., Beckert, S., Capesius, I., Malek, O., and Knoop, V. (1999) Plant mitochondrial RNA editing. *J Mol Evolution* **48**, 303–312.
56. Tillich, M., Lewahrk, P., Morton, B. R., and Maier, U. G. (2006) The evolution of chloroplast RNA editing. *Mol Biol Evol* **23**, 1912–1921.
57. Okuda, K., Myouga, M., Motohashi, R., Shinozaki, K., and Shikanai, T. (2007) Conserved domain structure of pentatricopeptide repeat proteins involved in chloroplast RNA editing. *Proc Nat Acad Sci USA* **104**, 8178–8183.
58. Kugita, M., Yamamoto, Y., Fujikawa, T., Matsumoto, T., and Yoshinaga, K. (2003) RNA editing in hornwort chloroplasts makes more than half the genes functional. *Nucleic Acids Res* **31**, 2417–2423.
59. Wolf, P. G., Rowe, C. A., and Hasebe, M. (2004) High levels of RNA editing in a vascular plant chloroplast genome: Analysis of transcripts from the fern *Adiantum capillus-veneris*. *Gene* **339**, 89–97.
60. Groth-Malonek, M., Pruchner, D., Grewe, F., and Knoop, V. (2005) Ancestors of *trans*-splicing mitochondrial introns support serial sister group relationships of hornworts and mosses with vascular plants. *Mol Biol Evol* **22**, 117–125.
61. Duff, R. J. (2006) Divergent RNA editing frequencies in hornwort mitochondrial *nad5* sequences. *Gene* **366**, 285–291.
62. Notsu, Y., Masood, S., Nishikawa, T., Kubo, N., Akiduki, G., Nakazono, M., Hirai, A., and Kadowaki, K. (2002) The complete sequence of the rice (*Oryza sativa* L.) mitochondrial genome: Frequent DNA sequence acquisition and loss during the evolution of flowering plants. *Mol Genet Genomics* **268**, 434–445.
63. Handa, H. (2003) The complete nucleotide sequence and RNA editing content of the mitochondrial genome of rapeseed (*Brassica napus* L.): Comparative analysis of the mitochondrial genomes of rapeseed and *Arabidopsis thaliana*. *Nucleic Acids Res* **31**, 5907–5916.
64. Schuster, W., Hiesel, R., Wissinger, B., and Brennicke, A. (1990) RNA editing in the cytochrome *b* locus of the higher plant *Oenothera berteriana* includes a U-to-C transition. *Mol Cell Biol* **10**, 2428–2431.
65. Rajasekhar, V. K., and Mulligan, R. M. (1993) RNA editing in plant mitochondria: α-phosphate is retained during C-to-U conversion in mRNAs. *Plant Cell* **5**, 1843–1852.
66. Bock, R., Hermann, M., and Fuchs, M. (1997) Identification of critical nucleotide positions for plastid RNA editing site recognition. *RNA* **3**, 1194–1200.
67. Reed, M. L., Lyi, S. M., and Hanson, M. R. (2001) Edited transcripts compete with unedited mRNAs for *trans*-acting editing factors in higher plant chloroplasts. *Gene* **272**, 165–171.
68. Verbitskiy, D., Takenaka, M., Neuwirt, J., van der Merwe, J. A., and Brennicke, A. (2006) Partially edited RNAs are intermediates of RNA editing in plant mitochondria. *Plant J* **47**, 408–416.
69. van der Merwe, J. A., Takenaka, M., Neuwirt, J., Verbitskiy, D., and Brennicke, A. (2006) RNA editing sites in plant mitochondria can share *cis*-elements. *FEBS Lett* **580**, 268–272.

70. Chateigner-Boutin, A. -L., and Hanson, M. R. (2002) Cross-competition in transgenic chloroplasts expressing single editing sites reveals shared *cis* elements. *Mol Cell Biol* **22**, 8448–8456.
71. Chateigner-Boutin, A. -L., and Hanson, M. R. (2003) Developmental co-variation of RNA editing extent of plastid editing sites exhibiting similar *cis*-elements. *Nucleic Acids Res* **31**, 2586–2594.
72. Hirose, T., and Sugiura, M. (1997) Both RNA editing and RNA cleavage are required for translation of tobacco chloroplast *ndhD* mRNA: A possible regulatory mechanism for the expression of a chloroplast operon consisting of functionally unrelated genes. *EMBO J* **16**, 6804–6811.
73. Shikanai, T. (2007) The NAD(P)H dehydrogenase complex in photosynthetic organisms: Subunit composition and physiological function. *Funct Plant Sci Biotechnol* **1**, 129–137.
74. Nakagawara, E., Sakuraba, Y., Yamasato, A., Tanaka, R., and Tanaka, A. (2007) Clp protease controls chlorophyll *b* synthesis by regulating the level of chlorophyllide *a* oxygenase. *Plant J* **49**, 800–809.
75. Bentolila, S., Chateigner-Boutin A. -L., and Hanson, M. R. (2006) Ecotype allelic variation in C-to-U editing extent of a mitochondrial transcript identifies RNA-editing quantitative trait loci in Arabidopsis. *Plant Physiol* **139**, 2006–2016.
76. Bock, R., Kössel, H., and Maliga, P. (1994) Introduction of a heterologous editing site into the tobacco plastid genome: The lack of RNA editing leads to a mutant phenotype. *EMBO J* **13**, 4623–4628.
77. Reed, M. L., and Hanson, M. R. (1997) A heterologous maize *rpoB* editing site is recognized by transgenic tobacco chloroplasts. *Mol Cell Biol* **17**, 6948–6952.
78. Miyamoto, T., Obokata, J., and Sugiura, M. (2002) Recognition of RNA editing sites is directed by unique proteins in chloroplasts: Biochemical identification of *cis*-acting elements and *trans*-acting factors involved in RNA editing in tobacco and pea chloroplasts. *Mol Cell Biol* **22**, 6726–6734.
79. Schmitz-Linneweber, C., Tillich, M., Herrmann, R. G., and Maier, R. M. (2001) Heterologous, splicing-dependent RNA editing in chloroplasts: Allotetraploidy provides *trans*-factors. *EMBO J* **20**, 4874–4883.
80. Tillich, M., Poltnigg, P., Kushnir, S., and Schmitz-Linneweber, C. (2006) Maintenance of plastid RNA editing activities independently of their target sites. *EMBO Rep* **7**, 308–313.
81. Barkan, A., Walker, M., Nolasco, M., and Johnson, D. (1994) A nuclear mutation in maize blocks the processing and translation of several chloroplast mRNAs and provides evidence for the differential translation of alternative mRNA forms. *EMBO J* **13**, 3170–3181.
82. Choury, D., and Araya, A. (2006) RNA editing site recognition in heterologous plant mitochondria. *Curr Genet* **50**, 405–416.
83. Blanc, V., Litvak, S., and Araya, A. (1995) RNA editing in wheat mitochondria proceeds by a deamination mechanism. *FEBS Lett* **373**, 56–60.
84. Yu, W., and Schuster, W. (1995) Evidence for a site-specific cytidine deamination reaction involved in C to U RNA editing of plant mitochondria. *J Biol Chem* **270**, 18227–18233.
85. Wedekind, J. E., Dance, G. S. C., Sowden, M. P., and Smith, H. C. (2003) Messenger RNA editing in mammals: New members of the APOBEC family seeking roles in the family business. *Trends Genet* **19**, 207–216.
86. Faivre-Nitschke, S. E., Grienenberger, J. -M., and Gualberto, J. -M. (1999) A prokaryotic-type cytidine deaminase from *Arabidopsis thaliana*. Gene expression and functional characterization. *Eur J Biochem* **263**, 896–903.
87. Takenaka, M., Verbitskiy, D., van der Merwe, J. A., Zehrmann, A., Plessmann, U., Urlaub, H., and Brennicke, A. (2007) In vitro RNA editing in plant mitochondria does not require added energy. *FEBS Lett* **581**, 2743–2747.
88. Hegeman, C. E., Hayes, M. L., and Hanson, M. R. (2005) Substrate and cofactor requirements for RNA editing of chloroplast transcripts in *Arabidopsis in vitro*. *Plant J* **42**, 124–132.
89. Nakamura, T., Ohta, M., Sugiura, M., and Sugita, M. (2001) Chloroplast ribonucleoproteins function as a stabilizing factor of ribosome-free mRNAs in the stroma. *J Biol Chem* **276**, 147–152.
90. Miyamoto, T., Obokata, J., and Sugiura, M. (2004) A site-specific factor interacts directly with its cognate RNA editing site in chloroplast transcripts. *Proc Nat Acad Sci USA* **101**, 48–52.
91. Williams, P. M., and Barkan, A. (2003) A chloroplast-localized PPR protein required for plastid ribosome accumulation. *Plant J* **36**, 675–686.
92. Ding, Y. -H., Liu, N. -Y., Tang, Z. -S., Liu, J., and Yang, W. -C. (2006) *Arabidopsis glutamine-rich protein23* is essential for early embryogenesis and encodes a novel nuclear PPR motif protein that interacts with RNA polymerase II subunit III. *Plant Cell* **18**, 815–830.

93. Cushing, D. A., Forsthoefel, N. R., Gestaut, D. R., and Vernon, D. M. (2005) *Arabidopsis emb175* and other *ppr* knockout mutants reveal essential roles for pentatricopeptide repeat (PPR) proteins in plant embryogenesis. *Planta* **221**, 424–436.
94. Kobayashi, K., Suzuki, M., Tang, J., Nagata, N., Ohyama, K., Seki, H., Kiuchi, R., Kaneko, Y., Nakazawa, M., Matsui, M., Matsumoto, S., Yoshida, S., and Muranaka, T. (2007) Lovastatin Insensitive 1, a novel pentatricopeptide repeat protein, is a potential regulatory factor of isoprenoid biosynthesis in *Arabidopsis*. *Plant Cell Physiol* **48**, 322–331.
95. Koussevitzky, S., Nott, A., Mockler, T. C., Hong, F., Sachetto-Martins, G., Surpin, M., Lim, J., Mittler, R., and Chory, J. (2007) Signals from chloroplasts converge to regulate nuclear gene expression. *Science* **316**, 715–719.
96. Prasad, A. M., Sivanandan, C., Resminath, R., Thakare, D. R., and Srinivasan, S. R. B. (2005) Cloning and characterization of a pentatricopeptide protein encoding gene (LOJ) that is specifically expressed in lateral organ junctions in *Arabidopsis thaliana*. *Gene* **353**, 67–79.
97. Oguchi, T., Sage-Ono, K., Kamada, H., and Ono, M. (2004) Genomic structure of a novel *Arabidopsis* clock-controlled gene, *AtC401*, which encodes a pentatricopeptide repeat protein. *Gene* **330**, 29–37.
98. Gothandam, K. M., Kim, E. S., Cho, H., and Chung, Y. Y. (2005) OsPPR1, a pentatricopeptide repeat protein of rice is essential for the chloroplast biogenesis. *Plant Mol Biol* **58**, 421–433.
99. Manthey, G. M., and McEwen J. E. (1998) The product of the nuclear gene PET309 is required for translation of mature mRNA and stability or production of intron-containing RNAs derived from the mitochondrial COX1 locus of *Saccharomyces cerevisiae*. *EMBO J* **14**, 4031–4043.
100. Coffin, J. W., Dhillon, R., Ritzel, R. G., and Nargang, F. E. (1997) The *Neurospora crassa cya-5* nuclear gene encodes a protein with a region of homology to the *Saccharomyces cerevisiae* PET309 protein and is required in a post-transcriptional step for the expression of the mitochondrially encoded COXI protein. *Curr Genet* **32**, 273–280.

PART II

FUNCTIONAL COORDINATION OF RNA EDITING WITH OTHER CELLULAR MECHANISMS

CHAPTER 5

TRANSFER RNA EDITING ENZYMES; AT THE CROSSROADS OF AFFINITY AND SPECIFICITY

Juan D. Alfonzo
F. Nina Papavasiliou

In all organisms, transfer RNAs (tRNAs) play a central role in hardwiring the genetic information found in DNA, with the protein synthesizing machinery at the far end of the genetic information cascade. Within cells, however, tRNAs are not inert molecules and in the process of maturation may undergo a number of changes, most notably the acquisition of a large variety of chemical groups collectively known as posttranscriptional modifications. The majority of modifications are apparently necessary to alter the structure of a tRNA to ensure proper folding. This chapter will focus on what is currently known about a subclass of modifications, grouped under the umbrella term of tRNA editing, which have direct bearings on a tRNA's function.

5.1 INTRODUCTION: STRUCTURAL VERSUS FUNCTIONAL tRNA EDITING

By far, tRNA undergoes the most varied types of posttranscriptional changes that qualify as RNA editing. Mechanistically, these changes may involve nucleotide deamination, polymerization, chemical modification, and so on. It is this aspect of tRNA biosynthesis that has led to a constant rephrasing and reinterpretation of the definition of RNA editing to accommodate the ever-growing number of editing examples found in tRNA.

As originally stated (and discussed throughout this book), "RNA editing" refers to the posttranscriptional alteration of sequence information in mRNA beyond what is encoded in the DNA genome from various organisms (1). Initially, this definition was sufficient to explain the mechanism of insertion and deletion of nucleotides into the pre-mRNAs of trypanosomatid mitochondria (see Chapter 3), as well as the single

RNA and DNA Editing: Molecular Mechanisms and Their Integration into Biological Systems,
Edited by Harold C. Smith

nucleotide changes to the coding regions of mammalian mRNAs (2–9). However, soon after the discovery of mRNA editing, the report of similar nucleotide changes in noncoding RNAs (mainly tRNAs) required the use of a broader term. Gray and co-workers first reported the posttranscriptional substitution of nucleotides to a noncoding RNA (10). They found that nucleotides were added to the acceptor stem of *Acanthamoeba castellani* tRNAs as a required step in their maturation. They then expanded the definition of editing to include any alteration in the sequence of an RNA (coding or noncoding) that leads to the introduction of one of the four canonical nucleotides (11). According to Covello and Gray, such "programmed alterations" result in the generation of transcripts whose sequence could have been potentially encoded in the DNA genome. Most recently, in an attempt to differentiate between RNA editing and modification, Grosjean has adopted a more strict definition of editing whereby, regardless of the mechanism, any sequence alteration that changes the genetic meaning of a transcript is called editing, while merely structural changes are called modification (12).

Arguably, when changes in structural information do not occur stochastically, these changes may also represent a form of programmed alteration of genetic meaning; thus in this chapter, rather than trying to establish arguments for or against either definition, we will use both. This chapter will divide tRNA editing into two major groups: functional editing (using Grosjean's definition) and structural editing (using Gray's definition). To this end, this chapter will summarize what is currently known about tRNA editing changes that have direct bearings on a tRNA's function by altering its identity (i.e., expanding its decoding properties), as strictly defined by Grosjean. These include A to I, C to U, and lysidine formation at the three-anticodon nucleotides as well as nucleotide additions at the ends of tRNAs. This chapter, however, will also include a number of canonical nucleotide changes occurring posttranscriptionally that may affect tRNA structure (indirectly affecting function) and that have been described in the literature as tRNA editing by using the broader definition originally coined by Covello and Gray.

5.2 TRANSFER RNA EDITING FOR STRUCTURE

A majority of editing examples, reported to date, do not directly affect the identity or change the decoding properties of a tRNA. These events encompass a number of different mechanisms that include single nucleotide transitions and transversions that mostly occur internally in a tRNA molecule. The unifying principle in this group is that the given editing events may not be needed for proper tRNA function but may instead play a more structural role, with the proviso that "not playing a direct role" in tRNA function may be nothing more than an inference as the enzymes responsible for these types of editing have yet to be identified.

5.2.1 C-to-U Editing of the tRNA Backbone

The tRNA backbone (i.e., sequences other than the anticodon arm) generally contains important information that determines the overall structure of a tRNA molecule.

Backbone sequences also play crucial roles in providing key recognition features to proteins that intimately interact with tRNAs (most notably amino acyl tRNA synthetases and a number of modification enzymes). RNA editing by changing single nucleotides in the tRNA backbone may influence the formation of these same structural features. To date, the best-studied editing changes in the backbone of a tRNA include a number of C-to-U substitutions in tRNAs from various plant species. Marechal-Drouard and co-workers discovered a single C-to-U change at position 28 of plant mitochondrial tRNAs (13) (Figure 5.1). Curiously, this editing event occurs at a site in the anticodon stem where pseudouridine (Ψ) is found in the mature tRNA. This discovery has led to the proposal of a two-step model for the formation of Ψ, where C-to-U editing occurs first, presumably by a deamination mechanism followed by isomerization of the U to form Ψ. Another interesting aspect of plant mitochondrial tRNA editing is the observation that tRNA end-trimming and editing are sequential events (14, 15). Editing of a mismatched nucleotide at the fourth position of the acceptor stem follows 5′ trimming but precedes 3′ maturation by ribonuclease Z. In this case, editing not only restores structures that might be important for synthetase recognition but may also serve as a checkpoint for complete maturation of tRNAs (Figure 5.1).

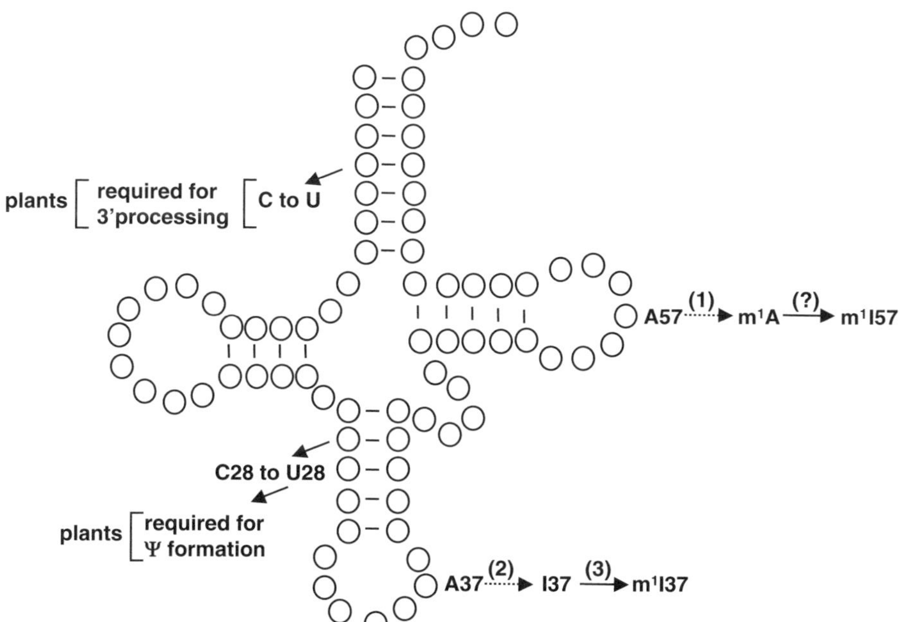

Figure 5.1 Structural editing by single nucleotide changes in the backbone of tRNAs in plants. In some instances the editing event occurs at a nucleotide position that is also the substrate for modification; in these cases the two processes are often interdependent. Solid arrows indicate the enzymatic reaction leading to the final product. Dashed arrows indicate the reaction that yields the required intermediate. (1) Reaction catalyze by the Trm I enzyme in Archaea, (2) reaction catalyzed by ADAT1p in eukaryotes, and (3) reaction catalyzed by Trm5p (?) Yet to be identified enzymatic activities.

Most recently, the Marechal-Drouard group has tested the fate of the unedited tRNA in potato mitochondria. They introduced, via direct uptake, a DNA template encoding a nonedited version of larch $tRNA^{His}$. They not only found that the potato mitochondria failed to edit this "foreign" tRNA, but also found that the unedited tRNA was quickly degraded (16). This again emphasizes the possibility that tRNA editing might play a role in ensuring quality control during processing. Evolutionarily, the idea that the potato editing machinery cannot recognize a foreign tRNA could also pose a hurdle for the horizontal transfer of tRNA genes from different plant species.

5.2.2 A-to-I Editing and Modification at Position 37 and 57 of tRNAs

Although inosine at the first position of the anticodon has been described in Eukarya and bacteria and their role in expanding the decoding capabilities of tRNA are clear, subtler is the role that inosines play at two other tRNA positions. In Eukarya a number of tRNAs undergo A-to-I editing at position 37 (Figure 5.1). This reaction is catalyzed by ADAT1p (adenosine deaminase acting on tRNA 1), which in terms of primary sequence has the conserved signature motifs found in *bona fide* adenosine deaminases (see Figure 5.9) (17). Therefore, this enzyme is closer in sequence to the ADAR family of enzymes than those enzymes that deaminate the first position of the anticodon (namely, ADATa and ADAT2/3). The biological significance of I37 is not yet clear, but it presumably plays a role in translational efficiency; however, unlike the wobble-specific deaminases, these enzymes are not essential for viability. The only other known case of inosine in tRNA outside the anticodon involves the formation of N-1 methylinosine 57 (m^1I57) in the TΨC loop of many archaeal tRNAs (18–20) (Figure 5.1).

Despite the rather disparate nature of these two seemingly unrelated modifications proceeding via a common chemistry, they do share a curious relationship. Both involve known examples of editing and modification that occur at a single site. In the case of I37 in all eukaryal $tRNA^{Ala}$, inosine is further methylated at the base to form m^1I. Grosjean and co-workers could show that deamination did not required methylation (17, 21), while in the case of m^1I57 in *Archaea*, Droogmans and co-corkers then elegantly showed a perfect precursor–product relationship that involved formation of a methylated adenosine (m^1A) first, followed by the subsequent deamination of m^1A to form the final reaction product, m^1I (19). This observation suggests an absolute requirement for methylation as a deamination determinant. This also raised the possibility that either one enzyme or alternatively a multienzyme complex could perform both reactions. To date, the enzyme responsible for m^1A formation has been identified (TRMI) and the recombinant enzyme can efficiently methylate a synthetic substrate to form m^1A (19). A robust deaminase activity has also been described in crude extracts from various Archaea. Grosjean and co-workers have suggested that anticodon-modifying enzymes are generally "architecture-dependent" (requiring a full-length tRNA for activity) whereas enzymes that act on the backbone of the tRNA are "architecture-independent" (22). Interestingly, comparing the eukaryal A37 to the archaeal m1A57 deaminase reveals a similar trend, where ADAT1 only works on a full-length substrate while the archaeal deaminase can utilize "minimalist" substrates (21).

The TΨC loop of tRNAs, like the anticodon loop, also forms a characteristic U-turn. If one was to compare the structure of the typical anticodon U-turn and that of the TΨC loop U-turn, the two are nearly indistinguishable and m^1I57 occurs at a nucleotide position equivalent to position 35 in the anticodon. Thus in thinking about substrate specificity, this raises the question as to what keeps the TΨC deaminase from editing the second position of the anticodon in Archaea. In other words, it provides for the possibility that A35 containing tRNAs in Archaea may undergo "functional editing." Evidently, it could be that the methylase is not able to deaminate position 35 of the anticodon and that the lack of it (being a prerequisite for deamination) will then act as a negative determinant for anticodon editing. It still, however, creates the problem as to what keeps the methylase from working on anticodon position 35, and thus implies differences in local structure as key contributors to editing and modification specificity.

5.3 TRANSFER RNA EDITING FOR FUNCTION

The degeneracy of the genetic code is implied in the need for 61 sense codons to specify 20 different amino acids and, with the exception of methionine and tryptophan, each amino acid is encoded by more than one codon. This discrepancy between codon and amino acid numbers was first explained by Crick's wobble hypothesis, which invoked the need for base-pairing flexibility between the first anticodon and third codon positions during decoding. Since the inception of the wobble rules, over 100 posttranscriptional modifications have been described with the largest number affecting the anticodon of tRNA. As anticodon modifications accrue, new findings lead to a constant redefinition of the wobble rules to include novel effects on tRNA function. A number of tRNA editing events permit decoding of many more codons than what is implicit in their gene sequence. In this realm, decoding changes imparted by tRNA editing provide a mechanism to effectively accommodate genetic code degeneracy. Although a number of editing events have direct effects in expanding a tRNA's decoding capacity, some editing events indirectly affect tRNA function by repairing otherwise nonfunctional tRNAs.

5.3.1 The Lysidine Story

Following the elucidation of the genetic code and proposal of the wobble hypothesis, scientists naturally proceeded to test the various assumptions suggested by the newly established code and the proposed pairing rules. Early studies reported the noticeable absence of a $tRNA^{Ile}$ that could decode the AUA isoleucine codons in *E. coli* (23). The only $tRNA^{Ile}$ sequenced at the time contained an unmodified G at position 34 (the wobble position). This tRNA could recognize both the U- and C-ending codons for isoleucine, but it could not decode the AUA codons (Figure 5.2A). However, it had been demonstrated, using an *E. coli in vitro* translation system, that indeed the codon AUA specified isoleucine (24). Nishimura and co-workers reasoned that perhaps the answer to this conundrum lied on the fact that if the AUA codons are rare, maybe they are decoded by a minor $tRNA^{Ile}$ species. This assumption led to the purification of a

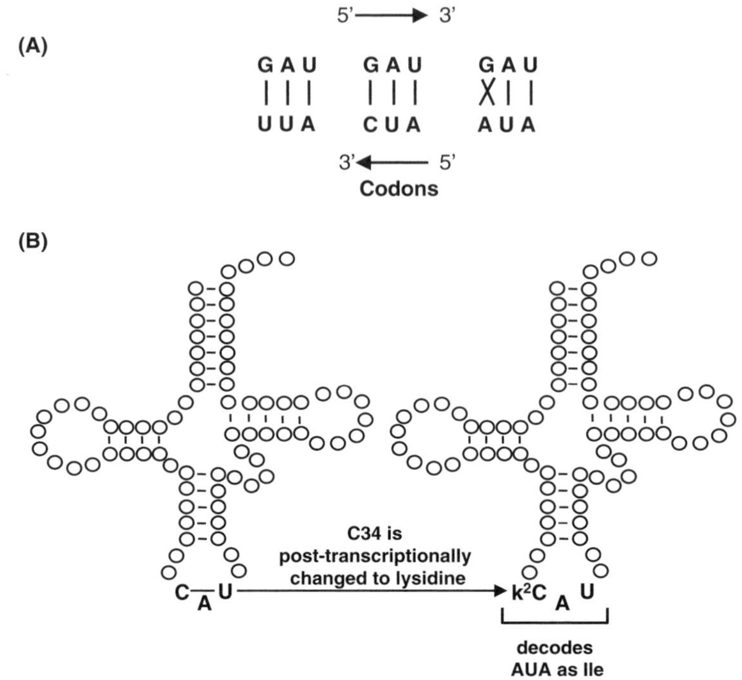

Figure 5.2 Lysidine formation at the first position of the anticodon of tRNA^{Met}. Lysidine formation changes both the decoding capacity and amino acid identity of the tRNA. (A) The three isoleucine codons of E. coli. The (X) denotes the absence of a tRNA that can decode the AUA codon. (B) C34 in tRNA^{Ile}(CAU) is changed to lysidine (k^{2}C), permitting decoding of the AUA codons for isoleucine.

minor tRNA from E. coli with an unusual modified nucleotide at position 34 (the wobble base) which they called N$^+$ (25). Although at the time these researchers demonstrated the ability of this minor species to support specific amino acylation with isoleucine, the chemical nature of the modification was not immediately elucidated.

At about the same time, sequences revealed the presence of a tRNA^{Ile} specific for the AUA codon (26). Surprisingly, the only tRNA in this genome that could recognize this codon had a CAU anticodon, which according to the coding rules should decode the canonical AUG codons for methionine (similar findings were also reported in chloroplasts). Yokoyama and co-workers, in collaboration with Nishimura, then used a combination of NMR and mass spectrometry to elucidate the chemical structure of the N$^+$ nucleotide, a cytosine derivative that could account for the apparently aberrant decoding. They found that this new nucleoside contained an unusual lysine side-chain at the C2 position of the pyrimidine ring and termed it "lysidine" (27) (Figure 5.2).

Functional studies revealed that the native lysidine-containing tRNA^{Ile}, despite the presence of a CAU anticodon, could be efficiently aminoacylated by isoleucine. In fact, this tRNA could not support charging with methionine, as its anticodon sequence might have suggested. In addition, replacement of lysidine for cytidine, in an otherwise native tRNA, led to a drastic reduction in isoleucine accepting activity and an increase in methionine accepting activity. Combined, these observations led Yokoyama and co-workers to conclude that the single change of cytidine to lysidine could lead to the simultaneous conversion of both the codon and amino acid specificity of $tRNA^{Ile}$ (28).

Although not proven at a chemical level, the presence of a lysidine-like change in the anticodon of $tRNA^{Ile}$(CAU) is suggested to occur in a number of other systems, including spinach chloroplast and bean mitochondria (29). However, it is only in potato mitochondria that the presence of a yet unidentified C34 modification also changes the identity of the $tRNA^{Ile}$(CAU) in a fashion similar to that described for the bacterial system (29). In addition, 13 different archaeal genomes contain $tRNA^{Ile}$ with a CAU anticodon, suggesting the possible use of lysidine (30). However, mass spectrometry analysis of total tRNA preparations from Archaea revealed the absence of a species with a molecular mass corresponding to that of lysidine. Therefore in these organisms, a different lysidine-like modification may exist for the reassignment of the CAU anticodon from Met to Ile.

A major breakthrough in the studies of lysidine biosynthesis comes from a recent report from Suzuki and co-workers, who, using a bioinformatics approach, identified the gene encoding lysidine synthetase in bacteria (31). By searching the Cluster of Orthologous Groups (COGs) database, they identify 48 predictably essential genes in E. coli, which at the time had no assigned function. Screening available conditional mutants for a number of these potential genes led to the identification of one locus responsible for lysidine formation (the yacA gene). In the absence of yacA expression, lysidine disappeared from $tRNA^{Ile}$ as determined by the lack of an activity for the direct incorporation of lysine into $tRNA^{Ile}$ in the conditional mutants, which was also confirmed by mass spectrometry (31–33).

The product of the yacA gene expressed recombinantly (lysidine synthetase, TilS) was by itself sufficient to synthesize lysidine in an ATP-dependent manner and only required *in vitro* transcribed $tRNA^{Ile}$ and lysine (31). These studies have led to a proposed two-step reaction mechanism by which TilS catalyzes first the activation of the C-2 position of C34 in $tRNA^{Ile}$, followed by a nucleophilic attack by the ε-amino group of the lysine side chain (33) (Figure 5.3). Interestingly, while lysidine formation involves the incorporation of an amino acid into the anticodon of $tRNA^{Ile}$, the proposed reaction mechanism does not follow that of aminoacyl-tRNA synthetases. These enzymes instead activate the amino acid by formation of an aminoacyladenylate followed by subsequent ligation of the amino acid to the tRNA substrate with the release of AMP. It is suggested that TilS owes its lysidine synthetase activity to its evolutionary path. The TilS sequence shares great similarity with a family of P-loop PPi synthetases (e.g., GMP synthetase) that analogously activate a nucleoside substrate by adenylation, followed by displacement of the adenylate by a secondary nucleophilic attack from the amino group of lysine (33).

In terms of substrate discrimination, it is of interest to note that the bacterial $tRNA^{Ile}$ contains a full set of positive determinants required for methionine incorporation by met-RS, including base pairs G2-C71, C3-G70, A73, and the CAU anticodon in addition to key determinants for Ile-RS. Recent results further confirmed that lysidine formation at C34 not only serves to create a strong determinant for IleRS recognition but also serves as a negative determinant for metRS (32). In this manner,

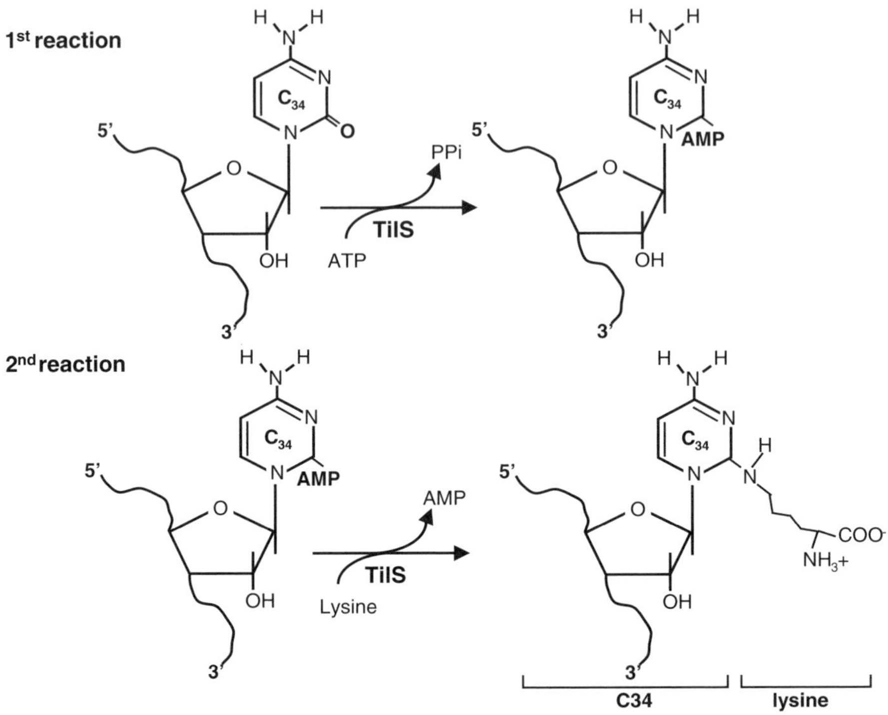

Figure 5.3 The proposed two-step mechanism for lysidine formation. TilS is the lysidine synthetase responsible for catalyzing both steps of the reaction.

the C34 to k^{2}C34 editing event is essential for converting a tRNA^{Ile}, with an encoded Met anticodon, to specify isoleucine by changing its amino acid acceptance. The presence of lysidine also affects the tRNA's decoding capacity while preventing the undesirable formation of a met-tRNA^{Ile} product that could have deleterious consequences in translation.

Curiously, the story of lysidine, a study of an unusual modification in a minor tRNA of E. coli that began as an attempt to explain the newly elucidated genetic code, now ends as a new type of tRNA editing. In terms of evolution, one can easily infer that for lysidine formation a conserved chemical mechanism utilized by nucleotide biosynthetic enzymes has been recruited by the newly discovered lysidine synthetase to provide the reassignment of a tRNA, with an unusual anticodon, in a process that is essential for cell viability.

5.3.2 Nucleotide Additions at the Ends of tRNAs

The first-ever example of tRNA editing reported occurs in the amoeboid protozoan *Acanthamoeba castellanii* (10). Lonergan and Gray showed that certain tRNAs in the mitochondria of these organisms have sequences that differ from those found in their encoding genes. These changes thus occur posttranscriptionally and help restore single or multiple mismatches in the acceptor stem of tRNAs in this organism

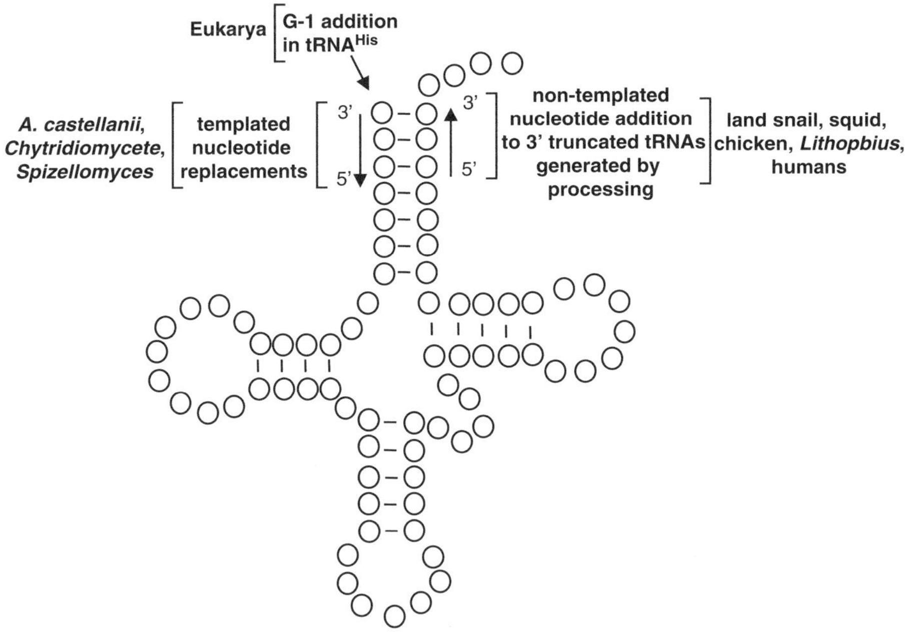

Figure 5.4 Nucleotide additions at tRNA ends. G-1 addition occurs at the 5′ end of all $tRNA^{His}$ in Eukarya but not in bacteria. In addition, editing by the addition of multiple nucleotides at the 5′ and 3′ ends of tRNAs have been described to repair the ends of tRNAs; but because of the nature of the additions, two mechanisms are inherently different. The arrows with 5′ and 3′ denote the direction of polymerization.

(Figure 5.4). Unlike the single nucleotide changes involving C-to-U or A-to-I conversions, editing in A. *castellanii* converts U to A, U to G, and A to G. Since some of these changes involve nucleotide transversions, mechanistically they cannot occur by a simple deamination as proposed for marsupial and trypanosomatid C-to-U editing and demonstrated for inosine formation. The observation of multiple nucleotide mismatches in a contiguous stretch of sequence led Gray and co-workers to propose a nucleotide addition mechanism, whereby the mismatched nucleotides are removed from the 5′ end of the tRNA and replaced by nucleotides that recreate canonical base pairs (34, 35). This mode of nucleotide insertion is of course unusual in that it requires the 3′-to-5′ templated addition of nucleotides, where the sequences on the 3′ side of the acceptor stem will provide the editing information (Figure 5.4).

With their unorthodox nucleotide addition mechanism in mind, the Gray laboratory set out to characterize this novel editing activity. They found that *A. castellanii* mitochondrial extracts could indeed edit a synthetic tRNA (35). Most recently, following the report of a similar type of editing in the *Chytridiomycete* fungus, *Spizellomyces punctatus*, these researchers also demonstrated the presence of a similar activity (34). Both activities require ATP or a 5′-triphosphate to allow nucleotide incorporation. Mechanistically, it is proposed that the 5′ residue to which an incoming nucleotide is to be added has to be first activated by AMP. This adenylated species is subsequently attacked by the incoming nucleotide with the concomitant release of AMP. This mechanism implies the need for an activity that can remove the mismatched nucleotides prior to nucleotide addition. Although removal of radioactively labeled 5′-mismatched nucleotides have not been conclusively demonstrated, a 5′-to-3′ exonuclease activity is a likely candidate for such an activity.

To date, the only example of naturally occurring addition of a nucleotide in the 3′-to-5′ direction occurs during the incorporation of guanosine in $tRNA^{His}$ (Figure 5.4). A reaction that creates an unpaired G at the 5′ end of $tRNA^{His}$ that is absolutely required for synthetase recognition (36). However, this reaction is nontemplated and mechanistically could be performed by a ligation-type mechanism with a requirement for ATP. Therefore, the apparent need for a reverse polymerase in the case of *A. castellani* seems, a priori, untenable as no polymerase meeting these characteristics has ever been described. However, Phizicky and co-workers, while studying the enzyme responsible for G-1 addition to $tRNA^{His}$ of yeast ($tRNA^{His}$ guanylyltransferase, Thg1), first made the connection that the nontemplated reaction was akin to the editing required of A. castellani (37, 38). These authors showed that during G-1 addition all that was required is the activation of the 5′ end of the tRNA by ATP to create a nucleotide triphosphate in which pyrophosphate will then serve as the leaving group in the subsequent addition of the incoming GTP (38) (Figure 5.4). They then argued that in fact thermodynamically there should be no barrier for the reverse polymerization of nucleotides, provided that a triphosphate is already present at the 5′ end of any polynucleotide. Jackman and Phizicky thus asked whether Thg1 could in fact function as a reverse polymerase. They proposed a model similar to that of Gray and co-workers for A castellani editing. They incubated Thg1 with a tRNA substrate containing a 5′ nucleotide triphosphate and found that indeed this enzyme could polymerize nucleotides in the 3′-to-5′ direction; but unlike the G-1 addition, reverse polymerization occurred in a templated fashion (37, 38). This observation then raises a number of questions: What is the biological significance for this reverse polymerase activity? The answer to this question is not immediately clear, but Jackman and Phizicky have suggested some role in DNA repair, following their observation that just like repair polymerases, Thg1 did not discriminate between ribonucleotide and deoxynucleotides. Consistent with the idea of a role in DNA repair and/or replication, Thg1 has been implicated in G2/M transition in yeast (39). A second major question, implicit in the former, is, What are the natural substrates for the reverse polymerase? Clearly, tRNAs could not be the natural targets for reverse polymerization, since tRNA molecules are universally process by RNAse P and all bear a nucleotide monophosphate at their 5′ end. Perhaps a more relevant question is, following G-1 addition, which generates the triphosphate needed for reverse polymerization, how is the activity controlled in $tRNA^{His}$? Evidently, these and other question will still await further experimentation.

Other than 5′ addition of nucleotides, in a number of systems, tRNAs with either truncations or mismatched nucleotides at their 3′ end are created in mitochondria during processing or synthesis. Generally, these tRNAs are repaired by the nontemplated addition of adenosine, presumably by poly A polymerase, as is the case of tRNA repair in snail, squid, and chicken mitochondria (40, 41) (Figure 5.4). In animal mitochondria, tRNA genes are often encoded as overlapping cassettes where the processing of the downstream tRNA yields an upstream tRNA with a shortened 3′ end.

Morl and co-workers showed that these tRNAs are repaired by a CCA-nucleotidyl transferase-like activity, but still in a nontemplated manner (42, 43).

Recently, genes that encode tRNAs with mismatches at the 3′ end of the acceptor stems have been reported in the mitochondria of the centipede *Lithobius forficatus* (44) and *Seculamonas ecuadoriensis* (a single-cell protist) (45). In these systems, mismatched nucleotides are replaced with nucleotides that regenerate canonical base pairs. However, neither the activity nor an *in vitro* assay has been established for this system. It has been proposed that due to the templated nature of this 3′ addition, an RNA-dependent RNA polymerase could represent a perfect candidate enzyme for this type of editing.

Editing mechanisms that repair or restore the structure of tRNAs represent by far the most common and varied types of tRNA editing. However, just like with single nucleotide substitutions, in most cases of tRNA repair either the enzyme has not been identified or an *in vitro* editing system does not exist. Thus one could only speculate on the mechanisms that operate in these types of editing events.

In terms of evolution, however, often the idea of sequestration of a preexisting activity is usually invoked in some scenario in which editing forces the cell to use an activity normally present for some other function to solve some major road block that a given cell may encounter in its evolution. Once the activity that solves the problem is sequestered, then the original accident becomes frozen and is incorporated into the normal pathway of tRNA maturation in the given cell. Perhaps a latent example of an activity ready for sequestration comes from the demonstration by Morl and co-workers that in yeast where overlapping tRNAs do not exist, the cells are still able to utilize a preexisting activity to properly process overlapping genes similar to those naturally found in human mitochondria (46). This observation highlights the fact that inevitably enzyme sequestration also implies a degree of substrate promiscuity by the activity to be sequestered.

5.3.3 C-to-U Editing in Marsupials and Trypanosomatids

To date, there are only two examples of C-to-U editing of tRNA at either of the three anticodon nucleotides: C-to-U editing of $tRNA^{Gly}$ in marsupial mitochondria (47), which affects the second position of the anticodon (C35), and C-to-U editing of $tRNA^{Trp}$ in trypanosomatid mitochondria, specific for the first position (C34) (48) (Figures 5.5 and 5.6).

In marsupial mitochondria, Janke and Paabo first observed the absence of a gene for a tRNA that could decode the mitochondrial aspartate codons. They found that at the position of the mitochondrial genome where most mammals encode $tRNA^{Asp}$, the gene had instead been replaced by a $tRNA^{Gly}$GCC. This tRNA could predictably decode two out of four mitochondrial codons for glycine (i.e., GGY) but would fail to decode the Asp codons. Sequence comparison with tRNA databases showed that, save for the three-anticodon nucleotides, the $tRNA^{Gly}$ was very similar to $tRNA^{Asp}$ from other organisms (47). These researchers raised the possibility, and later found, that this $tRNA^{Gly}$ undergoes a single C-to-U editing at the second position of the anticodon (C35) which changes the codon specificity of this tRNA from glycine to aspartate (47, 49–51) (Figure 5.5). They also showed that this editing event was widespread in marsupial

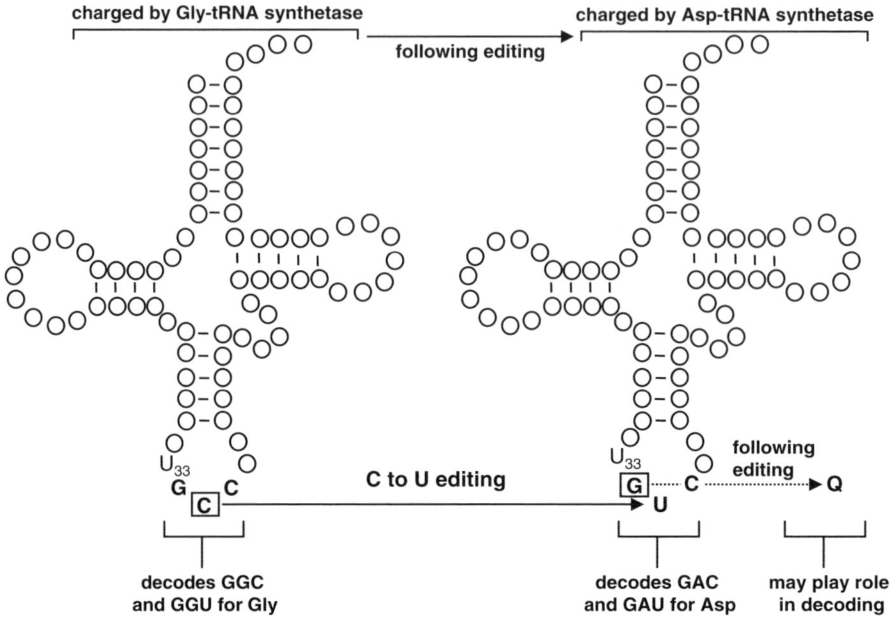

Figure 5.5 C-to-U editing in marsupial mitochondria. This editing event changes the amino acid and the decoding capacity of the tRNA and also affects the formation of queuosine (Q). The arrows indicate the various changes and their consequences.

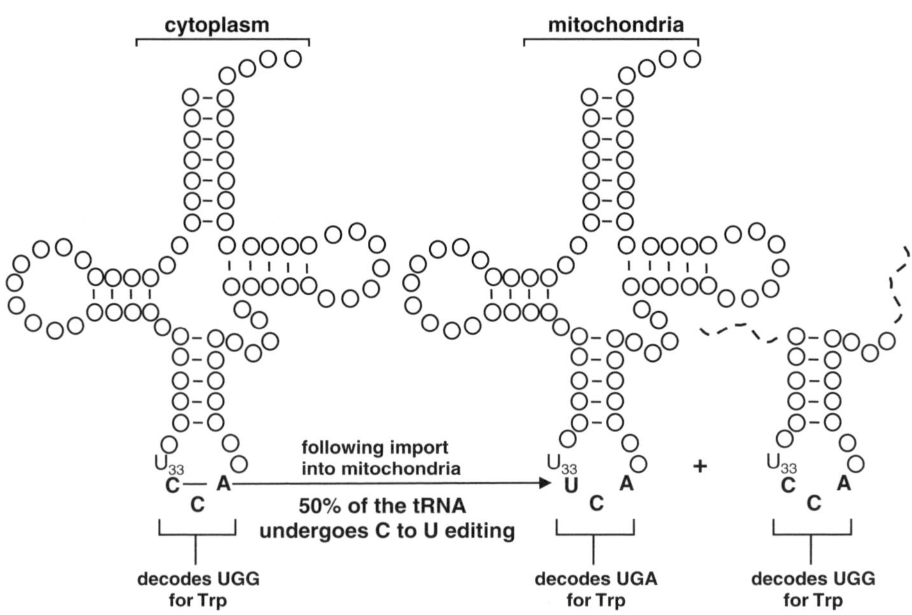

Figure 5.6 C-to-U editing in trypanosomatid mitochondria. Following import, the only $tRNA^{Trp}$ (CCA) encoded in the genome is imported into mitochondria where it is edited. This editing is thought to be essential for decoding the mitochondrial UGA codons as tryptophan.

mitochondria isolated from different tissues; however, the editing levels were fairly constant between tissues. Thus editing does not appear to play a role in tissue–specific gene expression. Interestingly, only 50% of the $tRNA^{Gly}$GCC was converted to $tRNA^{Asp}$ by editing (50), thereby raising the possibility that either both the unedited and edited tRNAs play a role in mitochondrial translation or only the edited species is functional but editing is inefficient. Further testing revealed that *in vitro* transcripts representing the two tRNA versions could be efficiently charged with synthetase fractions; however, the $tRNA^{Gly}$ could only be charged with glycine, while the edited tRNA could only accept aspartate (51). Use of an elegantly designed assay (OXOCIRC) where the ability of sodium periodate to oxidize adjacent free hydroxyl (i.e., those found at the 3′ of RNA) to form dialdehydes was exploited led to the assessment of whether or not both tRNAs were also charged *in vivo* (51). In this assay, the presence of an amino acid at the 3′-end of the tRNA blocks the accessibility of the end to oxidation and, when coupled with circularization by RNA ligase and RT-PCR, permits quantitation of the number of charged tRNAs regardless of their anticodon sequence identity. These study led to the conclusion that both the unedited and edited versions of the tRNA were functional. However, it is not clear what determines the 50% balance in terms of C-to-U conversion. Furthermore, the biological significance of this balance, although also unclear, perhaps suggests some function in decoding.

Besides affecting charging and codon recognition, C-to-U editing of $tRNA^{Gly}$ also plays a role as a modification determinant. The Paabo laboratory demonstrated that the C-to-U conversion at position 35 also led to the creation of the sequence motif (UGU, where the last U is created by C-to-U editing) needed for further conversion of G34 into the hypermodified nucleoside queosine (Q) (50). The nucleoside Q, in turn, has been implicated in affecting *in vivo* decoding by apparently stabilizing codon–anticodon interactions (Figure 5.7) (52, 53).

We discovered the only other example of C-to-U editing in the anticodon of a tRNA, which occurs in the mitochondria of trypanosomatids (48). In these organisms the mitochondrial genome does not encode a single tRNA gene. Thus for mitochondrial translation to take place, every tRNA is synthesized in the nucleus, transits through the cytoplasm, and is then imported into the mitochondria. The single nucleus-encoded tryptophanyl tRNA ($tRNA^{Trp}$) is transcribed with a CCA anticodon, posing the conundrum of how this tRNA is able to decode mitochondrial UGA codons. In trypanosomatid mitochondria, like in most eukaryotes, the UGA codons have been reassigned to tryptophan; in the cytoplasm, however, UGA functions as a stop codon (opal codon). Trypanosomatids have to be able to decode the UGA codons while preventing suppression of the same codon while the tRNA transits through the cytoplasm. To achieve this, a subpopulation of this tRNA is imported into the mitochondrion while still bearing a CCA anticodon. Following import, RNA editing of C_{34} creates the U_{34}CA anticodon required to translate the mitochondrial UGA codons as tryptophan (48, 54) (Figure 5.6). Despite the existence of other tRNAs that contain C34, only $tRNA^{Trp}$ undergoes editing, raising the question as to what determines specificity. Mass spectrometry analysis of native cytosolic and mitochondrial versions of $tRNA^{Trp}$ revealed that this tRNA undergoes a number of mitochondria-specific modifications following mitochondrial import (55). We have raised the possibility that these modifications occur in a cascade and only when they occur in the proper sequence, editing can be specified. In this proposal, although other tRNAs may contain the same modification set, only tRNAs in which modifications occur in the correct sequence combined with other determinants only present in $tRNA^{Trp}$ serve as some of the basis of editing specificity. Indeed in this system, we have identified a number of key determinants for tRNA editing, which include unique nucleotides at the anticodon loop as well as single base pairs at the acceptor stem. Nevertheless, the editing enzyme still remains elusive and neither in the marsupial system above nor in the trypanosomatid system a functional assay has been established for *in vitro* editing precluding identification of this important enzyme.

In the examples listed above, it is clear that a prevailing mechanism for reassigning the identity of a tRNA involves nucleotide transitions at the anticodon nucleotides. However, the chemical mechanism by which C-to-U transitions occur in the anticodon of tRNAs is not clear so far, because the enzyme(s) that perform this type of editing have remained elusive. Much work is still needed to identify the activities

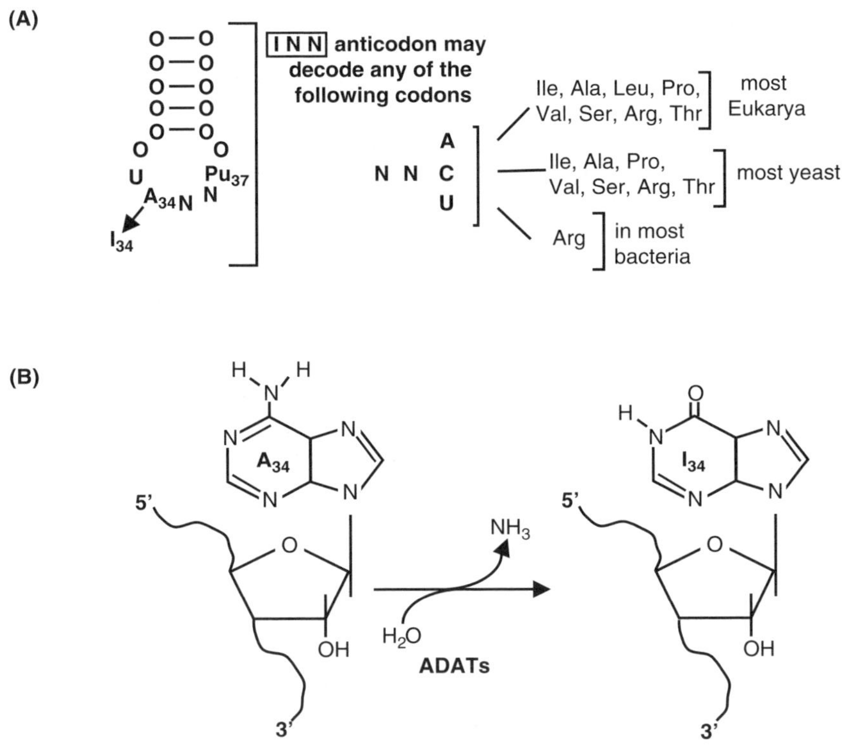

Figure 5.7 Inosine at the first position of the anticodon: significance and mechanism. (A) A34 is almost universally changed to inosine in most organisms and is necessary for expanding the decoding capacity of a tRNA to decode multiple codons by wobbling. INN refers to ino sine containing anticodon where N refers to any of the four canonical nucleotides. The arrow denotes the position of the A-to-I editing event in tRNA. (B) The mechanism of A-to-I editing in tRNAs, where the editing enzymes mediate the hydrolytic deamination of the A34 to form inosine; 5′ and (3′) refer to the ends of the tRNA, respectively. ADATs denotes adenosine deaminase acting on tRNA (according to accepted nomenclature).

and clarify the mechanism. Also, as seen in the case of C-to-U editing, although many other tRNAs contain encoded cytosines in the anticodon, the editing enzyme is able to discriminate and only edit the specific tRNA. Future work should also help answer the basis of specificity and discrimination of different substrate ensuring fidelity of the process.

5.3.4 A-to-I Editing of tRNAs in Yeast and Bacteria

The discovery of the nucleoside inosine (56, 57), by Holley and co-workers, over 40 years ago caused an immediate and now historical stir. Scientists at the time were amidst trying to explain how (or why) 20 different amino acids required 61 different codes for protein synthesis. The discovery of inosine in yeast $tRNA^{Ala}$(AUA) (because of its predicted base-pairing properties) led Crick to propose that the base-pair capabilities of this tRNA were not limited to the four canonical nucleotides. Crick reasoned that the newly discovered, inosine-containing $tRNA^{Ala}$(AUA) could pair with three different codons (ending in A, C, or U) to specify the same amino acid. Crick further elaborated this notion into his now famous "Wobble Hypothesis" (58), which provides explanation for (a) the presence of multiple codes for one amino acid, (b) the existence of more than one tRNA to specify a single amino acid, and (c) the implicit ability of a single tRNA to decode more than one code (as in the case of inosine). It is now well established that tRNAs encoded with an adenosine at position 34 are almost universally edited to inosine, expanding the decoding capacity of the given tRNA (Figure 5.7A).

Despite the early discovery of inosine, it was pioneering work by Grosjean and co-workers 30 years later that led to the identification of an enzymatic activity in S100 supernatants that could convert adenosine to inosine in tRNA (22). They showed that, *in vitro*, extracts from yeast, *Xenopus*, mammalian cells, and bacteria could specifically convert adenosine to inosine at the first anticodon position in tRNAs. These authors demonstrated that incubation of a tRNA substrate labeled at either the phosphate backbone or the base led to preservation of the label following inosine formation, ruling out possible mechanisms that involve complete replacement of the nucleotide (insertion/deletion-type mechanisms) or base replacement (transglycosylation-type mechanisms) (21). Furthermore, when unlabeled substrate was incubated in the presence of S100 supernatants in the presence of $[O^{18}]$ water, incorporation of labeled oxygen was observed concomitantly with the formation of inosine (21, 22). These experiments, of course, were guided by similar experiments performed earlier to unveil the mechanism of mRNA editing in *C. elegans* and mammalian cells (as discussed elsewhere in this book). Together, these findings led to the conclusion that inosine formation, regardless of the substrate (mRNA or tRNA), proceeded by a hydrolytic adenosine to inosine deamination mechanism (Figure 5.7B). To date, however, only the yeast and bacterial enzymes have been characterized to some extent.

Several years later, knowledge of a conserved deamination mechanism prompted Gerber and Keller to search the then newly sequenced yeast genome for open reading frames encoding putative deaminases. They found that in yeast the enzyme responsible for A to I conversion at the wobble position had two subunits, the products of the ADAT2 and ADAT3 genes (for adenosine deaminase acting on tRNA,

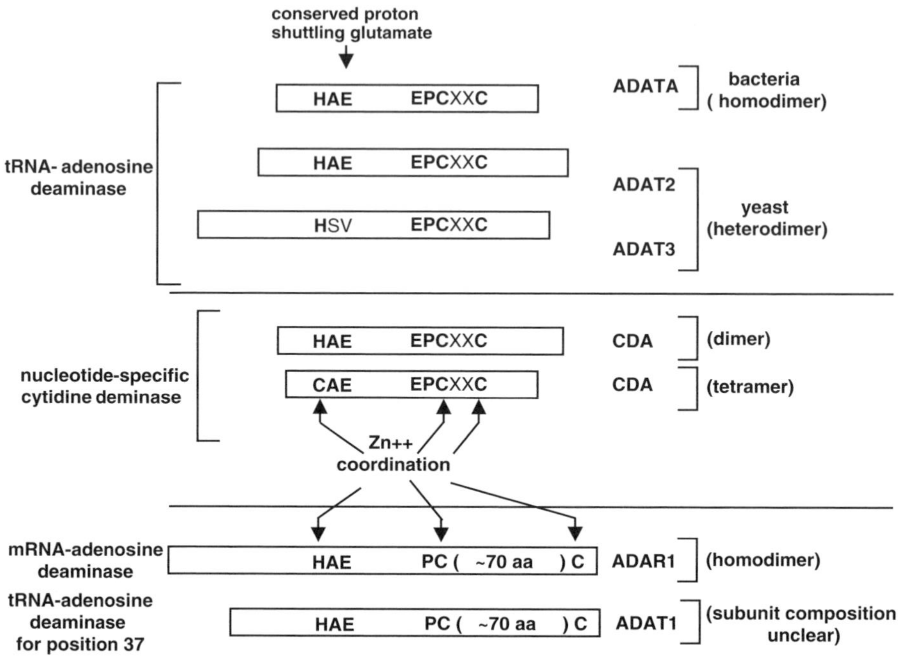

Figure 5.8 Different tRNA deaminases compared to their nucleotide- and mRNA-specific counterparts. Arrows denote the conserved glutamate (E) involved in proton shuttling and the conserved H/C and PCxxC motifs involved in Zn^{2+} coordination; both are essential for activity.

using the accepted editing nomenclature and replacing their former names of tad2p and tad3p) (59). Although both proteins are homologous, they are not identical, differing both in sequence and size. ADAT2 is a 24-kD protein that harbors all the conserved motifs required for deamination (Figure 5.8). ADAT3, by contrast, is a 30-kD polypeptide that contains a conserved zinc-binding motif, but lacks a highly conserved proton-shuttling domain found in most deaminases (Figure 5.8). Neither of the subunits, however, is by itself able to support editing of tRNAs (59). Therefore in yeast, the editing enzyme is formed by heterodimerization of two subunits that, upon association, create a functional enzyme that can then specifically deaminate the wobble base of all seven A-34-containing tRNAs in this organism (59).

In bacteria, only $tRNA^{Arg}$ contains inosine at the first anticodon position. The *E. coli* enzyme, the product of the adatA gene, is similar to the smaller subunit of the yeast enzyme (60) and catalyzes the same A-to-I editing, but, unlike the yeast enzyme, it can deaminate much smaller substrates, including molecules that are essentially short versions of the anticodon stemloop (ASL) (59–61). This enzyme, however, is very specific in that it is able to deaminate the cognate bacterial $tRNA^{Arg}$ but is unable to edit any of the eukaryotic A-34-containing tRNAs (60). However, the ADATA protein efficiently deaminates eukaryotic tRNAs that contain a transposed bacterial $tRNA^{Arg}$ ASL. Also, unlike yeast, and presumably other eukaryotes, tadAp forms a functional, albeit weak, homodimer in solution, which has been confirmed in the recently solved structure of the *Aquifex aeolicus* enzyme (62). This structure showed that the two subunits form a three-layered $\alpha/\beta/\alpha$ structure, with a dimerization interface so extensive that it buries over 16% of the total monomer surface area (or 1300 Å^2 out of a total of 8100 Å^2). Interestingly, an Asp-to-Glu change at a highly conserved residue within the dimerization interface of the bacterial proteins yields an enzyme that is inactive *in vitro* but fully active *in vivo*. This region either is important to stabilize the enzyme or is needed *in vivo* to interact with other proteins, which then leads to structural stability (62). This last observation is of special interest, given the finding by Grosjean and co-workers that the native mammalian enzyme is of a much larger size (>200 kD) (22) than that expected through homo- or heterodimerization of two putative subunits, perhaps suggesting that within cells the A-to-I tRNA editing enzyme is part of a much larger protein complex.

Although the enzymes involved in A-to-I editing of tRNA have been identified in yeast and bacteria and *in vitro* assays exist for a number of organisms (including *Xenopus*, trypanosomatids, humans, etc.), these activities are far from being fully characterized. To date, it is not known how these enzymes manage to specifically deaminate the first anticodon position or what is the basis for tRNA binding.

5.3.5 Double Editing in Trypanosomatids

Our laboratory has been studying A-to-I editing of tRNAs in trypanosomatids (*Trypanosoma* and *Leishmania*); like in mammals, eight different tRNAs containing an A at the first position of the anticodon can be identified by genomic database searches. These tRNAs undergo A-to-I editing at position 34 of the anticodon (63) (Figure 5.9). This editing is essential to decode the C-ending codons for the amino

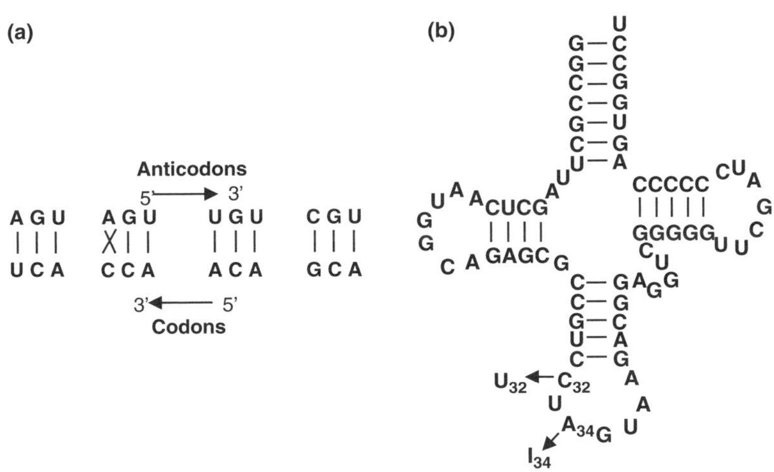

Figure 5.9 Double editing of tRNA^{Thr} in trypanosomatids. (A) The different threonine codons and their predicted anticodon. (X) denotes the absence of an anticodon in the genome that can decode the CCA codon. (B) In order to decode the CCA codons trypanosomatids edit tRNA^{Thr} (AGU), the arrows indicate the A-to-I change and the unexpected finding of a second editing event (C to U) in the same anticodon.

acids isoleucine (Ile), alanine (Ala), leucine (Leu), proline (Pro), valine (Val), serine (Ser), arginine (Arg), and threonine (Thr). We have investigated the A-to-I editing of threonyl tRNA ($tRNA^{Thr}$) based on the ease of folding of synthetic versions of this tRNA. Under *in vitro* conditions, synthetic $tRNA^{Thr}$ is efficiently aminoacylated with crude synthetase fractions from *Leishmania* and *Trypanosoma* (Alfonzo and Ibba, unpublished results). In *Trypanosoma brucei*, there are four codons and three genes encoding isoaccepting tRNAs for the amino acid threonine. Two of these isoacceptors (anticodon CGU and UGU) decode the ACG and ACA, respectively. The remaining tRNA (anticodon AGU) can decode the ACU codon but is unable to decode the ACC codon (Figure 5.9A). No tRNA is encoded in the *T. brucei* genome that may decode the ACC codon. The $tRNA^{Thr}$AGU must therefore undergo A-to-I_{34} editing to expand its decoding capacity (Figure 5.9B). Unexpectedly, we found that this tRNA also undergoes C-to-U conversion at position 32 of the same anticodon loop (Rubio and Alfonzo, unpublished results) (Figure 5.9B). Currently, the biological significance of the two editing events is not clear. Nevertheless, *in vitro* we could show that C-to-U formation at position 32 has a stimulatory effect in the further conversion of A to I at the wobble base. Additionally, both events appear to occur outside the organelle, indicating that tRNA editing by C-to-U conversion is more widespread than previously thought (Rubio, Gaston, Papavasiliou, and Alfonzo, unpublished results).

5.4 THE TRANSFER RNA EDITING ENZYMES OF TRYPANOSOMATIDS: A SPECIAL CASE OF CATALYTIC FLEXIBILITY

Keller and co-workers previously remarked that the anticodon-specific ADATs, despite performing adenosine deaminations, contained core signature sequences characteristic of cytidine deaminases (not adenosine deaminases) (Figure 5.8). They proposed an evolutionary path for these enzymes whereby these enzymes are derived from a gene duplication of an ancestor cytidine deaminase. This proposal thus suggests that through evolution, changes accumulated in the duplicated gene that led to the conversion of a C-to-U deaminase into an A-to-I-specific enzyme. This is in sharp contrast with the only other known tRNA deaminase, ADAT1, which forms inosine at position 37 in eukaryotic tRNAs. ADAT1 contains a set of conserved core sequence that resembles those of adenosine deaminase acting on RNA ADAR1 (an mRNA editing enzyme) and in theory appeared by a different evolutionary path.

From an evolutionary and mechanistic standpoint, it is unclear what events led to the conversion of cytidine deaminases into adenosine deaminases. Presumably, features of the individual proteins by themselves (in the case of *E. coli*; see below) or upon association of two different subunits (in the case of yeast) led to specificity changes from a pyrimidine to a purine deaminase. Evidently, merely scanning the conserved amino acids in anticodon ADATs is not sufficient to reveal what changes led to the differences in substrate specificity.

In light of the finding of two editing events in a single anticodon loop in trypanosomatids, we decided to explore the possibility that the trypanosome anticodon

ADATs could in fact play a role in both C-to-U and A-to-I editing. The sequences from yeast ADAT2 and ADAT3 were used to perform searches of the genomic database from *T. brucei* leading to the identification of the two homologs from trypanosomatids (tbADAT2 and tbADAT3). We then used RNA interference (RNAi) to reduce the levels of ADAT2 (which like in yeast is the catalytic subunit). We observed a significant reduction in both A-to-I and C-to-U editing events following induction of RNAi, suggesting that ADAT2 plays a role *in vivo* in specifying both editing events; whether directly or indirectly is not currently clear. Nonetheless, this observation reinforces the idea that ADAT2 might be able to perform both activities. However, recombinant expression of this two proteins showed that although TbAADAT2/3 could form a complex (like in yeast), they could only catalyze the formation of inosine at position 34 but not C to U. By modeling the sequence of ADAT2 to that of the structure of CDD1 (a nucleotide deaminase from yeast) and also comparing it to that of AID (the DNA cytidine deaminase), we observed some similarities which then led us to the hypothesis that perhaps tbADAT2/3 could perform deamination of DNA. In fact, we could demonstrate both by an *in vivo* screen and also *in vitro* that tbADAT2/3 can support a very robust C-to-U activity when single-stranded DNA is provided as a substrate (64). Thus the conservation in the cytidine deaminase motifs by ADATs is not accidental; in fact at least in the case of trypanosomatid, these enzymes still maintain the ability to perform both reactions, albeit in different substrates (DNA versus RNA). This in our view provides the first line of experimental evidence in supporting the path for the evolution of editing deaminases originally proposed by Keller and co-workers. In turn, it also raises the following question: What, if any, could be the biological significance of DNA deamination? We have speculated that the mutagenic activity of TbADAT2/3 may be related to the mechanism of antigenic variation. Indeed, deaminase-initiated class switch recombination in the immunoglobulin locus in vertebrate cells and gene conversion in the variable region glycoprotein (VSG) locus in *T. brucei* are very similar in mechanistic requirements: Both loci are telomeric, both are flanked by highly repetitive regions, and in both cases the locus must be transcribed for recombination to occur (65–68). However, whether cytidine deamination within DNA is important for VSG switching and antigenic variation in *T. brucei* will remain an open question.

5.5 COMPLEX FORMATION BY TRANSFER RNA EDITING ENZYMES: A MODEL FOR THE REGULATION OF EDITING ACTIVITY

Our forays in the trypanosomatid tRNA editing systems has naturally led us to the question of subcellular localization of the different editing events and activities (C to U and A to I) in $tRNA^{Thr}$. Given the possibility that the DNA deamination reaction we have described is of some biological significance, it follows logically that a least ADAT2 should localize to the nucleus of these cells. Analysis of nuclear and cytosolic RNA fractions from *T. brucei* and *L. tarentolae* have revealed that in $tRNA^{Thr}$ the C-to-U editing occurs in the nucleus while A-to-I editing occurs in the cytoplasm (Gaston, Rubio, Alfonzo and Papavasiliou, unpublished results). This observation is also supported by preliminary analysis with antibodies against ADAT2, also suggesting a dual localization for this enzyme, in line with the finding that it plays a role in both nuclear and cytoplasmic editing. We have thus proposed a model by which subcellular localization of the eukaryotic editing deaminases may alter their specificity. In the trypanosomatid example, this model suggests that ADAT2 may have different specificity depending on two variables: its subcellular localization (i.e., nuclear, cytoplasmic, and maybe even mitochondrial) or its association with different protein subunits. The first part of the model is supported by our nuclear localization experiments but needs further corroboration by more elaborate immunofluorescence approaches. The second part of the model is more difficult to test because it requires prior knowledge of who associates with whom within a cell and it will of course require lengthy and rigorous experiments. However, for the second part of the model in a way nature has already performed the experiment for us. ADAT2 and ADAT3, cyitidine deaminases in terms of primary sequence, pair up in the form of a heterodimer and indeed act as adenosine deaminases in tRNA.

5.6 CONCLUDING REMARKS: EVOLUTION OF TRANSFER RNA EDITING DEAMINASES: AFFINITY VERSUS SPECIFICITY

In this chapter, we have tried to summarize what is currently known about a plethora of editing events that occurs in tRNAs. Ideally, our wish would be to expand our comments into enzymatic activities that may illustrate difference between different tRNA editing systems, but unfortunately only a handful of enzymes have been identified. Therefore the tRNA editing world is not as well understood as other editing systems (e.g., mRNA editing in mammals). Nonetheless, current knowledge is already more than sufficient to get creativity rolling and to propose a number of very testable models for the evolution of this family of editing enzymes.

The story of the tRNA deaminases presents a case study on the evolution of substrate specificity by these enzymes that in fact might be applicable to all known editing deaminases, be it those that work on RNA or DNA. A look at the editing deaminase of bacteria reveals a single substrate enzyme with sufficient affinity for its target that in fact it can fairly efficiently utilize shorter substrates *in vitro* (i.e., stemloops representing the anticodon arm of $tRNA^{Arg}$). In the bacterial system, $tRNA^{Arg}$ is the only tRNA encoded in the genome containing A at the first position of the anticodon. Predictably, in this system, high affinity for its substrate, however minimal that substrate may be, implies that the active site of ADATa has enough binding functions built into it that is sufficient to recognize a single substrate. Eukaryotic tRNA deaminases, on the other hand, recognize 7–8 different tRNAs, depending on the system. Curiously, all of the ADAT2/3 studied so far (namely, those from yeast and trypanosomatids) are architecture-dependent and do require a full-length tRNA for catalysis. This leaves us with the proposal that perhaps in the process of evolution and the rising need to accommodate different substrates, the eukaryotic deaminases acquired a number of amino acid changes. These changes led to a certain degree of active-site relaxation at the expense of active site affinity. In our view, this relaxation process has then led to the need to create a binding function away from the active site to be able to now switch the specificity for multiple substrates. In this realm, it is predicted that the requirement for a full-length tRNA is nothing more than a reflection of the need to move the binding domain of these deaminases away from the active site, permitting accommodation of a larger variety of substrates. A corollary of this process of active site relaxation could indeed be the driving force that has created tRNA editing deaminases with the ability to perform additional reactions such as the observed ability to deaminate DNA. Evidently, the fact that such activities still remain today in a group of enzymes that are widely accepted to only perform one task may be telling us something about the different requirements set forth by different biological systems.

The ever-changing landscape that once led to genetic code degeneracy has throughout evolution set the stage for the recruitment of preexisting activities and has no doubt exerted some selective pressure in the evolution of editing systems. In the case of tRNA editing enzymes, their evolutionary path and the dynamics of decoding have played a major role for a group of enzymes caught between the need to edit single substrates, as they may arise, and the need to accommodate similar variants of these substrates. Thus it is at the crossroads of affinity and specificity that the rules were set for how many of these editing activities have appeared and continue to evolve.

REFERENCES

1. Benne, R., Van den Burg, J., Brakenhoff, J. P., Sloof, P., Van Boom, J. H., and Tromp, M. C. (1986) *Cell* **46**, 819–826.
2. Polson, A. G., Crain, P. F., Pomerantz, S. C., McCloskey, J. A., and Bass, B. L. (1991) *Biochemistry* **30**, 11507–11514.
3. Hurst, S. R., Hough, R. F., Aruscavage, P. J., and Bass, B. L. (1995) *RNA* **1**, 1051–1060.
4. Rueter, S. M., Burns, C. M., Coode, S. A., Mookherjee, P., and Emeson, R. B. (1995) *Science* **267**, 1491–1494.
5. Wagner, R. W., Smith, J. E., Cooperman, B. S., and Nishikura, K. (1989) *Proc Natl Acad Sci USA* **86**, 2647–2651.
6. Lai, F., Chen, C. X., Carter, K. C., and Nishikura, K. (1997) *Mol Cell Biol* **17**, 2413–2424.
7. Powell, L. M., Wallis, S. C., Pease, R. J., Edwards, Y. H., Knott, T. J., and Scott, J. (1987) *Cell* **50**, 831–840.
8. Higuchi, K., Monge, J. C., Lee, N., Law, S. W., Brewer, H. B. Jr., Sakaguchi, A. Y., and Naylor, S. L. (1987) *Biochem Biophys Res Commun* **144**, 1332–1339.
9. Chen, S. H., Habib, G., Yang, C. Y., Gu, Z. W., Lee, B. R., Weng, S. A., Silberman, S. R., Cai, S. J., Deslypere, J. P., Rosseneu, M., et al. (1987) *Science* **238**, 363–366.
10. Lonergan, K. M., and Gray, M. W. (1993) *Science* **259**, 812–816.
11. Covello, P. S., and Gray, M. W.(1998). In: Grosjean, H., and Benne, R., *Modification and Editing of tRNA*, ASM Press, Washington, DC., p. 596.
12. Grosjean, H., and Bjork, G. R. (2004) *Trends Biochem Sci* **29**, 165–168.
13. Fey, J., Weil, J. H., Tomita, K., Cosset, A., Dietrich, A., Small, I., and Marechal-Drouard, L. (2002) *Gene* **286**, 21–24.
14. Marechal-Drouard, L., Cosset, A., Remacle, C., Ramamonjisoa, D., and Dietrich, A. (1996) *Mol Cell Biol* **16**, 3504–3510.
15. Kunzmann, A., Brennicke, A., and Marchfelder, A. (1998) *Proc Natl Acad Sci USA* **95**, 108–113.
16. Placido, A., Gagliardi, D., Gallerani, R., Grienenberger, J. M., and Marechal-Drouard, L. (2005) *J Biol Chem* **39**, 33573–33579.
17. Gerber, A., Grosjean, H., Melcher, T., and Keller, W. (1998) *EMBO J* **17**, 4780–4789.

18. Droogmans, L., Roovers, M., Bujnicki, J. M., Tricot, C., Hartsch, T., Stalon, V., and Grosjean, H. (2003) *Nucleic Acids Res* **31**, 2148–2156.
19. Roovers, M., Wouters, J., Bujnicki, J. M., Tricot, C., Stalon, V., Grosjean, H., and Droogmans, L. (2004) *Nucleic Acids Res* **32**, 465–476.
20. Grosjean, H., Constantinesco, F., Foiret, D., and Benachenhou, N. (1995) *Nucleic Acids Res* **23**, 4312–4319.
21. Grosjean, H., Auxilien, S., Constantinesco, F., Simon, C., Corda, Y., Becker, H. F., Foiret, D., Morin, A., Jin, Y. X., Fournier, M., and Fourrey, J. L. (1996) *Biochimie* **78**, 488–501.
22. Auxilien, S., Crain, P. F., Trewyn, R. W., and Grosjean, H. (1996) *J Mol Biol* **262**, 437–458.
23. Takemura, S., Murakami, M., and Miyazaki, M. (1969) *J Biochem (Tokyo)* **65**, 489–491.
24. Gardner, R. S., Wahba, A. J., Basilio, C., Miller, R. S., Lengyel, P., and Speyer, J. F. (1962) *Proc Natl Acad Sci USA* **48**, 2087–2094.
25. Harada, F., and Nishimura, S. (1974) *Biochemistry* **13**, 300–307.
26. Scherberg, N. H., and Weiss, S. B. (1972) *Proc Natl Acad Sci USA* **69**, 1114–1118.
27. Muramatsu, T., Yokoyama, S., Horie, N., Matsuda, A., Ueda, T., Yamaizumi, Z., Kuchino, Y., Nishimura, S., and Miyazawa, T. (1988) *J Biol Chem* **263**, 9261–9267.
28. Muramatsu, T., Nishikawa, K., Nemoto, F., Kuchino, Y., Nishimura, S., Miyazawa, T., and Yokoyama, S. *Nature* **336** (1988) 179–181.
29. Weber, F., Dietrich, A., Weil, J. H., and Marechal-Drouard, L. (1990) *Nucleic Acids Res* **18**, 5027–5030
30. Sprinzl, M., and Vassilenko, K. S. (2005) *Nucleic Acids Res* **33** (database issue), D139–D140.
31. Soma, A., Ikeuchi, Y., Kanemasa, S., Kobayashi, K., Ogasawara, N., Ote, T., Kato, J., Watanabe, K., Sekine, Y., and Suzuki, T. (2003) *Mol Cell* **12**, 689–698.
32. Ikeuchi, Y., Soma, A., Ote, T., Kato, J., Sekine, Y., and Suzuki, T. (2005) *Mol Cell* **19**, 235–246
33. Nakanishi, K., Fukai, S., Ikeuchi, Y., Soma, A., Sekine, Y., Suzuki, T., and Nureki, O. (2005) *Proc Natl Acad Sci USA* **102**, 7487–7492.
34. Bullerwell, C. E., and Gray, M. W. (2005) *J Biol Chem* **280**, 2463–2470.
35. Price, D. H., and Gray, M. W. (1999) *RNA* **5** 302–317.
36. Cooley, L., Appel, B., and Soll, D. (1982) *Proc Natl Acad Sci USA* **79** 6475–6479.
37. Jackman, J. E., and Phizicky, E. M. (2006) *RNA* **12** 1007–1014.
38. Jackman, J. E., and Phizicky, E. M. (2006) *Proc Natl Acad Sci USA* **103** 8640–8645.
39. Rice, T. S., Ding, M., Pederson, D. S., and Heintz, N. H. (2005) *Eukaryot Cell* **4**, 832–835.
40. Yokobori, S. I., and Paabo, S. (1995) *Nature* **377**, 490.
41. Yokobori, S., and Paabo, S. (1995) *Proc Natl Acad Sci USA* **92**, 10432–10435.
42. Reichert, A., Rothbauer, U., and Morl, M. (1998) *J Biol Chem* **273**, 31977–31984.
43. Reichert, A. S., and Morl, M. (2000) *Nucleic Acids Res* **28**, 2043–2048.
44. Lavrov, D. V., Brown, W. M., and Boore, J. L. (2000) *Proc Natl Acad Sci USA* **97**, 13738–13742.
45. Leigh, J., and Lang, B. F. (2004) *RNA* **10** 615–621.
46. Schuster, J., Betat, H., and Morl, M. (2005) *EMBO Rep* **6** 367–372.
47. Janke, A., and Paabo, S. (1993) *Nucleic Acids Res* **21**, 1523–1525.
48. Alfonzo, J. D., Blanc, V., Estevez, A. M., Rubio, M. A., and Simpson, L. (1999) *EMBO J* **18**, 7056–7062.
49. Janke, A., Feldmaier-Fuchs, G., Thomas, W. K., von Haeseler, A., and Paabo, S. (1994) *Genetics* **137**, 243–256.
50. Morl, M., Dorner, M., and Paabo, S. (1995) *Nucleic Acids Res* **23**, 3380–3384.
51. Borner, G. V., Morl, M., Janke, A., and Paabo, S. (1996) *EMBO J* **15**, 5949–5957.
52. Urbonavicius, J., Stahl, G., Durand, J. M., Ben Salem, S. N., Qian, Q., Farabaugh, P. J., and Bjork, G. R. (2003) *RNA* **9**, 760–768.
53. Urbonavicius, J., Qian, Q., Durand, J. M., Hagervall, T. G., and Bjork, G. R. (2001) *EMBO J* **20**, 4863–4873.
54. Simpson, L., Thiemann, O. H., Savill, N. J., Alfonzo, J. D., and Maslov, D. A. (2000) *Proc Natl Acad Sci USA* **97**, 6986–6993.
55. Crain, P. F., Alfonzo, J. D., Rozenski, J., Kapushoc, S. T., McCloskey, J. A., and Simpson, L. (2002) *RNA* **8**, 752–761.
56. Holley, R. W., Apgar, J., Everett, G. A., Madison, J. T., Marquisee, M., Merrill, S. H., Penswick, J. R., and Zamir, A. (1965) *Science* **147**, 1462–1465.
57. Holley, R. W., Everett, G. A., Madison, J. T., and Zamir, A. (1965) *J Biol Chem* **240**, 2122–2128

58. Crick, F. H. (1966) *J Mol Biol* **19**, 548–555.
59. Gerber, A. P., and Keller, W. (1999) *Science* **286**, 1146–1149.
60. Wolf, J., Gerber, A. P., and Keller, W. (2002) *EMBO J* **21**, 3841–3851.
61. Grosjean, H., Edqvist, J., Straby, K. B., and Giege, R. (1996) *J Mol Biol* **255**, 67–85.
62. Kuratani, M., Ishii, R., Bessho, Y., Fukunaga, R., Sengoku, T., Shirouzu, M., Sekine, S., and Yokoyama, S. (2005) *J Biol Chem* **280**, 16002–16008.
63. Rubio, M. A., Ragone, F. L., Gaston, K. W., Ibba, M., and Alfonzo, J. D. (2006) *J Biol Chem* **281**, 115–120.
64. Rubio, M. A., Pastar, I., Gaston, K. W., Ragone, F. L., Janzen, C. J., Cross, G. A., Papavasiliou, F. N., and Alfonzo, J. D. (2007) *Proc Natl Acad Sci USA* **104**, 7821–7826.
65. McCulloch, R., Rudenko, G., and Borst, P. (1997) *Mol Cell Biol* **17**, 833–843.
66. Cross, G. A., Wirtz, L. E., and Navarro, M. (1998) *Mol Biochem Parasitol* **91**, 77–91.
67. Pays, E., Vanhamme, L., and Perez-Morga, D. (2004) *Curr Opin Microbiol* **7**, 369–374.
68. Dreesen, O., Li, B., and Cross, G. A. (2007) *Nat Rev Microbiol* **5**, 70–75.

CHAPTER 6

A-TO-I EDITING AS A CO-TRANSCRIPTIONAL RNA PROCESSING EVENT

Marie Öhman

THIS CHAPTER will focus on the potential of A-to-I editing as a modifier of nuclear pre-mRNA and the possible effects this type of modifications can have on the mRNA maturation process. The power of site-selective editing within coding sequence will be discussed as well as editing in untranslated regions that may have an influence on RNA processing and export. Other pre-mRNA processing events are believed to occur co-transcriptionally, the possibility of RNA editing as a co-transcriptional event will also be discussed.

6.1 INTRODUCTION

Eukaryotic messenger RNA maturation is a complex, multistep process that includes 5′-end capping, splicing, 3′-end cleavage, and polyadenylation. Some of these transcripts, particularly in the brain, have also been shown to be A-to-I edited. Each of these RNA processing events has been well-characterized on the biochemical level and can be reproduced *in vitro*. In recent years it has been shown that these processes are interconnected and often occur during ongoing transcription. They are coupled to each other by a network of molecular interactions involving some of the factors that participate in the different processing events. It is therefore no longer possible to consider these processes as separate events since the coordination during pre-mRNA maturation can influence the efficiency and regulate gene expression that can be both cell-type-specific and organism-specific. In this perspective it is also important to see A-to-I RNA editing as an event that participates in the network of coupled interactions in gene expression.

RNA and DNA Editing: Molecular Mechanisms and Their Integration into Biological Systems,
Edited by Harold C. Smith

6.1.1 Overview of Co-transcriptional Pre-mRNA Processing

To understand the complexity of co-transcriptional RNA processing, the coordination of the different steps during maturation of an mRNA by the transcription machinery needs to be appreciated. It is important to note that the RNA substrate for these processing events is not a static molecule. Rather, it is a target for the processing machineries as the RNA is being synthesized and folded. It is common knowledge that the general RNA processing events during mRNA maturation are linked and dependent on each other. Co-transcriptional pre-mRNA processing can be pictured as an "mRNA factory" combining the initiating, elongating, and terminating transcription machinery together with RNA processing factors (Figure 6.1) (reviewed in reference 1).

In this "factory," RNA–protein and protein–protein interactions occur simultaneously on the growing RNA. The large subunit of RNA polymerase II (pol II) plays a major role for efficiency and coordination of these RNA processing events. More specifically, the C-terminal domain (CTD) of this large pol II subunit has been shown to allosterically regulate capping, splicing, and cleavage/polyadenylation during its different stages of phosphorylation. The extreme flexibility in the structure of the CTD

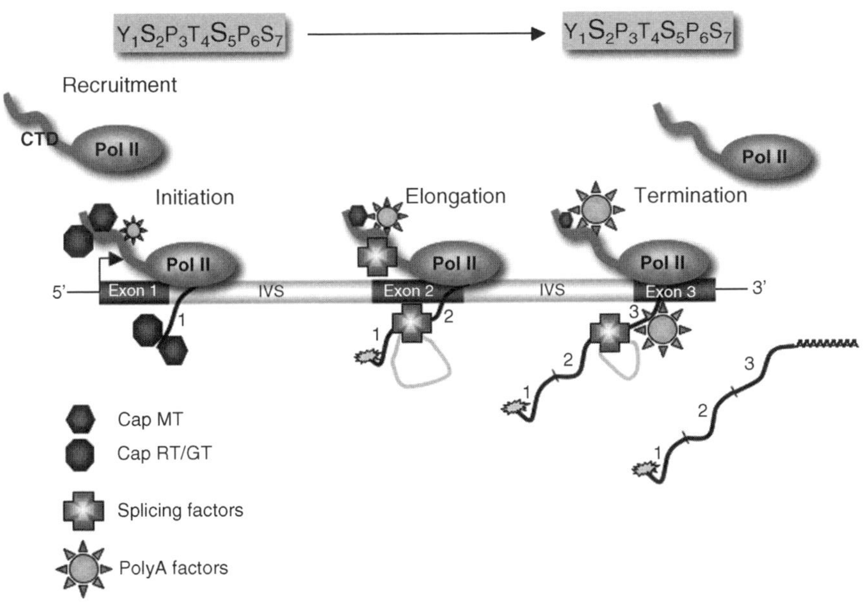

Figure 6.1 Co-transcriptional mRNA processing. Pre-mRNA processing factors are recruited to the C-terminal domain (CTD) of the largest subunit of RNA polymerase II. The capping enzymes RNA triphosphatase (RT), guanylyltransferase (GT), and 7-methyltransferase (MT), along with 3′-end cleavage/polyadenylation factors, are recruited to the genes during the initiation of transcription. As pol II is transcribing the gene, splicing factors, associate with the transcription complex. Phosphorylation of Ser2 (S_2) and Ser5 (S_5) residues in the CTD heptad repeats are indicated in bold, and increased size indicates extent of phosphorylation. Increased size of the symbols for RNA processing factors represent increased number of factors bound.

helps it to bind so many different factors. In mammals the CTD consists of 52 heptad repeats with the consensus sequence $Y_1S_2P_3T_4S_5P_6S_7$ that are phosphorylated on serine 2 and serine 5. Proteins involved in capping are the first mRNA processing enzymes to be recruited by the CTD (Figure 6.1). The RNA triphosphases (guanylyltransferase and 7-methyltransferase) capping enzyme activities depend on the CTD to efficiently cap the pre-mRNA. During transcription initiation, the CTD is phosphorylated at serine 5 (S_5) in the heptad repeats; this phosphorylation permits recruitment of the capping factors and co-transcriptional capping. Capping of the 5′end of the pre-mRNA is an addition of a methylated guanosine. The cap structure is important for mRNA stability but will also stimulate splicing and 3′-end formation (reviewed in reference 2). The cap structure has even been suggested to be important for nuclear export (3).

Pre-mRNA splicing is performed by a large multicomponent complex called the spliceosome. In the spliceosome, small nuclear ribonucleoproteins (snRNPs), together with a number of other splicing factors, interact with the nascent RNA in a specific order. Even though it is not clear if there is a direct contact between factors involved in splicing and the CTD of pol II, it is evident that splicing factors are recruited during transcription elongation and that splicing can be CTD-dependent (4). Introns are also removed co-transcriptionally or marked during the process of transcription for posttranscriptional removal (5). During this phase, the CTD of the elongating pol II is dephosphorylated at Ser5 and phosphorylated at Ser2 progressively to reach the highest state of phosphorylation at the time of transcription termination.

The cleavage/polyadenylation process at the 3′ end of the mRNA is important for export of the mature mRNA out of the nucleus and is also important for translation. The 3′ end of an mRNA is produced by a cleavage of the pre-mRNA between a conserved AAUAAA and a G/U-rich sequence. These regions are recognized by the cleavage and polyadenylation-specific factor (CPSF) and the cleavage stimulation factor (CstF). The recruitment of these cleavage/polyadenylation factors to the CTD occurs not only at the 3′ end of the transcript. Some factors required for 3′-end processing interact with the CTD at the time of transcription initiation, and factors are progressively recruited throughout transcription elongation (6, 7). It is also clear that Ser2 CTD phosphorylation, which occurs increasingly toward the end of transcription, is important for 3′-end maturation. In summary, a dynamic change in phosphorylation of the CTD seems to play an important role in the coordination of different RNA processing events to occur at the right time and place during pol II transcription.

6.1.2 Localization of the ADAR Proteins

A-to-I RNA editing by the ADAR enzymes is believed to be performed both in the nucleus and in the cytoplasm in mammals. The long form of ADAR1 (p150) is the only ADAR enzyme found in the cytoplasm, but this protein has also been found in the nucleus (8). The p150 ADAR1 is interferon-inducible and has been suggested to play a role in viral defense by deaminating viral double-stranded RNA (dsRNA). This protein shuttles between the nucleus and the cytoplasm because it contains both a nuclear export and a nuclear import signal (9). It is therefore possible that it can target dsRNA viruses in both compartments.

The shorter form of ADAR1 (p110) exists exclusively in the nucleus. All isoforms of ADAR2 contain a nuclear localization signal but lack nuclear export signals and are therefore nuclear proteins. Moreover, both ADAR1 and ADAR2 have been found in the nucleolus (10, 11). The localization of the ADAR proteins to the nucleolus is dependent on RNA, probably through binding to ribosomal RNA in a nonspecific manner. The function of this nonspecific binding might be to keep the enzyme away from "accidental" editing. Once a specific editing substrate appears, the ADAR proteins can relocate from the nucleolus to the nucleoplasm suggesting that the nucleolus is a place for storage to prevent nonspecific editing (10).

6.1.3 A-to-I Editing as a Pre-mRNA Processing Event

The majority of the known selectively edited sites in coding sequence are edited by ADAR2, although some sites, even within the same transcript, are exclusively edited by ADAR1 (12, 13). In the nucleus, both pre-mRNA and mRNA are potential targets for A-to-I editing. Among the known substrates, editing at the Q/R site in the AMPA glutamate receptor subunit B (GluR-B) is perhaps the most well known. A dramatic negative effect on editing at the Q/R site of the GluR-B pre-mRNA has been seen in ADAR2$^{-/-}$ knockout mice. Another, more obvious reason to believe that ADAR targets this substrate on the pre-mRNA level is that both exonic and intronic sequence are required for editing at this site. As discussed in Chapters 1, 9, and 12, both ADAR1 and ADAR2 only edit adenosines within RNA molecules that are largely double-stranded. ADAR1 contains three double-stranded RNA binding motifs while ADAR 2 has two that recognize and bind to dsRNA in a nonspecific manner.

Editing of up to 50% of the adenosines is seen on artificial substrates consisting of long duplexes of double-stranded RNA (14, 15). Natural substrates are of two categories; site-selectively edited and hyper-edited with many edited adenosines. Selectively edited substrates like the GluR-B transcript consist of a hairpin structure with a stem that is interrupted by bulges and loops. The edited Q/R site is situated in exon 11 while the editing complementary sequence (ECS), required for ADAR2 recognition, is situated in the downstream intron. It is therefore evident that this transcript is edited as pre-mRNA.

6.2 MAIN QUESTIONS IN THE FIELD AND APPROACHES

6.2.1 Why Are Edited Sites Often Situated Close to Exon/Intron Border?

Substrates for A-to-I editing often consist of both exon and intron sequence, with the editing site within the coding region (Figure 6.2). What is the advantage of having the editing complementary sequence within an intron sequence? The ADAR enzymes require an RNA structure that is largely double-stranded in order to interact with the substrate. There is no clear consensus sequence that the ADAR enzymes (ADAR1 and ADAR2) recognize, but both structure and sequence in the vicinity of the edited site are often extremely conserved between species (see

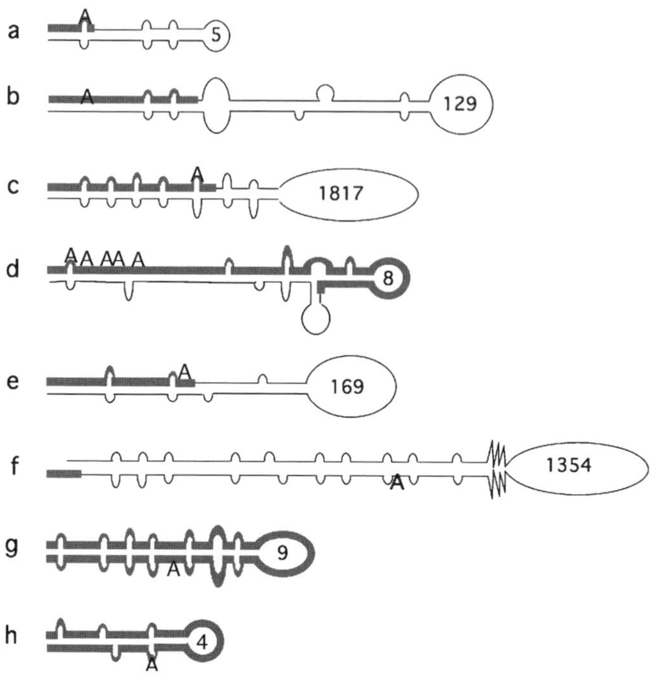

Figure 6.2 Predicted secondary structures of substrates for mammalian ADAR enzymes. The substrates are: (a) GluR-B site R/G; (b) GluR-B site Q/R; (c) GluR-6 site Q/R; (d) 5-HT_{2C} site A-E; (e) FlnA site Q/R; (f) Adar2 site −1; (g) Kv1.1 site I/V; and (h) GABRA3 site I/M. Edited sites are indicated with an A in bold. Exon sequence is in gray, while intron sequence is indicated as a thin black line. Numbers represent number of nucleotides in the loops.

Chapters 1, 12, and 15). Therefore, there may be very little flexibility allowed to achieve efficient editing.

Within the coding sequence, changes are usually limited to the third codon position in order to preserve the amino acid sequence. The intron allows the sequence to be much more flexible so that changes can occur without consequences on the proteome. In Figure 6.2, some examples of site-selectively edited substrates are shown. In the first five substrates, the edited site is located close to the exon–intron border, with the editing complementary sequence in the downstream intron. As illustrated in these examples, the ECS can be separated from the edited site by over a thousand nucleotides. Thus, even though the hairpin appears to be unbranched, to qualify as an editing substrate there is no loop size requirement and efficiency of editing is not dependent on the distance between the edited site and the ECS.

Most known sites of selective editing utilize intron sequence in order to create an efficiently edited site. Nevertheless, there are a few exceptions. In at least two known sites of selective editing, both the targeted A and the ECS are within coding sequence. One of these is in the transcript of the intronless human potassium channel gene *hK$_v$1.1* (16). In this substrate the putative hairpin RNA structure required for editing is disrupted by several internal loops of 2–10 nucleotides. The double-stranded sequence in the vicinity of the edited site is only 6 base pairs long (Figure 6.2g). Still this site is edited to almost 50% in the mouse brain, indicating that a short continuous stretch of base pairs is sufficient for efficient editing when it is in the vicinity of other regions that are base-paired. Another example where the entire sequence required for editing is situated in an exon is in the transcript coding for the alpha3 subunit of the $GABA_A$ receptor. This receptor subunit was recently found to be edited to nearly 100% in the adult mouse brain (17). In this case the putative editing complementary sequence is located 15 bases upstream of the edited site in a short stem loop of 54 nucleotides within exon 9 (Figure 6.2h). With a loop of 4 nucleotides, this is the shortest stem known to be efficiently edited.

6.2.2 The Potential of A-to-I Editing in Changing the Transcriptome

There are several different possibilities for how editing can change the transcriptome and the downstream proteome summarized in Figure 6.3. Perhaps the most obvious effect the editing event can have on an mRNA is to make new protein isoforms by changing a codon so that it codes for another amino acid. At the Q/R editing site in the GluR-B transcript a CAG codon for glutamine (Q) is changed to CIG upon editing. The CIG codes for arginine (R), since I is read as a G by the ribosome (18). More than 99% of the GluR-B subunits are in the edited R form. There are several other single sites of editing that give rise to changes in the codon usages, most of them with functional consequences. Another, more complex example is the transcript of the G-protein-coupled receptor 5-HT_{2C}, which undergoes A-to-I editing at 5 sites: A, B, E, C, and D, which are situated in close proximity to each other (Figure 6.2d). Editing at the A site or at the A and B sites converts an isoleucine into a valine at position 156 of the human receptor. Editing solely at the B site generates a methionine at the same position. Asparagine 158 is changed to serine upon editing at the C site and editing at both C and E generates a glycine. Finally, editing at the D site results in the substitution of a valine for an isoleucine at position 160. Theoretically, 24 different receptor variants can be produced from

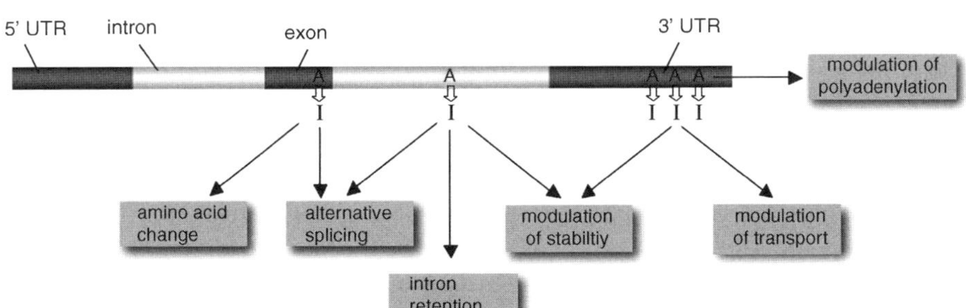

Figure 6.3 Possible effects of A-to-I editing of pre-mRNAs in the cell nucleus. The location of sites for editing are indicated on the pre-mRNA, and their potential consequences are in boxes. Exon sequence is in dark gray, while intron and untranslated regions are in light gray.

different combinations of the edited sites. Indeed, at least 10 different isoforms have been shown to be produced in various amounts in different parts of the rat brain and 12 isoforms in different human brain regions (19, 20). However, the editing pattern differs considerably between rodents and humans, suggesting that the isoforms specific to the different species may have distinct functional roles (21).

Although most sites of selective editing have been found to give rise to codon sense changes, isoform diversity can also be created by editing induced changes in splicing patterns. The mammalian pre-mRNA transcript contains several sequences that are required for the splicing process. At the 5′ splice site (marked with |) the consensus sequence AG|**G**URAGU (R = purine) can be created or disrupted by a functional A to G change caused by editing. A novel 3′ splice site, defined by **YAG**|R (Y = pyrimidine), can also be created upon editing. Further, within the intron the branchpoint CUR**A**Y is positioned close to the 3′ splice site, followed by a pyrimidine-rich track. The highly conserved adenosine at the branchpoint is also a potential target for editing to change splicing.

There are several eukaryotic viruses that are edited. Hepatitis delta virus (HDV) uses editing to vary its proteome during the infection. HDV is a virus with a negative-strand RNA genome of about 1700 nucleotides that replicates through a positive-strand intermediate called the antigenome. Editing occurs on the antigenome, and ADAR1 is primarily responsible for this editing event (22). Human HDV is edited at a single adenosine, converting an amber stop codon into a tryptophan (W) coding triplet (23). This amber/W site enables the virus to express a short and a long form of the viral protein delta antigen (HDAg), the only protein encoded by the virus. Both isoforms are required for the viral life cycle. The short form (HDAg-S) is used during viral replication, whereas the long form (HDAg-L) is involved in packaging of new virus particles.

With so much potential for editing to affect RNA processing, what limits the editing site utilization? The major limit for A-to-I editing is the requirement of an RNA structure that is largely double-stranded. However, the secondary structure in the vicinity of single sites of editing is rarely completely double-stranded since the helix structure often is interrupted by bulges and internal loops. Currently we do not know how short a continuous double-stranded region can be in order to be seen by the ADAR enzymes (see Chapters 1, and 12 for additional discussion). In addition to constraints on ADAR binding to RNA, editing may be regulated by the storage of ADAR proteins in the nucleolus.

6.2.3 RNA Editing, the Influence on Pre-mRNA Splicing and Vice Versa

Why is it important that editing happens prior to splicing? As discussed earlier, several of the known site-selectively edited substrates have the putative editing complementary sequence situated in the downstream intron (Figure 6.2). If splicing is a co-transcriptional event (see Figure 6.1) occurring as soon as the intron is transcribed, it is important that editing also occurs co-transcriptionally in a way so that it happens before the splicing machinery performs its action. In the GluR-B transcript, both the

Q/R and the R/G editing site are dependent on the downstream introns for editing to occur (Figure 6.2). Therefore, editing has to happen prior to or in concert with pre-mRNA splicing. The Q/R site in exon 11 is situated 24 nucleotides upstream of the 5′ splice site of intron 11, but transcription has to continue for another 325 nucleotides, at least, to express the editing complementary sequence situated in the intron in order to produce a substrate for editing.

Although there seems to be no physical interference between ADAR2 binding to the Q/R site and the interactions by the splicing machinery, these two processes have to be coordinated because editing cannot occur after the intron has been removed. At the R/G editing site in exon 13 of GluR-B, it is even more important that editing and splicing are synchronized. The R/G site is located only one nucleotide upstream of the 5′ splice site of intron 13. Therefore, the U1 snRNP or other spliceosomal components could compete with ADAR2 for binding at this position. *In vivo*, however, a mechanism appears to exist to ensure that co-transcriptional editing and splicing occur sequentially. Editing at the R/G site can also influence splicing since the first A in the recognition sequence for the 5′ splice site is edited. Jantsch and co-workers have recently shown that a change from AG|GU to IG|GU at the 5′ splice site has a negative effect on splicing of the intron (24).

Editing has also been suggested to influence splicing in a more indirect way. The pre-mRNA transcript of the *para* gene in *Drosophila* is A-to-I edited at several sites (25). Downstream of the exonic editing site there is a 5′ splice site within the double-stranded structure required for editing. It is presumed that the double-stranded structure prevents the 5′ splice site from being accessible to the splicing machinery. Therefore it is predicted that a double-stranded RNA helicase is required to resolve the double-stranded structure. Indeed the mle^{napts} RNA helicase mutation has been shown to cause exon-skipping by using alternative splice donors upstream of the edited exon. The exon skipping phenomenon can be explained if the normal splicing requires the resolution of the structure by a dsRNA helicase.

There are other examples where editing and splicing requires sophisticated coordination. In the Adar2 pre-mRNA the RNA structure required for editing is entirely intronic (Figure 6.2f). After editing, an alternative 3′ splice site is created, since an AA dinucleotide is edited to AI, read as an AG by the splicing machinery (26). This site is efficiently used as an alternative 3′ splice, 47 nucleotides upstream of the normal splice site (Figure 6.4, bottom left). We know that this novel 3′ splice site competes very well with the normal acceptor site (27). However, at this site, editing has to be a rapid, co-transcriptional event in order for the alternative splicing to be able to compete with the normal splice site. The use of a mutant polymerase with a lower rate of transcription could, in theory, favor editing, because this would increase the window for the editing event to happen. Nevertheless, no increase in alternative splicing could be seen when the Adar2 pre-mRNA was transcribed by a mutant pol II with a slower transcription rate, indicating that the editing is very rapid once the required RNA structure has been transcribed (Källman and Öhman, unpublished).

In $ADAR2^{-/-}$ knockout mice only a minor amount of editing at the GluR-B Q/R site can be detected (12). In the brain of these mice, the lack of Q/R editing, leads to inefficient excision of intron 11. It is not known if the absence of ADAR2 only

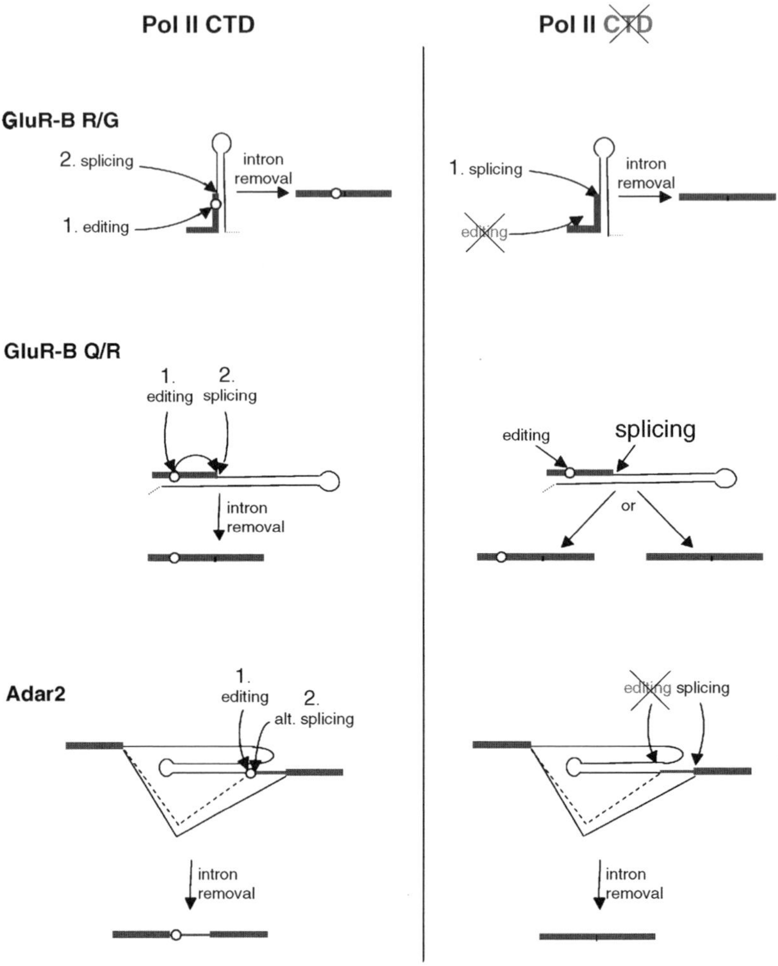

Figure 6.4 Functional consequences of transcription by a pol II lacking the CTD on mammalian editing/splicing substrates. The numbers indicate the order of the events. Exon sequence is in gray while intron sequence is indicated as a thin black line. Editing is shown as an open circle. Increased size of letters indicate increased processing.

influences splicing of intron 11 or if it has a negative effect on splicing in general. However, recent results in our laboratory suggest that it is the A-to-I modification at the Q/R site that induce splicing rather than the presence of the ADAR2 protein per se (28). Further, editing of an intronic hotspot is required for efficient splicing (24). Thus, editing induced splicing might be a way to ensure that only properly edited mRNAs are produced.

6.2.3.1 What Is the Role of RNA Polymerase II in Coordination of Editing and Pre-mRNA Splicing?

Deletion of the CTD of pol II inhibits other RNA processing events like capping and cleavage/polyadenylation (4, 29). However, editing in a context where it is not dependent on splicing is not affected by the CTD deletion (28). Editing often has to occur during a relatively short time window before the process of splicing is completed. This can be achieved if ADAR binding has priority over and/or stalls the splicing machinery. Analysis of *in vitro* editing and splicing of a pre-made GluR-B transcript has revealed that editing and splicing can interfere with each other when their sites of recognition are in proximity to each other, while co-transcriptional editing and splicing during pol II transcription *in vivo* of the same substrate is synchronized as sequential events (30).

We have investigated the coordination between ADAR2 editing, splicing, and the role of the CTD as a coordinator of these two RNA processing events. From these analyses, models have been proposed for synchronized editing and splicing where the CTD coordinates these two RNA processing events in the GluR-B and Adar2 transcripts (Figure 6.4). The CTD is required for the coordination when sites of editing and splicing are in close proximity—as is the case for the GluR-B R/G site, where the editing site is located only one nucleotide upstream of a 5′ splice site. This coordination is dependent on the CTD of pol II because editing is severely decreased when transcribed by a polymerase that lacks the CTD. The Q/R site of editing in the GluR-B pre-mRNA is more distant to the splicing donor site (Figure 6.2b). At this site, the CTD also has a less prominent effect on editing. However, the coordination between editing and splicing seems to be disrupted without the CTD, causing an increase in splicing efficiency and a slight reduction in editing efficiency (28). Unexpectedly, the A-to-I modification has a stimulatory effect on splicing. Thus splicing and editing influence each other and the CTD helps restrain splicing to facilitate editing at the Q/R site.

In summary, the CTD of pol II helps coordinate editing and splicing in different ways at the R/G and the Q/R site of GluR-B (Figure 6.4). In both cases splicing is retained and occurs after editing in presence of the CTD. During transcription without the CTD, editing at the R/G site is inhibited and many of the transcripts have escaped editing after intron removal. At the Q/R site, splicing is more efficient without the CTD, but this only has a moderate effect on editing and a mixture of edited and unedited transcripts are detected after intron removal (Figure 6.4).

During the editing-induced alternative splicing of the Adar2 pre-mRNA, it is crucial that splicing is restrained until editing takes place and the new 3′ splice site is formed. The CTD is not required for splicing of this pre-mRNA because both normal and alternative splicing occur independently of the CTD. The CTD is required for efficient co-transcriptional editing of the pre-mRNA. In absence of the CTD, the decrease in editing efficiency more frequently leads to an mRNA with normal splicing (Figure 6.4). Upon editing, the alternative 3′ splice site competes well with the normal acceptor site, and the majority of the edited transcripts are alternatively spliced after intron removal.

Taken together, this indicates that editing is dependent on the CTD for the coordination with other RNA processing events during the maturation of an mRNA. These results also indicate that the CTD has a role as synchronizer rather than a pure inducer and that the coordination is solved differently, depending on location of editing and splice sites.

6.3 CAN EDITING INFLUENCE THE FATE OF A MESSENGER RNA IN OTHER WAYS?

There is great potential that A-to-I editing can fine-tune and diversify the proteome. There are, however, other processes in the eukaryotic cell that can be affected by RNA editing. Without changes in coding sequence, editing can still have a large impact on gene expression by influencing nuclear export and mRNA stability. Sites of editing for this type of regulation have been found in the 3′ UTR of the transcript and also within introns (see Chapters 9 and 12). Below are some examples that indicate that RNA editing and export as well as editing and RNA stability are coupled.

6.3.1 Editing and Its Potential Effect on RNA Export

The CAT2 protein is a plasma membrane receptor that facilitates the uptake of L-arginine, a substrate for synthesis of nitric oxide (NO). The NO pathway is stress-induced and an important component of the cellular defense program. The mCAT2 and the CTN-RNA are both transcribed from the *mouse cationic amino acid transporter2* (*mCAT2*) gene using alternative promoters and poly(A) sites (31). The poly(A) site in the CTN-RNA is situated 4000 bases downstream of the cleavage/polyadenylation signal in mCAT2. Thus, the CTN-RNA has a 4-kilobase 3′-untranslated region (UTR) that is not present in the mCAT2 transcript. The 3′-UTR contains inverted repeats of SINE elements that can form duplex RNA subjected to A-to-I editing at several sites. This edited RNA binds a nuclear complex, containing p54/nrb, PSF, and matrin 3, known to be responsible for nuclear retention (32) (see also Chapter 9).

The nuclear retained CTN-RNA is more abundant than the mCAT2 transcript. Furthermore, specific knock-down of the CTN-RNA downregulates not only the CTN-RNA but also the mCAT2 RNA, indicating that the CTN-RNA has a role in stabilizing its protein coding partner (31). During normal conditions, only the mCAT2 mRNA is exported to the cytoplasm and translated into CAT2 protein. Upon stress, the more abundant CTN-RNA is cleaved within its 3′ UTR, thereby eliminating the nuclear retention element. The cleaved mCAT2-like RNA can then be transported out to the cytoplasm where it is translated. Due to a rapid cleavage of the highly abundant CTN-RNA, high levels of translation-competent cytoplasmic mCAT2-like CTN-RNA can be achieved upon stress. The exact position of the cleavage site is not known, or if editing is required for targeting of the cleavage. Nevertheless, A-to-I editing seems to help storing transcripts that can rapidly be expressed into CAT2 protein upon stress.

6.3.2 Editing as a Modulator of RNA Stability

Scadden and co-workers have shown that hyper-edited dsRNA can induce cleavage at specific sites and thereby affect the stability of an RNA (33, 34). The staphylococcal nuclease Tudor (Tudor-SN) is a subunit of RISC, the RNA-induced silencing complex, involved in RNA interference by microRNAs. It has been shown that this enzyme interacts with and cleave inosine containing RNA (35) (see also Chapter 9). Interestingly, the first edited natural RNA substrate shown to be degraded by Tudor-SN is the primary microRNA 142 (pri-miRNA 142) (36). In the previous section, nuclear retention of the hyper-edited CTN-RNA was discussed. One possibility is that it is Tudor-SN that targets the inosine containing 3′-UTR of the CTN-RNA and thereby induces the transport out to the cytoplasm. Other candidates that may be targets for this type of cleavage are DNA or RNA viruses with intermediates of long dsRNA. However, perhaps the most likely targets for Tudor-SN-induced cleavage in man are transcripts that are hyper-edited in their 3′-UTRs due to inverted repeats of Alu sequence elements.

Several computational methods have led to the discovery of over 2000 genes with predicted A-to-I RNA editing in their UTRs (37–40) (see Chapters 1, and 12). It is likely that editing and subsequent nuclease cleavage induces degradation of at least some of these transcripts. However, not all inosine-containing RNAs are cleaved by Tudor-SN. Cleavage occurs preferentially in a long dsRNA sequence that contains both IU and UI base pairs (34). A 295-base-pair-long perfect duplex of RNA is hyper-edited by both ADAR1 and ADAR2, preferentially creating this multiple IU and UI pairs. This result indicates that ADAR can contribute to the creation of recognition sites for Tudor-SN cleavage and might therefore induce degradation of edited mRNAs.

6.3.2.1 Can Editing Induce Nonsense Mediated Decay? In the pre-mRNA of ADAR2 and PTPN6, editing causes alternative splicing that leads to a frame shift. The change of open reading frame creates a premature stop codon that might lead to the expression of a truncated protein. Another possibility is that the RNA is subjected to nonsense-mediated decay (NMD). By inducing alternative splicing, editing has the potential to stimulate mRNA degradation by the NMD pathway. It is believed that the mammalian cell routinely utilizes alternative splicing to trigger NMD in order to regulate protein levels (reviewed in reference 41). It is therefore possible that editing-induced alternative splicing is also used to regulate gene expression. Furthermore, when the NMD pathway was downregulated by siRNA-mediated depletion of hUpf1 in HeLa cells, a two-fold induction of ADAR2 mRNA was observed by microarray analysis comparing depleted with undepleted cells (42). This result indicates that the ADAR2 transcript is stabilized in NMD-deficient cells. In line with this observation, less than 5% of the ADAR2 proteins correspond to the truncated protein, although 80% of ADAR2 transcripts were subjected to editing and subsequent alternative splicing in rat brain (26), suggesting that the editing-induced alternatively spliced ADAR2 pre-mRNA is subjected to NMD.

6.3.3 Editing and Its Influence on Polyadenylation

The mouse polyomavirus has a circular DNA genome. The expression is divided into the early and the late transcription unit, transcribed in opposite directions. As in many eukaryotic viruses, the early transcript is required for viral replication while the late transcript codes for viral proteins. The early and late polyadenylation sites overlap with at least 45 nucleotides. The overlapping RNAs create a double-stranded RNA that is a substrate for editing (43). The polyadenylation site of both the early and the late primary transcripts are edited, but their fates differ. The early transcript is degraded after being edited, while the late transcript is stabilized. In fact, the late antisense RNA only accumulates after replication initiation. RNA editing will therefore regulate the early–late switch during the polyomavirus infection by altering the sequence required for polyadenylation of the two transcripts.

6.4 PROSPECTIVES FOR FUTURE RESEARCH

Can altered editing affect other RNA processing events and then be a cause of disease? The tyrosine phosphatase PTPN6 is a cytoplasmic protein that downregulates several receptors that promote growth during the development of hematopoietic cells by dephosphorylating them. PTPN6 is believed to play a tumor-suppressing role while alterations in PTPN6 expression have an oncogenic potential. The pre-mRNA of PTPN6 is edited at several sites, with the major site for A-to-I editing at the branchpoint A^{7866} in intron 3 (44). Editing at A^{7866} leads to an altered splicing pattern and retention of intron 3. Retention of the 251 nucleotides of intron sequence changes the open reading frame of the translated protein creating a premature stop codon. The branchpoint is less efficiently edited in patients with acute myeloid leukemia. Therefore, editing might have a regulatory role in downregulating of the PTPN6 protein expression. Posttranscriptional processing of PTPN6, due to a decrease in editing in acute myeloid leukemia patients, might play a role in the leukemogenesis.

Another substrate for editing that has the potential to play a role in a disease was found in the pre-mRNA coding for the endothelin B receptor (ETB). Editing was found in healthy individuals but occurred at a much higher frequency in patients with Hirschsprung disease (45). A-to-I editing was detected at A^{950} in exon 4 of the ETB RNA. This editing event results in a glutamine to arginine (Q/R) substitution after translation. Editing at the Q/R site of ETB was only observed in transcripts that were subjected to alternative splicing by exon-skipping of exon 5. These findings suggest the possibility that the processes of alternative splicing and editing influence each other. Exon-skipping of exon 5 leads to a frame shift, which results in premature stop codons. This alternative mRNA product has not been detected, possibly due to rapid degradation through the NMD pathway.

Since A-to-I editing has the potential to change the sequences required for splicing at several sites, we might expect to find alterations in splicing caused by an increase or decrease in editing in many more transcripts yet to be discovered. The discovery of hyper-editing in untranslated regions revealed that massive editing occurs within mRNAs containing inverted repetitive elements (see Chapter 9). If editing is so abundant in these regions, it is possible that also other noncoding regions of the pre-mRNA, like the introns, are abundantly edited. Editing that creates or disrupts excision of introns has the potential to be the answer to many alternative splicing events. However, editing-induced alternative splicing might be hard to detect if ADAR targets the pre-mRNA within intron sequence that is excised and therefore not visible in the expressed sequence tag (EST). To find these, new computational methods for detection of novel sites of editing are required that are not based on comparisons between the genomic sequence and the cDNA or EST. In combination with experimental analyses, future analysis of sequence conservation in noncoding sequence along with predictions of double-stranded RNA structure could possibly detect editing-induced splicing in both healthy individuals and diseased ones.

REFERENCES

1. Bentley, D. L. (2005) Rules of engagement: co-transcriptional recruitment of pre-mRNA processing factors. *Curr Opin Cell Biol* **17**, 251–256.
2. Proudfoot, N. J., Furger, A., and Dye, M. J. (2002) Integrating mRNA processing with transcription. *Cell* **108**, 501–512.
3. Cheng, H., Dufu, K., Lee, C. S., Hsu, J. L., Dias, A., and Reed, R. (2006) Human mRNA export machinery recruited to the 5′ end of mRNA. *Cell* **127**, 1389–1400.
4. McCracken, S., Fong, N., Yankulov, K., Ballantyne, S., Pan, G., Greenblatt, J., Patterson, S. D., Wickens, M., and Bentley, D. L. (1997) The C-terminal domain of RNA polymerase II couples mRNA processing to transcription. *Nature* **385**, 357–361.
5. Bauren, G., and Wieslander, L. (1994) Splicing of Balbiani ring 1 gene pre-mRNA occurs simultaneously with transcription. *Cell* **76**, 183–192.
6. Licatalosi, D. D., Geiger, G., Minet, M., Schroeder, S., Cilli, K., McNeil, J. B., and Bentley, D. L. (2002) Functional interaction of yeast pre-mRNA 3′ end processing factors with RNA polymerase II. *Mol Cell* **9**, 1101–1111.
7. Kim, M., Ahn, S. H., Krogan, N. J., Greenblatt, J. F., and Buratowski, S. (2004) Transitions in RNA polymerase II elongation complexes at the 3′ ends of genes. *EMBO J* **23**, 354–364.
8. Patterson, J. B., and Samuel, C. E. (1995) Expression and regulation by interferon of a double-stranded-RNA-specific adenosine deaminase from human cells: Evidence for two forms of the deaminase. *Mol Cell Biol* **15**, 5376–5388.
9. Poulsen, H., Nilsson, J., Damgaard, C. K., Egebjerg, J., and Kjems, J. (2001) CRM1 mediates the export of ADAR1 through a nuclear export signal within the Z-DNA binding domain. *Mol Cell Biol* **21**, 7862–7871.
10. Desterro, J. M., Keegan, L. P., Lafarga, M., Berciano, M. T., O'Connell, M., and Carmo-Fonseca, M. (2003) Dynamic association of RNA-editing enzymes with the nucleolus. *J Cell Sci* **116**, 1805–1818.
11. Sansam, C. L., Wells, K. S., and Emeson, R. B. (2003) Modulation of RNA editing by functional nucleolar sequestration of ADAR2. *Proc Natl Acad Sci USA* **100**, 14018–14023.
12. Higuchi, M., Maas, S., Single, F. N., Hartner, J., Rozov, A., Burnashev, N., Feldmeyer, D., Sprengel, R., and Seeburg, P. H. (2000) Point mutation in an AMPA receptor gene rescues lethality in mice deficient in the RNA-editing enzyme AD AR2. *Nature* **406**, 78–81.
13. Hartner, J. C., Schmittwolf, C., Kispert, A., Muller, A. M., Higuchi, M., and Seeburg, P. H. (2004) Liver disintegration in the mouse embryo caused by deficiency in the RNA-editing enzyme AD AR1. *J Biol Chem* **279**, 4894–4902.
14. Bass, B. L., and Weintraub, H. (1988) An unwinding activity that covalently modifies its double-stranded RNA substrate. *Cell* **55**, 1089–1098.
15. Nishikura, K., Yoo, C., Kim, U., Murray, J. M., Estes, P. A., Cash, F. E., and Liebhaber, S. A. (1991) Substrate specificity of the dsRNA unwinding/modifying activity. *EMBO J* **10**, 3523–3532.
16. Bhalla, T., Rosenthal, J. J., Holmgren, M., and Reenan, R. (2004) Control of human potassium channel inactivation by editing of a small mRNA hairpin. *Nat Struct Mol Biol* **11**, 950–956.
17. Ohlson, J., Skou Pedersen, J., Haussler, D., and Öhman, M. (2007) Editing modifies the GABA-A receptor subunit alpha3. *RNA* **13**, 698–703.
18. Higuchi, M., Single, F. N., Kohler, M., Sommer, B., Sprengel, R., and Seeburg, P. H. (1993) RNA editing of AMPA receptor subunit GluR-B: A base-paired intron–exon structure determines position and efficiency. *Cell* **75**, 1361–1370.

19. Burns, C. M., Chu, H., Rueter, S. M., Hutchinson, L. K., Canton, H., Sanders-Bush, E., and Emeson, R. B. (1997) Regulation of serotonin-2C receptor G-protein coupling by RNA editing. *Nature* **387**, 303–308.
20. Wang, Q., O'Brien, P. J., Chen, C. X., Cho, D. S., Murray, J. M., and Nishikura, K. (2000) Altered G protein-coupling functions of RNA editing isoform and splicing variant serotonin2C receptors. *J Neurochem* **74**, 1290–1300.
21. Niswender, C. M., Copeland, S. C., Herrick-Davis, K., Emeson, R. B., and Sanders-Bush, E. (1999) RNA editing of the human serotonin 5-hydroxytryptamine 2C receptor silences constitutive activity. *J Biol Chem* **274**, 9472–9478.
22. Jayan, G. C., and Casey, J. L. (2002) Inhibition of hepatitis delta virus RNA editing by short inhibitory RNA-mediated knockdown of ADAR1 but not ADAR2 expression. *J Virol* **76**, 12399–12404.
23. Casey, J. L., and Gerin, J. L. (1995) Hepatitis D virus RNA editing: Specific modification of adenosine in the antigenomic RNA. *J Virol* **69**, 7593–7600.
24. Schoft, V. K., Schopoff, S., Jantsch, M. F., (2007). Regulation of glutamate receptor B pre-mRNA splicing by RNA editing. *Nucleic Acids Res* **35**, 3723–3732.
25. Reenan, R. A., Hanrahan, C. J., and Barry, G. (2000) The mle(napts) RNA helicase mutation in drosophila results in a splicing catastrophe of the para Na^+ channel transcript in a region of RNA editing. *Neuron* **25**, 139–149.
26. Rueter, S. M., Dawson, T. R., and Emeson, R. B. (1999) Regulation of alternative splicing by RNA editing. *Nature* **399**, 75–80.
27. Laurencikiene, J., Källman, A. M., Fong, N., Bentley, D. L., and Öhman, M. (2006) RNA editing and alternative splicing: the importance of co-transcriptional coordination. *EMBO Rep* **7**, 303–307.
28. Ryman, K., Fong, N., Bratt, E., Bentley, D. L., and Öhman, M. (2007) The C-terminal domain of RNA pol II helps ensure that editing precedes splicing of the GluR-B transcript. *RNA* **13**, 1071–1078.
29. McCracken, S., Fong, N., Rosonina, E., Yankulov, K., Brothers, G., Siderovski, D., Hessel, A., Foster, S., Shuman, S., and Bentley, D. L. (1997) 5′-Capping enzymes are targeted to pre-mRNA by binding to the phosphorylated carboxy-terminal domain of RNA polymerase II. *Genes Dev* **11**, 3306–3318.
30. Bratt, E., and Öhman, M. (2003) Coordination of editing and splicing of glutamate receptor pre-mRNA. *RNA* **9**, 309–318.
31. Prasanth, K. V., Prasanth, S. G., Xuan, Z., Hearn, S., Freier, S. M., Bennett, C. F., Zhang, M. Q., and Spector, D. L. (2005) Regulating gene expression through RNA nuclear retention. *Cell* **123**, 249–263.
32. Zhang, Z., and Carmichael, G. G. (2001) The fate of dsRNA in the nucleus: a p54(nrb)-containing complex mediates the nuclear retention of promiscuously A-to-I edited RNAs. *Cell* **106**, 465–475.
33. Scadden, A. D., and Smith, C. W. (2001) Specific cleavage of hyper-edited dsRNAs. *EMBO J* **20**, 4243–4252.
34. Scadden, A. D., and O'Connell, M. A. (2005) Cleavage of dsRNAs hyper-edited by ADARs occurs at preferred editing sites. *Nucleic Acids Res* **33**, 5954–5964.
35. Scadden, A. D. (2005) The RISC subunit Tudor-SN binds to hyper-edited double-stranded RNA and promotes its cleavage. *Nat Struct Mol Biol* **12**, 489–496.
36. Yang, W., Chendrimada, T. P., Wang, Q., Higuchi, M., Seeburg, P. H., Shiekhattar, R., and Nishikura, K. (2006) Modulation of microRNA processing and expression through RNA editing by ADAR deaminases. *Nat Struct Mol Biol* **13**, 13–21.
37. Athanasiadis, A., Rich, A., and Maas, S. (2004) Widespread A-to-I RNA editing of Alu-containing mRNAs in the human transcriptome. *PLoS Biol* **2**, e391.
38. Blow, M., Futreal, P. A., Wooster, R., and Stratton, M. R. (2004) A survey of RNA editing in human brain. *Genome Res* **14**, 2379–2387.
39. Kim, D. D., Kim, T. T., Walsh, T., Kobayashi, Y., Matise, T. C., Buyske, S., and Gabriel, A. (2004) Widespread RNA editing of embedded alu elements in the human transcriptome. *Genome Res* **14**, 1719–1725.
40. Levanon, E. Y., Eisenberg, E., Yelin, R., Nemzer, S., Hallegger, M., Shemesh, R., Fligelman, Z. Y., Shoshan, A., Pollock, S. R., Sztybel, D., Olshansky, M., Rechavi, G., and Jantsch, M. F. (2004) Systematic identification of abundant A-to-I editing sites in the human transcriptome. *Nat Biotechnol* **22**, 1001–1005.
41. Lejeune, F., and Maquat, L. E. (2005) Mechanistic links between nonsense-mediated mRNA decay and pre-mRNA splicing in mammalian cells. *Curr Opin Cell Biol* **17**, 309–315.

42. Mendell, J. T., Sharifi, N. A., Meyers, J. L., Martinez-Murillo, F., and Dietz, H. C. (2004) Nonsense surveillance regulates expression of diverse classes of mammalian transcripts and mutes genomic noise. *Nat Genet* **36**, 1073–1078.
43. Gu, R., Zhang, Z., and Carmichael, G. G. (2007) How a small DNA virus uses dsRNA but not RNAi to regulate its life cycle. *Cold Spring Harbor Symposia on Quantitative Biology* **LXXI** 1–7.
44. Beghini, A., Ripamonti, C. B., Peterlongo, P., Roversi, G., Cairoli, R., Morra, E., and Larizza, L. (2000) RNA hyperediting and alternative splicing of hematopoietic cell phosphatase (PTPN6) gene in acute myeloid leukemia. *Hum Mol Genet* **9**, 2297–2304.
45. Tanoue, A., Koshimizu, T. A., Tsuchiya, M., Ishii, K., Osawa, M., Saeki, M., and Tsujimoto, G. (2002) Two novel transcripts for human endothelin B receptor produced by RNA editing/alternative splicing from a single gene. *J Biol Chem* **277**, 33205–33212.

CHAPTER 7

STUDYING AND WORKING WITH RIBONUCLEOPROTEINS THAT CATALYZE H/ACA GUIDED RNA MODIFICATION

U. Thomas Meier

THE MAJORITY of H/ACA ribonucleoproteins (RNPs) are small nucleolar (sno)RNPs, which form the largest class of RNPs in the nucleus of every cell. Most H/ACA RNPs function in the pseudouridylation of target RNAs such as ribosomal and spliceosomal small nuclear RNAs. However, most crucial to a cell may the function of a minority of H/ACA RNPs in ribosomal RNA processing, telomerase RNA stabilization, and yet-to-be-determined processes. The challenge of H/ACA RNP study– for example, their role in the inherited bone marrow failure syndrome dyskeratosis congenita— is to distinguish not only between their many functions but also between the ~150 virtually identical RNPs differing solely in their RNAs.

7.1 INTRODUCTION

Recently, several excellent reviews have covered many aspects of H/ACA RNP biology (1–5). Here, I will focus on mammalian complexes and highlight methods and challenges in the analysis of H/ACA RNPs. H/ACA RNAs are generally short, ~140-nucleotide-long RNAs with a characteristic "hairpin–hinge–hairpin–tail" secondary structure (Figure 7.1A). They derive their name from a nearly invariant ACA triplet exactly three nucleotides from the 3′end and from the hinge region with a similar sequence (ANANNA). In a major bulge of each hairpin (pseudouridylation pocket), both strands at the bottom of the upper stem can exhibit 3 to 10 nucleotide complementarity to a target RNA (Figure 7.1A, gray RNA). Hybridization of these nucleotides to sequences flanking a uridine in the target RNA specifies it for

RNA and DNA Editing: Molecular Mechanisms and Their Integration into Biological Systems, Edited by Harold C. Smith

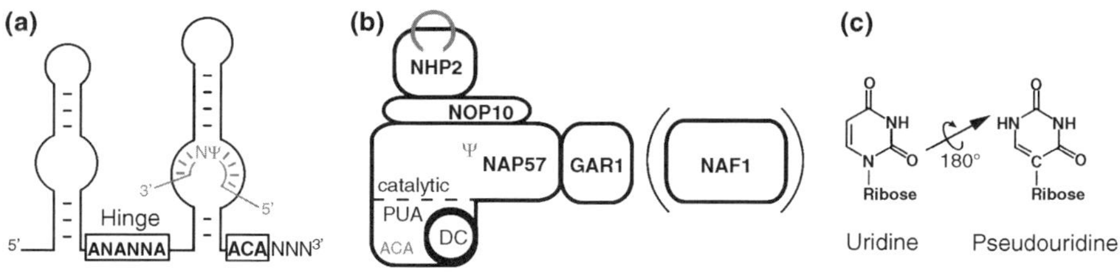

Figure 7.1 (**A**) Structure of uridine and pseudouridine. Note the 180° rotation of uracil and C-glycosidic bond in pseudouridine. (**B**) Secondary structure of a generic H/ACA RNA pointing out the hallmark "hairpin–hinge–hairpin-tail" organization and the conserved sequence elements in the hinge region and the tail. A target RNA hybridizing to nucleotides in the bulge (pseudouridylation pocket) of the 3′ hairpin is indicated in gray. Note that the unpaired uridine at the bottom of the terminal stem is converted to pseudouridine (Ψ). Either one or both hairpins can guide pseudouridylation. (**C**) Schematic representation of the H/ACA core protein complex. The sketch is based on the crystal structure of archaeal H/ACA RNPs, and the relative sizes of the proteins are based on that of mammalian RNPs. Note the catalytic and PUA domains of NAP57 with the site of catalysis (gray Ψ); the location of the tethering of a single H/ACA RNA hairpin via its terminal loop and its ACA (in gray); and the assembly factor NAF1, which is not part of mature particles but binds to NAP57 at the same site as GAR1.

pseudouridylation. Therefore, these H/ACA RNAs are also referred to as guide RNAs. In the mammalian nucleolus, most of these RNAs guide the pseudouridylation of ~100 uridines in ribosomal RNA. In Cajal bodies (subnuclear organelles with a ~1-μm diameter), the modification of an additional ~30 uridines in spliceosomal snRNAs is guided by H/ACA RNAs that carry a four-nucleotide Cajal body targeting sequence in the terminal loop of one or both of their hairpins. Therefore, these H/ACA RNAs belong to the small Cajal body (sca)RNA class of guide RNAs. Perhaps most interesting is a growing number of orphan H/ACA RNAs (>20) that lack complementarity to noncoding RNAs and therefore may target unknown RNAs for pseudouridylation or serve novel functions altogether. Two H/ACA RNAs, U17/E1 (snR30 in yeast) and mammalian telomerase RNA (hTR), stand out as they exhibit complementarity to target RNAs without guiding their pseudouridylation. U17 is an essential H/ACA RNA that is required for a site-specific cleavage during pre-rRNA processing, whereas hTR (which ends in a 3′ H/ACA domain) provides the template for telomere length maintenance.

Every one of these H/ACA RNAs assembles with the same four core proteins to generate a metabolically stable RNP (Figure 7.1B). In addition to the 57-kD pseudouridine synthase NAP57 (aka dyskerin or Cbf5 in yeast and archaea), these proteins include the 8 to 23-kD NOP10, NHP2 (L7Ae in archaea), and GAR1. These five components, one function-defining H/ACA RNA and four core proteins, are thought to form the basic functional unit of every single H/ACA RNP. The structure and assembly of these particles has been studied extensively (see below) and revealed the requirement for transacting factors. Such factors include NAF1, SHQ1, SMN, Nopp140 (Srp40 in yeast), p50/p55 (Rvb1 and 2 in yeast), and Has1, which interact with H/ACA RNP components without being an integral part of the functional particle. Since their discovery 10 years ago, these simple five-component RNPs have yielded an astounding array of complexity in biogenesis and function, yet many questions remain unanswered.

7.2 DISCOVERY OF COMPLEX (RNA-GUIDED) PSEUDOURIDINE SYNTHASES

Several independent lines of investigation lead to the appreciation of H/ACA RNPs. Identification of H/ACA RNAs and their role in guiding pseudouridylation was driven by the earlier discovery of C/D RNAs and their role in guiding 2′-*O*-methylation (see Chapter 8) (6, 7). The pseudouridine synthase of H/ACA RNPs was originally identified in yeast as the centromere binding factor Cbf5 and subsequently in mammalian cells as the nucleolar Nopp140 associated protein NAP57 with homology to an at-the-time-uncharacterized bacterial open reading frame (8, 9). Soon after, the protein encoded by this open reading frame was purified based on its pseudouridylation activity of the universally conserved U55 in elongator t-RNA and termed TruB (10). Purified TruB alone proved highly active toward t-RNA and was sufficient for specific pseudouridylation of U55. However, attempts to use bacterially expressed NAP57 in the same assay were unsuccessful (U.T. Meier and J. Ofengand, unpublished results). The discovery of H/ACA guide RNAs explained our failure. Thus, unlike single-protein pseudouridine synthases, which recognize the target uridine and catalyze the pseudouridylation (and which, even in eukaryotes, are responsible for most of the tRNA modifications), NAP57 pseudouridylates its targets only when guided by an H/ACA RNA. This concept and the composition of H/ACA RNPs was originally deciphered taking advantage of the ease of manipulation of the yeast *Saccharomyces cerevisiae*. For example, genetic depletion of Cbf5, the yeast NAP57 ortholog, or mutation of its catalytic aspartate resulted in a global loss of pseudouridines in rRNA (11, 12). Moreover, purification of the H/ACA RNA snR30 (U17/E1 in metazoans) by gradient fractionation or of H/ACA RNPs in general by using a tagged Gar1 [which was originally recognized as associated with H/ACA RNAs based on its sequence relationship with fibrillarin (13–15), the methylase of C/D RNPs (16)] revealed the full complement of H/ACA core proteins, Cbf5, Nop10, Nhp2, and Gar1 (17–19). Depletion of all core proteins, except Gar1, causes a loss of all H/ACA RNAs and of the other core proteins, whereas all core proteins are essential for cell viability and for site-specific pseudouridylation of rRNA *in vivo*. Important determinants for the requirement of H/ACA RNPs for cell viability are the roles of certain RNPs in pre-rRNA cleavage (snR30) and in pre-mRNA splicing (see Chapter 14). The fact that the same four core proteins are associated with all H/ACA RNAs, even when they differ in function, complicates their molecular and functional analysis.

7.3 APPROACHES AND CHALLENGES

Identification of components of individual H/ACA RNPs is tricky. Using affinity purification for one of the core proteins has been helpful in identifying H/ACA RNAs and the major core proteins as outlined above. However, if one or few of H/ACA RNPs differed in their composition from the bulk, such particle-specific components might escape detection when the ~150 H/ACA RNPs are co-isolated. For example, the small number of composite scaRNAs (consisting of an H/ACA appended to a C/D domain) will harbor core proteins from both types of RNPs, H/ACA and C/D (20, 21). Thus, individual H/ACA RNPs may differ by more than just their H/ACA RNA. To identify such H/ACA RNP-specific components, individual H/ACA RNPs need to be isolated and analyzed. The most logical solution to this problem is to tag specific H/ACA RNAs and use them as handles to pull out their associated RNPs. This of course leaves us with a choice of ~150 different H/ACA RNAs! Nevertheless, there are a few obvious candidates—that is, mammalian telomerase RNA, the rRNA processing-specific U17/E1, the tissue-specifically expressed HBI-36 [exclusive to the choroid plexus of the brain (22)], and perhaps one of the orphan H/ACA RNAs (because their number is lower). But what about the rest of the H/ACA RNAs? Is it possible that the modification of certain uridines requires separate transacting factors, such as helicases (23)?

Tagging and/or affinity purification of H/ACA RNAs entails its own problems because an extended RNA tag and/or hybridization to an endogenous H/ACA RNA may interfere with RNP function, biogenesis, and/or integrity. Moreover, without having the pure RNP actually in hand, there is no way of judging if the particle is fully active and contains the full complement of proteins. The purification procedure itself may be problematic, because loosely associated proteins may be lost and because other proteins associate nonspecifically. The latter was illustrated when the H/ACA RNP assembly factor NAF1, which is not a part of the mature RNPs, associated with RNPs only after cell lysis (24). The obvious choice of organism for RNA tagging is yeast where the endogenous copy of an H/ACA RNA can easily be replaced by a tagged one. While this allows *in vivo* testing for functionality of this H/ACA RNA, the selection of RNA will be limited to guide RNAs but exclude telomerase or tissue-specific H/ACA RNAs, which are mammalian-specific. However, the advent of RNA interference approaches now allows similar technology in mammalian cells, although the relatively elaborate procedure and optimization associated with it raises again the question of which H/ACA RNA is worth the effort.

Having raised all these issues, it should be pointed out that recently a very elegant, activity-based approach was successful in isolating specifically active telomerase RNPs from tissue culture cells (25). Surprisingly, in addition to the telomerase RNA, this active RNP only contained the telomerase reverse transcriptase TERT and NAP57 but none of the other H/ACA core proteins (25). The absence of the other core proteins, which have been identified as hTR associated (26, 27) and which (except for Gar1) are required for stable expression of hTR in yeast (28), could indicate that active telomerase contains NAP57 as the only H/ACA core protein. Indeed, NAP57 is the protein least exchangeable off H/ACA RNPs, whereas the other core proteins are less tightly associated (24). Alternatively, the other core proteins may have been lost during purification or escaped detection.

7.4 RNP RECONSTITUTION

The need for functional and structural analysis of individual H/ACA RNPs is not only restricted to the specialized telomerase and U17/E1 H/ACA RNAs but also extends to the entire family of RNPs involved in rRNA modification. That the individual H/ACA RNAs can impart slightly different integrity to their RNPs is perhaps best illustrated by their different resistance to salt during elution from a cellular mixture of H/ACA RNPs that were isolated and immobilized via a tagged core protein (29). Therefore, there is an imperative need for pure individual particles. Given the only five core components, the obvious approach to obtain such individual H/ACA RNPs is to reconstitute the particles using specific H/ACA RNAs. Several approaches were successful in this regard, each with its own set of advantages and disadvantages.

Reconstitution of eukaryotic H/ACA RNPs proved particularly challenging. Based on the experience from several groups, expression of useful amounts of the full-length pseudouridine synthase in soluble form in bacteria and other systems was unsuccessful (30, 31). However, all core proteins, including NAP57, could be expressed by *in vitro* transcription/translation in rabbit reticulocyte lysate (31). Incorporation of ^{35}S-methionine produced sufficient amounts of protein for detection and analysis by immunoprecipitation. Using this approach, we determined that, unusually for an RNP, the four core proteins can form a complex independent of an H/ACA RNA. Specifically, NAP57 binds NOP10 as a prerequisite for association of NHP2 (Figure 7.1B). Together these three proteins form a core trimer that is sufficient for specific recognition of H/ACA RNAs. GAR1 binds NAP57 (and the core trimer) independently (Figure 7.1B) (31). Although these interactions are supported by studies in yeast (29) and by crystal structures of archaeal H/ACA RNPs (see below), RNPs reconstituted in reticulocyte lysate are inactive. This could be due to missing factors and/or to insufficient sensitivity of the *in vitro* pseudouridylation assay. Regardless, active individual RNPs can be reconstituted in cell extracts by the addition of specific H/ACA RNAs (21, 32). Of course, as in the case of the reticulocyte lysate, while these are individual complexes, they are not pure.

In this context, another obstacle in H/ACA RNP study needs to be pointed out—that is, measuring pseudouridylation activity. Pseudouridine is an isomer of uridine with its base attached via a *C*- instead of an *N*-glycosidic bond and cannot be identified by a simple labeling technique (Figure 7.1C). However, the two isomers exhibit different physicochemical properties that most commonly are distinguished by thin-layer chromatography and by modification with *N*-cyclohexyl-*N*′-β-(4-methylmorpholinium)ethylcarbodiimide *p*-tosylate (CMC) (33, 34). Applied to site-specifically labeled substrates (in itself a feat) or to unlabeled rRNA, respectively, both methods are tricky and labor-intensive (21, 32, 33, 35). There is a clear need for more quantitative and higher throughput assays. Some alternative approaches are starting to emerge, such as HPLC and mass spectrometry-based assays (36–38).

Several cell-based assays have been developed for studying individual H/ACA RNPs. They all rely on the introduction/expression of a specific H/ACA RNA in a homologous or heterologous cell system. For example, when injected into *Xenopus* oocytes, *in vitro* transcribed H/ACA RNAs are efficiently incorporated into RNPs and, if fluorescently labeled, their subnuclear targeting can be traced (39, 40). In mammalian cells, most H/ACA RNAs are expressed from introns of host genes (1). Therefore, transient and stable expression of constructs harboring a specific H/ACA RNA in an intron of a homologous or a heterologous gene has been employed to analyze the biogenesis of H/ACA RNPs (41–43). While these approaches provided invaluable insight into the life of H/ACA RNPs, biochemical analysis of individual RNPs nevertheless requires reconstitution from pure components. Taking a step backward in evolution, archaea provided the solution to this problem.

7.5 LESSONS FROM ARCHAEAL H/ACA RNPs

Archaea, like eukaryotes, use, in addition to single-protein enzymes, H/ACA guide RNAs and four core proteins to modify their rRNA. The main structural and sequence features of their H/ACA RNAs are conserved, except that they contain a kink–turn motif in one or several of their terminal stem-loops (44). The core proteins are also conserved but truncated by about one-third at their amino and/or carboxyl termini relative to their eukaryal counterparts (45). Interestingly, the archaeal NHP2 homolog, L7Ae, specifically binds to the kink-turn and is shared between ribosomes and C/D and H/ACA RNPs, marking it as a common ancestor of the homologous but different proteins in each of these three eukaryotic RNPs (44). Most important for this discussion, all core proteins are readily expressed in soluble form and can be reconstituted together with an H/ACA RNA into pseudouridylation competent RNPs (46, 47). From such reconstitution studies we learned that Cbf5 (the archaeal NAP57), Nop10, and an H/ACA RNA are sufficient for minimal pseudouridylation of rRNA, although addition of the other core proteins dramatically increases the activity (47). Indeed, taking a page out of the book of archaea, truncated (i.e., archaea-like) yeast Cbf5 could be expressed in recombinant form and associated specifically with a single hairpin of an H/ACA RNA (30). These data are consistent with each hairpin of an H/ACA RNA associating with its own complement of core proteins and is supported by the crystal structures of archaeal H/ACA RNPs.

Exploiting the soluble expression of archaeal core proteins, the crystal structures of subcomplexes and, most recently, of an entire archaeal H/ACA RNP have been solved (48–51). These structures supported the biochemically derived interactions within the RNP (29, 31, 46, 47). Specifically, archaeal GAR1 and NOP10 independently interact with Cbf5 (NAP57), whereas L7Ae (NHP2) is held in place next to NOP10 via its specific interaction with the kink-turn of the H/ACA RNA (Figure 7.1B). A single hairpin of an H/ACA RNA is draped over the relatively flat platform provided by the NAP57-NOP10-NHP2 core trimer, with the terminal stem loop holding NHP2 in its place (Figure 7.1B, gray loop) and the ACA anchored at the conserved pseudouridine archaeosine transglycosylase (PUA) domain of NAP57 (gray ACA), thereby loosely positioning the pseudouridylation pocket next to the catalytic aspartate of NAP57 (gray Ψ) (49). The pinning down of the H/ACA RNA by only two conserved elements, the ACA sequence and the terminal hairpin loop, provides an elegant explanation for how the same protein complex can accommodate many different H/ACA RNAs (52). Despite this tremendous structural insight, many questions regarding mammalian H/ACA RNPs remain because archaeal RNPs differ substantially in their mass, mode of assembly, and RNA composition, to name a few.

7.6 BIOGENESIS OF EUKARYOTIC H/ACA RNPs

Many of the eukaryotic H/ACA RNP associated factors are not conserved in archaea, and some not even between mammals and yeast, e.g., SMN (53). These proteins may directly attest to the more complicated assembly pathway of eukaryotic RNPs. One such factor, originally identified in yeast, is the nuclear assembly factor NAF1 (54–56). NAF1 is essential for the metabolic stability of H/ACA RNAs without being an integral part of the RNPs (41, 57). It shares a conserved central domain with GAR1 and binds NAP57 competitively with GAR1 (41, 55). This suggests that they bind with their conserved domain to the same part of NAP57 (Figure 7.1B). NAF1 is a low abundant nuclear protein that shuttles between the nucleus and cytoplasm (24, 41). Using chromatin immunoprecipitation (ChIP) and a stable cell line with multiple tandem copies of an exogenous H/ACA RNA stably integrated into the genome, factors recruited to the H/ACA RNA transcription site were identified by co-immunoprecipitation and fluorescent visualization (41, 58, 59). Whereas NAF1 (and the NAP57-NOP10-NHP2 core trimer) can be observed at the site of transcription, GAR1 is absent but concentrates in Cajal bodies and nucleoli from which NAF1 is excluded. An interaction of NAF1 with the C-terminal domain of the large subunit of RNA polymerase II may be important for the recruitment to the transcription site (55, 58, 59). These and other data suggest that NAF1 escorts NAP57 (possibly in form of the core trimer) to the site of H/ACA RNA transcription for RNA association. Subsequently, GAR1 replaces NAF1 to yield functional H/ACA RNPs in Cajal bodies and nucleoli. Apparently, NAF1 is required as a chaperone for the unstable NAP57 before association with the RNA.

Additional factors may play a role in this process—for example, in the exchange and release of RNPs from the site of transcription as well as for the targeting within the nucleus. Candidates for such proteins are SHQ1, Nopp140, and SMN, all of which bind to one or the other H/ACA core protein (1, 4). Given the surprising complexity of assembly of these simple five component particles, more factors may yet be discovered and their precise function analyzed. Stay tuned.

7.7 DEBATE ON DYSKERATOSIS CONGENITA

Mutations in the human NAP57 gene *DKC1* cause a rare bone marrow failure syndrome, dyskeratosis congenita (DC) (60). DC mutations extend into regions of *DKC1* that are not conserved, thus providing another reason to study specifically the more elaborate mammalian H/ACA RNPs (61). DC is often recognized in childhood by the three mucocutaneous features: abnormal skin pigmentation, nail dystrophy, and mucosal leukoplakia. Patients frequently die in their third decade of life due to bone marrow failure and pulmonary complications but also are predisposed to malignancies in rapidly dividing tissues. DC is inherited in three forms: X-linked, and autosomal

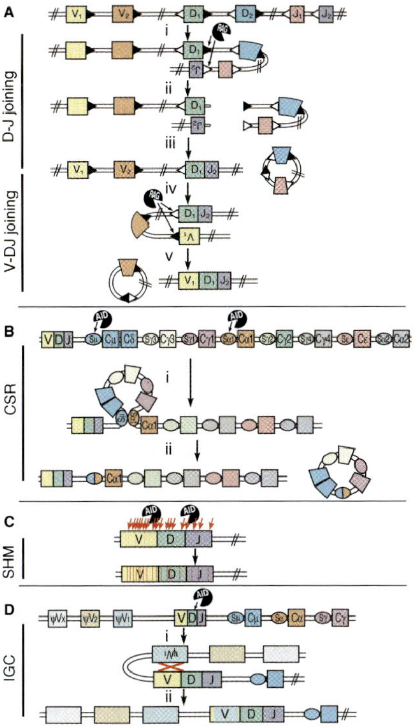

Figure 2.3 Similarities and differences between the mechanisms of V(D)J recombination, CSR, SHM, and IGC. (A) V(D)J recombination. (B) Class switch recombination. (C) Somatic hypermutation. (D) Immunoglobulin gene conversion. (*See text for full caption.*)

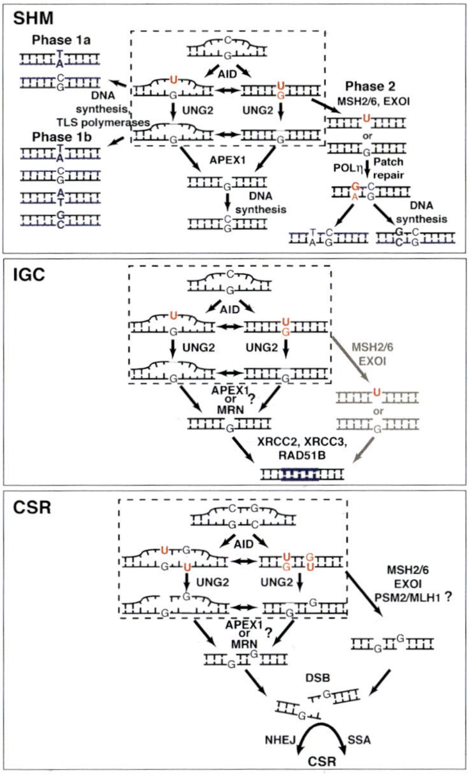

Figure 2.4 The molecular events leading to SHM, IGC and CSR. (*See text for full caption.*)

Figure 3.1a RNA editing by U-insertion and U-deletion. (*See text for full caption.*)

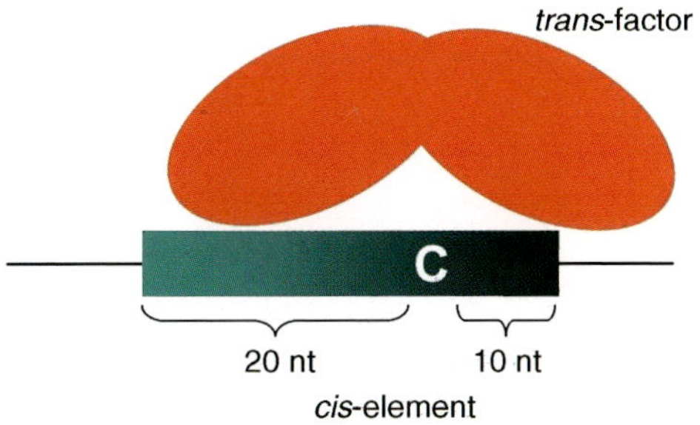

Figure 4.1 An early model of RNA editing in plant organelles. Site recognition for RNA editing requires fewer than 20 upstream nt and fewer than 10 nt downstream in some cases (*cis*-element). This *cis*-element is recognized by a site-specific factor (*trans*-factor).

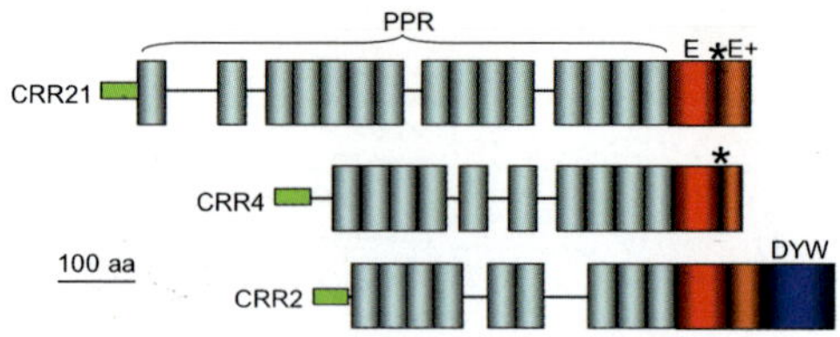

Figure 4.2 Comparison of domain structures among the PPR proteins present in *Arabidopsis* plastids. (*See text for full caption.*)

AtndhF-2 GAUCCACUUACUUCUAUUAUGU--CAAUAUUA
AtndhB-3 GGGGCAAGCUCUUCUAUUCUGGUUCAUGGUUU
AtndhD-1 CCUGGUGUAUCUUGUCUUUACCA-CGAAUGAU
NtndhD-1 CCAAGUGUAUCUUGUCUUUACUA-CGAAUUAU
OsndhD AUCUUGUCUUUAUCACUUUAUCA-UGAGUUCU

Figure 4.3 Alignment of three putative *cis*-elements for RNA editing in *Arabidopsis*. (*See text for full caption.*)

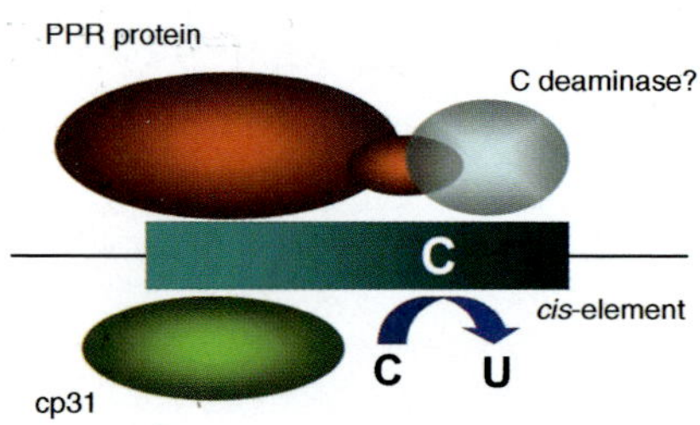

Figure 4.5 Current model of RNA editing in plastids. The PPR protein is a *trans*-factor that binds to a *cis*-element. The C-terminal region of the PPR protein has a common function in the RNA editing machinery, such as binding to an editing enzyme. (*See text for full caption.*)

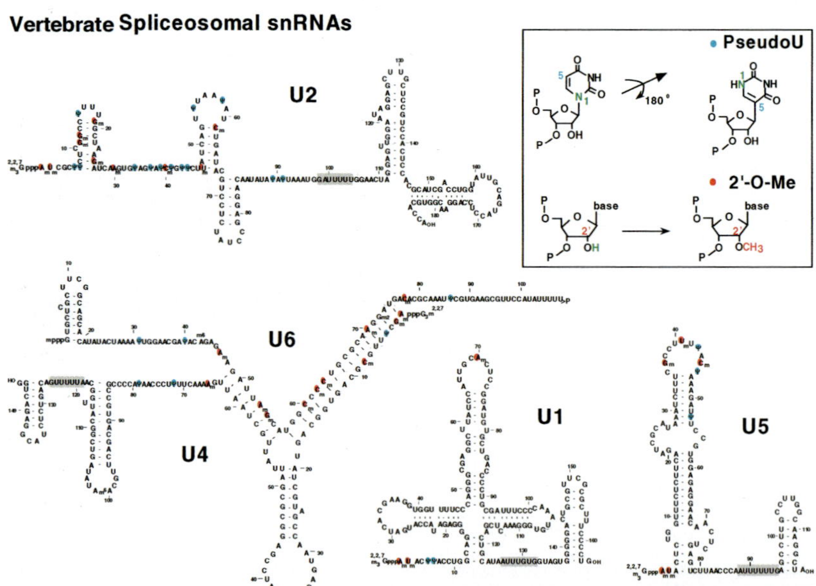

Figure 8.1 Primary sequences and secondary structures of vertebrate major spliceosomal snRNAs. (*See text for full caption.*)

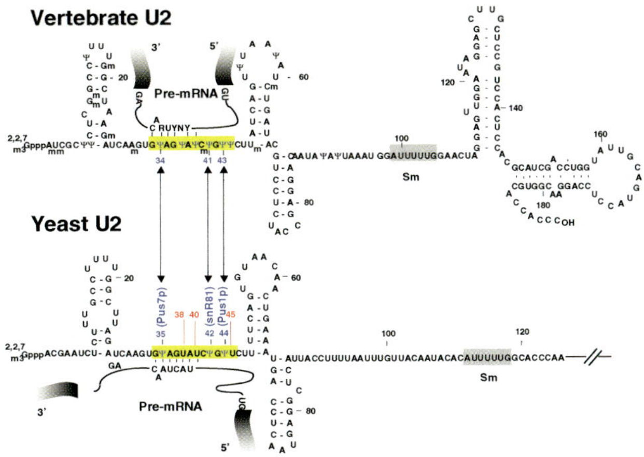

Figure 8.2 Base-pairing interaction between U2 and the branch site of pre-mRNA. (*See text for full caption.*)

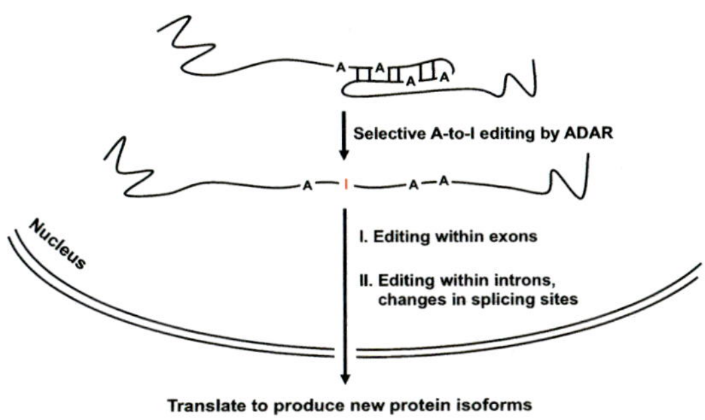

Figure 9.1 Editing of short, imperfect RNA duplexes in the nucleus. (*See text for full caption.*)

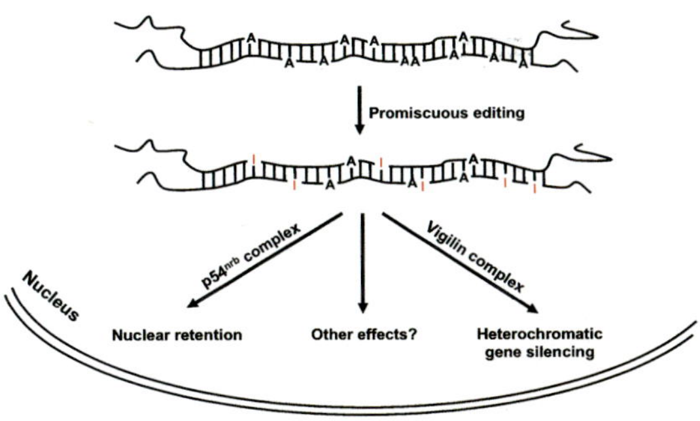

Figure 9.2 Editing of long, perfect RNA duplexes in the nucleus. (*See text for full caption.*)

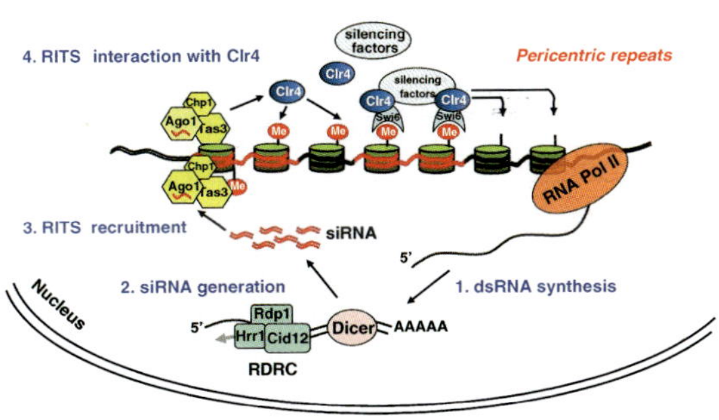

Figure 9.3 RNAi-induced heterochromatin formation in fission yeast. (*See text for full caption.*)

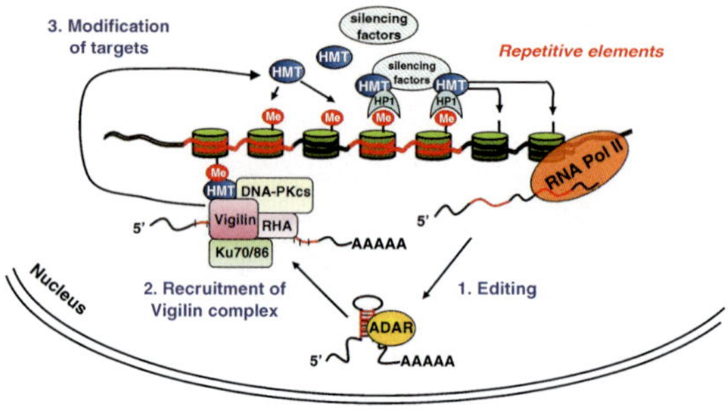

Figure 9.4 A speculative model for I-RNA-induced heterochromatic gene silencing in mammalian cells. (*See text for full caption.*)

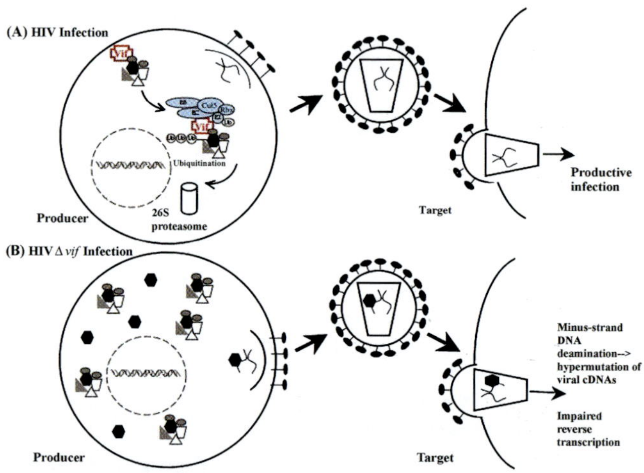

Figure 11.2 Current model of APOBEC3G function. (*See text for full caption.*)

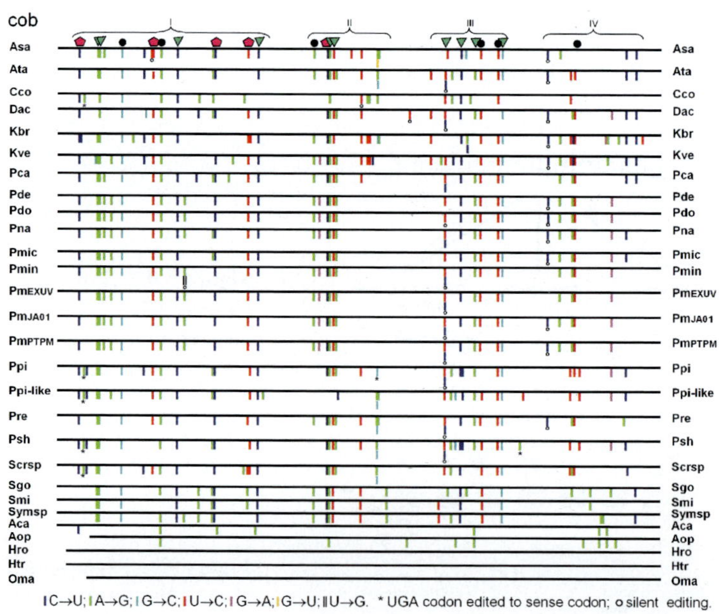

Figure 13.1 Schematic diagram of distribution of editing events in mitochondrial cytochrome *b* (*cob*) in dinoflagellates. (*See text for full caption.*)

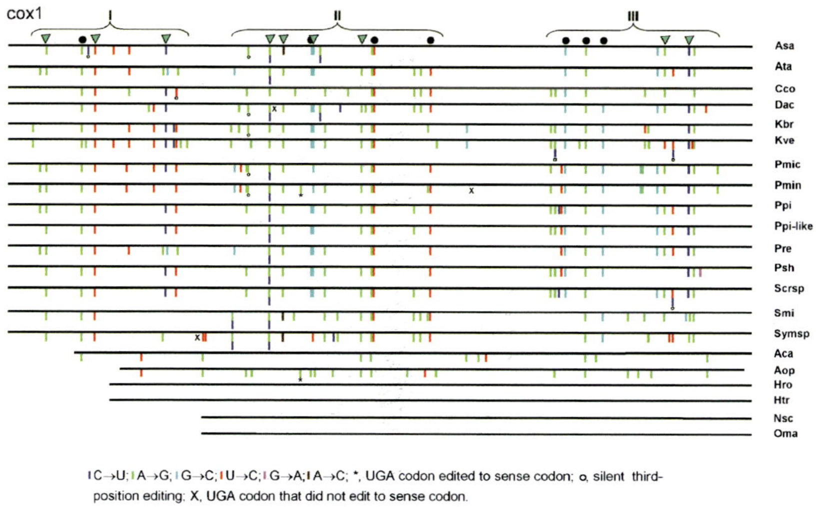

Figure 13.2 Schematic diagram of distribution of editing events in mitochondrial cytochrome *c* oxidase subunit I (*cox1*) in dinoflagellates. (*See text for full caption.*)

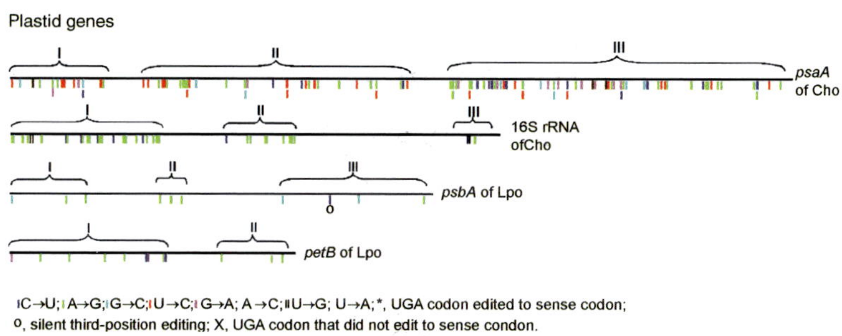

Figure 13.3 Schematic diagram of distribution of editing events in plastid genes in dinoflagellates. (*See text for full caption.*)

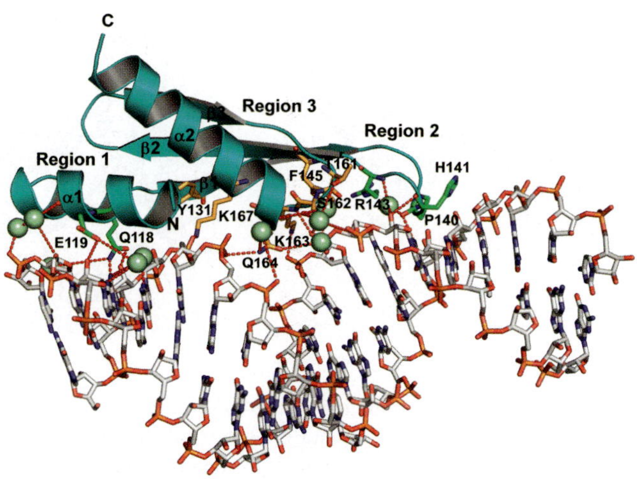

Figure 15.3 Crystal structure of the second *Xenopus laevis* XIrpba double-stranded RNA-binding domain at 1.9 Å. (*See text for full caption.*)

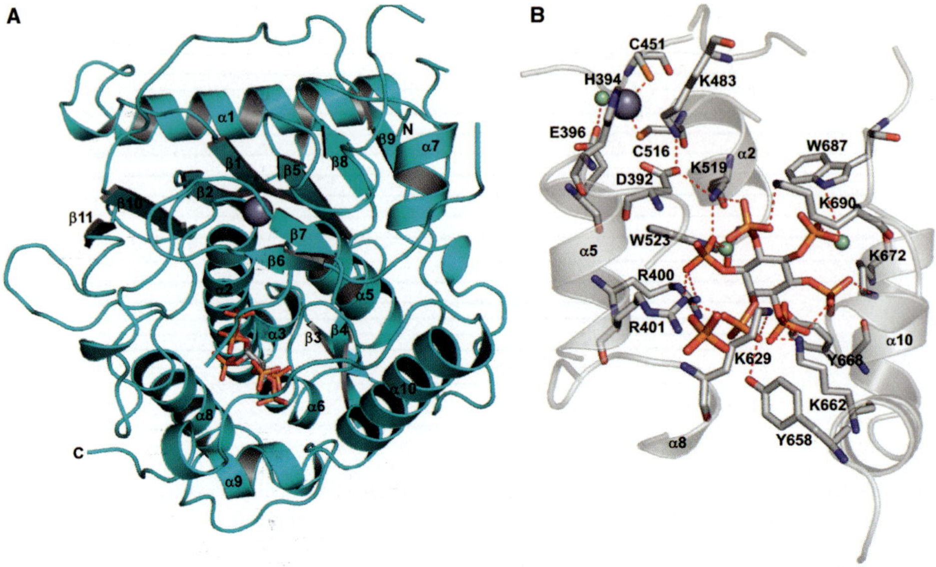

Figure 15.4 Crystal structure of the ADAR2 catalytic domain at a resolution of 1.9 Å. (**A**) The deaminase domain has a spherical shape constructed by 10 α-helices and 11 β-strands. A blue-gray sphere represents the active-site zinc ion. (**B**) Residue interactions at the active site. Zinc and inositol hexakisphosphate (IP6) are shown. The zinc ion and E396 coordinate the nucleophilic water (green sphere). The hydrogen-bond relay between zinc and IP6 is shown as dashes, as are hydrogen bonds between conserved residues and IP6 (gray sticks). IP6 interactions with W523 and W687 are mediated by waters (green spheres).

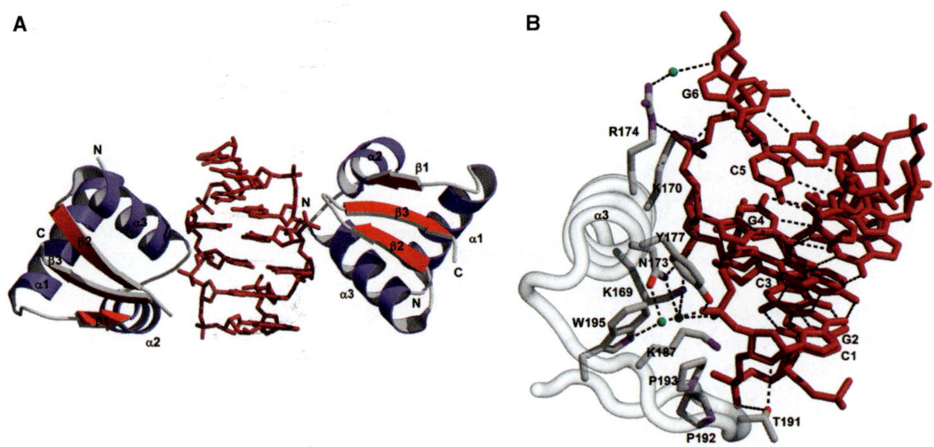

Figure 15.5 Overview of the Zα domain bound to left-handed Z-DNA. (**A**) Residues 134–198 of two symmetry-related Zα monomers and the 6-base-pair left-handed DNA duplex d (CGCGCG)2 are shown. Labels indicate N- and C-termini of the proteins as well as helices (α1–α3) and strands (β1–β3). (**B**) Protein residues involved in DNA contact. Hydrogen bonds are indicated by dashed lines. Water molecules in key positions are indicated by green spheres.

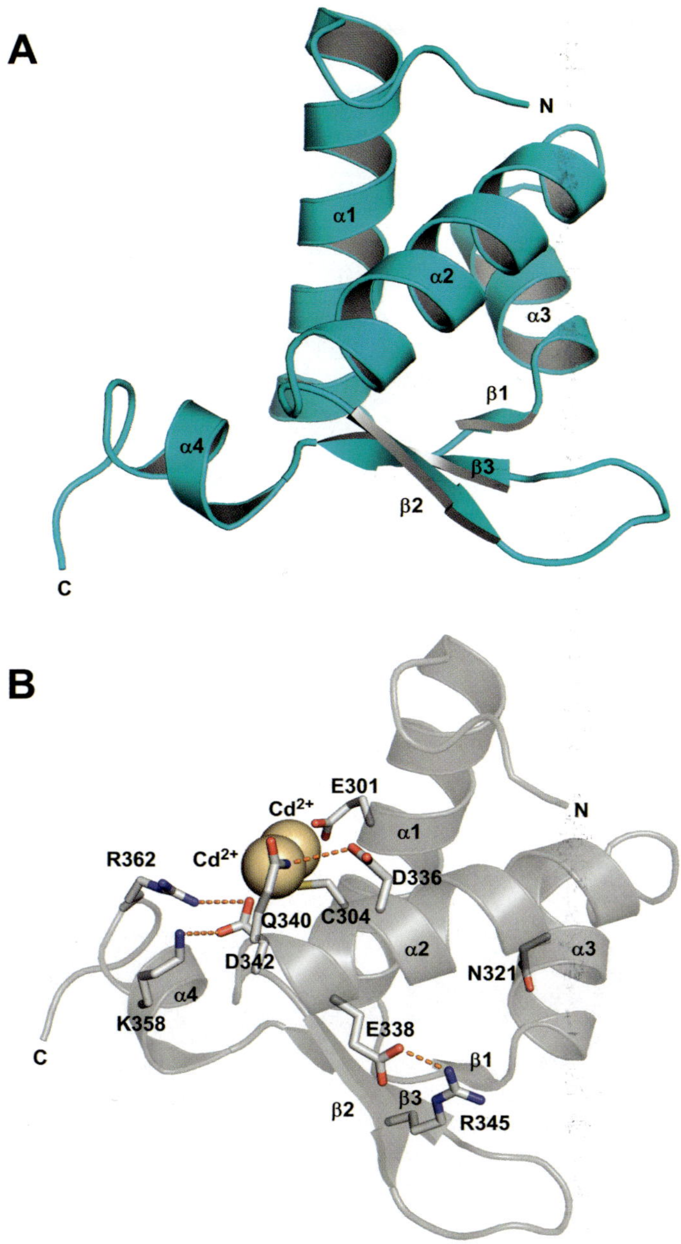

Figure 15.6 The ADAR1 Zβ domain. (**A**) Ribbon diagram of the ADAR1 Zβ domain. (**B**) Conserved residues and those involved in metal-binding residues are shown. Cadmium ions are depicted as spheres. Residues E301 and C304 are involved in metal binding. Residue Q340 forms a hydrogen bond with D336. The conserved D342 forms salt bridges with R362 and K358. E338 forms two hydrogen bonds with R345. A hydrogen bond is formed between N321 and the backbone carbonyl moiety of N317.

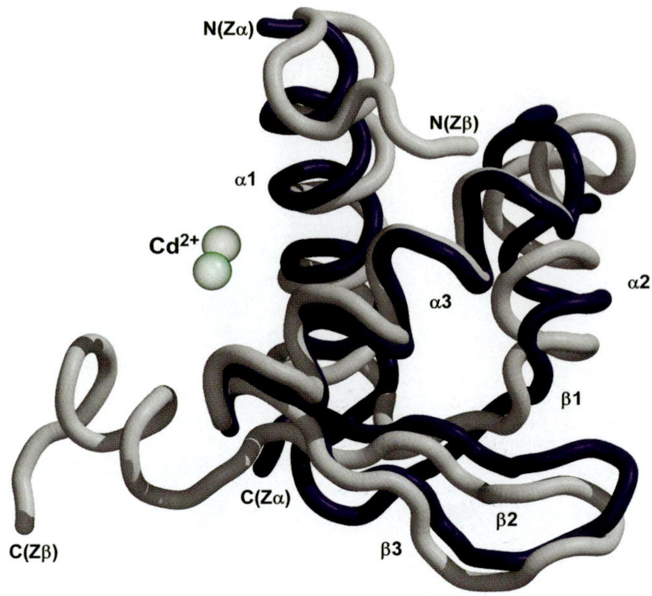

Figure 15.7 Superposition of the Zα and Zβ domains. The ADAR1 Zα and Zβ domains have the same topology and overall structure. Metal ions identified in the Zβ structure are shown as spheres.

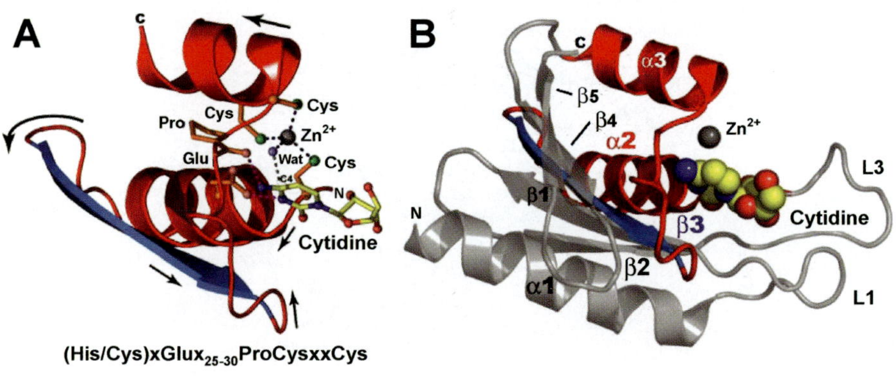

Figure 16.2 Ribbon diagrams of the representative zinc-dependent deaminase motif bound to cytidine. The coordinates were derived from the mouse cytidine deaminase crystal structure (43). (**A**) The helix–strand–helix structure of the ZDD motif indicating the spatial orientation of key residues. Dashed lines (blue) indicate ionic coordination to Zn^{2+}; pink lines indicate hydrogen bonds. The activated water makes a close contact to atom C4 of the cytidine ring (gray line). Black arrows indicate the progression of the polypeptide chain from N- to C-terminus. (**B**) The ZDD signature motif in the context of the CDA domain (gray, semitransparent). The cytidine substrate is depicted as a space-filling model.

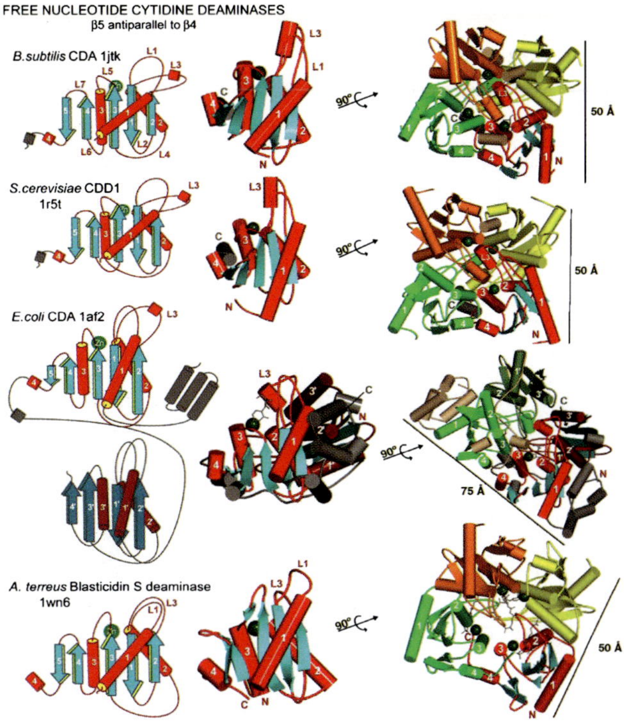

Figure 16.5 Free nucleotide cytidine deaminases (fnCDAs). (*See text for full caption.*)

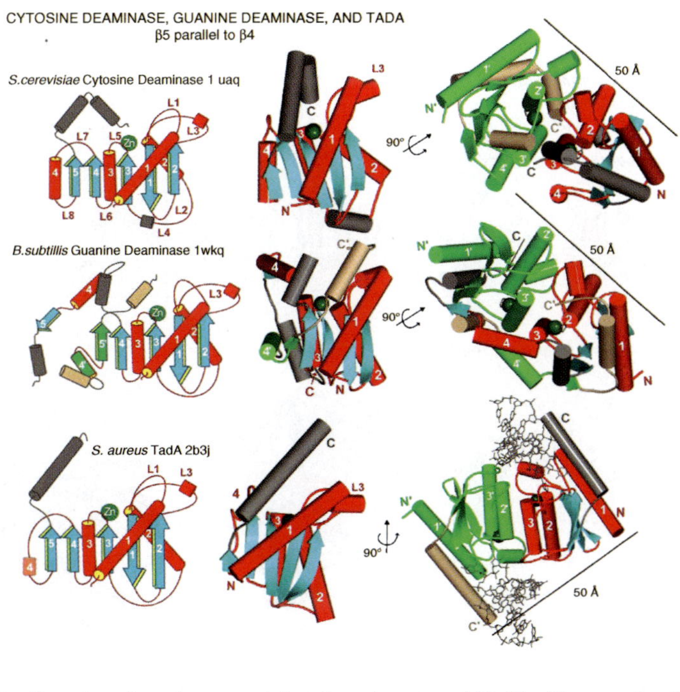

Figure 16.6 Cytosine deaminase, guanine deaminase, and TadA. (*See text for full caption.*)

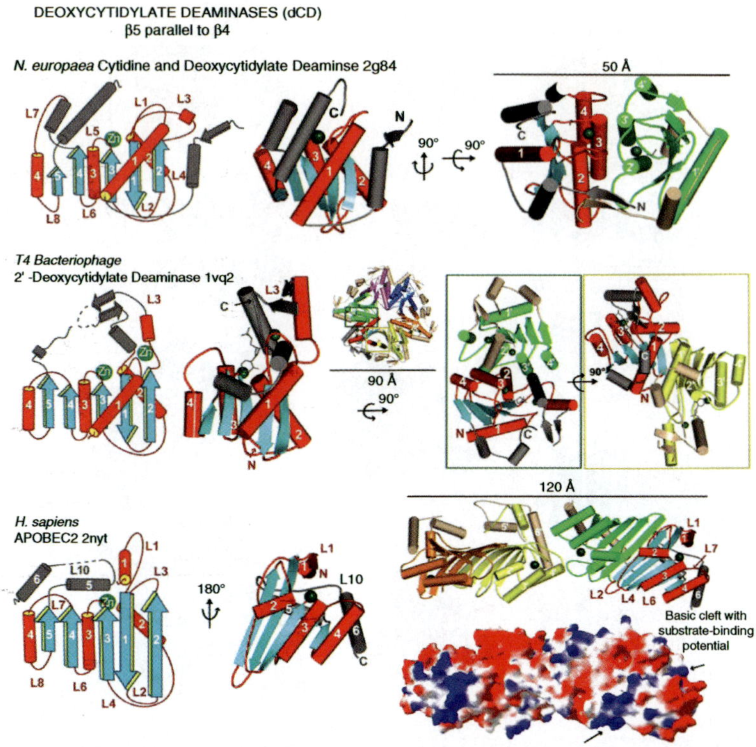

Figure 16.7 Deoxycytidylate Deaminases and APOBEC2. (*See text for full caption.*)

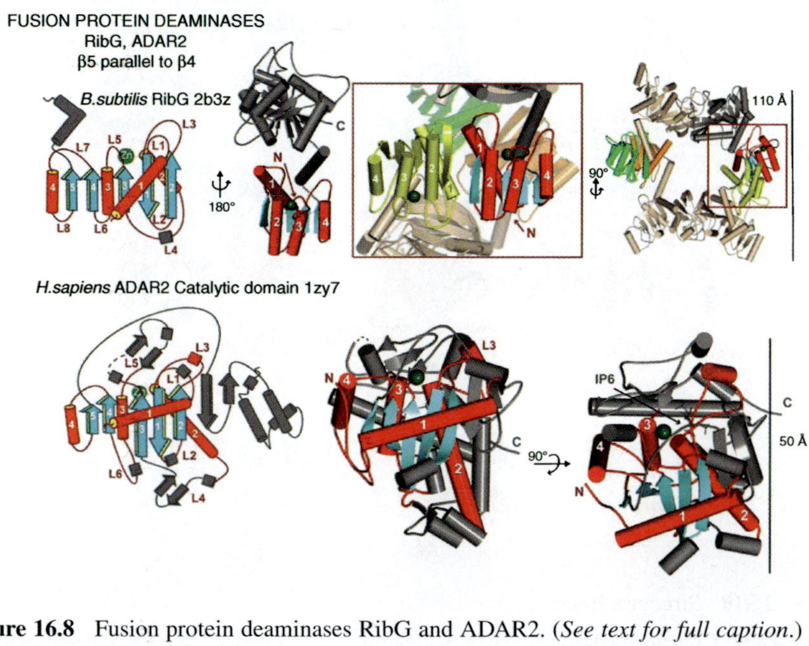

Figure 16.8 Fusion protein deaminases RibG and ADAR2. (*See text for full caption.*)

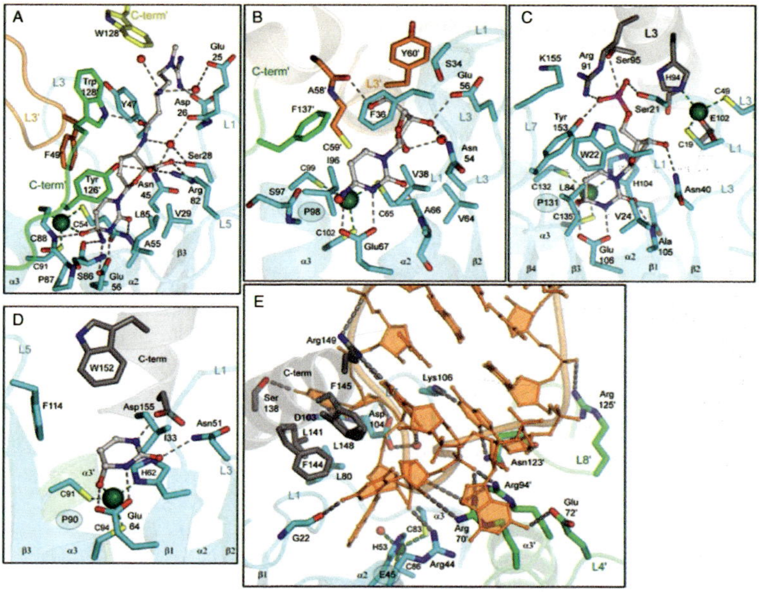

Figure 16.9 Schematic diagrams of substrate binding in the active sites of representative CDA superfamily members. (*See text for full caption.*)

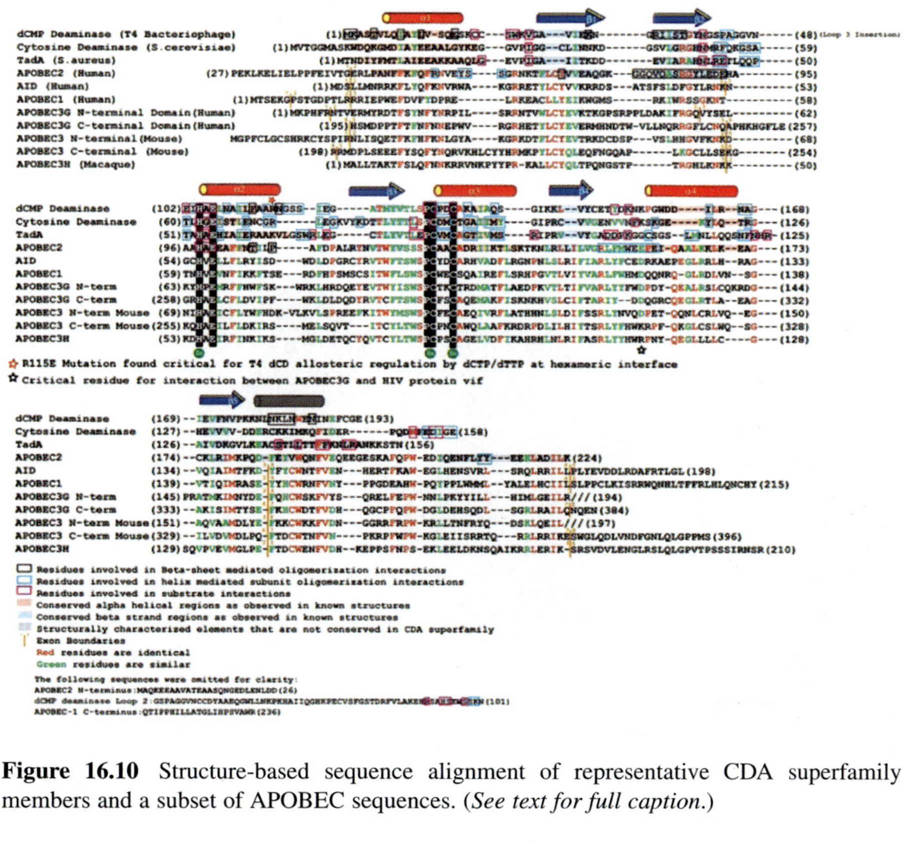

Figure 16.10 Structure-based sequence alignment of representative CDA superfamily members and a subset of APOBEC sequences. (*See text for full caption.*)

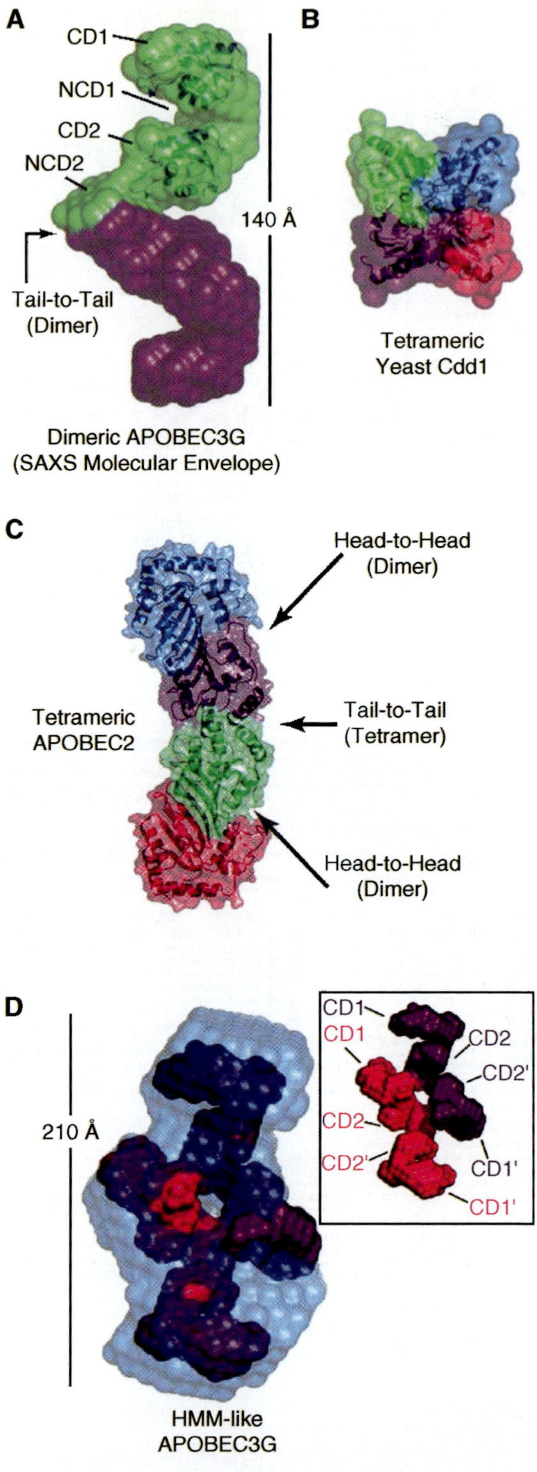

Figure 16.12 The global molecular envelope for APOBEC3G based on shape restorations from SAXS, and subunit inter- actions of crystallographically defined cytidine deaminases. (*See text for full caption*.)

dominant and recessive. X-linked DC is the most common and severe form and is caused by mutations in *DKC1* (60). Autosomal dominant DC is the least frequent and mildest form and is due to mutations in the telomerase RNA and reverse transcriptase (62, 63). Autosomal recessive DC proofs to be genetically heterogeneous and in one family has been associated with a mutation in NOP10 (64). The discovery of hTR as an H/ACA RNA started the debate if X-linked DC was a deficiency of telomere maintenance and/or of one or several of the other H/ACA RNP functions—for example, pre-rRNA modification and processing (protein synthesis) and snRNA modification (pre-mRNA splicing).

The answer may lie somewhere in between. For example, shortened telomeres could explain bone marrow failure through impairment of hematopoietic renewal. In fact, reduced levels of hTR appear to be responsible for shortened telomeres observed in cells of patients with X-linked DC (65, 66). In contrast, studies in mouse models of X-linked DC indicate that pseudouridylation defects caused by mutations in *DKC1* are not necessarily accompanied by telomere deficiencies (67, 68). Moreover, a specific impairment of IRES (internal ribosome entry site)-dependent translation was identified in *DKC1* mutant mice and in cells from X-linked DC patients (69). Given the tight association of NAP57 with H/ACA RNAs (24), mutations in NAP57 could impair the function of any H/ACA RNA. Nevertheless, based on the sequence variability among H/ACA RNAs, that impact could be dramatically different for individual H/ACA RNPs—for example, the telomerase RNP. Molecular dissection of possibly minor differences between human H/ACA RNPs requires pure complexes, which leads us back to a need for an *in vitro* reconstitution system and the problems associated with it (see above). In particular, expression of soluble human NAP57 poses a major obstacle to overcome.

The approximately 40, mutations (mostly missense) identified to date in X-linked DC patients distribute to the first and fourth one-fifth of the 514-amino-acid-long NAP57 (61). However, when the structure of human NAP57 is modeled based on that of its archaeal counterparts, these mutations cluster in a single three-dimensional domain (49, 51). Although many mutations cannot be modeled because the amino and carboxyl termini of human NAP57 are missing in the shorter archaeal orthologs, the DC mutation cluster in NAP57 falls on one side of the PUA domain (Figure 7.1B, DC circle). Surprisingly, this locates most of the DC mutations on a surface of NAP57 that is exposed and does not contact any of the other components of the ribonuclear complex—that is, H/ACA RNAs and the other core proteins (Figure 7.1B) (49). This suggests that the mutations cause allosteric effects or impact contacts of NAP57 with yet-to-be-identified factors. The hunt for such factors may prove challenging as the effects of the often conservative missense mutations are likely subtle. In contrast, a NAP57 knockout in mouse is embryonically lethal (70). Obviously, there is still a long way to go before we understand DC on a molecular level.

7.8 IMPORTANCE AND FUTURE OF H/ACA RNPs

As illustrated by the debate on the molecular basis of DC, the function of H/ACA RNPs impacts many cellular and consequently organismal aspects. Despite the simplicity of these five component ribonuclear complexes, many questions remain. Some of these questions and points of interest have been raised throughout this chapter and some are pointed out here.

Are we alone? Do the four core proteins and an H/ACA RNA constitute simply the heart of complexes that routinely perform in the company of additional factors? For example, does the small Cajal body targeting element of certain H/ACA RNAs bind a specific protein responsible for its particular localization—for example, Sm proteins (71)? Alternatively, is the core particle much more dynamic than assumed and most of the core proteins exchange off the particle and only meet for catalysis of pseudouridylation and/or other H/ACA RNP functions? In this regard, only NAP57, the least exchangeable of core proteins (24), together with hTR and TERT but not the other H/ACA core proteins, forms active telomerase (25). How many of the already identified H/ACA RNP interacting proteins (in addition to NAF1) are required for RNP biogenesis and how many for RNP function?

Although all pseudouridine synthases are structurally related (regardless if they function independently or in the context of an RNP and regardless of primary sequence differences), their mode of catalysis may vary (3, 72). For example, 5-fluorouracil, a general inhibitor of pseudouridine synthases, forms a covalent adduct with some synthases but not with others (73). These data indicate that the precise mechanism of catalysis for individual enzymes remains to be elucidated. Additionally, do the RNA-guided, complex pseudouridylases require helicases for the release of substrate RNAs? The identification of the DEAD box helicase Has1p in yeast suggests yes (23). How many more such catalysis-associated factors are out there?

Finally, how many H/ACA RNAs (and consequently RNPs) remain to be identified? Despite fully sequenced genomes and sophisticated search tools, the full complement of mammalian H/ACA RNAs remains to be uncovered (74–80). In addition, the functionality of genomically identified H/ACA RNAs needs to be proven experimentally, as complementarity to a target RNA within the pseudouridylation pocket may not be sufficient to specify a uridine for isomerization. What is the function of the many orphan H/ACA RNAs? Does NAP57 function in a capacity other than pseudouridine synthase, as possibly in the case of the telomerase RNP, and are there other similar examples?

As indicated by these many questions, the next few years of H/ACA RNP research may bring us a bounty of interesting answers. Additionally, these questions have not even addressed the function(s) of pseudouridines, which in large part is (are) poorly understood. For a role of pseudouridines in pre-mRNA splicing, you are referred to the Chapter 8.

ACKNOWLEDGMENTS

I thank Petar Grozdanov and the rest of the members in my laboratory for their helpful comments on the manuscript. The work in the author's laboratory is supported by grants from the National Heart Lung and Blood Institute and the March of Dimes Birth Defects Foundation.

REFERENCES

1. Kiss, T., Fayet, E., Jady, B. E., Richard, P., and Weber, M. (2006) Biogenesis and Intranuclear Trafficking of Human Box C/D and H/ACA RNPs. *Cold Spring Harbor Symp Quant Biol* **71**, 407–417.
2. Meier, U. T. (2005) The many facets of H/ACA ribonucleoproteins. *Chromosoma* **114**(1), 1–14.
3. Reichow, S. L., Hamma, T., Ferre-D'Amare, A. R., and Varani, G. (2007) The structure and function of small nucleolar ribonucleoproteins. *Nucleic Acids Res* **35**(5), 1452–1464.
4. Terns, M., and Terns, R. (2006) Noncoding RNAs of the H/ACA Family. *Cold Spring Harbor Symp Quant Biol* **71**, 395–405.
5. Yu, Y. T., Terns, R. M., and Terns, M. P. (2005) Mechanisms and functions of RNA-guided RNA modification. In: Grosjean, H.,(ed.), Fine-Tuning of RNA Functions by Modification and Editing, Springer-Verlag, New York.
6. Ganot, P., Bortolin, M. -L., and Kiss, T. (1997) Site-specific pseudouridine formation in preribosomal RNA is guided by small nucleolar RNAs. *Cell* **89**, 799–809.
7. Ni, J., Tien, A. L., and Fournier, M. J. (1997) Small nucleolar RNAs direct site-specific synthesis of pseudouridine in ribosomal RNA. *Cell* **89**, 565–573.
8. Jiang, W., Middleton, K., Yoon, H. -J., Fouquet, C., and Carbon, J. (1993) An essential yeast protein, CBF5p, binds *in vitro* to centromeres and microtubules. *Mol Cell Biol* **13**, 4884–4893.
9. Meier, U. T., and Blobel, G. (1994) NAP57, a mammalian nucleolar protein with a putative homolog in yeast and bacteria. J Cell Biol (correction appeared in **140**; 447) **127**, 1505–1514.
10. Nurse, K., Wrzesinski, J., Bakin, A., Lane, B. G., and Ofengand, J. (1995) Purification, cloning, and properties of the tRNA Ψ55 synthase from *Escherichia coli*. *RNA* **1**, 102–112.
11. Lafontaine, D. L. J., Bousquet-Antonelli, C., Henry, Y., Caizergues-Ferrer, M., and Tollervey, D. (1998) The box H+ACA snoRNAs carry Cbf5p, the putative rRNA pseudouridine synthase. *Genes Dev* **12**, 527–537.
12. Zebarjadian, Y., King, T., Fournier, M. J., Clarke, L., and Carbon, J. (1999) Point mutations in yeast CBF5 can abolish *in vivo* pseudouridylation of rRNA. *Mol Cell Biol* **19**(11), 7461–7472.
13. Ganot, P., Caizergues-Ferrer, M., and Kiss, T. (1997) The family of box ACA small nucleolar RNAs is defined by an evolutionarily conserved secondary structure and ubiquitous sequence elements essential for RNA accumulation. *Genes Dev* **11**, 941–956.
14. Bousquet-Antonelli, C., Henry, Y., Gélugne, J. -P., Caizergues-Ferrer, M., and Kiss, T. (1997) A small nucleolar RNP protein is required for pseudouridylation of eukaryotic ribosomal RNAs. *EMBO J* **16**, 4770–4776.
15. Girard, J. -P., Lehtonen, H., Caizergues-Ferrer, M., Amalric, F., Tollervey, D., and Lapeyre, B. (1992) GAR1 is an essential small nucleolar RNP protein required for pre-rRNA processing in yeast. *EMBO J* **11**, 673–682.
16. Niewmierzycka, A., and Clarke, S. (1999) S-Adenosylmethionine-dependent methylation in *Saccharomyces cerevisiae*. Identification of a novel protein arginine methyltransferase. *J Biol Chem* **274**(2), 814–824.
17. Henras, A., Henry, Y., Bousquet-Antonelli, C., Noaillac-Depeyre, J., Gelugne, J. P., and Caizergues-Ferrer, M. (1998) Nhp2p and Nop10p are essential for the function of H/ACA snoRNPs. *EMBO J* **17**(23), 7078–7090.
18. Lübben, B., Fabrizio, P., Kastner, B., and Lührmann, R. (1995) Isolation and characterization of the small nucleolar ribonucleoprotein particle snR30 from Saccharomyces cerevisiae. *J Biol Chem* **270** (19), 11549–11554.
19. Watkins, N. J., Gottschalk, A., Neubauer, G., Kastner, B., Fabrizio, P., Mann, M., and Lührmann, R. (1998) Cbf5p, a potential pseudouridine synthase, and Nhp2p, a putative RNA- binding protein, are present together with Gar1p in all box H/ACA-motif snoRNPs and constitute a common bipartite structure. *RNA* **4**(12), 1549–1568.
20. Darzacq, X., Jady, B. E., Verheggen, C., Kiss, A. M., Bertrand, E., and Kiss, T. (2002) Cajal body-specific small nuclear RNAs: A novel class of 2′-*O*-methylation and pseudouridylation guide RNAs. *EMBO J* **21**(11), 2746–2756.
21. Jady, B. E., and Kiss, T. (2001) A small nucleolar guide RNA functions both in 2′-O-ribose methylation and pseudouridylation of the U5 spliceosomal RNA. *EMBO J* **20**(3), 541–551.

22. Cavaille, J., Buiting, K., Kiefmann, M., Lalande, M., Brannan, C. I., Horsthemke, B., Bachellerie, J. P., Brosius, J., and Huttenhofer, A. (2000) Identification of brain-specific and imprinted small nucleolar RNA genes exhibiting an unusual genomic organization. *Proc Natl Acad Sci USA* **97**(26), 14311–14316.
23. Liang, X. H., and Fournier, M. J. (2006) The helicase Has1p is required for snoRNA release from pre-rRNA. *Mol Cell Biol* **26**(20), 7437–7450.
24. Kittur, N., Darzacq, X., Roy, S., Singer, R. H., and Meier, U. T. (2006) Dynamic association and localization of human H/ACA RNP proteins. *RNA* **12**(12), 2057–2062.
25. Cohen, S. B., Graham, M. E., Lovrecz, G. O., Bache, N., Robinson, P. J., and Reddel, R. R. (2007) Protein composition of catalytically active human telomerase from immortal cells. *Science* **315**(5820) 1850–1853.
26. Dragon, F., Pogacic, V., and Filipowicz, W. (2000) *in vitro* assembly of human H/ACA small nucleolar RNPs reveals unique features of U17 and telomerase RNAs. *Mol Cell Biol* **20**(9) 3037–3048.
27. Pogacic, V., Dragon, F., and Filipowicz, W. (2000) Human H/ACA small nucleolar RNPs and telomerase share evolutionarily conserved proteins NHP2 and NOP10. *Mol Cell Biol* **20**(23), 9028–9040.
28. Dez, C., Henras, A., Faucon, B., Lafontaine, D., Caizergues-Ferrer, M., and Henry, Y. (2001) Stable expression in yeast of the mature form of human telomerase RNA depends on its association with the box H/ACA small nucleolar RNP proteins Cbf5p, Nhp2p and Nop10p. *Nucleic Acids Res* **29**(3), 598–603.
29. Henras, A. K., Capeyrou, R., Henry, Y., and Caizergues-Ferrer, M. (2004) Cbf5p, the putative pseudouridine synthase of H/ACA-type snoRNPs, can form a complex with Gar1p and Nop10p in absence of Nhp2p and box H/ACA snoRNAs. *RNA* **10**(11), 1704–1712.
30. Normand, C., Capeyrou, R., Quevillon-Cheruel, S., Mougin, A., Henry, Y., and Caizergues-Ferrer, M. (2006) Analysis of the binding of the N-terminal conserved domain of yeast Cbf5p to a box H/ACA snoRNA. *RNA* **12**(10), 1868–1882.
31. Wang, C., and Meier, U. T. (2004) Architecture and assembly of mammalian H/ACA small nucleolar and telomerase ribonucleoproteins. *EMBO J* **23**(8) 1857–1867.
32. Wang, C., Query, C. C., and Meier, U. T. (2002) Immunopurified small nucleolar ribonucleoprotein particles pseudouridylate rRNA independently of their association with phosphorylated Nopp140. *Mol Cell Biol* **22**(24), 8457–8466.
33. Bakin, A., and Ofengand, J. (1993) Four newly located pseudouridylate residues in *Escherichia coli* 23S ribosomal RNA are all at the peptidyltransferase center: Analysis by the application of a new sequencing technique. *Biochemistry* **32**, 9754–9762.
34. Nishimura, S. (1972) Minor components in transfer RNA: Their charcterization, location, and function. *Prog Nucleic Acid Res Mol Biol* **12**, 49–85.
35. Ma, X., Zhao, X., and Yu, Y. T. (2003) Pseudouridylation (Psi) of U2 snRNA in *S. cerevisiae* is catalyzed by an RNA-independent mechanism. *EMBO J* **22**(8), 1889–1897.
36. Buchhaupt, M., Peifer, C., and Entian, K. D. (2007) Analysis of 2'-*O*-methylated nucleosides and pseudouridines in ribosomal RNAs using DNAzymes. *Anal Biochem* **361**(1), 102–108.
37. Emmerechts, G., Herdewijn, P., and Rozenski, J. (2005) Pseudouridine detection improvement by derivatization with methyl vinyl sulfone and capillary HPLC-mass spectrometry. *J Chromatogr* **825**(2), 233–238.
38. Pomerantz, S. C., and McCloskey, J. A. (2005) Detection of the common RNA nucleoside pseudouridine in mixtures of oligonucleotides by mass spectrometry. *Anal Chem* **77**(15) 4687–4697.
39. Lange, T. S., Ezrokhi, M., Amaldi, F., and Gerbi, S. A. (1999) Box H and box ACA are nucleolar localization elements of U17 small nucleolar RNA. *Mol Biol Cell* **10**(11), 3877–3890.
40. Narayanan, A., Lukowiak, A., Jady, B. E., Dragon, F., Kiss, T., Terns, R. M., and Terns, M. P. (1999) Nucleolar localization signals of box H/ACA small nucleolar RNAs. *EMBO J* **18**(18), 5120–5130.
41. Darzacq, X., Kittur, N., Roy, S., Shav-Tal, Y., Singer, R. H., and Meier, U. T. (2006) Stepwise RNP assembly at the site of H/ACA RNA transcription in human cells. *J Cell Biol* **173**(2), 207–218.
42. Kiss, T., and Filipowicz, W. (1995) Exonucleolytic processing of small nucleolar RNAs from pre-mRNA introns. *Genes Dev* **9**(11) 1411–1424.
43. Richard, P., Kiss, A. M., Darzacq, X., and Kiss, T. (2006) Cotranscriptional recognition of human intronic box H/ACA snoRNAs occurs in a splicing-independent manner. *Mol Cell Biol* **26**(7), 2540–2549.

44. Rozhdestvensky, T. S., Tang, T. H., Tchirkova, I. V., Brosius, J., Bachellerie, J. P., and Huttenhofer, A. (2003) Binding of L7Ae protein to the K-turn of archaeal snoRNAs: a shared RNA binding motif for C/D and H/ACA box snoRNAs in Archaea. *Nucleic Acids Res* **31**(3), 869–877.
45. Watanabe, Y., and Gray, M. W. (2000) Evolutionary appearance of genes encoding proteins associated with box H/ACA snoRNAs: cbf5p in Euglena gracilis, an early diverging eukaryote, and candidate Gar1p and Nop10p homologs in archaebacteria. *Nucleic Acids Res* **28**(12), 2342–2352.
46. Baker, D. L., Youssef, O. A., Chastkofsky, M. I., Dy, D. A., Terns, R. M., and Terns, M. P. (2005) RNA-guided RNA modification: functional organization of the archaeal H/ACA RNP. *Genes Dev* **19**(10), 1238–1248.
47. Charpentier, B., Muller, S., and Branlant, C. (2005) Reconstitution of archaeal H/ACA small ribonucleoprotein complexes active in pseudouridylation. *Nucleic Acids Res* **33**(10), 3133–3144.
48. Hamma, T., Reichow, S. L., Varani, G., and Ferre-D'Amare, A. R. (2005) The Cbf5-Nop10 complex is a molecular bracket that organizes box H/ACA RNPs. *Nat Struct Mol Biol* **12**(12), 1101–1107.
49. Li, L., and Ye, K. (2006) Crystal structure of an H/ACA box ribonucleoprotein particle. *Nature* **443** (7109), 302–307.
50. Manival, X., Charron, C., Fourmann, J. B., Godard, F., Charpentier, B., and Branlant, C. (2006) Crystal structure determination and site-directed mutagenesis of the *Pyrococcus abyssi* aCBF5-aNOP10 complex reveal crucial roles of the C-terminal domains of both proteins in H/ACA sRNP activity. *Nucleic Acids Res* **34**(3), 826–839.
51. Rashid, R., Liang, B., Baker, D. L., Youssef, O. A., He, Y., Phipps, K., Terns, R. M., Terns, M. P., and Li, H. (2006) Crystal structure of a Cbf5-Nop10-Gar1 complex and implications in RNA-guided pseudouridylation and dyskeratosis congenita. *Mol Cell* **21**(2), 249–260.
52. Meier, U. T. (2006) How a single protein complex accommodates many different H/ACA RNAs. *Trends Biochem Sci* **31**(6), 311–315.
53. Matera, A. G., Terns, R. M., and Terns, M. P. (2007) Non-coding RNAs: Lessons from the small nuclear and small nucleolar RNAs. *Nat Rev Mol Cell Biol* **8**(3), 209–220.
54. Dez, C., Noaillac-Depeyre, J., Caizergues-Ferrer, M., and Henry, Y. (2002) Naf1p, an essential nucleoplasmic factor specifically required for accumulation of box H/ACA small nucleolar RNPs. *Mol Cell Biol* **22**(20), 7053–7065.
55. Fatica, A., Dlakic, M., and Tollervey, D. (2002) Naf1p is a box H/ACA snoRNP assembly factor. *RNA* **8** (12), 1502–1514.
56. Yang, P. K., Rotondo, G., Porras, T., Legrain, P., and Chanfreau, G. (2002) The Shq1p.Naf1p complex is required for box H/ACA small nucleolar ribonucleoprotein particle biogenesis. *J Biol Chem* **277**(47), 45235–45242.
57. Hoareau-Aveilla, C., Bonoli, M., Caizergues-Ferrer, M., and Henry, Y. (2006) hNaf1 is required for accumulation of human box H/ACA snoRNPs, scaRNPs, and telomerase. *RNA* **12**(5), 832–840.
58. Ballarino, M., Morlando, M., Pagano, F., Fatica, A., Bozzoni, I., (2005) The cotranscriptional assembly of snoRNPs controls the biosynthesis of H/ACA snoRNAs in *Saccharomyces cerevisiae*. *Mol Cell Biol* **25**(13), 5396–5403.
59. Yang, P. K., Hoareau, C., Froment, C., Monsarrat, B., Henry, Y., and Chanfreau, G. (2005) Cotranscriptional recruitment of the pseudouridylsynthetase Cbf5p and of the RNA binding protein Naf1p during H/ACA snoRNP assembly. *Mol Cell Biol* **25**(8) 3295–3304.
60. Heiss, N. S., Knight, S. W., Vulliamy, T. J., Klauck, S. M., Wiemann, S., Mason, P. J., Poustka, A., and Dokal, I. (1998) X-linked dyskeratosis congenita is caused by mutations in a highly conserved gene with putative nucleolar functions. *Nat Genet* **19**(1), 32–38.
61. Vulliamy, T. J., Marrone, A., Knight, S. W., Walne, A., Mason, P. J., and Dokal, I. (2006) Mutations in dyskeratosis congenita: Their impact on telomere length and the diversity of clinical presentation. *Blood* **107**(7), 2680–2685.
62. Armanios, M., Chen, J. L., Chang, Y. P., Brodsky, R. A., Hawkins, A., Griffin, C. A., Eshleman, J. R., Cohen, A. R., Chakravarti, A., Hamosh, A., and Greider, C. W. (2005) Haploinsufficiency of telomerase reverse transcriptase leads to anticipation in autosomal dominant dyskeratosis congenita. *Proc Natl Acad Sci USA* **102**(44), 15960–15964.
63. Vulliamy, T., Marrone, A., Goldman, F., Dearlove, A., Bessler, M., Mason, P. J., and Dokal, I. (2001) The RNA component of telomerase is mutated in autosomal dominant dyskeratosis congenita. *Nature* **413** (6854), 432–435.

64. Walne, A. J., Vulliamy, T., Marrone, A., Beswick, R., Kirwan, M., Masunari, Y., Al-Qurashi, F. H., Aljurf, M., and Dokal, I. (2007) Genetic heterogeneity in autosomal recessive dyskeratosis congenita with one subtype due to mutations in the telomerase-associated protein NOP10. *Hum Mol Genet* **16**(13), 1619–1629.
65. Mitchell, J. R., Wood, E., and Collins, K. (1999) A telomerase component is defective in the human disease dyskeratosis congenita. *Nature* **402**(6761), 551–555.
66. Wong, J. M., and Collins, K. (2006) Telomerase RNA level limits telomere maintenance in X-linked dyskeratosis congenita. *Genes Dev* **20**(20), 2848–2858.
67. Mochizuki, Y., He, J., Kulkarni, S., Bessler, M., and Mason, P. J. (2004) Mouse dyskerin mutations affect accumulation of telomerase RNA and small nucleolar RNA, telomerase activity, and ribosomal RNA processing. *Proc Natl Acad Sci USA* **101**(29), 10756–10761.
68. Ruggero, D., Grisendi, S., Piazza, F., Rego, E., Mari, F., Rao, P. H., Cordon-Cardo, C., and Pandolfi, P. P. (2003) Dyskeratosis congenita and cancer in mice deficient in ribosomal RNA modification. *Science* **299**(5604), 259–262.
69. Yoon, A., Peng, G., Brandenburger, Y., Zollo, O., Xu, W., Rego, E., and Ruggero, D. (2006) Impaired control of IRES-mediated translation in X-linked dyskeratosis congenita. *Science* **312**(5775), 902–906.
70. He, J., Navarrete, S., Jasinski, M., Vulliamy, T., Dokal, I., Bessler, M., and Mason, P. J. (2002) Targeted disruption of Dkc1, the gene mutated in X-linked dyskeratosis congenita, causes embryonic lethality in mice. *Oncogene* **21**(50), 7740–7744.
71. Fu, D., and Collins, K. (2006) Human telomerase and Cajal body ribonucleoproteins share a unique specificity of Sm protein association. *Genes Dev* **20**(5), 531–536.
72. Ferre-D'Amare, A. R. (2003) RNA-modifying enzymes. *Curr Opin Struct Biol* **13**(1), 49–55.
73. Spedaliere, C. J., and Mueller, E. G. (2004) Not all pseudouridine synthases are potently inhibited by RNA containing 5-fluorouridine. *RNA* **10**(2), 192–199.
74. Luo, Y., and Li, S. (2007) Genome-wide analyses of retrogenes derived from the human box H/ACA snoRNAs. *Nucleic Acids Res* **35**(2), 559–571.
75. Xie, J., Zhang, M., Zhou, T., Hua, X., Tang, L., and Wu, W. (2007) Sno/scaRNAbase: A curated database for small nucleolar RNAs and cajal body-specific RNAs. *Nucleic Acids Res* **35**(database issue), D183–187.
76. Yang, J. H., Zhang, X. C., Huang, Z. P., Zhou, H., Huang, M. B., Zhang, S., Chen, Y. Q., and Qu, L. H. (2006) snoSeeker: an advanced computational package for screening of guide and orphan snoRNA genes in the human genome. *Nucleic Acids Res* **34**(18) 5112–5123.
77. Hüttenhofer, A., Kiefmann, M., Meier-Ewert, S., O'Brien, J., Lehrach, H., Bachellerie, J. P., and Brosius, J. (2001) RNomics: An experimental approach that identifies 201 candidates for novel, small, non-messenger RNAs in mouse. *EMBO J* **20**(11), 2943–2953.
78. Kiss, A. M., Jady, B. E., Bertrand, E., and Kiss, T. (2004) Human box H/ACA pseudouridylation guide RNA machinery. *Mol Cell Biol* **24**(13), 5797–5807.
79. Lestrade, L., and Weber, M. J. (2006) snoRNA-LBME-db, a comprehensive database of human H/ACA and C/D box snoRNAs. *Nucleic Acids Res* **34**(database issue), D158–162.
80. Schattner, P., Barberan-Soler, S., and Lowe, T. M. (2006) A computational screen for mammalian pseudouridylation guide H/ACA RNAs. *RNA* **12**(1), 15–25.

CHAPTER 8

FUNCTIONAL ROLES OF SPLICEOSOMAL snRNA MODIFICATIONS IN PRE-mRNA SPLICING

David Stephenson
John Karijolich
Yi-Tao Yu

THE SPLICEOSOMAL snRNAs, which are essential for pre-mRNA splicing, are all posttranscriptionally modified by pseudouridylation and 2′-*O*-methylation. Many of the modified nucleotides within the spliceosomal snRNAs are conserved across species and are almost always clustered in regions that are functionally important for pre-mRNA splicing. Splicesomal snRNA modifications can be catalyzed by both RNA-independent (protein-only) and RNA-dependent (RNA–protein complexes) mechanisms. Recent studies have indicated that modifications of spliceosomal snRNAs are required for pre-mRNA splicing.

8.1 INTRODUCTION

Eukaryotic messenger RNA precursors (pre-mRNAs) must be efficiently and accurately spliced before the resulting mature mRNAs are transported to the cytoplasm where they direct the translation of proteins. Pre-mRNA splicing is catalyzed by the spliceosome, a multicomponent complex containing a large number of proteins and five small nuclear (sn) RNAs (U1, U2, U4, U5, and U6). Over the years, the mechanism of pre-mRNA splicing has been studied extensively, and it is well-documented that the five spliceosomal snRNAs play crucial roles in orchestrating the splicing of pre-mRNAs (1–3).

Interestingly, all five spliceosomal snRNAs are extensively posttranscriptionally modified (4, 5). Aside from the 5′ cap modification, there are essentially two types

RNA and DNA Editing: Molecular Mechanisms and Their Integration into Biological Systems,
Edited by Harold C. Smith

of internal modifications within the spliceosomal snRNAs, namely, 2′-*O*-methylation and pseudouridylation. While 2′-*O*-methylation is an RNA backbone-targeting reaction that introduces a methyl group to the sugar ring at the 2′-*O* position, pseudouridylation is a uridine-base-specific modification whereby the uridine is isomerized to pseudouridine (Ψ) (Figure 8.1, inset).

Over the years, great efforts have been made toward understanding the nature of these two types of modification, and the progress concerning many aspects of these modifications, especially the mechanism of modifications, has been remarkable (6). We now know that spliceosomal snRNA modifications, including 2′-*O*-methylation and pseudouridylation, are catalyzed almost exclusively by box C/D sno/scaRNPs (small nucleolar/small Cajal body-specific ribonucleoproteins) and box H/ACA sno/scaRNPs, respectively, in high eukaryotes. Interestingly, in *S. cerevisiae*, both the RNA-dependent mechanism (box H/ACA sno/scaRNP) and a protein-only mechanism (RNA-independent), depending on the site, catalyze spliceosomal snRNA modification (at least for pseudouridylation in U2 snRNA) (7–9).

With regard to the function of these modifications, progress has been relatively slow due primarily to a lack of proper assays and experimental systems. Recently, especially over the past 10 years, the pace on functional modification research has quickened, resulting in the accumulation of a large amount of data (6). A clear picture regarding the function of spliceosomal snRNA modification is emerging. In this review, we discuss the functional aspect of spliceosomal snRNA modification, with a focus on the U2 snRNA because it has been the most extensively studied.

8.2 MODIFIED NUCLEOTIDES IN SPLICEOSOMAL snRNAS

The abundant spliceosomal snRNAs U1 and U2 were first identified in mammals about 40 years ago (10–12). When sequenced (by fingerprinting and other rather laborious techniques that were the only ones available at the time), these snRNAs were found to be extensively modified by 2′-*O*-methylation and pseudouridylation (for review, see reference 4). The other major spliceosomal snRNAs, including U4, U5, and U6, were identified later (13), and they, too, were found to contain multiple 2′-*O*-methylated residues and pseudouridines (4). However, although identified long ago, these modified nucleotides had been largely ignored for decades.

In 1977, the "split gene" was discovered (14, 15), leading to the discovery that the U snRNAs are in fact involved in pre-mRNA splicing (for review, see reference 16). Upon careful inspection of the sequence, it was noted that modified nucleotides in the spliceosomal snRNAs are concentrated almost exclusively in the 5′ half of the molecule that was known to be functionally important (4, 16). For instance, the majority of modified nucleotides in U1 are concentrated in its 5′ end sequence, which plays an important role in 5′ splice site recognition during splicing (Figure 8.1). Similarly, all six uridines in the U2 branch site recognition region located near the 5′

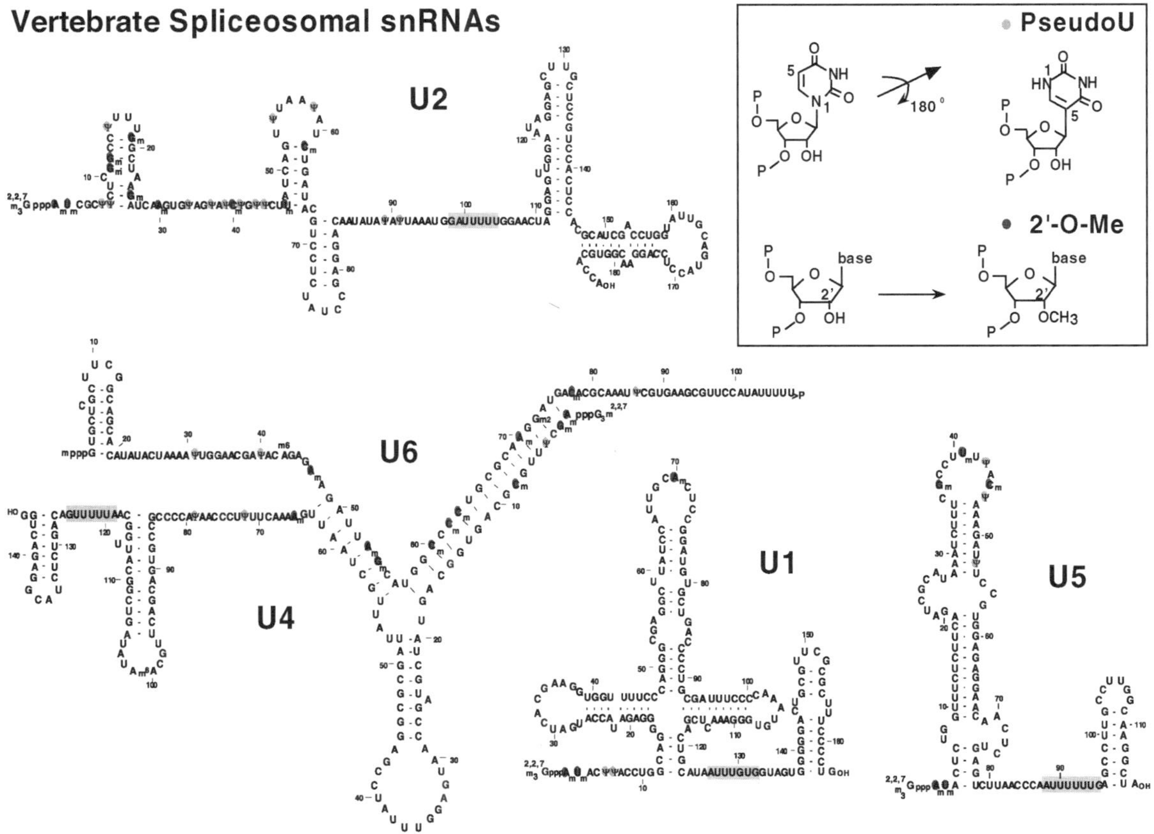

Figure 8.1 Primary sequences and secondary structures of vertebrate major spliceosomal snRNAs. Internal modified nucleotides are color-coded: blue for pseudouridines (Ψ) and red for 2′-*O*-methylated residues. The gray boxes indicate the Sm-binding sites present in U1, U2, U4, and U5 snRNAs. The 5′ caps (2,2,7-trimethylated guanosine cap for U1, U2, U4, and U5, and γ-methylated guanosine for U6) are also shown. U-to-Ψ isomerization (blue dot) and 2′-*O*-methylation (red dot) are schematized in the inset. (*See color insert.*)

end (nucleotides 33–44), which base-pairs with the branch site during splicing, are converted to pseudouridines following initial transcription (Figures 8.1 and 8.2). This clustering phenomenon was rather intriguing since it appeared to be unique to spliceosomal snRNAs—no clustering was observed in tRNAs and rRNAs, both of which contain extensive modifications that are scattered throughout the RNA molecules (4). Thus, it was anticipated that the unusual location of the modified nucleotides in spliceosomal snRNAs is functionally relevant. Yet, given a lack of effective assays and experimental systems, functional analysis of snRNA modifications was almost impossible.

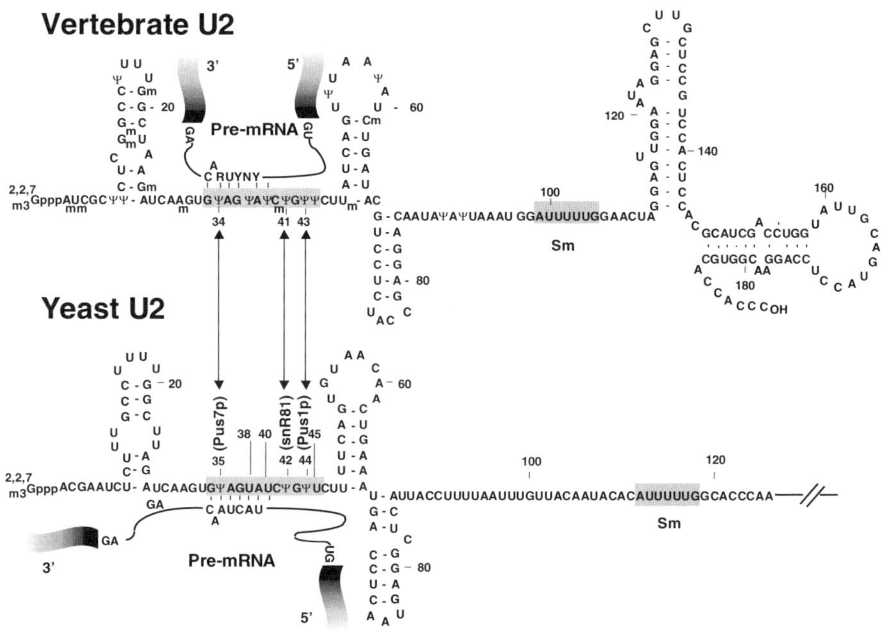

Figure 8.2 Base-pairing interaction between U2 and the branch site of pre-mRNA. Primary sequences (including modified nucleotides) and secondary structures of both vertebrate U2 snRNA and yeast U2 snRNA are shown. The branch site sequences of vertebrate (YNYURAC) and yeast (UACUAAC) pre-mRNAs are also shown. The fading gray ribbons represent exons, and the lines stand for introns (starting with the dinucleotide GU and ending with the dinucleotide AG). The branch site recognition region of vertebrate and yeast U2 is indicated by yellow boxes. The short lines between the U2 branch site recognition region and the pre-mRNA branch site indicate the base-pairing interactions during splicing. Pseudouridines (Ψ) within the U2 branch site recognition region are highlighed in blue. The conserved pseudouridines (Ψ34, Ψ41, Ψ43 in vertebrate U2 and their counterparts Ψ35, Ψ42, and Ψ44 in yeast U2) are indicated by the blue numbers with arrows. The three pseudouridylases responsible for the formation of Ψ35, Ψ42 and Ψ44 in yeast U2 are also indicated. The three unmodified uridines in the yeast U2 branch site recognition region are indicated as well. The gray boxes indicate the Sm binding sites. (*See color insert.*)

8.3 FUNCTIONAL ANALYSIS OF SPLICEOSOMAL snRNA MODIFICATIONS

In the early 1990s, Patton developed an *in vitro* system using HeLa cell S100 and nuclear extracts in which spliceosomal snRNA pseudouridylation could be assayed (17, 18). Interestingly, when incubated with the modification extracts, U2 and the other spliceosomal snRNAs were pseudouridylated. In contrast, when uridines were substituted with 5-fluorouridines (5FU), the resulting 5FU-containing U2 snRNA was not pseudouridylated.

In an effort to assess whether pseudouridylation is important for snRNP assembly, Patton carried out additional experiments using an *in vitro* assembly-modification system (18). To address the effects of 5FU incorporation on snRNP assembly, he incubated synthetic U2 containing or lacking 5FU in the assembly/modification extracts and subsequently analyzed snRNP assembly by cesium-sulfate buoyant density gradient centrifugation. Strikingly, while U2 RNA lacking 5FU was assembled into a U2 snRNP, practically all of the 5FU-containing U2 snRNAs were unable to form salt-resistant complexes, suggesting that no stable U2 snRNP formed. Patton was next interested in determining whether a snRNP is capable of forming on 5FU-containing U2 snRNA in the absence of a high concentration of salt. To do so, he incubated 5FU-containing U2 or U2 lacking 5FU in assembly/modification extracts and analyzed snRNP assembly by velocity sedimentation on 10–30% glycerol gradients. In both cases, velocity sedimentation revealed the formation of an U2 snRNP. Thus, the data clearly indicated that 5FU-containing U2 snRNPs, although capable of forming *in vitro*, were surprisingly susceptible to salt dissociation. These data, coupled with the observation that pseudouridylation was inhibited on 5FU-containing snRNA, clearly hinted at an association between the presence of pseudouridine and U2 snRNP biogenesis.

Around the same period, a number of functional reconstitution systems were developed, permitting a direct assessment of the function of spliceosomal snRNAs in pre-mRNA splicing (19–33). All of these systems involved specific depletion of one of the endogenous spliceosomal snRNAs followed by the supplement of a respective spliceosomal snRNA synthesized *in vitro*. The ability of the supplemented spliceosomal snRNA to reconstitute pre-mRNA splicing could then be monitored. Because *in vitro*-synthesized/transcribed spliceosomal snRNAs contained no modifications, the ability (or lack thereof) of the RNA to support pre-mRNA splicing could be potentially linked to the functionality of the modified nucleotides (assuming *in vitro*-transcribed spliceosomal snRNAs are no longer modified in the systems).

Surprisingly, however, most of these studies indicated that *in vitro* transcribed spliceosomal snRNAs were able to support pre-mRNA splicing. For instance, *in vitro* transcribed U1 (19), U4 (20), and U6 (25–28) can effectively reconstitute pre-mRNA splicing in mammalian nuclear extracts depleted of respective endogenous spliceosomal snRNAs. Also, *in vitro* synthesized U6 can rescue pre-mRNA splicing in U6-depleted cell-free extracts prepared from developing *Ascaris* embryos (29, 30). Additionally, it has been demonstrated that *in vitro* transcribed U2 and U6 were capable of restoring splicing activity in yeast cell extracts (22–24, 32, 33). Taken at face value, these results all suggested that spliceosomal snRNA modifications were not important for pre-mRNA splicing. However, it should be noted that this conclusion relies on the assumption that supplemented spliceosomal snRNAs could no longer be modified in the systems. This is certainly still an open question, because none of the above-mentioned work monitored modification of the *in vitro*-transcribed snRNAs following their addition to the systems. In fact, based on the experience of others (McPheeters, personal communication) and our own (unpublished data), *in vitro*-transcribed U2 snRNA, when added to yeast splicing extracts, is readily modified. In this regard, Lenz et al. showed that a 5-fluorouridine (5FU)-substituted U2, which can no longer be pseudouridylated, failed to reconstitute splicing in U2-depleted yeast cell extracts (34). Thus, at least for U2, it was premature to conclude that modified nucleotides were functionally irrelevant.

This view was further bolstered by independent work carried out later in 1995, in which both synthetic U2 and U5 were analyzed in the HeLa reconstitution system (21). Interestingly, although synthetic U5 was capable of restoring splicing, *in vitro* transcribed U2 completely failed to reconstitute pre-mRNA splicing. Further analysis indicated that neither U2 nor U5 was pseudouridylated in the reconstitution system (21). The failure to reconstitute splicing was also observed earlier when *in vitro* transcribed U2 snRNA was assayed for activity in the *Xenopus* oocyte reconstitution system (31). (Presumably, no modification occurred in the supplemented U2 snRNA under the conditions they used.) Taken together, these results suggested that at least for U2, modified nucelotides (e.g., pseudouridines) may have contributed to the function in pre-mRNA splicing.

8.4 MODIFIED NUCLEOTIDES OF U2 snRNA ARE IMPORTANT FOR PRE-mRNA SPLICING

Encouraged by the possible relationship between U2 snRNA modification and pre-mRNA splicing, Yu et al examined, in a direct and systematic way, the function of U2 snRNA modification in *Xenopus* oocytes (35). An antisense U2 deoxyoligonucleotide was injected into *Xenopus* oocytes, resulting in the formation of a hybrid duplex between the DNA oligonucleotide and the endogenous U2 snRNA. The formation of this hybrid activated an endogenous RNase H activity, resulting in the rapid degradation of endogenous U2 snRNA (the RNA strand of the hybrid). By the time U2 was completely degraded, an endogenous DNase activity degraded the remaining antisense U2 deoxynucleotide. Thus, the initial injection of antisense U2 deoxyoligonucleotide resulted in a complete depletion of endogenous U2 snRNA. *In vitro* transcribed U2 (unmodified) or cellularly derived U2 (modified) was then injected into the U2-depleted oocytes. After reconstitution, a radiolabeled adenovirus standard splicing substrate was injected, and the removal of the intron was monitored through denaturing gel analysis.

Yu et al. (35) observed that following a short reconstitution time, *in vitro* transcribed U2 was unable to rescue splicing. In contrast, reconstitution with cellularly derived U2 for the same amount of time was sufficient to reconstitute splicing, suggesting that U2 modification may indeed play a role in pre-mRNA splicing.

Interestingly, it was also observed that upon longer reconstitution periods, *in vitro* transcribed U2 was able to regenerate splicing activity. In an effort to understand the nature of this delayed activity, Yu et al. (35) monitored the modification status of *in vitro* transcribed U2 after its injection into the oocytes. Strikingly, they found that modification did indeed occur after prolonged reconstitution in the oocytes and that the slow rate of modification in U2 can account for the delayed restoration of splicing. Thus, an excellent correlation was established between the level of U2 modification and the ability of U2 to function in splicing (35).

To further verify the requirement of U2 pseudouridylation in pre-mRNA splicing, Yu et al. (35) took advantage of the fact that *in vitro* transcribed 5FU-containing U2 was unable to reconstitute splicing (34, 35) and that the 5FU-containing U2 functioned as a potent inhibitor that blocked the pseudouridylation of *in vitro* transcribed U2 snRNA containing no 5FU (35, 36). It is widely believed that 5FU-containing U2 can irreversibly bind to the pseudouridylases, thus making the enzymes unavailable for modification (35).

Using this scheme, Yu et al. further assessed the role of U2 pseudouridylation in splicing (35). Following injection of 5FU-containing U2 into the U2-depleted oocytes, either *in vitro* transcribed U2 (containing no 5FU) or cellularly derived U2 was injected. Remarkably, in the presence of 5FU-containing U2, *in vitro* transcribed U2 could no longer reconstitute splicing, even after prolonged reconstitution. In contrast, 5FU-containing U2 had no effect on the ability of cellularly derived U2 to rescue pre-mRNA splicing. These results clearly demonstrated that pseudouridylation alone is important for U2 function in splicing.

In light of concluding that internal modifications are essential for snRNP assembly and pre-mRNA splicing, Yu et al. (35) sought to determine the location of these critical modifications within U2 snRNA. Using site-specific RNaseH cleavage directed by 2′-*O*-methyl RNA–DNA chimeric oligonucleotides (37, 38), they constructed chimeric U2 snRNA molecules with either the 5′ or 3′ half of the molecule derived from cellular U2, while the other half was *in vitro* transcribed. Injection of these chimeric U2 snRNAs indicated that the internal modifications required for pre-mRNA splicing resided within the 27 5′-most nucleotides (35) (Figures 8.1 and 8.2). The functional importance of these modified nucleotides was recently confirmed in the HeLa reconstitution system by the Luhrmann group (39). In that study, chemically synthesized chimeric U2 containing various types and amounts of modifications at the 5′ end was added to U2-depleted HeLa nuclear extracts that contained a radiolabeled pre-mRNA splicing substrate. Since the HeLa nuclear extracts used in the study were unable to modify *in vitro* transcribed U2 snRNA (21, 39), the failure of an unmodified or undermodified U2 to rescue splicing would reflect the importance of the modification at the respective sites. Their results demonstrated that U2 RNA lacking pseudouridines at the 5′ end was incapable of supporting pre-mRNA splicing (39). Thus, these results once again indicate that modified nucleotides within the U2 5′ end region are required for pre-mRNA splicing.

Most recently, further dissection of U2 modifications indicated that the pseudouridines within the U2 branch site recognition region are also required for pre-mRNA splicing in *Xenopus* oocytes (40) (see Figure 8.2). Initial studies failed to identify these pseudouridines as important modifications (35), due to the fact that in the system used, pseudouridylation within the branch site recognition region occurred rather fast—in fact, pseudouridylation had been completed even before the splicing assay was performed (40). This rapid modification precludes the possibility of analyzing the functionality of the pseudouridines within this region using the conventional *Xenopus* oocyte reconstitution system described above (35). Therefore, Zhao and Yu took advantage of the fact that injection of oocytes with synthetic U2 snRNA containing 5FU only in the branch site recognition region specifically inhibits pseudouridylation in the same region of *in vitro* transcribed U2 snRNA injected at a later time (40). The reconstitution results demonstrate that prior injection of 5FU-containing U2 into U2-depleted oocytes almost completely abrogated the ability of *in vitro* transcribed U2 to rescue splicing, whereas full rescue was achieved with either cellular U2 or U2 containing pseudouridines only in the branch site recognition region (40). Taken together, the data accumulated thus far have clearly demonstrated that most modified nucleotides in U2 snRNA, including those residing within the 5′ end region and the branch site recognition region (a total of nine pseudouridines and five 2′-*O*-methylated residues; see Figure 8.2), are functionally important.

8.5 U2 MODIFICATIONS CONTRIBUTE TO snRNP BIOGENESIS AND SPLICEOSOME ASSEMBLY

In an effort to delineate the stage at which modifications contribute to U2 function, Yu et al. examined snRNP and spliceosome assembly in *Xenopus* oocytes using native gel electrophoresis and glycerol gradient sedimentation (35). Following a short reconstitution with *in vitro* transcribed U2 or cellularly derived U2, the oocyte nuclei were isolated and homogenized. The nuclear extract was then directly electrophoresized on a native gel. The result showed that while cellularly derived U2 snRNA was competent in forming splicing complexes A, B, and C, U2 snRNA with no or a low level of modification in its 5′ end region was unable to form higher-order complexes—in fact, no complexes (A, B, or C) were observed. Further analyses using anti-snRNP immunoprecipitation and glycerol gradient sedimentation indicate that U2 modifications contribute to snRNP biogenesis. Specifically, without modifications, U2 can form only a nonfunctional 12S U2 snRNP; modifications are required for the transition from the 12S snRNP to a functional 17S U2 snRNP. Using similar approaches, Zhao and Yu found that the pseudouridines in the branch site recognition region may also contribute to snRNP biogenesis (40). Taken together, these results suggest that the modified nucleotides of U2 snRNA are required at the stage of snRNP biogenesis, prior to spliceosomal assembly, a finding that is consistent with the earlier observations made by Patton (17, 18) that U2 pseudouridylation might contribute to the formation of a stable and salt-resistant snRNP (see above).

Similarly, the Luhrmann group (39) examined which stage of splicing was affected by the modifications in the 5′-end region of U2 in HeLa nuclear extracts. Chimeric U2 RNAs lacking some of the modifications in the U2 5′-end region were used for native gel analysis. Consistent with the Yu study, splicing complexes A, B, and C were undetectable. However, further analysis argued that the modified nucleotides in the U2 5′-end region may directly contribute to the formation of complex E, an early complex required for committing splicing, rather than the formation of functional snRNP, which had been reported earlier (35). The reason for this discrepancy is presently unclear. However, this may be a result of differences between the experimental systems—the *Xenopus* oocyte system (close to an *in vivo* system) (35) versus the HeLa nuclear extracts (an *in vitro* system) (39).

8.6 GENETIC ANALYSIS OF U2 MODIFICATION IN YEAST

Although not as extensively modified as their vertebrate counterparts, yeast spliceosomal snRNAs contain posttranscriptional modifications as well. There are at least three pseudouridines (Ψ) in the yeast U2 branch site recognition region, namely, Ψ35, Ψ42, and Ψ44 (equivalent to Ψ35, Ψ41, and Ψ43 in vertebrate U2; see Figure 8.2) (5). Recently, all pseudouridylases responsible for the pseudouridylation at these three sites (positions 35, 42, and 44) were identified (7–9). While pseudouridylation at positions 35 and 44 is catalyzed by Pus7p (7) and Pus1p (9), respectively, uridine-to-pseudouridine conversion at position 42 is catalyzed by the snR81 box H/ACA RNP, an RNA–protein complex (5) (see also Chapter 7). Given that these three pseudouridines are absolutely conserved across species from yeast to human, it is highly likely that they each play an important role or function synergistically in pre-mRNA splicing. Since all enzymes are known, it becomes possible to use yeast genetics to study the function of each of these pseudouridines.

Our lab first examined the function of pseudouridine 35 (Ψ35) by deleting the gene responsible for this modification, *PUS7*. Although viable in rich media, the *pus7*-deletion strain is growth-disadvantaged under certain conditions (7), such as high salt concentrations or when in competition with the wild-type strain. To clarify the function of Ψ35 in yeast, we used the *pus7*-deletion strain to screen a collection of mutant U2 snRNAs, each containing a point mutation near the branch site recognition region, for a synthetic growth defect phenotype (41). The screen identified two U2 mutants, one containing a U40 to G40 substitution and the other having a U40 deletion. Yeast strains carrying either of these U2 mutations grew as well as the wild-type strain in the selection medium; however, both displayed a temperature-sensitive growth defect phenotype when coupled with the *pus7*-deletion. A subsequent temperature-shift assay and a conditional *pus7*-depletion (via GAL promoter shutoff) in the U2-U40 mutant genetic background resulted in pre-mRNA accumulation, suggesting that Ψ35 is indeed required for pre-mRNA splicing under certain conditions (41).

We then assessed the role of the other two pseudouridines, Ψ42 and Ψ44, by deleting their pseudouridylase genes *SNR81* and *PUS1*, respectively (Stephenson and Yu, unpublished data). Just as what was observed for *PUS7*, deletion of either *SNR81* or *PUS1* exhibited no obvious growth defect phenotype in rich media. However, when the two pseudouridylase genes (*SNR81* and *PUS1*) were deleted simultaneously, a clear temperature-sensitive growth defect phenotype was observed (no growth at 37°C).

To further unravel the molecular mechanism by which the simultaneous deletion of the two pseudouridylases resulted in a growth defect, we took advantage of the intron-containing Cup1p-reporter system, in which the ability of cells to grow in the copper medium correlates directly to the efficiency of splicing that gives rise to a functional *CUP1* message and therefore a functional Cup1p, a protein conferring copper resistance in yeast. As expected, when both *PUS1* and *SNR81* were deleted, cells exhibited a growth defect phenotype in the copper-containing medium even at 30°C, linking the phenotype directly to the splicing defect that fails to produce the functional *CUP1* reporter mRNA. Indeed, detailed molecular analyses showed that the level of mature *CUP1* mRNA was greatly reduced in the *Pus1*-deletion and *Snr81*-deletion strain when compared to the wild type strain. Remarkably, when the adjacent uridines (positions 38 and 40, see Figure 8.2) were converted to pseudouridines by an artificial sno/scaRNP, the growth phenotype was rescued. Hence, it appears that the pseudouridines in the U2 branch recognition region function synergistically during pre-mRNA splicing. However, the way the pseudouridines support splicing is rather sloppy—pseudouridylation within a specific region rather than at a specific site is necessary.

8.7 CYTOTOXICITY ASSOCIATED WITH 5FU TREATMENT IS A RESULT OF INHIBITION ON PSEUDOURIDYLATION AND SPLICING

Although it has been extensively studied, the mechanism of action of 5FU on cancer cells has yet to be established (42–45). It is known that once taken up by the cell, a fraction of 5FU is converted to 5FUTP (5-fluorouridine triphosphate), which is a perfect ribonucleotide analog that can be readily incorporated into RNA chains. Based on our results showing that 5FU-containing U2 blocks U2 pseudouridylation and consequently inhibits splicing, Zhao and Yu postulated that 5FU incorporation into U2 snRNA could be a factor in the therapeutic effect of 5FU in chemotherapy (46).

Zhao and Yu (46) found that upon culturing HeLa cells in a 5FU-containing medium, 5FU was indeed efficiently incorporated into U2 snRNA at natural pseudouridylation sites within the branch site recognition region. Remarkably, 5FU incorporation effectively blocked the formation of important pseudouridines in U2 snRNA, because only a trace of pseudouridine was detected when cells had been exposed to a low dose of 5FU for 5 days. U2 snRNAs recovered from HeLa cells were further analyzed for their ability to reconstitute splicing in U2-depleted *Xenopus* oocytes. The results indicated that U2 snRNA isolated from 5FU-cultured cells failed to restore pre-mRNA splicing. In contrast, U2 snRNA isolated from control cells fully rescued pre-mRNA splicing. Further analysis using RT-PCR indicated that pre-mRNA splicing was indeed inhibited in HeLa cells exposed to 5FU, because an accumulation of unspliced pre-mRNAs was observed.

Given that 5FU, when converted into 5FUTP in cells, can also be incorporated into pre-mRNA, it is possible that 5FU-containing pre-mRNA is defective in splicing. To address this hypothesis, Zhao and Yu (46) incorporated 5FU into adenovirus standard pre-mRNA splicing substrate or human β-globin pre-mRNA via *in vitro* transcription and injected the pre-mRNA into *Xenopus* oocytes to assess whether it could be spliced *in vivo*. No effect on splicing was observed regardless of the degree of 5FUTP substitution. Taken together, we concluded that 5FU incorporation into pre-mRNA does not interfere with pre-mRNA splicing; the accumulation of pre-mRNAs observed in 5FU-treated cells is a result of the inability of hypo-pseudouridylated U2 snRNA to support pre-mRNA splicing. This effect could very well contribute to cell death and, therefore, explain at least some of the therapeutic effect of 5FU in chemotherapy.

8.8 BIOPHYSICAL ANALYSIS OF U2 snRNA MODIFICATION

Although it is now clear that U2 modifications play an important role in snRNP biogenesis and pre-mRNA splicing, the mechanism by which the modified nucleotides act is still unknown. It is well established that both 2′-*O*-methylation and pseudouridylation can change the chemical properties of nucleotide residues. Such changes may in turn bring further alterations in the local and/or global RNA structure, thermal stability, and chemical interactions, thus leading to changes in function (47). With regard to pseudouridylation, it has been shown that pseudouridine creates a more stable conformation in a short RNA fragment through changes in base stacking (48, 49). Along the same line, it has been shown that tRNA structure can be stabilized by pseudouridylation (50) (see also Chapter 5). Unlike uridine, pseudouridine coordinates a water molecule through the addition of an extra hydrogen bond, as exemplified by the crystal structure of $tRNA^{Gln}$, in which a water molecule forms a local structure that may be critical for stabilizing the tRNA (51).

In an attempt to understand the molecular mechanism by which U2 pseudouridylation contributes to function in splicing, the Greenbaum group carried out a structural analysis (52, 53). In that study, a short RNA duplex, mimicking the base-pairing interaction between the U2 branch site recognition region and the intron branch site (see Figure 8.2), was subjected to NMR analysis. Strikingly, the results showed that a single pseudouridine substitution at position 35 (naturally catalyzed by Pus7p in yeast, see Figure 8.2) in the U2 strand drastically changed the local structure surrounding the modified nucleotide. Specifically, the branch point adenosine in the intron strand was bulged out, establishing a configuration that favors the first step of splicing: The 2′-hydroxyl group of the bulged-out adenosine nucleophilically attacks the phosphate at the 5′-exon–intron junction. In support of this finding, Valadkhan and Manley showed that the change of a single uridine in the branch site recognition region to pseudouridine greatly enhanced the production of X-RNA, a product generated by a splicing-related branching reaction in a cell- and protein-free system (54, 55), suggesting that at least one pseudouridine in the branch site recognition region plays a critical role in splicing.

To date, no direct biophysical data are available with regard to spliceosomal snRNA 2′-*O*-methylation. The molecular mechanism of the action of 2′-*O*-methylation in U2 function remains unclear. However, it has been suggested that 2′-*O*-methylation has the ability to alter the characteristics of a nucleotide residue, thus leading to more stable RNA conformations. For instance, 2′-*O*-methylation alters the hydration sphere around the oxygen, resulting in the blockage of sugar edge interactions (49, 56–58). In addition, the hydrophobic methyl group modifies the ability of the ribose to engage in hydrogen bonding (59). It is conceivable that such alterations would have an impact on RNA structures. In this regard, thermophilic organisms show a direct correlation between their optimal growth temperature and the number of methylations present in rRNA (60), as was reported for tRNA (47, 59, 61, 62), suggesting that 2′-*O*-methylations may indeed help stabilize the RNA structures (59). Finally, it is known that 2′-*O*-methylation also plays a role in protecting the RNA from hydrolysis by alkali and nucleases, thus probably helping increase RNA stability in the cell.

8.9 CONCLUDING REMARKS

Recent progress on spliceosomal snRNA modification, including the mechanisms and, to some extent, the functions, has been remarkable. However, compared to the progress made on DNA and protein modifications, the study of spliceosomal snRNA modification or RNA modification in general, especially the function of modified nucleotides, remains lagging behind. Although we now know that U2 modifications play an important role in snRNP biogenesis and splicing, many fundamental questions remain. For instance, how do the modified nucleotides in U2 snRNA contribute to function? To answer this question, a detailed structural mechanistic analysis seems to be necessary. With regard to the modifications of the other spliceosomal snRNAs, there are even more outstanding questions. Currently, we still do not know whether modified nucleotides in U1, U4, U5, and U6 play any roles in splicing. It has been discussed above that *in vitro* transcribed snRNA, either U1, U4, U5, or U6, can reconstitute splicing in a number of reconstitution systems (19–30), suggesting that modified nucleotides may be functionally irrelevant. However, with the exception of U5 (21), none of these RNAs were monitored for modification during the course of reconstitution. It remains possible that *in vitro* transcribed snRNAs, when added to the reconstitution systems, become modified. If true, the use of modification inhibitors (e.g., 5FU incorporation to inhibit pseudouridylation) would be a desirable approach to this problem.

On the other hand, it should be noted that all the above-discussed reconstitution experiments utilized a strong splicing substrate, either actin pre-mRNA (for the yeast system), adenovirus standard pre-mRNA (for the higher systems), or human β-globin pre-mRNA (for the higher systems). It is possible that splicing of such optimal splicing substrates may not need the additional help from modified nucleotides in U1, U4, U5, or U6, given that most RNA modifications are only involved in the fine-tuning of RNA functions. Therefore, to fully understand the functional roles of these modifications, it is desirable to use some suboptimal splicing substrates. With a growing attention to RNA modifications, we expect that a clear picture concerning whether and how spliceosomal snRNA modifications contribute to function will be established soon.

REFERENCES

1. Burge, C. B., Tuschl, T., and Sharp, P. A. (1999) In: Gesteland, R. F., Cech, T. R., and At-kins, J. F. (eds.), *The RNA World*, 2nd ed., Cold Spring Harbor Laboratory Press, Cold Spring Harbor, NY, pp. 525–560.

2. Yu, Y. T., Scharl, E. C., Smith, C. M., and Steitz, J. A. (1999) In: Gesteland, R. F., Cech, T. R., and Atkins, J. F. (eds.), *The RNA World*, 2nd ed., Cold Spring Harbor Laboratory Press, Cold Spring Harbor, NY, pp. 487–524.
3. Nilsen, T. W. (1998) In: Simons, R. W., and Grunberg-Manago, M. (eds.), *RNA Structure and Function*, Cold Spring Harbor Laboratory Press, Cold Spring Harbor, NY, pp. 279–307.
4. Reddy, R., and Busch, H. (1988) In: Birnsteil, M. L. (ed.), *Structure and Function of Major and Minor Small Nuclear Ribonucleoprotein Particles*, Springer-Verlag Press, Heidelberg, pp. 1–37.
5. Massenet S., Mougin, A., and Branlant, C. (1998) In: Grosjean, H. (ed.), *Modification and Editing of RNA*, ASM Press, Washington, DC, pp. 201–228.
6. Yu, Y. T., Terns, R. M., and Terns, M. P. (2005) In: Grosjean H. (ed.), *Topics in Current Genetics*, **Vol. 12**, Springer-Verlag, New York, pp. 223–262.
7. Ma, X., Zhao, X., and Yu, Y. T., (2003) Pseudouridylation (Psi) of U2 snRNA in S. cerevisiae is catalyzed by an RNA-independent mechanism. *EMBO J* **22**, 1889–1897.
8. Ma, X., Yang, C., Alexandrov, A., Grayhack, E. J., Behm-Ansmant, I., and Yu, Y. T. (2005) Pseudouridylation of yeast U2 snRNA is catalyzed by either an RNA-guided or RNA-independent mechanism. *EMBO J* **24**, 2403–2413.
9. Massenet, S., Motorin, Y., Lafontaine, D. L., Hurt, E. C., Grosjean, H., and Branlant, C. (1999) Pseudouridine mapping in the *Saccharomyces cerevisiae* spliceosomal U small nuclear RNAs (snRNAs) reveals that pseudouridine synthase pus1p exhibits a dual substrate specificity for U2 snRNA and tRNA. *Mol Cell Biol* **19**, 2142–2154.
10. Hodnett, J. L., and Busch, H. (1968) Isolation and characterization of uridylic acid-rich 7 S ribonucleic acid of rat liver nuclei. *J Biol Chem* **243**, 6334–6342.
11. Weinberg, R. A., and Penman, S. (1968) Small molecular weight monodisperse nuclear RNA. *J Mol Biol* **38**, 289–304.
12. Muramatsu, M., and Busch, H. (1965) Studies on the nuclear and nucleolar ribonucleic acid of regenerating rat liver. *J Biol Chem* **240**, 3960–3966.
13. Lerner, M. R., and Steitz, J. A. (1979) Antibodies to small nuclear RNAs complexed with proteins are produced by patients with systemic lupus erythematosus. *Proc Natl Acad Sci USA* **76**, 5495–5499.
14. Berget, S. M., Moore, C., and Sharp, P. A. (1977) Spliced segments at the 5′ terminus of adenovirus 2 late mRNA. *Proc Natl Acad Sci USA* **74**, 3171–3175.
15. Chow, L. T., Gelinas, R. E., Broker, T. R., and Roberts, R. J. (1977) An amazing sequence arrangement at the 5′ ends of adenovirus 2 messenger RNA. *Cell* **12**, 1–8.
16. Steitz, J. A., black, D. L., Gerke, V., Parker, K. A., Kramer, A., Frendewey, D., and Keller, W. (1988) In: Birnstiel, M. L. (ed.), *Small Nuclear Ribonucleoprotein Particles*, Springer-Verlag, Berlin, pp. 115–154.
17. Patton, J. R. (1993) Multiple pseudouridine synthase activities for small nuclear RNAs. *Biochem J* **290** (Pt 2), 595–600.
18. Patton, J. R. (1993) Ribonucleoprotein particle assembly and modification of U2 small nuclear RNA containing 5-fluorouridine. *Biochemistry* **32**, 8939–8944.
19. Will, C. L., Rumpler, S., Klein Gunnewiek, J., van Venrooij, W. J., and Luhrmann, R. (1996) In vitro reconstitution of mammalian U1 snRNPs active in splicing: The U1-C protein enhances the formation of early (E) spliceosomal complexes. *Nucleic Acids Res* **24**, 4614–4623.
20. Wersig, C., and Bindereif, A. (1992) Reconstitution of functional mammalian U4 small nuclear ribonucleoprotein: Sm protein binding is not essential for splicing in vitro. *Mol Cell Biol* **12**, 1460–1468.
21. Segault, V., Will, C. L., Sproat, B. S., and Luhrmann, R. (1995) In vitro reconstitution of mammalian U2 and U5 snRNPs active in splicing: Sm proteins are functionally interchangeable and are essential for the formation of functional U2 and U5 snRNPs. *EMBO J* **14**, 4010–4021.
22. Fabrizio, P., McPheeters, D. S., and Abelson, J. (1989) In vitro assembly of yeast U6 snRNP: A functional assay. *Genes Dev* **3**, 2137–2150.
23. Fabrizio, P., and Abelson, J. (1990) Two domains of yeast U6 small nuclear RNA required for both steps of nuclear precursor messenger RNA splicing. *Science* **250**, 404–409.
24. Fabrizio, P., and Abelson, J. (1992) Thiophosphates in yeast U6 snRNA specifically affect pre-mRNA splicing in vitro. *Nucleic Acids Res* **20**, 3659–3664.
25. Wolff, T., and Bindereif, A. (1992) Reconstituted mammalian U4/U6 snRNP complements splicing: A mutational analysis. *EMBO J* **11**, 345–359.

26. Wolff, T., and Bindereif, A. (1995) Mutational analysis of human U6 RNA: Stabilizing the intramolecular helix blocks the spliceosomal assembly pathway. *Biochim Biophys Acta* **1263**, 39–44.
27. Wolff, T., Menssen, R., Hammel, J., and Bindereif, A. (1994) Splicing function of mammalian U6 small nuclear RNA: conserved positions in central domain and helix I are essential during the first and second step of pre-mRNA splicing. *Proc Natl Acad Sci USA* **91**, 903–907.
28. Wolff, T., and Bindereif, A. (1993) Conformational changes of U6 RNA during the spliceosome cycle: An intramolecular helix is essential both for initiating the U4-U6 interaction and for the first step of slicing. *Genes Dev* **7**, 1377–1389.
29. Yu, Y. T., Maroney, P. A., and Nilsen, T. W. (1993) Functional reconstitution of U6 snRNA in nematode cis- and trans-splicing: U6 can serve as both a branch acceptor and a 5′ exon. *Cell* **75**, 1049–1059.
30. Yu, Y. T., Maroney, P. A., Darzynkiwicz, E., and Nilsen, T. W. (1995) U6 snRNA function in nuclear pre-mRNA splicing: A phosphorothioate interference analysis of the U6 phosphate backbone. *RNA* **1**, 46–54.
31. Pan, Z. Q., and Prives, C. (1989) U2 snRNA sequences that bind U2-specific proteins are dispensable for the function of U2 snRNP in splicing. *Genes Dev* **3**, 1887–1898.
32. McPheeters, D. S., and Abelson, J. (1992) Mutational analysis of the yeast U2 snRNA suggests a structural similarity to the catalytic core of group I introns. *Cell* **71**, 819–831.
33. McPheeters, D. S., Fabrizio, P., and Abelson, J. (1989) In vitro reconstitution of functional yeast U2 snRNPs. *Genes Dev* **3**, 2124–2136.
34. Lenz, H. J., Manno, D. J., Danenberg, K. D., and Danenberg, P. V. (1994) Incorporation of 5-fluorouracil into U2 and U6 snRNA inhibits mRNA precursor splicing. *J Biol Chem* **269**, 31962–31968.
35. Yu, Y. T., Shu, M. D., and Steitz, J. A. (1998) Modifications of U2 snRNA are required for snRNP assembly and pre-mRNA splicing. *EMBO J* **17**, 5783–5795.
36. Patton, J. R., Jacobson, M. R., and Pederson, T. (1994) Pseudouridine formation in U2 small nuclear RNA. *Proc Natl Acad Sci USA* **91**, 3324–3328.
37. Yu, Y. T., Shu, M. D., and Steitz, J. A. (1997) A new method for detecting sites of 2′-*O*-methylation in RNA molecules. *RNA* **3**, 324–331.
38. Lapham, J., and Crothers, D. M. (2000) Site-specific cleavage of transcript RNA. *Methods Enzymol* **317**, 132–139.
39. Donmez, G., Hartmuth, K., and Luhrmann, R. (2004) Modified nucleotides at the 5′ end of human U2 snRNA are required for spliceosomal E-complex formation. *RNA* **10**, 1925–1933.
40. Zhao, X., and Yu, Y. T. (2004) Pseudouridines in and near the branch site recognition region of U2 snRNA are required for snRNP biogenesis and pre-mRNA splicing in *Xenopus* oocytes. *RNA* **10**, 681–690.
41. Yang, C., McPheeters, D. S., and Yu, Y. T. (2005) Psi35 in the branch site recognition region of U2 small nuclear RNA is important for pre-mRNA splicing in Saccharomyces cerevisiae. *J Biol Chem* **280**, 6655–6662.
42. Heidelberger, C., Chaudhuri, N. K., Danneberg, P., Mooren, D., Griesbach, L., Duschinsky, R., Schnitzer, R. J., Pleven, E., and Scheiner, J. (1957) Fluorinated pyrimidines, a new class of tumour-inhibitory compounds. *Nature* **179**, 663–666.
43. Longley, D. B., Harkin, D. P., and Johnston, P. G. (2003) 5-fluorouracil: Mechanisms of action and clinical strategies. *Nat Rev Cancer* **3**, 330–338.
44. Parker, W. B., and Cheng, Y. C. (1990) Metabolism and mechanism of action of 5-fluorouracil. *Pharmacol Ther* **48**, 381–395.
45. Ghoshal, K., and Jacob, S. T. (1997) An alternative molecular mechanism of action of 5-fluorouracil, a potent anticancer drug. *Biochem Pharmacol* **53**, 1569–1575.
46. Zhao, X., and Yu, Y. T. (2007) Incorporation of 5-fluorouracil into U2 snRNA blocks pseudouridylation and pre-mRNA splicing in vivo. *Nucleic Acids Res* **35**, 550–558.
47. Agris, P. F. (1996) The importance of being modified: Roles of modified nucleosides and Mg^{2+} in RNA structure and function. *Prog Nucleic Acid Res Mol Biol* **53**, 79–129.
48. Davis, D. R. (1995) Stabilization of RNA stacking by pseudouridine. *Nucleic Acids Res* **23**, 5020–5026.
49. Davis, D. R. (1998) In: Grosjean H. and Benne R. (eds.), *Modification and Editing of RNA*, ASM Press, Washington, DC, pp. 85–102.
50. Cabello-Villegas, J., and Nikonowicz, E. P. (2005) Solution structure of psi32-modified anticodon stem-loop of *Escherichia coli* tRNAPhe. *Nucleic Acids Res* **33**, 6961–6971.

51. Arnez, J. G., and Steitz, T. A. (1994) Crystal structure of unmodified tRNA(Gln) complexed with glutaminyl-tRNA synthetase and ATP suggests a possible role for pseudo-uridines in stabilization of RNA structure. *Biochemistry* **33**, 7560–7567.
52. Newby, M. I., and Greenbaum, N. L. (2002) Sculpting of the spliceosomal branch site recognition motif by a conserved pseudouridine. *Nat Struct Biol* **9**, 958–965.
53. Newby, M. I., and Greenbaum, N. L. (2001) A conserved pseudouridine modification in eukaryotic U2 snRNA induces a change in branch-site architecture. *RNA* **7**, 833–845.
54. Valadkhan, S., and Manley, J. L. (2003) Characterization of the catalytic activity of U2 and U6 snRNAs. *RNA* **9**, 892–904.
55. Valadkhan, S., and Manley, J. L. (2001) Splicing-related catalysis by protein-free snRNAs. *Nature* **413**, 701–707.
56. Auffinger, P., and Westhof, E. (1997) Rules governing the orientation of the 2′-hydroxyl group in RNA. *J Mol Biol* **274**, 54–63.
57. Auffinger, P., and Westhof, E. (1998) Hydration of RNA base pairs. *J Biomol Struct Dyn* **16**, 693–707.
58. Helm, M. (2006) Post-transcriptional nucleotide modification and alternative folding of RNA. *Nucleic Acids Res* **34**, 721–733.
59. Lapeyre, B. (2005) In: Grosjean H. (ed.), *Topics in Current Genetics*, **Vol. 12**, Springer-Verlag, New York, pp. 263–284.
60. Noon, K. R., Bruenger, E., and McCloskey, J. A. (1998) Posttranscriptional modifications in 16S and 23S rRNAs of the archaeal hyperthermophile *Sulfolobus solfataricus*. *J Bacteriol* **180**, 2883–2888.
61. Agris, P. F., Koh, H., and Soll, D. (1973) The effect of growth temperatures on the in vivo ribose methylation of *Bacillus stearothermophilus* transfer RNA. *Arch Biochem Biophys* **154**, 277–282.
62. Kowalak, J. A., Dalluge, J. J., McCloskey, J. A., and Stetter, K. O. (1994) The role of posttranscriptional modification in stabilization of transfer RNA from hyperthermophiles. *Biochemistry* **33**, 7869–7876.

CHAPTER 9

A ROLE FOR A-TO-I EDITING IN GENE SILENCING

Jing Zhou
Ling-Ling Chen
Gordon G. Carmichael

DOUBLE-STRANDED RNA (dsRNA) is expressed at significant levels in the nuclei of all eukaryotic cells. In lower eukaryotes, this RNA has been shown to trigger heterochromatic gene silencing through the RNAi machinery. But evidence is weak for a similar mechanistic connection in higher organisms. In mammalian cells, nuclear dsRNAs may be very rapidly and efficiently edited by ADAR activity, thus preventing them from being acted upon by the RNAi machinery. We argue here that in cells with robust editing activity, there may exist an alternative pathway from dsRNA to heterochromatin, involving both A-to-I editing by ADAR and the highly conserved vigilin class of proteins.

9.1 EXPRESSION OF DOUBLE-STRANDED RNA IN CELLS

Double-stranded RNAs (dsRNAs) are produced continually in cells. The fates of these dsRNAs depend on both their location and their length. In higher eukaryotes, long cytoplasmic dsRNAs are often the result of viral infection. These long dsRNAs (>30 bp) can either (a) activate the potent interferon and protein kinase R (PKR) antiviral pathways that lead to non-sequence-specific effects including apoptosis (see reference (1) for review) or (b) trigger antimicrobial immune responses by activation of NF-κB and induction of beta interferon through Toll-like receptor 3 (TLR3) (2, 3).

In contrast to cytoplasmic dsRNAs, nuclear dsRNAs are almost always produced within cells and not introduced from without. They can arise from antisense transcription or from bidirectional transcription in which both strands of the DNA are transcribed and anneal to form dsRNA. Antisense RNAs that are transcribed from the opposite strand of the same genomic locus as the sense RNA have a long (and perfect)

RNA and DNA Editing: Molecular Mechanisms and Their Integration into Biological Systems,
Edited by Harold C. Smith

overlap with the sense transcripts. On the other hand, antisense RNAs that are transcribed from a genomic locus different from the sense RNA may give rise to imperfect duplexes with sense transcripts (1, 4). Antisense RNAs have been observed in various organisms, and their expression is considered to be a conserved feature within the genomes of all species from archaebacterials to humans. Employing appropriate computational tools, it has been reported that up to 22% of human transcripts have the potential to form sense–antisense pairs based on analysis of the available human gene database (5).

A significant fraction of the long terminal repeat retrotransposon LINE elements, which occupy about 17% of genomic DNA, may be transcribed in an antisense orientation and may therefore have the potential to form dsRNA in such a way (1). Interestingly, a large fraction of the LINE element sequences (as well as tandem sequence arrays that can express sense–antisense pairs) appear to be localized to heterochromatic regions.

A second way in which dsRNAs can arise is by base pairing of complementary sequences that exist as parts of the same molecule (intramolecular base pairing). For example, about 90% of human genes contain in their introns multiple copies of *Alu* sequences of the SINE family of transposable elements. Since these *Alu* sequences are highly conserved in humans, transcription of inverted *Alu* elements on the same strand can lead to the formation of intramolecular dsRNA (see reference 6 and references therein).

9.2 THE ACTIVITY OF ADAR IN THE NUCLEUS

As discussed above, there is a large amount of dsRNA expressed in the nucleus. What is the fate of such RNA? One important fate of nuclear dsRNA is to be edited by ADARs (adenosine deaminases that act on dsRNAs). These enzymes convert adenosines to inosines via hydrolytic deamination (7). The editing activity is sensitive to the length of dsRNAs. Duplexes less than 15 base pairs are poorly modified, while long perfectly matched dsRNAs of at least 25–30 bp are hyperedited in which many A's can be edited into I's (8, 9). Optimal editing activity is seen in dsRNAs of more than 100 bp and 50% or more of the A's on each strand of long perfect duplexes can be converted into I's (10). Editing produces I:U mismatches that replace A:U base pairs, and inosine preferentially base pairs with cytosine. Therefore, ADAR editing dramatically alters the structure of RNAs, importantly reducing secondary structure.

In short imperfect duplexes, only a few A's are edited to I's. Site-selective editing is directed to specific adenosines embedded in favorable secondary structures and has been reported primarily in mRNAs important for the function of the nervous system (11). Discovery of large number of ADAR substrates by combining comparative genomics and experimental approaches has uncovered a general feature of this type of editing: It is directed by the interaction of exonic sequences with partner sequences that lie in adjacent introns (12). Such pre-mRNA editing must be quite rapid in the nucleus: It occurs cotranscriptionally and prior to (or simultaneous with) splicing (Figure 9.1) (see Chapters 1 and 6).

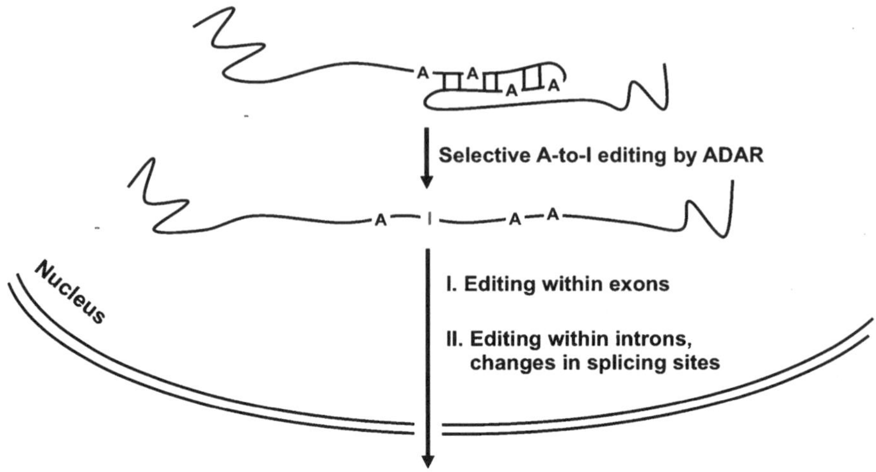

Figure 9.1 Editing of short, imperfect RNA duplexes in the nucleus. In this situation, A-to-I editing is selective, often involving only one or a few editing sites within exon sequences that base-pair with intron sequences. Editing within an intron could also in some cases generate novel splice sites, leading to new mRNAs. Selectively edited RNAs can be exported to the cytoplasm and translated into new protein isoforms. (*See color insert.*)

9.3 ALTERNATIVE FATES OF EDITED RNAs IN THE NUCLEUS

What are the consequences of promiscuous and site selective editing in the nucleus? Since inosine preferentially base pairs with cytosine, the edited A's are recognized as G's by the translation machinery. Thus, site-selective editing can introduce missense codons but not nonsense codons into mRNAs when the editing occurs in the coding regions. This leads to alterations of protein primary sequence and/or structure and expands the diversity of the cellular proteome. Also, site-selective editing has the potential to affect splicing when it occurs in the intronic sequences by either removing a splicing recognition sequence (AG $\rightarrow$ IG $\rightarrow$ GG) or creating a new splice acceptor site (AA $\rightarrow$ AI $\rightarrow$ AG) or splice donor site (AU $\rightarrow$ IU $\rightarrow$ GU) (13–15) (see Chapters 1 and 6).

Unlike selective editing, promiscuous editing in the coding regions would produce mutant proteins that may have detrimental effects on cells (Figure 9.2). A common fate of these promiscuously edited RNAs appears to be nuclear retention, mediated by binding to a protein complex containing $p54^{nrb}$, the splicing factor PSF and the nuclear matrix structural protein Matrin 3 (16). $p54^{nrb}$ shows strong preference for inosine containing RNA and such specificity is conserved between its orthologues in mouse, *Drosophila melanogaster*, *Chironomus tentans,* and *Xenopus laevis* (Q. Wang and G. Carmichael, unpublished). Thus, one natural function of the $p54^{nrb}$ complex may be to prevent the export of promiscuously edited

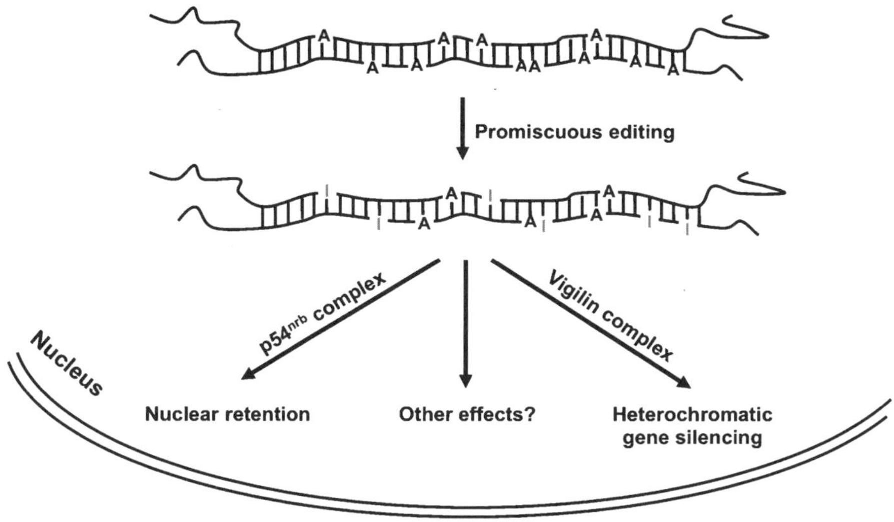

Figure 9.2 Editing of long, perfect RNA duplexes in the nucleus. In this situation, A-to-I editing is promiscuous, with up to 50% or more of the adenosines on each strand deaminated to inosines. As discussed in the text, such RNAs can be retained in the nucleus in complexes containing p54^{nrb}, used to induce heterochromatic gene silencing via the vigilin pathway or to exert other, yet unknown effects. (*See color insert.*)

RNAs (and thereby prevent the translation of such RNAs into mutated proteins), thus providing a mechanism in the quality control of mRNA expression and gene regulation. This control may act co-transcriptionally since both PSF and p54^{nrb} associate strongly with the C-terminal domain of RNA polymerase II (17) (see Chapter 6).

9.4 A POSSIBLE CONNECTION BETWEEN RNA EDITING AND GENE SILENCING

Nuclear retention may not be the only fate of promiscuously edited RNA in the nucleus. We have recently discovered a protein, vigilin, that binds avidly to promiscuously edited RNAs, but which also appears to participate directly in the establishment of heterochromatin. In order to understand how this might occur at the molecular level, it is instructive to briefly review heterochromatin and what is known about how the RNAi machinery has been reported to function in silencing.

9.4.1 Heterochromatin

Heterochromatin is cytologically visible nuclear material that typically remains condensed throughout the cell cycle. It is generally transcriptionally inactive and largely composed of tandem arrays of repetitive sequences and transposable elements. Constitutive heterochromatin is always located within pericentric and telomeric regions, both of which are important structures for the maintenance of intact chromosomes and faithful transmission of their genetic information. The molecular markers of heterochromatin include (a) methylation of histone 3 on lysine 9, which is catalyzed by SUV39H1 (SU(VAR)3-9 in *Drosophila* and Clr4 in fission yeast), and (b) the binding of heterochromatin protein 1 (HP1, Swi6 in fission yeast) to the methylation site.

9.4.2 RNAi-Directed Heterochromatin Formation

In the past several years a growing body of evidence has pointed to dsRNA and the RNAi machinery in the establishment of some heterochromatin. Involvement of RNAi in the establishment of heterochromatin has been best studied in the pericentromeric region of the fission yeast *S. pombe* (see Figure 9.3). The central core of *S. pombe* centromeres are flanked by two different types of inverted repeats, the innermost (imr) and outer (otr) repeats. The outer repeat is composed of *dg* and *dh* tandem repeats, which are classical constitutive heterochromatin. In wild-type cells, transcripts of only one strand, the reverse strand of the repeats, can be detected and only at very low level. However, when components of the RNAi machinery are mutated, transcripts from both strands of *dg-dh* sequences accumulate, with reduced amounts of H3K9 methylation and repeat-associated Swi6. These cells exhibit chromosome segregation defects. Two different complexes have been identified that play important roles in heterochromatin

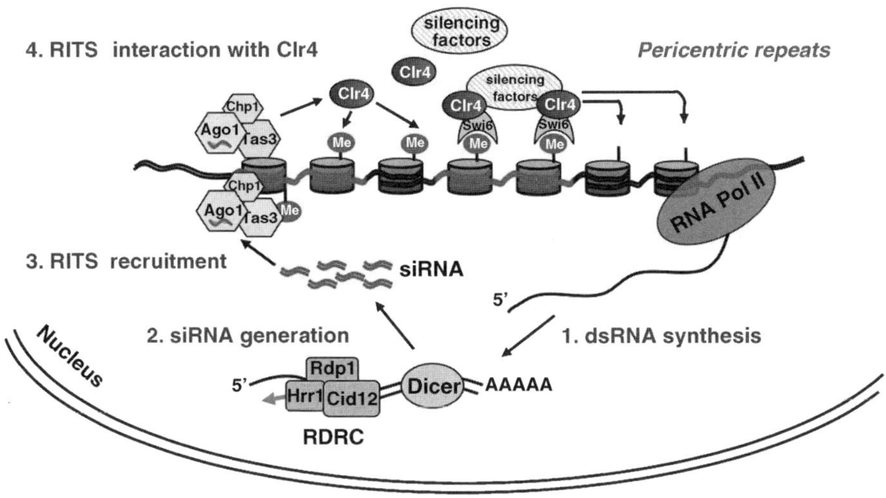

Figure 9.3 RNAi-induced heterochromatin formation in fission yeast. RNA polymerase II transcribes the reverse strand of the pericentromeric repeats. dsRNA is then synthesized from the pericentric ssRNA by Rdp1, one of the components from the RNA-directed RNA polymerase complex (RDRC). Then the dsRNA is acted on by Dicer to generate siRNAs, which are incorporated into RNA-induced initiation of transcriptional silencing (RITS) complexes. These RITS complexes can recruit Clr4, which methylates H3K9, leading to the binding of Swi6. Clr4 interacts with Swi6, thus promoting further methylation and spreading of silencing in the pericentromeric region (50–53). (*See color insert.*)

formation in the pericentric region of *S. pombe*: the RNA-induced initiation of transcriptional silencing (RITS) complex and the RNA-directed RNA polymerase complex (RDRC). Clr4 (the H3K9 methyltransferase) and Dcr1 (having Dicer activity) are required for the interaction and localization of these two complexes to the pericentric region (18). In this model, the RDRC could generate dsRNAs from an ssRNA template, the reverse strand of the *dg–dh* repeats transcribed by RNA pol II. These dsRNAs are processed by Dicer into siRNAs, which are then incorporated into RITS. This RNAi effector complex is possibly also involved in recruiting the histone methyltransferase Clr4. Through RNA–RNA or RNA–DNA homology, the RITS complex is directed to target sequences that become methylated on lysine 9 of histone H3 (H3K9Me). Chromodomain proteins Swi6 (the *S. pombe* homolog of HP1) and Chp1 are then recruited to the modified histones. Clr4 may also interact with Swi6. This interaction may facilitate the spreading of H3K9Me and creating additional binding site for Chp1, thereby spreading the silent state to the nearby regions (Figure 9.3).

It should be pointed out that the above model is not a universal model and may not apply to other organisms. While similarities in heterochromatin composition and formation may exist between organisms, some of the key components in this model appear to be missing in different organisms. For example, the establishment of heterochromatin in *Neurospora* does not require RNAi components at all, and homologs of the RNA-dependent RNA polymerase have not been identified in either *Drosophila* or in mammals (19).

Although evidence for the involvement of RNAi in nuclear gene silencing is strong in lower organisms, in mammals, the contribution of RNAi in heterochromatin formation has remained largely elusive. Cells lacking Dicer showed delocalization of HP1 and Rad21, premature sister-chromatid separation, and chromosome mis-segregation, the phenotype caused by failure of heterochromatin formation in the pericentromeric region (20). In Dicer-null mouse embryonic stem cells, one group observed similar phenotypes (21), while another group reported no defects in DNA methylation and heterochromatic silencing (22). Further, no mammalian centromeric small RNAs have been isolated even though longer transcripts of both strands of centromeric satellite repeats exist (23). Recently, RNAi-independent heterochromatin formation and gene silencing has been demonstrated in mammalian cells (24). In this experiment, neither Dicer knockdown-induced transcriptional reactivation or decondensation of heterochromatin that had been formed in a transgenic repeat locus nor the siRNA corresponding to the repeat sequences was found in these transgenic cell lines. Furthermore, when the inducible transgene was activated, overexpression of exogenous siRNA directed to the transgene did not facilitate reconstruction of heterochromatin in the transgenic locus. Interestingly, overexpression HP1 caused repression of the expression of the transgene in a Dicer-independent manner. These findings are in contrast with the reports that antisense strand-specific siRNA to a promoter directs DNA methylation/histone modification (25). However, a prerequisite for this apparent siRNA-mediated transcriptional gene silencing is the active delivery of siRNA to the nucleus. So far, the natural pathway of nuclear delivery of siRNA is unknown and small RNAs originating from natural antisense transcripts have not been identified either. In fact, natural antisense transcripts of IGF2R have been shown to regulate gene expression, but do not form dsRNA that enter the RNAi pathway (26). Thus, it is possible that another RNA mediated pathway distinct from RNAi exists and directs heterochromatin formation in the nucleus of mammalian cells.

9.4.3 Connections Between RNAi and dsRNA Editing

In the nucleus, ADAR can edit dsRNA, and it does so quite efficiently. Further, the editing machinery can actually suppress RNAi in mammalian cells. For example, 1:U mismatches can disrupt the structure of RNA duplexes to such an extent that they can no longer be processed by Drosha (27). Also, hyperedited RNA might in some cases be degraded by Tudor-SN, which is a component of RISC in some species (28). Finally, short siRNAs, which are too short to trigger editing activity of ADAR, still can be bound by ADAR with high affinity and are prevented from entering the RNAi pathway (29). Thus, cells with robust ADAR activity may be compromised, at least in the nucleus, in their ability to respond to long dsRNAs through the RNAi pathway.

9.4.4 Vigilin

Affinity chromatography using inosine-containing RNA led to the discovery of the protein vigilin, which may provide an alternative RNA-dependent gene silencing pathway in the nucleus of mammalian cells (30). Vigilin is a ubiquitous and highly conserved protein of about 160 kDa. Vigilin was first cloned as a high-density lipoprotein binding protein (HDLBP). However, the predicted structure of HDLBP does not conform to that of any known receptor for HDL, suggesting that it does not function as a classic plasma membrane receptor (31). Later studies showed that this protein could bind specific RNAs and perhaps play a role in the control of mRNA metabolism and translation in the cytoplasm. It can be induced by estrogen and binds to the 3′-UTR of vitellogenin mRNA with high affinity (32). This binding protects the mRNA from cleavage by the mRNA endonuclease polysomal ribonuclease 1 *in vitro* and stabilizes it (33). Vigilin homologs have been found in many species including yeast, worms, flies, zebrafish, frogs, chickens, mouse and human. Its homolog in *Drosophila* is DDP1 and in budding yeast is Scp160p. Vigilin and all its homologs consist primarily of 14 tandem-related but nonidentical type I KH (hnRNP K homology) domains. The KH domain is a common nucleic acid binding motif but also may be involved in protein–protein interactions. Each KH domain consists of about 70 amino acids with a conserved hydrophobic core (VIGXXGXXI).

Consistent with its role in maintaining mRNA stability and facilitating translation, in mammalian cells, vigilin is localized mostly in the cytoplasm, associated with ER and polyribosomes (34). However, it is also found in the nucleus in regions of higher DNA content. In *Drosophila* S2 cells, nuclear DDP1 is concentrated in the region of the chromocenter and almost completely colocalizes with heterochromatin protein 1 (HP1) (30, 35). Mutation or depletion experiments of vigilin homologs demonstrated its role in establishing heterochromatin in the nucleus. Disruption of Scp160p in *S. cerevisiae* is not lethal but results in cells of decreased viability, abnormal morphology, and increased DNA ploidy (36). Ectopic expression of DDP1 complements a Scp160p deletion (37). DDP1 knockdown S2 cells grow slowly and show increased average DNA content, a phenotype similar to the chromosome segregation defects in yeast Scp160p mutants (30). This phenotype appears to be the result of aberrant pericentromeric heterochromatin formation. In *Drosophila*, a DDP1 mutant was generated by transposon insertion at the second intron of the gene in the 5′-UTR region. The homozygous mutant was viable but with reduced viability that affects the female more dramatically. The extent of the reduction of viability was correlated with the reduction in DDP1 protein level. Heterochromatin-induced position effect variegation was suppressed in the DDP1 mutant. Further, histone H3–K9 methylation and HP1 deposition at chromocenter heterochromatin was also strongly reduced, and chromosome condensation and segregation were compromised (35). These results demonstrated that DDP1 contributes to the structural and functional properties of heterochromatin.

9.4.5 Recognition of RNA by Vigilins

In addition to the demonstration that vigilin and DDP1 bind strongly to ADAR-edited RNAs (Q. Wang, J. Zhou, and G. Carmichael, unpublished), several other studies have shed light on how these proteins interact with their targets. Both vigilin and DDP1 prefer long nucleic acids devoid of secondary structure. In the vitellogenin mRNA 3′-UTR, deletions analysis showed that efficient binding of vigilin requires an unusually long, 75-nucleotide RNA sequence (38). When a series of mutations were made in the identified binding site in order to identify consensus mRNA sequences and important structures for vigilin binding, it was found that the mutants having higher binding affinity exhibited hypermutation of G residues leading to a largely unstructured single-stranded region containing multiple conserved (A)nCU and UC(A)n motifs. On the other hand, mutations that created small stem-loop structures that moved the conserved sequence from a single-stranded region of the RNA to a double-stranded region severely decreased binding. Consistent with a model for sequence-independent but length-dependent requirement for efficient binding of RNA by vigilin, the nucleic acid binding affinity of DDP1 was also found to be severely affected by the length and secondary structure of its substrates (39). DDP1 was originally identified and purified based on its high affinity to the C-strand of *Drosophila* dodeca-satellite, a highly repeated DNA sequence that is localized to the pericentric heterochromatin (37). The dodecasatellite has a remarkable purine–pyrimidine strand asymmetry. The purine-rich G-strand forms intramolecular hairpin structures that are stabilized by the formation of non-Watson–Crick GA pairs as well as regular Watson–Crick GC pairs, which leaves the pyrimidine rich C-strand as an unstructured single-stranded DNA (40). DDP1 also shows strong affinity for the unstructured pyrimidine-rich strand of centromeric *Drosophila* AAGAG satellite. Like vigilin, a minimal length about 75–100 nucleotides is required for efficient DDP1 binding, and this binding is also facilitated by the lack of secondary structure of the substrate (39).

Consistent with the above results, hyperedited RNAs are relatively unstructured because A:U base pairs are replaced by I:U mismatches in the edited regions. We have recently observed that not only vigilin and DDP1, but even the yeast homolog Scp160p all bind specifically to promiscuously edited RNA (Q. Wang and G. Carmichael, unpublished results). The binding of hyper-edited RNA by Scp160p strongly suggests that this protein recognizes edited RNA through its lack of secondary structure and not through inosines, since no ADAR activity has been found in yeast.

9.4.6 The Vigilin Complex

Following the discovery of vigilin as a protein that binds to edited RNAs, it was shown that this protein in fact exists in the nucleus as a part of a larger complex that contains not only the editing enzyme ADAR but also the Ku86/70 autoantigen and RNA helicase A (RHA, ref.30). In the presence of RNA, DNA-PKcs is recruited and a discrete set of targets including RHA, HP1, and H2AX are phosphorylated. Importantly, all of these targets are known to participate in chromatin silencing (30).

Ku86 and Ku70 form a stable heterodimer in cells and play an important role in DNA repair of double-stranded breaks (DSB) and telomere maintenance (41). Following DNA damage, the Ku70/86 heterodimer binds to the free ends of DSB and recruits the enzyme DNA-PKcs, which subsequently phosphorylates histone H2AX, leading to the recruitment of repair factors and chromatin silencing around the sites of DNA damage (42).

RHA is another vigilin partner and an RNA-dependent phosphorylation target of DNA-PKcs. This protein plays multiple roles in gene expression (including chromatin) and can unwind both dsRNA and dsDNA (43). Its nucleic acid unwinding activity can be regulated by the nucleic acid binding domains flanking both sides of the helicase's catalytic center. Both DNA and RNA binding stimulate the phosphorylation of RHA by the catalytic subunit of DNA-PK. Hence both the DNA- and RNA-related activities of RHA may be regulated by DNA-PK (44). It was reported that RHA could act together with the RNA editing enzyme ADAR for a coordinated editing and splicing of the glutamate receptor pre-mRNA (13). It is possible that there is a general cooperation between RHA and ADAR in the processing of cellular or viral dsRNAs.

Another vigilin partner and phosphorylation substrate is the heterochromatin protein HP1 (30). HP1 homologs can be found in almost all eukaryotes from *S. pombe* to mammals. At least three isoforms of HP1 (HP1α, HP1β, and HP1γ) exist in mammals. HP1α and HP1β are involved in the formation and maintenance of heterochromatin, while HP1γ is associated with euchromatin. Recent studies showed that RNA components may be involved in targeting HP1 to heterochromatin because RNase treatment of nuclei leads to a release of HP1 in mammalian cells (45, 46). This view is also supported by the identification of the RNA binding activity of HP1 that could be assigned to a conserved region within the hinge domain (46) and further strengthened by the observation that mutations in components of RNAi machinery prevent heterochromatin formation mediated by SWi6/HP1 in fission yeast (47). The role of HP1 in heterochromatin is also tightly regulated by phosphorylation. Mutations in the phosphorylation site of HP1 can reduce or completely eliminate its silencing activity. On the other hand, constitutive phosphorylation introduced by site mutation can also abolish its silencing activity. This indicates that phosphorylation of HP1 is a dynamic process, and both phosphorylated and unphosphorylated isoforms are functionally important for their role in heterochromatin formation (48). Both HP1β and SUV39H1 associate with DNA methyltransferase *in vivo*. Abolishment of histone methylation enzymatic activity leads to decrease of DNA methylation, but not vice versa (49). But methylated DNA can also recruit SUV39H1 through MeCP2 that binds to the methylated DNA. These sequential events thereby create a self-propagating epigenetic cycle for the spreading and maintenance of the heterochromatin status.

A final and important vigilin partner is the histone methyl transferase SUV39H1. This enzyme methylates lysine 9 of histone H3 and is a key player in heterochromatic gene silencing. In recent work (J. Zhou, L. Chen, and G. Carmichael, unpublished results) we have found that this protein interacts strongly with vigilin in the nucleus, thus providing a direct link between this RNA-binding protein and heterochromatin. Further, this important interaction appears to be influenced by a conformational change in vigilin, perhaps as a result of RNA binding or posttranslational modification (unpublished results).

9.5 A MODEL FOR THE NUCLEAR FUNCTION OF VIGILIN

Although both p54^{nrb} and vigilin bind to promiscuously edited RNA, they form distinct complexes in response to nuclear dsRNAs. We speculate that the p54^{nrb} pathway deals with dsRNAs made by transcriptional errors by editing the aberrant RNAs and preventing them from being exported and translated in the cytoplasm (Figure 9.4). However, when more serious problems occur, such as the expression of a large amount of dsRNA from one genomic locus, the vigilin pathway may be

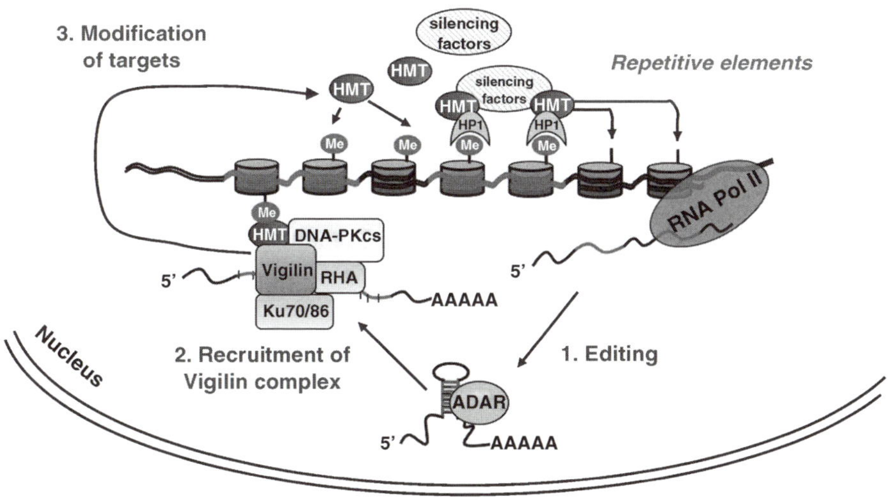

Figure 9.4 A speculative model for I-RNA-induced heterochromatic gene silencing in mammalian cells. In this model, repetitive elements lead to transcripts that can form RNA duplexes either intramolecularly (shown) or intermolecularly. After ADAR editing (producing inosine-containing RNA, or I-RNA), the vigilin complex binds, leading to the recruitment and activation of gene silencing machinery, including at least one protein kinase and a histone methyltransferase. This leads to the methylation of H3K9, and H3K9Me may interact with HP1 to facilitate the spreading of methylation and further gene silencing. (*See color insert.*)

mobilized presumably by a higher concentration of dsRNA generated in a local region, such as centromeric region or areas rich in repetitive elements. These dsRNAs could be edited by ADAR, which works coordinately with RHA that can bind and unwind them. Vigilin could then bind the edited RNAs in the vicinity of the targeted chromatin. In the presence of RNA, conformational changes may allow the recruitment of DNA-PKcs, which then phosphorylates H2AX, RHA, HP1α, and possibly other proteins. The phosphorylation of these proteins could then promote the establishment of heterochromatin in an epigenetic way. At the same time, either the association with RNA or proteins or other posttranslational modification might trigger a conformational change in vigilin, allowing the binding of SUV39H1, which methylates H3K9 to create the binding site for HP1α.

In conclusion, we propose a model in which edited RNA can lead to heterochromatin formation through the nuclear vigilin silencing complex. While the evidence for the participation of vigilin in heterochromatic gene silencing is strong, the evidence that ADAR-edited RNAs are, in fact, *in vivo* triggers for vigilin-mediated silencing remains indirect. In *in vitro* experiments, vigilin binds tightly not only to edited RNAs, but also to RNAs with low secondary structure. Thus, we speculate that it is also possible for many RNAs to trigger heterochromatin through the vigilin pathway. ADAR-edited RNAs are proposed to constitute a subset of these triggers. The identification and sequencing of RNAs bound to vigilin in the nucleus may provide important insight into this issue.

REFERENCES

1. Wang, Q., and Carmichael, G. G. (2004) Effects of length and location on the cellular response to double-stranded RNA. *Microbiol Mol Biol Rev* **68**, 432–452.
2. Alexopoulou, L., Holt, A. C., Medzhitov, R., and Flavell, R. A. (2001) Recognition of double-stranded RNA and activation of NF-kappaB by Toll-like receptor 3. *Nature* **413**, 732–738.
3. Matsumoto, M., Kikkawa, S., Kohase, M., Miyake, K., and Seya, T. (2002) Establishment of a monoclonal antibody against human Toll-like receptor 3 that blocks double-stranded RNA-mediated signaling. *Biochem Biophys Res Commun* **293**, 1364–1369.
4. Kumar, M., and Carmichael, G. G. (1998) Antisense RNA: Function and fate of duplex RNA in cells of higher eukaryotes. *Microbiol Mol Biol Rev* **62**, 1415–1434.
5. Chen, J., Sun, M., Kent, W. J., Huang, X., Xie, H., Wang, W., Zhou, G., Shi, R. Z., and Rowley, J. D. (2004) Over 20% of human transcripts might form sense–antisense pairs. *Nucleic Acids Res* **32**, 4812–4820.
6. DeCerbo, J., and Carmichael, G. G. (2005) SINES point to abundant human editing. *Genome Biol* **6**, 216–217.
7. Bass, B. L. (2002) RNA editing by adenosine deaminases that act on RNA. *Annu Rev Biochem* **71**, 817–846.
8. Nishikura, K. (1992) Modulation of double-stranded RNAs in vivo by RNA duplex unwindase. *Ann NY Acad Sci* **660**, 240–250.
9. Bass, B. L., and Weintraub, H. (1988) An unwinding activity that covalently modifies its double-stranded RNA substrate. *Cell* **55**, 1089–1098.
10. Polson, A. G., and Bass, B. L. (1994) Preferential selection of adenosines for modification by double-stranded RNA adenosine deaminase. *EMBO J* **13**, 5701–5711.
11. Reenan, R. A. (2001) The RNA world meets behavior: A–I pre-mRNA editing in animals. *Trends Genet* **17**, 53–56.
12. Hoopengardner, B., Bhalla, T., Staber, C., and Reenan, R. (2003) Nervous system targets of RNA editing identified by comparative genomics. *Science* **301**, 832–836.

13. Bratt, E., and Ohman, M. (2003) Coordination of editing and splicing of glutamate receptor pre-mRNA. *RNA* **9**, 309–318.
14. Raitskin, O., Cho, D. S., Sperling, J., Nishikura, K., and Sperling, R. (2001) RNA editing activity is associated with splicing factors in lnRNP particles: The nuclear pre-mRNA processing machinery. *Proc Natl Acad Sci USA* **98**, 6571–6576.
15. Rueter, S. M., Dawson, T. R., and Emeson, R. B. (1999) Regulation of alternative splicing by RNA editing. *Nature* **399**, 75–80.
16. Zhang, Z., and Carmichael, G. G. (2001) The fate of dsRNA in the nucleus. A p54(nrb)-containing complex mediates the nuclear retention of promiscuously A-to-I edited RNAs. *Cell* **106**, 465–475.
17. Emili, A., Shales, M., McCracken, S., Xie, W., Tucker, P. W., Kobayashi, R., Blencowe, B. J., and Ingles, C. J. (2002) Splicing and transcription-associated proteins PSF and p54nrb/nonO bind to the RNA polymerase II CTD. *RNA* **8**, 1102–1111.
18. Morris, K. V., Chan, S. W., Jacobsen, S. E., and Looney, D. J. (2004) Small interfering RNA-induced transcriptional gene silencing in human cells. *Science* **305**, 1289–1292.
19. Freitag, M., Lee, D. W., Kothe, G. O., Pratt, R. J., Aramayo, R., and Selker, E. U. (2004) DNA methylation is independent of RNA interference in *Neurospora*. *Science* **304**, 1939.
20. Fukagawa, T., Nogami, M., Yoshikawa, M., Ikeno, M., Okazaki, T., Takami, Y., Nakayama, T., and Oshimura, M. (2004) Dicer is essential for formation of the heterochromatin structure in vertebrate cells. *Nat Cell Biol* **6**, 784–791.
21. Kanellopoulou, C., Muljo, S. A., Kung, A. L., Ganesan, S., Drapkin, R., Jenuwein, T., Livingston, D. M., and Rajewsky, K. (2005) Dicer-deficient mouse embryonic stem cells are defective in differentiation and centromeric silencing. *Genes Dev* **19**, 489–501.
22. Murchison, E. P., Partridge, J. F., Tam, O. H., Cheloufi, S., and Hannon, G. J. (2005) Characterization of Dicer-deficient murine embryonic stem cells. *Proc Natl Acad Sci USA* **102**, 12135–12140.
23. Bernstein, E., and Allis, C. D. (2005) RNA meets chromatin. *Genes Dev* **19**, 1635–1655.
24. Wang, F., Koyama, N., Nishida, H., Haraguchi, T., Reith, W., and Tsukamoto, T. (2006) The assembly and maintenance of heterochromatin initiated by transgene repeats are independent of the RNA interference pathway in mammalian cells. *Mol Cell Biol* **26**, 4028–4040.
25. Weinberg, M. S., Villeneuve, L. M., Ehsani, A., Amarzguioui, M., Aagaard, L., Chen, Z. X., Riggs, A. D., Rossi, J. J., and Morris, K. V. (2006) The antisense strand of small interfering RNAs directs histone methylation and transcriptional gene silencing in human cells. *RNA* **12**, 256–262.
26. Faghihi, M. A., and Wahlestedt, C. (2006) RNA interference is not involved in natural antisense mediated regulation of gene expression in mammals. *Genome Biol* **7**, R38.
27. Yang, W., Chendrimada, T. P., Wang, Q., Higuchi, M., Seeburg, P. H., Shiekhattar, R., and Nishikura, K. (2006) Modulation of microRNA processing and expression through RNA editing by ADAR deaminases. *Nat Struct Mol Biol* **13**, 13–21.
28. Scadden, A. D. (2005) The RISC subunit Tudor-SN binds to hyper-edited double-stranded RNA and promotes its cleavage. *Nat Struct Mol Biol* **12**, 489–496.
29. Nishikura, K. (2006) Editor meets silencer: crosstalk between RNA editing and RNA interference. *Nat Rev Mol Cell Biol* **7**, 919–931.
30. Wang, Q., Zhang, Z., Blackwell, K., and Carmichael, G. G. (2005) Vigilins bind to promiscuously A-to-I-edited RNAs and are involved in the formation of heterochromatin. *Curr Biol* **15**, 384–391.
31. McKnight, G. L., Reasoner, J., Gilbert, T., Sundquist, K. O., Hokland, B., McKernan, P. A., Champagne, J., Johnson, C. J., Bailey, M. C. Holly, R., *et al.* (1992) Cloning and expression of a cellular high density lipoprotein-binding protein that is up-regulated by cholesterol loading of cells. *J Biol Chem* **267**, 12131–12141.
32. Dodson, R. E., and Shapiro, D. J. (1997) Vigilin, a ubiquitous protein with 14 K homology domains, is the estrogen-inducible vitellogenin mRNA 3′-untranslated region-binding protein. *J Biol Chem* **272**, 12249–12252.
33. Cunningham, K. S., Dodson, R. E., Nagel, M. A., Shapiro, D. J., and Schoenberg, D. R. (2000) Vigilin binding selectively inhibits cleavage of the vitellogenin mRNA 3′-untranslated region by the mRNA endonuclease polysomal ribonuclease 1. *Proc Natl Acad Sci USA* **97**, 12498–12502.
34. Klinger, M. H., and Kruse, C. (1996) Immunocytochemical localization of vigilin, a tRNA-binding protein, after cell fractionation and within the exocrine pancreatic cell of the rat. *Anat Anz* **178**, 331–335.

35. Huertas, D., Cortes, A., Casanova, J., and Azorin, F. (2004) *Drosophila* DDP1, a multi-KH-domain protein, contributes to centromeric silencing and chromosome segregation. *Curr Biol* **14**, 1611–1620.
36. Wintersberger, U., Kuhne, C., and Karwan, A. (1995) Scp160p, a new yeast protein associated with the nuclear membrane and the endoplasmic reticulum, is necessary for maintenance of exact ploidy. *Yeast* **11**, 929–944.
37. Cortes, A., Huertas, D., Fanti, L., Pimpinelli, S., Marsellach, F. X., Pina, B., and Azorin, F. (1999) DDP1, a single-stranded nucleic acid-binding protein of *Drosophila*, associates with pericentric heterochromatin and is functionally homologous to the yeast Scp160p, which is involved in the control of cell ploidy. *EMBO J* **18**, 3820–3833.
38. Kanamori, H., Dodson, R. E., and Shapiro, D. J. (1998) In vitro genetic analysis of the RNA binding site of vigilin, a multi-KH-domain protein. *Mol Cell Biol* **18**, 3991–4003.
39. Cortes, A., and Azorin, F. (2000) DDP1, a heterochromatin-associated multi-KH-domain protein of *Drosophila melanogaster*, interacts specifically with centromeric satellite DNA sequences. *Mol Cell Biol* **20**, 3860–3869.
40. Ferrer, N., Azorin, F., Villasante, A., Gutierrez, C., and Abad, J. P. (1995) Centromeric dodeca-satellite DNA sequences form fold-back structures. *J Mol Biol* **245**, 8–21.
41. Koike, M. (2002) Dimerization, translocation and localization of Ku70 and Ku80 proteins. *J Radiat Res (Tokyo)* **43**, 223–236.
42. Collis, S. J., DeWeese, T. L., Jeggo, P. A., and Parker, A. R. (2005) The life and death of DNA-PK. *Oncogene* **24**, 949–961.
43. Zhang, S., and Grosse, F. (2004) Multiple functions of nuclear DNA helicase II (RNA helicase A) in nucleic acid metabolism. *Acta Biochim Biophys Sinica* **36**, 177–183.
44. Zhang, S., Schlott, B., Gorlach, M., and Grosse, F. (2004) DNA-dependent protein kinase (DNA-PK) phosphorylates nuclear DNA helicase II/RNA helicase A and hnRNP proteins in an RNA-dependent manner. *Nucleic Acids Res* **32**, 1–10.
45. Maison, C., Bailly, D., Peters, A. H., Quivy, J. P., Roche, D., Taddei, A., Lachner, M., Jenuwein, T., and Almouzni, G. (2002) Higher-order structure in pericentric heterochromatin involves a distinct pattern of histone modification and an RNA component. *Nat Genet* **30**, 329–334.
46. Muchardt, C., Guilleme, M., Seeler, J. S., Trouche, D., Dejean, A., and Yaniv, M. (2002) Coordinated methyl and RNA binding is required for heterochromatin localization of mammalian HP1alpha. *EMBO Rep* **3**, 975–981.
47. Hall, I. M., Shankaranarayana, G. D., Noma, K., Ayoub, N., Cohen, A., and Grewal, S. I. (2002) Establishment and maintenance of a heterochromatin domain. *Science* **297**, 2232–2237.
48. Norwood, L. E., Grade, S. K., Cryderman, D. E., Hines, K. A., Furiasse, N., Toro, R., Li, Y., Dhasarathy, A., Kladde, M. P., Hendrix, M. J., Kirschmann, D. A., and Wallrath, L. L. (2004) Conserved properties of HP1(Hsalpha). *Gene* **336**, 37–46.
49. Fuks, F., Hurd, P. J., Deplus, R., and Kouzarides, T. (2003) The DNA methyltransferases associate with HP1 and the SUV39H1 histone methyltransferase. *Nucleic Acids Res* **31**, 2305–2312.
50. Volpe, T. A., Kidner, C., Hall, I. M., Teng, G., Grewal, S. I., and Martienssen, R. A. (2002) Regulation of heterochromatic silencing and histone H3 lysine-9 methylation by RNAi. *Science* **297**, 1833–1837.
51. Verdel, A., Jia, S., Gerber, S., Sugiyama, T., Gygi, S., Grewal, S. I., and Moazed, D. (2004) RNAi-mediated targeting of heterochromatin by the RITS complex. *Science* **303**, 672–676.
52. Motamedi, M. R., Verdel, A., Colmenares, S. U., Gerber, S. A., Gygi, S. P., and Moazed, D. (2004) Two RNAi complexes, RITS and RDRC, physically interact and localize to noncoding centromeric RNAs. *Cell* **119**, 789–802.
53. Huisinga, K. L., Brower-Toland, B., and Elgin, S. C. (2006) The contradictory definitions of heterochromatin: Transcription and silencing. *Chromosoma* **115**, 110–122.

CHAPTER 10

BIOLOGICAL IMPLICATIONS AND BROADER-RANGE FUNCTIONS FOR APOBEC-1 AND APOBEC-1 COMPLEMENTATION FACTOR (ACF)

Valerie Blanc
Nicholas O. Davidson

BASE-MODIFICATION RNA editing is an important mechanism in regulating and diversifying gene expression, allowing distinct protein products to be generated from a single nuclear gene (1). One of the most widely described examples of base modification RNA editing in mammals is C-to-U conversion of the nuclear RNA encoding the lipid transporting protein apolipoprotein B (apoB). ApoB mRNA undergoes a single, site-specific cytidine deamination in the spliced nuclear transcript and represents a highly adapted enzymatic mechanism for RNA modification (2). Two gene products are necessary and sufficient for C-to-U apoB RNA editing: namely apobec-1, the catalytic deaminase, and apobec-1 complementation factor (ACF) (3, 4). The overarching biological importance of these evolutionarily adapted functions of apobec-1 and ACF likely extends beyond a restricted role in mammalian lipid metabolism.

10.1 OVERVIEW

Enzymatic C-to-U modification of the genomically encoded cytidine converts a glutamine codon (CAA) into a premature STOP codon (UAA), resulting in translational termination of a truncated protein referred to as apoB48 (2). This posttranscriptional modification of apoB mRNA occurs in the small intestine of all mammals, and in the liver

RNA and DNA Editing: Molecular Mechanisms and Their Integration into Biological Systems,
Edited by Harold C. Smith

of certain species (rat, mouse) (5, 6). Humans however, express the edited form of apoB mRNA only in the small intestine (7, 8). By contrast, human liver expresses the product of the *unedited* apoB mRNA, referred to as apoB100 (9). The net result is that all mammals express two distinct isoforms of apoB, each with a distinct functional itinerary in lipid homeostasis. The evolutionary significance of this adaptation is the facilitation and regulation of dual, tissue-specific pathways for mammalian lipid transport and tissue uptake in which there is an important dialogue between hepatic lipoprotein formation and uptake (endogenous pathway) and intestinal lipoprotein secretion (exogenous or dietary-dependent pathway) (reviewed in reference (10)). While the essential elements of apoB RNA editing have been well-described and its functional biological implications elaborately detailed, many important questions remain concerning the mechanisms that restrict C-to-U RNA editing to a single nucleotide in one of the largest mammalian exons. These unanswered questions in turn raise important implications for the broader range functions for apobec-1 and ACF.

C-to-U editing of apoB mRNA is mediated by a core holoenzyme complex whose known subunits (apobec-1 and ACF) have a defined developmental and tissue-specific expression pattern and which exhibit metabolic regulation with respect to abundance and compartmentalization. With the cloning of the core components of the apoB RNA editing enzyme, considerable advances have taken place in our understanding of the biochemical and genetic mechanisms that facilitate and regulate C-to-U RNA editing. In the following paragraphs, we will outline the background to our current understanding of the composition, molecular genetic, and biochemical features of the core components of the apoB RNA editing holoenzyme. We will then discuss the major questions and experimental approaches underway to examine the range of known functions of these novel genes. Finally, we will speculate about possible future directions the field may take as we consider some possible alternative functions for apobec-1 and ACF.

10.2 BACKGROUND TO OUR CURRENT UNDERSTANDING OF C-TO-U EDITING OF ApoB mRNA: CANONICAL FUNCTIONS FOR APOBEC-1 AND ACF

10.2.1 Role of Cis-Acting Elements

ApoB mRNA editing is an exquisitely precise process that targets a single cytidine at nucleotide position 6666, which is located in the middle of a 14-kb spliced nuclear RNA and more specifically in the middle of exon 26, which spans over 7 kb (reviewed in reference (2)). Using *in vitro* transcripts prepared from subclones of cDNA fragments surrounding the targeted base, the key sequences have been functionally mapped to a minimum cassette of approximately 50 nucleotides flanking the edited base, which is embedded in an AU-rich context (11, 12). In particular, there is a highly conserved 11-nucleotide sequence, referred to as a "mooring sequence," located 5 nucleotides downstream of the edited base (Figure 10.1) and which has been demonstrated using *in vitro* C-to-U RNA editing assays as being absolutely required

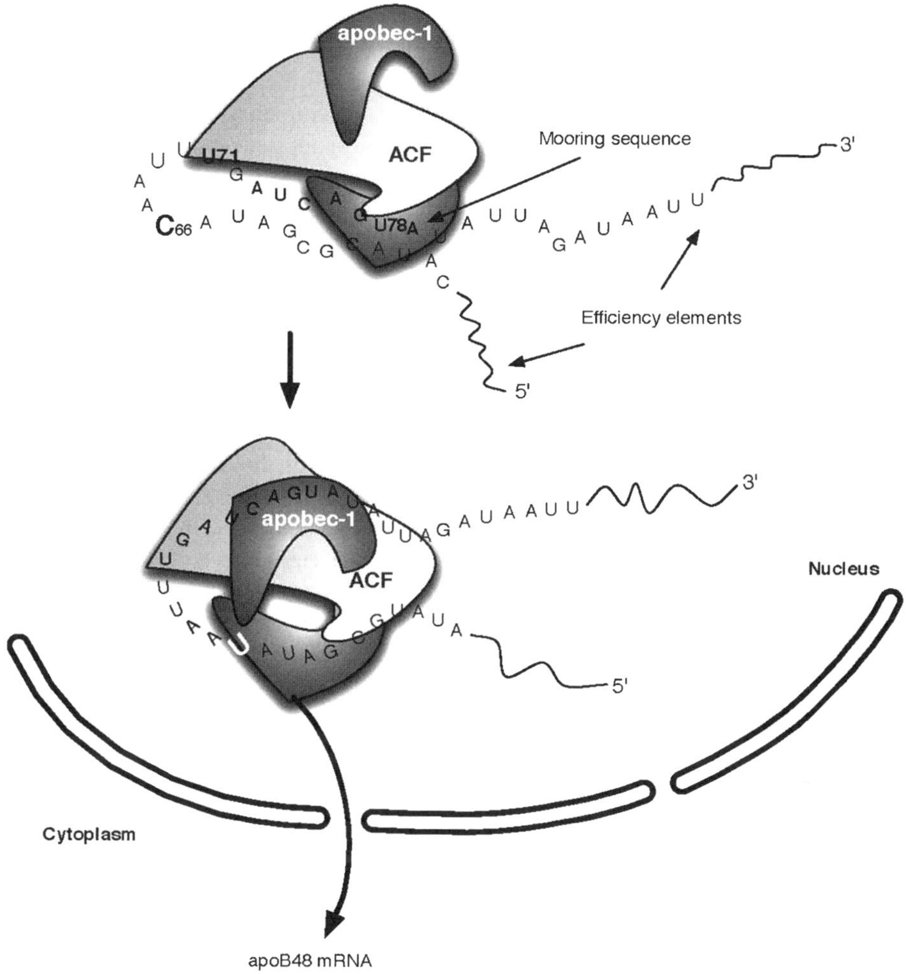

Figure 10.1 Model for C-to-U RNA editing of apoB mRNA and the role of apobec-1 and ACF. This model is based on predictions of stem-loop structure of apoB mRNA flanking the edited nucleotide at cDNA position 6666 (**C66**, upper panel) and incorporates the importance of the 11 nucleotide mooring sequence [UGAUCAGUAUA] located 3′ of the targeted cytidine. In addition, the model incorporates a role for *cis*-acting elements both 5′ and 3′ of the edited base that function as "efficiency elements." The predictions are further based on the results of NMR studies conducted by Allain and colleagues (44), demonstrating an interaction between ACF and the mooring sequence that is suggested to introduce a conformational change through contacts with U71 and U78 (upper panel). Following the interaction of ACF with the target RNA, the proposed mechanism of C-to-U deamination likely involves positioning of the targeted C within the active site of one of the subunits of the dimeric apobec-1 complex, while the other subunit interacts with the mooring sequence (lower panel). This proposal has implicit implications for the recognition of other targets of both ACF and apobec-1 with the suggestion that ACF may provide both target recognition information as well as a mechanism to constrain promiscuous C-to-U RNA editing.

for targeted deamination of the appropriate cytidine residue (13, 14). Single mutations in the conserved mooring sequence generally abolished or greatly abrogated *in vitro* C-to-U RNA editing activity (11, 12). These findings, confirmed by several laboratories using apoB templates derived from a large number of mammalian species, suggest that there are tight constraints on the *cis*-acting elements for appropriate recognition by the RNA-binding subunit of the holoenzyme (ACF) and for optimal substrate orientation with respect to the catalytic subunit, apobec-1. It should be emphasized that under normal physiological circumstances, C-to-U RNA editing of the apoB transcript is constrained by both *cis*-acting elements and also the abundance and stoichiometry of the *trans*-acting factors that mediate the enzymatic deamination (15).

In attempting to understand the properties of the *cis*-acting elements within apoB RNA that constrain C-to-U RNA editing, possible secondary structures have been considered that may theoretically be adopted by the apoB mRNA itself. Several models have been proposed in which apoB RNA adopts a stem-loop structure, with the editing site being exposed in the single-stranded part of the structure (14, 16). A detailed phylogenetic analysis has revealed substitutions in the flanking sequences both 5′ and 3′ of the edited base that influence folding and modulate the efficiency of *in vitro* C-to-U RNA editing (Figure 10.1). In particular, a comparison of 32 mammalian apoB cDNA sequences flanking the edited base revealed a divergence in guinea pig apoB in the 3′ flanking efficiency element region that reduced the efficiency of *in vitro* C-to-U RNA editing (13). Single-nucleotide revertants of this 3′ efficiency region to the mammalian consensus restored the efficiency of *in vitro* C-to-U RNA editing and also altered the predicted secondary structure of the stem loop containing the targeted C as well as the positioning of a downstream bulged U that is predicted to represent the binding site of apobec-1 in its optimal configuration. The functional importance of this predicted stem-loop structure of apoB RNA in relation to the biochemical properties of the deaminase machinery will be considered further in a later section.

10.2.2 Identification and Characterization of *Trans*-Acting Factors

Even though an intact apoB RNA holoenzyme remains to be purified biochemically, a minimal functional complex has been identified that contains (a) an RNA-specific deaminase, apobec-1, that represents the catalytic subunit (17) and (b) an RNA-binding subunit, apobec-1 complementation factor, ACF (3, 4).

10.2.2.1 Apobec-1: Identification and Key Functions Apobec-1 was cloned by Teng and colleagues in 1993 using functional complementation in *Xenopus* oocytes injected with sib-selected fractions of rat intestinal cDNAs, from which extracts were prepared and C-to-U editing of an *in vitro* transcribed apoB RNA template examined (17). The species- and tissue-specific regulation of apobec-1 expression will be discussed below, but two important features should be emphasized at the outset. First, apobec-1 is expressed almost ubiquitously in rodents such as mice and rats and in particular in the two major lipid-exporting tissues, the liver and the small intestine (6). Secondly, apobec-1 expression in humans appears to be confined to the luminal gastrointestinal tract, the stomach, the small intestine, and the colon (7). By contrast, with rats and mice, human liver does not express apobec-1 (18). Among the important biochemical and functional features of apobec-1 was the observation that, like all cytidine deaminases, apobec-1 is an obligate dimer (8, 19, 20). The dimerization of apobec-1 involves carboxy-terminal residues within a leucine-rich repeat domain (20). A catalytically inactive mutant apobec-1 protein was shown to act in a dominant negative manner *in vivo* in mouse liver, resulting in greater than 50% decrease in apoB mRNA editing (21). A further key observation was that while absolutely required for C-to-U editing of apoB RNA, apobec-1 alone was insufficient (17, 22). While it is capable alone of mediating deamination of a monomeric cytidine or deoxycytidine substrate, the physiological function of apobec-1 as an RNA-specific cytidine deaminase— in the context of cytidine 6666 of apoB mRNA— occurs exclusively in the presence of an additional protein factor. This intrinsic property distinguishes apobec-1 from members of the adenosine deaminase family acting on RNA (ADAR family), where ADAR gene products biochemically fulfill both RNA binding and A-to I conversion properties without the need for additional cofactor(s). Earlier assay systems utilized a tissue-based complementation extract (derived from chicken small intestine, a species in which apoB RNA is not edited) to promote C-to-U editing activity of recombinant apobec-1 (23) and pointed strongly to the likelihood of an identifiable auxiliary component. The cloning of apobec-1 and the recognition that other factor(s) were required for C-to-U RNA editing opened an era of intense investigation into the identification of these additional auxiliary factors and culminated in the definitive isolation of one such cofactor, ACF, described below.

10.2.2.2 Apobec-1 Complementation Factor, ACF: Identification and Key Functions As alluded to above, ACF was identified as a required biochemical component (presumptively the RNA binding subunit) of the apoB RNA C-to-U editing holoenzyme, simultaneously by two groups, each using different approaches. In one approach, Mehta et al. used RNA affinity chromatography in which an immobilized truncated apoB RNA fragment containing the mooring sequence was used as bait to isolate a 65-kD protein from liver extracts and which was further demonstrated using *in vitro* C-to-U RNA editing assays to complement apobec-1 activity (3). Independently, Lellek et al. using a biochemical enrichment strategy obtained a fraction from rat nuclear extracts that was enriched with apobec-1 complementation activity and which contained two proteins, one of which was identified as ASP (apobec-1 supplementing protein) and is identical to the major splice variant of ACF (4). An additional protein, KSRP, was co-purified with ASP/ACF in the isolation procedure used by Lellek and colleagues but appears dispensable for apoB RNA editing. KSRP may play a role in modulating apoB RNA editing as a result of its physical interaction with ACF. This possibility is discussed below. Both groups showed that recombinant ACF and apobec-1 are together capable of mediating efficient *in vitro* RNA editing of a synthetic apoB template and these findings collectively support the proposal that apobec-1 and ACF represent a minimal functional core C-to-U RNA editing enzyme (3, 4). In both mice and humans, ACF mRNA and protein is developmentally regulated (24, 25) and is expressed in high abundance in liver and small intestine (the major sites of apoB RNA expression) as well as many tissues, including lung, kidney, heart and brain, some of which express almost no apoB mRNA (3). The functional consequences of this widespread expression of ACF will be amplified in a later section in relation to its possible targets.

10.2.2.3 Additional Putative Trans-Acting Factors: Key Players or Supporting Actors? Even though apobec-1 and ACF most plausibly represent the minimal editing-competent core, sucrose gradient fractionation of rat liver extracts allowed enrichment of a macromolecular complex that contained C-to-U RNA editing activity, functionally identified as an "editosome," and which fractionated with a size of 27S, that is an ~450-kD protein complex (26). This observation suggested the possibility that the activity of core C-to-U RNA editing enzyme may be functionally modulated through recruitment of additional protein factors. Among the various scenarios considered was that these factors could function to enhance or limit access of the core enzyme to its substrate or to sequester one or both of the subunits within an intracellular compartment. Several factors have been identified through strategies based on their capacity for binding to or interacting with either apobec-1 and/or apoB mRNA. Among these additional factors, studies have characterized several interacting proteins including GRY-RBP, CUGBP2, hnRNP C1, ABBP1, ABBP2, KSRP, BAG4, and Aux240 (4, 27, 28). Virtually all these interacting proteins were identified before the cloning of ACF, and they were originally identified during a search for genes that would complement apobec-1 and enhance C-to-U RNA editing of apoB. In addition, almost all these interacting proteins—if they had any effect at all—functioned to *decrease*, rather than increase C-to-U editing of apoB mRNA, when added to an editing competent extract. Interestingly, several of these factors were originally shown to participate in splicing, mRNA turnover, and translation (29, 30). This observation further underscores the observed relationship between RNA editing and mRNA splicing, an issue directly established in studies using fractionated rat liver nuclear extracts in which it was conclusively demonstrated that C-to-U RNA editing is a nuclear event only (31). These latter studies further demonstrated that the ideal template for C-to-U editing is the spliced and polyadenylated, nuclear apoB mRNA (31). GRYRBP, also known as hnRNP Q (32), is highly homologous to ACF and binds apoB RNA and also interacts with apobec-1 (27). GRY-RBP has been demonstrated to inhibit C-to-U editing in a dose-dependent manner *in vitro* and *in vivo*, most probably by sequestering both ACF and apobec-1 (27). Similarly, CUG-BP2 binds to and colocalizes with both ACF and apobec-1 in the nucleus and also binds to apoB RNA, but again inhibits apoB mRNA editing (33), possibly by modulating the stoichiometry of the core complex components through sequestration of either ACF and/or apobec-1. Alternatively, since many of these additional proteins demonstrate RNA binding activity, it is possible that binding to the apoB RNA template exerts a steric effect in hindering further binding by either apobec-1 and/or ACF. These possibilities remain speculative however and the specific role of each individual factor, their functional relevance, and their organization within the holoenzyme remain to be more fully clarified. It is conceivable that the interactions of apobec-1 and/or ACF with these additional *trans*-acting proteins plays an important role in the regulation of their canonical functions in C-to-U RNA editing of apoB. More plausibly, these interactions are likely to be important in the longer-range functions of apobec-1 and ACF. It is worth noting that other members of the *APOBEC* superfamily also demonstrate high molecular mass complex formation in association with other proteins involved in mRNA splicing and translation (Chapter 12).

10.3 CURRENT UNDERSTANDING OF APOBEC-1 AND ACF: STRUCTURE–FUNCTION AND GENETIC REGULATION

10.3.1 Apobec-1: Structure–Function Relationships

Apobec-1 is an RNA-dependent cytidine deaminase. As a member of the superfamily of cytidine deaminases, apobec-1 is highly conserved and shares with the other cytidine deaminase members several residues that constitute a zinc-binding active (deaminase) site with the consensus His-X-Glu-$X_{(\sim 20-\sim 28-40)}$-Pro-Cys-$X_{(2-4)}$-Cys (19). As will be reviewed in Chapters 2 and 16 in this volume, both apobec-1 and its archeal ancestor AID contain a single deaminase motif as do several other members of the related larger gene family, including apobec-2/ARCD1, apobec-3A, apobec-3C, apobec-3D, and apobec-3H (2). The human *APOBEC1* gene is located on chromosome 12p13.1 approximately 1 MB from the locus encoding AID (33, 34); and based on the expression pattern in avian and other lower species of AID but not apobec-1, the most reasonable conclusion is that apobec-1 arose from a recent gene duplication at this locus (33, 35). As will be reviewed in Chapter 16 of this volume, other family members such as the *APOBEC3* cluster on chromosome 22, including apobec-3B, apobec-3F, and apobec-3G, all contain duplicated active sites (34).

In addition to the deaminase motif, apobec-1 contains (a) a bipartite nuclear localization sequence (NLS) within its amino terminus and (b) a nuclear export sequence (NES) in its carboxyl terminus (36, 37). Work from Smith and colleagues has demonstrated that neither consensus sequence—either the NLS or the NES—functions as an authentic targeting motif for a chimeric cargo protein, suggesting that these are atypical domains, although the NES domain appears to override the functional effects of an authentic SV-40-type NLS when expressed *in trans* (38). These findings were complemented by studies using deletion mutants of the carboxyl terminus of apobec-1 that demonstrated nuclear localization of epitope tagged apobec-1 following removal of the leucine-rich NES (37). Those results, taken in conjunction with other studies, again using epitope-tagged apobec-1, have provided evidence for both a nuclear and cytoplasmic distribution (27, 36, 39) along with the demonstration that apobec-1 shuttles between the nucleus and cytoplasm (27, 37).

Complicating the interpretation of these data, however, is the absence of experimental data validating subcellular localization of *endogenous* apobec-1, most likely the result of its low abundance in mammalian cells. The limited immunohistochemical data available in human tissue (39) demonstrates both nuclear and cytoplasmic staining of apobec-1 within intestinal enterocytes, but comparable data for murine tissue have yet to be convincingly demonstrated. Accordingly, a major unresolved question and focus of current effort is an unambiguous resolution of the intracellular localization and itinerary of endogenous apobec-1 in both human and murine tissues.

In addition to the role of functional domains involved in its subcellular itinerary, apobec-1 is an RNA binding protein. Several residues have been implicated in this function, including the zinc coordinating residues, two Phe residues located between the two active site domains, and residues within the carboxyl-terminus leucine-rich region (40, 41). However, no canonical RNA binding motif has been identified. Mutations introduced into recombinant apobec-1 (other than at the catalytic/zinc coordinating sites) that eliminate RNA binding still retain cytidine deaminase activity as inferred by its activity on a monomeric cytidine or deoxycytidine substrate (42). By contrast, RNA binding-defective mutants of apobec-1 exhibit greatly reduced or undetectable C-to-U apoB RNA editing activity (40, 41). These findings strongly suggest that apobec-1 acquired RNA-specificity in association with its evolving role in targeting a specific cytidine residue(s) in the context of a larger template—that is, spliced nuclear apoB RNA. By corollary, one of the major unresolved questions was how apobec-1 mediates targeted deamination of a single cytidine base in the 14,000-nucleotide apoB transcript. This question was initially addressed in early studies based on the information that the minimal sequence requirements for C-to-U RNA editing were contained in a 50 to 100-nucleotide cassette that was ~70% AU-rich (12–14). Recombinant apobec-1 binds apoB RNA with low specificity (40, 41); and modeling with the E. coli cytidine deaminase suggested a model for molecular recognition of the region flanking the edited base in apoB RNA, based on the ability of the stem loop to fit within the catalytic pocket of the active site of one of the two apobec-1 subunits (19). Modeling with the yeast cytidine deaminase (CDD1) provided further functional analysis on the molecular mechanism that allows apobec-1 to accommodate long mRNA substrats. The yeast CDD model suggests that apobec-1 functions as a head-to-head dimer with the two active sites on opposite faces. Thus each dimer could theoretically deaminate two mRNA molecules (43). Prior to the cloning of ACF, one of the proposed models for the site-specific deamination of apoB mRNA was that binding of the dimeric apobec-1 complex, perhaps augmented by interactions with auxiliary proteins, resulted in the optimal configuration of enzyme and substrate. With the characterization of ACF and with the recent description of a nuclear magnetic resonance (NMR) structure prediction for the apoB RNA substrate, it appears more likely that ACF provides the major targeting and guidance mechanism with respect to substrate organization and that its higher order interactions with apobec-1 and with the RNA together restrict C-to-U deamination (44) (Figure 10.1). These predictions notwithstanding, apobec-1 itself exhibits RNA binding activity and an important focus of current efforts is to define the range of its potential targets.

10.3.2 Functions of Apobec-1 Beyond apoB mRNA Editing

The demonstration that apobec-1 binds to an AU-rich region of the apoB RNA flanking the edited base led to a series of studies in which the binding of recombinant apobec-1 to several AU-rich templates was examined. A predicted consensus motif was identified with the sequence UUUN[A/U]U, and a hierarchy of RNA substrate/templates was constructed based on RNA binding affinity to a range of potential physiologic targets (45). Direct binding of recombinant apobec-1 was demonstrated to a series of AU-rich targets using UV crosslinking and electrophoretic mobility shift assays and supershifting experiments (45). These AU-rich targets included a number of cytokine and cell signaling molecules (including the protooncogene *c-myc*) whose 3′ untranslated region (3′ UTR) contains the apobec-1 consensus in the context of other AU-rich elements (AREs). The function of AREs is well-established as a regulatory motif in controlling mRNA stability and in targeting transcripts for selective translation versus degradation (46).

In order to establish a plausible foundation for the exploration of alternative RNA targets for apobec-1, gain-of-function experiments were undertaken in which apobec-1 was stably transfected into F442A pre-adipocytes, a mammalian cell line in which apobec-1 is not expressed, and mRNA stability of *c-myc* examined. The findings demonstrated that the half-life of *c-myc* mRNA in apobec-1 transfected cells was increased from 90 to 240 minutes, suggesting that as a result of its RNA binding activity, apobec-1 stabilizes *c-myc* mRNA in this cell line (45). Control experiments using apobec-1 transfectants in which the RNA binding (but not cytidine deaminase activity) of apobec-1 was eliminated through mutation of the phenylalanine residues revealed no alteration in *c-myc* mRNA half-life compared to control cells. These results demonstrate that the RNA binding activity alone of apobec-1 is necessary and sufficient for conferring mRNA binding and stabilization of other candidate mRNAs.

Apobec-1 has been demonstrated to participate in nonsense-mediated decay (NMD) of an edited apoB-chimeric RNA where its function is possibly related to the introduction of premature termination codons close to the 3′ terminus of exon junction complexes (37). In this context the role of apobec-1 in NMD is dependent on apobec-1's catalytic activity (a catalytically inactive mutant of apobec-1 failed to induce NMD) (37). The role of apobec-1 in NMD will require further exploration in other model systems and will be discussed further below. However, in regard to the role of apobec-1 in mediating RNA stabilization, it should be emphasized that RNA binding of apobec-1 to the alternative targets identified above (for example, *c-myc*) occurs without evidence of cytidine deamination in the region bound. In other words, despite similarities in the apobec-1 consensus binding site to that of the canonical mooring sequence and notwithstanding the observation that this consensus motif overlaps the 5′ end of the mooring sequence, the requirements for apobec-1 mediated C-to-U deamination are more stringent than for RNA binding alone.

That said, there are at least two exceptions to the paradigm that site-specific RNA binding and C-to-U deamination by apobec-1 are mechanistically coupled in direct proximity. The first exception is within apoB mRNA itself, where in addition to the known RNA targets for apobec-1-mediated cytidine deamination at the canonical site (nucleotide 6666 in the cDNA) there is another site at 6802, downstream of the canonical site that is edited at ~10% efficiency (47). There is no functional consequence of C-to-U RNA editing at this alternative site within endogenous mammalian apoB mRNA since editing at the 6666 site is more than 90% efficient in the small intestine. However, the mooring sequence divergence at these two sites strongly suggests that additional sequence information is required for C-to-U editing at this downstream site. Studies by Hersberger and Innerarity (16) strongly suggest that an AU-rich region ~50 nucleotides 5′ to the 6802 site is required. The second exception is the human neurofibromatosis 1 (NF1) mRNA, which also undergoes C-to-U RNA editing (48, 49). Within the human NF1 transcript, an alternatively spliced exon 23A creates a template in which a single nucleotide is inefficiently (<20%) edited by apobec-1 and undergoes C-to-U RNA editing of nucleotide 3916, changing a CGA (arginine) to a UGA termination codon (49). Two groups have demonstrated that this process occurs in a subset of peripheral nerve sheath tumors from subjects with NF1 and *in vitro* assays have confirmed that this processs is mediated by apobec-1 in a reaction process that creates a truncated protein that represents a loss of function allele (49). mRNA amplification confirmed that both apobec-1 and ACF were expressed in the subset of peripheral nerve sheath tumors in which C-to-U RNA editing was detected. The finding of RNA sequence divergence between the region of NF1 RNA flanking the edited base and the corresponding region of apoB RNA again strongly points to the likelihood that additional sequence and/or conformational signals are required for C-to-U RNA editing of this template. The precise mechanisms by which C-to-U RNA editing is constrained to a small number of physiological targets therefore still remain incompletely understood.

A further functional characteristic of apobec-1 is its functional organization as a higher order multimer. Using immunoprecipitation of mixtures of *in vitro* translated epitope tagged apobec-1 and native apobec-1 protein, Lau and colleagues demonstrated that apobec-1 likely exists as a homodimer (8). Similar conclusions were reached in regard to apobec-1 by Navaratnam and colleagues, who used a yeast two-hybrid approach to characterize protein–protein interaction (19). In addition, these workers characterized the range of residues that facilitate and promote dimerization, including residues in the carboxyl-terminus leucine-rich region (19). The most reasonable prediction is that apobec-1 binds to the apoB RNA template as a dimer in which one subunit accommodates the stem loop and targeted cytidine base in its active site cleft, while the paired subunit binds to a consensus site in the mooring sequence (Figure 10.1) (50). As outlined below, important unresolved questions remain concerning this hypothetical model. These include the key issue of whether the apobec-1 dimer binds to one or more molecules of ACF and whether a higher-order complex assembles *in vivo* (as strongly suggested by the work of Smith and colleagues (50) and, if so, with what kinetics and with what composition?

10.3.3 Apobec-1: Genetic Regulation and Gain- and Loss-of-Function

10.3.3.1 Apobec-1: Genetic Regulation The regulation of apobec-1 gene expression has been extensively examined in relation to the known physiologic stimuli that modulate C-to-U RNA editing of apoB. Examples of the physiologic regulation of apobec-1 gene expression include developmental regulation in both mammalian liver and small intestine (24), where a temporal increase in the extent of C-to-U RNA editing of apoB has been demonstrated to coincide with increased expression of apobec-1 mRNA and protein. A second example is the metabolic response in rat liver to prolonged fasting and carbohydrate refeeding, where there is a dramatic increase in hepatic lipogenesis accompanied by an increase in apobec-1 mRNA abundance (51, 52). The promoter elements that might mediate this transcriptional increase in apobec-1 expression have not been identified. In addition to these examples of developmental and nutritional modulation of apobec-1 gene expression, there are many examples of hormonal and other environmental stimuli that modulate C-to-U RNA editing of apoB mRNA, some of which involve alterations of expression of ACF and/or the localization and/or activity of the core enzyme complex (52, 53). Selected examples will be reviewed below, but the reader is referred to recent reviews for a comprehensive accounting of experimental models of altered C-to-U editing of apoB mRNA (53).

10.3.3.2 Apobec-1: Gain-of-Function Emerging from the early descriptions that C-to-U RNA editing of apoB was confined to a single nucleotide, studies addressed the question of whether forced overexpression of apobec-1 could lead to loss of specificity of C-to-U RNA deamination. Using an immortalized rat hepatoma cell culture model in which hepatic apoB mRNA demonstrates a constitutive but low-level (10–20%) C-to-U RNA editing, Smith and colleagues demonstrated that overexpression of apobec-1 was associated with promiscuous but "mooring sequence-dependent" deamination of cytidine residues located *upstream* of the canonical cytidine 6666 in apoB mRNA (54). By contrast, Innerarity and colleagues, using forced transgenic overexpression of apobec-1 in mouse and rabbit liver, demonstrated aberrant editing of cytidine residues *downstream* of the canonical site, a phenomenon they referred to as hyperediting (55). Interestingly, the conclusions reached by Innerarity and colleagues was that hyperediting was "mooring sequence independent" but that site selection for targeted deamination was constrained by auxiliary protein(s) (56). It is possible that C-to-U editing at the noncanonical sites may reflect subtle changes in the stoichiometry of apobec-1 and the auxiliary factors. In this scenario, a single mooring site/editosome complex interaction potentially functions to position additional cytidine bases in a favorable configuration relative to apobec-1, which is expressed at supraphysiological levels. This suggestion is consistent with the findings of Smith et al. (54). It bears emphasis that the reports concerning gain-of-function phenotypes associated with overexpression of apobec-1 preceded the identification of ACF and much remains to be learned concerning the mechanisms that constrain apobec-1-mediated C-to-U RNA editing to a single nucleotide in the apoB transcript *in vivo*.

A major observation emerging from the transgenic apobec-1 overexpression studies reported by Innerarity and colleagues was that mice and rabbits expressing high copy numbers of the rabbit apobec-1 transgene developed hepatic dysplasia with large aberrant nodules and hepatocellular carcinomas (57). The presumed mechanism for this phenotype of tumor development was that other RNAs such as tumor suppressor genes underwent aberrant C-to-U RNA editing. These workers identified other candidate genes with a conserved mooring sequence or with a single nucleotide difference, yet validated only one, a protein kinase (Tec) that revealed barely detectable levels (~1%) of C-to-U RNA editing. These findings reinforce the fact that the mooring sequence by itself is not sufficient to support RNA editing.

However, a bona fide additional candidate substrate for C-to-U RNA editing by apobec-1—referred to as novel apobec-1 target 1 (NAT1)—was identified through the forced transgenic overexpression studies alluded to above, using differential display

(58). In apobec-1 transgenic liver RNA, NAT1 revealed multiple C-to-U deamination events along the entire coding sequence with the introduction of multiple translational termination stop codons (58). The frequency of editing at each cytidine ranged from 4% to 71% with a preference for sites where mooring-like sequences were clustered. However, apobec-1 alone failed to support editing of these sites in an *in vitro* editing assay, emphasizing the importance of both *cis*-acting elements and auxiliary factors in target selection. NAT1 protein demonstrates homology to the carboxyl terminus of the translation inhibitor eIF4G whose activity controls the assembly and function of the initiation complex containing eIF4E, eIF4A, and eIF3. NAT1 physically interacts with eIF4A and eIF3 and has been presumed to function as an inhibitor of translation. By contrast, overexpression of eIF4G has been associated with transformation in cell culture (59), suggesting a plausible link to the observed gain-of-function phenotype and the implication that translation initiation factors can function as cellular oncogenes. Accordingly, loss of NAT1 function by either alteration of its synthesis or by the production of a truncated form via aberrant editing could account for tumorigenesis. NAT1 also plays an essential role in embryonic development as evidenced by the embryonic lethality at E7.5-E11.5 following targeted deletion by homologous recombination (58). In summary, the gain-of-function studies demonstrate complexity in the observed patterns of C-to-U editing of noncanonical sites. Site selection appears distinctive and not random, only partially constrained by *cis*-acting elements but largely dependent on the presence of auxiliary factors.

10.3.3.3 Apobec-1: Loss-of-Function As alluded to above, apobec-1 is one member of a multigene family with homologs expressed in a tissue and cell-specific distribution in all mammals. In order to explore the physiological functions for apobec-1, its targeted deletion was undertaken using homologous recombination in mice. Apobec-1$^{-/-}$ mice are fertile and manifest no obvious developmental or age-related abnormalities in growth or nutrition. However, since they lack the capacity for C to U RNA editing of apoB, apobec-1$^{-/-}$ mice synthesize and secrete only apoB100 from both the liver and from the small intestine (60, 61). This observation demonstrates that there is no functional redundancy among the other members of the APOBEC gene family (cf. Chapters 2 and 16) with regard to C-to-U editing of the canonical target of apobec-1. The major phenotypic consequence for the loss-of-function of this allele is the presence of exclusively apoB100 in apobec-1$^{-/-}$ mice. The loss of apobec-1 and the absence of intestinal apoB48 secretion were associated with a subtle defect in lipid absorption in congenic mice expressing apoB100 (62, 63). Another important phenotype, however, emerged when apobec-1$^{-/-}$ mice were crossed into LDLR$^{-/-}$ mice, since the combined deletion of these two alleles yielded a mouse model of the human disease familial hypercholesterolemia, and while consuming a low-fat chow diet the combined double knockout mice [apobec-1$^{-/-}$, LDLR$^{-/-}$] manifested serum cholesterol levels greater than 400 mg/dL and extensive atherosclerosis (64). The longer-range implications for loss-of-function of apobec-1 will be considered below.

10.3.4 ACF: Structure–Function Relationships

ACF is a 64.5–kD protein member of the large family of RNA-binding proteins, of which *HuR* and the *Drosophila* protein *elav* are also members. Mutagenesis studies of

ACF have revealed the importance of three nonidentical RNA recognition motifs (RRM) located in the amino-terminal half of the protein (27, 65, 66). These motifs are well-characterized and have been identified in many proteins involved in RNA processing and metabolism (66). *In vitro* RNA-binding studies have shown that recombinant ACF binds to apoB RNA with a general sequence preference surrounding the editing site and partially overlapping the 5′ end of the mooring sequence (25). However, an important objective of future studies will be to map with greater specificity a consensus binding site for ACF and to determine whether RNA binding is quantitatively augmented in the presence of apobec-1. This is an important objective since apobec-1 alone has been shown to bind apoB RNA with low affinity (K_d in the range of 400 nM) while ACF binds apoB RNA with a K_d in the range of <50 nM (25, 65). In other words, does ACF bind the substrate and direct deamination site selection alone or in conjunction with apobec-1? The most recent modeling predictions with recombinant ACF and apobec-1 suggest that ACF mediates the initial site recognition of the target region and assist in aligning the optimal configuration of the targeted cytidine with respect to the active site of the deaminase (44). This prediction is represented schematically in Figure 10.1.

Deletion analysis of recombinant ACF demonstrated that apoB RNA binding activity requires the second and third RRM domains (27, 65). By contrast, apobec-1 interaction relies exclusively on RRM2 (27). Among other predicted structural motifs within ACF was a double-stranded RNA-binding domain and six arginine–glycine (RG) motifs identified at the carboxy end of the protein. Complete deletion or point mutations within these motifs failed to abolish either apobec-1 interaction, or apoB RNA binding or C-to-U apoB editing, suggesting that these domains may be dispensable for the function of ACF in C-to-U RNA editing. In support of this suggestion, a truncated ACF protein containing only the first 380 amino acids appeared to interact with both apoB RNA and apobec-1, allowing efficient editing of apoB template *in vitro* (our own unpublished data). It remains possible that ACF functions in others aspects of RNA processing or that there are additional functions of ACF beyond apoB editing for which these carboxyl terminus motifs would be required. In rat (but not human or mouse) liver, two additional splice variant mRNAs give rise to ACF43 and ACF45 proteins, both sharing the amino-terminal two-thirds of ACF65. However, these two isoforms appear less efficient in complementing apobec-1 for apoB mRNA editing (67). Smith and colleagues showed that the metabolic regulation of rat hepatic apoB mRNA editing observed with fasting/refeeding is associated with a change in the relative proportions of these individual isoforms (ACF65 versus ACF43/ACF45), with lower levels of apoB mRNA editing correlating with elevated level of ACF43/ACF45 (67).

Analysis of the primary sequence of ACF revealed the presence of an SV-40-type nuclear localization motif in the RRM2 domain (3). In addition, FLAG-tagged ACF localized in the nucleus of transfected cells, suggesting that ACF may be primarily a nuclear protein (27). However, mutagenesis of this putative NLS did not affect the nuclear distribution of ACF, suggesting that ACF may contain another motif responsible for its nuclear targeting. Indeed a novel nuclear localization signal was identified in the carboxy half of the ACF (2). This atypical nuclear localization sequence functioned to target a chimeric reporter gene (beta-galactosidase) into the nucleus, suggesting that ACF contains a functional NLS with atypical sequence characteristics (2). Heterokaryon assays confirmed the nature of this new motif and indicated that ACF was able to shuttle between nucleus and cytoplasm. ACF was further proposed to function in the directed translocation of apobec-1 from the cytoplasm to the nucleus. This prediction was experimentally tested by co-transfecting cells with both ACF and apobec-1 expression plasmids. Upon co-transfection, apobec-1 redistributed to the nucleus, the site where apoB RNA editing occurs (2). It bears emphasis that these results are divergent to the findings of Navaratnam and colleagues, who reported that apobec-1, rather than ACF, directs nuclear import and nucleocytoplasmic shuttling activity (37). Resolution of these issues will require access to genetically targeted strains of mice in which either ACF and/or apobec-1 has been deleted. Some, but not all, of these lines have been developed.

10.3.5 Intersections of Apobec-1 and ACF Regulation in the Modulation of C-to-U RNA Editing

In relation to the developmental regulation of mammalian intestinal C-to-U RNA editing and the expression of the *trans*-acting factors, Chen et al. have recently undertaken to determine the relationship between the level of apoB mRNA editing and the expression of the different auxiliary factors. These authors confirmed most of the earlier observations that showed the developmental upregulation of apoB mRNA editing accompanied by an increase in apobec-1 and ACF mRNA expression (68). These authors also extended these earlier findings by demonstrating a developmental decrease of CUG-BP2 mRNA expression followed by an increase in ACF and apobec-1 mRNA expression and a later decrease in GRY-RBP and hnRNP C1 mRNA levels. Changes in auxiliary factor expression were found in both the small intestine (an abundant source of apoB mRNA) and also in the colon (where endogenous apoB mRNA is barely detectable using RT-PCR). In contrast, none of the others putative auxiliary factors including KSRP, ABBP1, ABBP2, BAG4, and AUX240 showed altered gene expression profiles over this developmental window (68).

10.3.5.1 Physiological Regulation of apoB mRNA Editing The study by Chen et al. (68) highlights some of the key questions concerning the physiological function of apobec-1 and ACF. These questions include the mechanisms by which C-to-U RNA is regulated in both site selection of the targeted base and the extent of deamination and also the role of putative auxiliary factors in modulating the stoichiometry of the core components of the apoB mRNA editing enzyme. Experimental immunoprecipitation of nuclear or cytoplasmic ACF and or apobec-1 followed by proteomic analysis could be undertaken to clearly identify the exact composition of the editing complexes at these different stages of development, and this strategy may identify alterations in the composition of editing competent extracts or editosomes that correlate with differences in C-to-U editing activity. The likely mechanisms at play in regulating C-to-U RNA editing activity presumably involve regulation of the core enzyme stoichiometry (i.e., apobec-1 and ACF) and their interactions with additional proteins that are represented within a larger complex or functional editosome.

As alluded to above, modulating the composition of the holoenzyme constitutes an important, but poorly understood, molecular pathway for the regulation of apoB mRNA editing. Studies undertaken over the past 20 years have shown that several physiologic, hormonal, and dietary or nutritonal stimuli can alter *in vivo* apoB mRNA editing and also the composition of editing competent cell extracts. More specifically, a range of hormonal treatments (estrogen, thyroid hormone, growth hormone) (69), dietary interventions (fasting, carbohydrate refeeding) (70), treatment with ethanol (71, 72), and phosphorylation status (68) contribute to the metabolic regulation of C-to-U apoB RNA editing. The focus of these earlier studies largely centered on possible alterations in apobec-1 expression or localization. The identification of ACF has introduced another potential level of regulation. Smith and colleagues have demonstrated that alcohol and insulin supplementation of the media of hepatoma cells results in an increase in ACF translocation to the nucleus, suggesting one possible level of control of C-to-U RNA editing activity (71, 73). More recently, Smith and colleagues extended these findings with the demonstration that regulation of ACF trafficking may be regulated by serine phosphorylation. In these studies, Smith and colleagues demonstrated that ethanol treatment of hepatocytes resulted in a simultaneous increase in ACF phosphorylation and its distribution into the 27S fraction of editosomal particles, namely the functionally active component (72). More recent studies by this same group have identified two putative serine phosphorylation sites, residues 154 and 368, as likely targets for this posttranslational modification (74). Serine to aspartic acid mutations at these sites yielded extracts with enhanced editing activity that were comparable to those seen following ethanol treatment, suggesting that phosphorylation may be a relevant regulatory mechanism in controlling the distribution and activity of ACF (72).

10.3.5.2 Thyroid Hormone Regulation of apoB RNA Editing Another animal model has also provided novel insights into the metabolic regulation of C-to-U apoB RNA editing and its integration with ACF and apobec-1 gene expression. A novel strain of mutant, hypothyroid mice (Pax $8^{-/-}$), has been developed that lack thyroid follicles and which display reduced hepatic apoB mRNA editing and a concomitant decrease in ACF mRNA and protein abundance (75). Upon rescue of these neonatal mice with exogenous thyroid hormone, hepatic apoB mRNA editing is restored to wild-type levels in association with increased expression of ACF mRNA and a shift in ACF protein abundance to a predominantly nuclear localization (75). Similarly, treatment of isolated primary rat hepatocytes with insulin or ethanol has been shown to stimulate apoB mRNA editing (76, 77). Once again the principal mechanism underlying the increase in apoB mRNA editing was increased abundance of ACF in the nucleus (50). One hypothesis emerging from these observations is that the shift in ACF from cytoplasmic to predominantly nuclear in turn leads to increased recruitment of apobec-1 for nuclear import—the site where C-to-U RNA editing occurs. This hypothesis will require further exploration however.

In summary, regulation of apoB RNA editing and the biological actions of apobec-1 and ACF involve several distinct "check points." First, C-to-U RNA editing can be modulated by both tissue and cell-specific transcriptional regulation of the core enzyme complex as well as through alterations in the expression of additional auxiliary (inhibitory) factors (CUG-BP2, GRY-RBP, hnRNP C1, etc.). Second, the subcellular distribution of the constituents of the core enzyme appear to be dynamically regulated and may represent an additional level of control, along with subtle alterations in protein–protein interactions among the components of the holoenzyme or functional editosome. Taken together, the existing data strongly suggest that mammals have developed multiple strategies to facilitate targeted and efficient cytidine deamination of precisely selected targets and to simultaneously abrogate the deleterious effects of unconstrained or promiscuous nucleotide attack. The longer-range implications for this machinery will be considered below.

10.4 IMPLICATIONS AND BROAD-RANGE FUNCTION FOR APOBEC-1 AND ACF: FUTURE DIRECTIONS AND OVERARCHING QUESTIONS

10.4.1 Apobec-1

10.4.1.1 Apobec-1 and AU-Rich mRNAs As alluded to above, an apobec-1 consensus binding site was predicted as a result of RNA-binding studies performed using recombinant GST-apobec-1 and a panel of synthetic RNA templates. The consensus motif identified through this approach UUUN[A/U]U is similar to the consensus of AU-rich elements frequently found in the U-rich regions of the 3′-UTR of short-lived mRNAs and which are known to control the half-life of an mRNA by stimulating or inhibiting degradation (78). An important question emerging from these original approaches is whether the consensus binding sequence is influenced by ACF-apobec-1 interactions as a result of conformational changes in the RNA binding domains. The regulation of mRNA stability and degradation by apobec-1 is a subject of intense investigation, and some additional discussion of the role of AU-rich RNA motifs as signals for regulating cytoplasmic mRNA turnover is merited.

AU-rich elements (AREs) are characterized by the presence of an AUUUA pentamer surrounded by a variable number of uridine residues. AREs were initially identified in mRNAs encoding proteins (principally cytokines and growth factors) involved in the regulation of cell growth or in the response to inflammatory stimuli and other environmental factors. The spatial and temporal expression of these genes is highly regulated and involves multiple levels of control including mRNA transcription, stability, and translation (79, 80). In the resting state or in the absence of certain environmental stimuli (radiation or heat shock), AU-rich RNAs are rapidly degraded and their protein products are expressed at a very low level. The functional significance of this mode of gene regulation through rapid mRNA turnover has a counterpart in human disease states, where dysregulation of the stability of ARE-containing mRNAs has been described in cancers, as well as chronic inflammatory and autoimmune diseases (reviewed in references 77 and 81).

The list of mRNAs containing such AREs continues to expand, but includes among other genes *c-myc*, *c-fos*, granulocyte-macrophage colony-stimulating factor (GM-CSF), interferon, interleukin 2 and 3, and tumor necrosis factor (TNF-α) (for review see reference 46). Studies undertaken to identify apobec-1 putative additional RNA targets have shown that apobec-1 exhibits binding affinity for *c-myc*, TNF-α, interleukin 2, and GM-CSF (45). In addition, more recently, recombinant apobec-1 have been demonstrated to bind *in vitro* to the 3′UTR of COX-2 mRNA as well as to the 3′UTRs of interleukin-8 and vascular endothelial growth factor, suggesting that apobec-1 may have physiological functions in relation to a range of potential RNA targets (82).

10.4.1.2 Lessons from Apobec-1$^{-/-}$ Mice: Apobec-1: The Good and the Bad In terms of the possible long-range effects of apobec-1 expression and its role in the response to injury, important insights have emerged from experiments using *apobec*-1$^{-/-}$ mice. We examined the response to acute intestinal radiation injury in wild type congenic and apobec-1$^{-/-}$ mice (82). These strains of mice were compared in their capacity to regenerate intestinal stem cells upon administration of one dose of sublethal γ-irradiation. This is a useful model of development and repair since the entire intestinal crypt–villus unit undergoes proliferation, maturation, and migration from base to tip every three days. The γ-irradiation model of intestinal injury is well-established; and after irradiation, proliferating intestinal cells stop proliferating, undergo apoptosis, and are extruded off the villus tip. The surviving stem cells proliferate and form the regenerative crypts, which then eventually repopulate the entire intestinal epithelium. Apobec-1$^{-/-}$ mice demonstrated a 50% decrease in the number of regenerating crypts and in the number of proliferative cells (82). These findings strongly suggest that apobec-1 provides a survival advantage for the regeneration of the small intestinal epithelium following radiation injury.

In exploring the potential mechanisms underlying this response, it was earlier reported that administration of lipopolysaccharide (LPS) prior to sublethal irradiation provides a protective mechanism for animals as a result of increased prostaglandin production (83). In this model, LPS administration induces cyclooxygenase 2 (COX-2) protein synthesis, which in turn leads to increased production of prostaglandin E2 (PGE2) which protects intestinal cells from irradiation. These same effects can be achieved by administration of the synthetic prostaglandin dimethyl PGE2. One important question emerging from the radiation experiments was whether COX-2 might represent a valid RNA target for apobec-1. The 3′UTR of COX-2 mRNA has been shown to contain multiple AU-rich sequence elements that alter either mRNA stability or the efficiency of translation of the mRNA (84, 85). Western blot analysis of intestinal protein extracts after γ-irradiation revealed that COX-2 protein was less abundant in the extracts from apobec-1$^{-/-}$ mice compared to wild-type congenic mice. In addition, LPS-mediated production of PGE2 failed to occur in the apobec-1$^{-/-}$ mice. These findings suggest that apobec-1 may mediate the LPS-dependent induction of COX-2 expression and strongly suggest that apobec-1 may play an important role in the physiological regulation of COX-2 gene expression. The interaction of apobec-1 and COX-2 mRNA in murine intestine was confirmed *in vivo* by analyzing RNA immunopurified with an anti-apobec-1 antibody. These data strongly suggest that apobec-1 physically interacts with the target mRNA. Furthermore, this physical interaction of apobec-1 with COX-2 AU-rich 3′ UTR resulted in a 2.5-fold increase of the half-life of COX-2 mRNA as inferred from luciferase chimeric plasmid constructs containing different regions of the COX-2 3′ UTR. Importantly, the dominant interacting domain appeared to be within the first 60 nucleotides of the COX-2 3′ UTR, the site of a cluster of AUUUA motifs (82). Taken together, these data indicated that apobec-1 protects murine small intestine from radiation injury through its ability to bind to the AU-rich 3′UTR of COX-2 mRNA, and this interaction results in increased stability and augmented translation.

These findings raise important questions concerning the mechanisms by which apobec-1 might exert such a dominant role *in vivo*. In the first place, the data support previous observations showing a cytoplasmic localization of apobec-1 in transfected cells, reinforcing the proposal that apobec-1 shuttles between nucleus and cytoplasm. An important question arising from this prediction is whether binding of apobec-1 to these AU-rich cytoplasmic RNAs in turn alters the dynamics of its nuclear–cytoplasmic shuttling.

A further long-range possibility for the biological functions of apobec-1 stems directly from the findings concerning COX-2 mRNA binding in the irradiated mouse intestine. As noted above, PGE2 is a pro-proliferative and anti-apoptotic agent in the intestine, and its production is highly dependent on the expression of COX-2 (86). By corollary, it has been proposed that overexpression of PGE2 might be a feature of proliferative diseases of the intestinal epithelium, including colorectal adenomas and polyps, which are important precursor lesions to adenocarcinomas. In support of this possibility, observations made earlier demonstrated that COX-2 overexpression is a signature finding in human colorectal carcinoma and in their adenomatous precursor lesions (87). There is a predictable series of mutations and genetic defects that accompany transition from normal intestinal mucosa to adenomatous polyp and from there to large, dysplastic adenomas with carcinoma in-situ, among which COX-2 overexpression is one of the earliest findings (88, 89). In view of the demonstrated effects of apobec-1 on cox-2 mRNA stabilization (82), we hypothesized that the absence of apobec-1 would have a protective effect on intestinal polyposis and colorectal cancer formation by virtue of the abrogation of COX-2 expression as a result of decreased mRNA stability. Indeed, compound [$Apc^{min/+}$, $apobec\text{-}1^{-/-}$] mice show a significant reduction of small intestinal adenoma formation when compared to the parental $Apc^{min/+}$strain (90). This phenotype was associated with increased small intestinal apoptosis, decreased proliferation, a twofold decrease in cox-2 mRNA abundance and reduced prostaglandin E2 synthesis compared to tumors from $Apc^{min/+}$ mice. Interestingly, several other mRNAs, including TNF-α, GMCSF, and c-myc, all containing the apobec-1 consensus binding site in their 3′UTR, showed reduced expression in [$Apc^{min/+}$, $apobec\text{-}1^{-/-}$] mice compared to the parental $Apc^{min/+}$ line. These findings not only underscore the complex contribution of apobec-1 in the development of intestinal adenomas in $Apc^{min/+}$ mice but also reaffirm how crucial it is to tightly regulate apobec-1 expression and cellular distribution. Considering that colorectal cancer is one of the most common malignancies, these findings open a vast field of investigative possibilities.

10.4.2 ACF

To date, most (if not all) the studies concerning ACF have focused on its characterization in the context of apoB mRNA C-to-U editing. They include, on the one hand, analysis of the functional domains involved in both ACF-apoB RNA and ACF-apobec-1 interactions and, on the other hand, the role of ACF in mediating the nuclear localization of the core editing complex. These latter studies have addressed questions concerning the influence of metabolic or other challenges in the context of C-to-U RNA editing (72, 75). However, relatively little is known concerning the possibility of other RNA targets for ACF and whether additional or alternative protein partners exist for ACF. It is important to recall that in humans as well as in mice, ACF protein is readily detectable in tissues that do not manifest apoB editing activity and which do not even express detectable apoB mRNA. The obvious question is, therefore, What roles does ACF fulfill in such tissues?

10.4.2.1 ACF Loss-of-Function To address this fundamental question, our laboratory has undertaken the targeted deletion of the murine ACF gene, using homologous recombination. Germ-line deletion of ACF resulted in early embryonic lethality, a phenotype discussed in more detail below. Heterozygous ACF knockout animals, however, showed no apparent phenotype. These heterozygous knockout mice developed normally and behaved similarly to wild-type mice with respect to feeding and reproductive outcomes. ACF protein was reduced in all tissues analyzed, including liver, kidney, and small intestine (25). Considering that apobec-1 and ACF constitute *in vitro* the minimal core editing-enzyme complex, endogenous apoB editing in tissues from heterozygous ACF knockout mice was examined with the anticipation that heterozygous ACF knockout mice might exhibit reduced C-to-U RNA editing.

To our surprise, the extent of apoB mRNA C-to-U editing was statistically *increased* in the liver and the kidney whereas small intestinal editing remained unaltered. This unexpected result is similar to what Chen et al had earlier observed in the mouse liver-derived cell line BNLCL2 (68). These workers reported low but detectable levels (5%) of C-to-U apoB RNA editing despite undetectable levels of ACF by RT-PCR, raising the possibility that an alternative gene product may function in apobec-1 complementation. It bears emphasis that other studies using antisense oligonucleotides to knock down expression of ACF in rat hepatoma cells demonstrated virtual elimination of C-to-U RNA editing, so there may be unresolved questions concerning the obligate genetic requirements for ACF *in vivo*.

That said, the simplest explanation for the increase in C-to-U RNA editing with the ACF heterozygous knockout mice would be a compensatory effect leading to the substitution of ACF by another factor. So far, none of the candidate proteins identified (outlined above) has been shown to complement apobec-1 in an *in vitro* C-to-U RNA editing assay. A testable hypothesis could be proposed to explain the increased editing observed in the heterozygous ACF knockout mice. Specifically, one might speculate that the relative abundance of one or more of the known inhibitor factors, such as CUG-BP2, GRY-RBP, or hnRNP C1, may be decreased either via the reduction of their expression or by changes in their capacity to associate with the holoenzyme. These inhibitory factors have all been proposed and shown to exert their inhibitory effect by sequestering ACF, apobec-1, and apoB. One might speculate that a reduction in ACF abundance might alter the stoichiometry of the holoenzyme, releasing some of these inhibitory factors from their association with the core complex and rendering the resulting complex more active in C-to-U RNA editing of apoB.

Interestingly, even though the efficiency of C-to-U RNA editing at the canonical cytidine residue was increased in the liver of heterozygous ACF knockout mice, none of the other sites in apoB RNA, previously identified as promiscuous editing targets upon apobec-1 overexpression, were edited. These findings demonstrate that even with reduced expression of ACF, the constraints on site selection of C-to-U RNA editing of apoB mRNA are retained. These findings further raise the possibility that ACF alone is not sufficient to constrain C-to-U RNA editing to the canonical site C6666, but this possibility will require the development of homozygous ACF deletor mice. Taken together, these observations reinforced the idea that subtle changes of the stoichiometry of the elements constituting the editosome determine the amplitude and specificity of apoB editing. Further studies involving immunoprecipitation of apobec-1 from liver extracts from wild-type and ACF heterozygous knockout mice may shed light on the composition of the holoenzyme.

10.4.2.2 Role of ACF in Early Development As mentioned above, heterozygous ACF knockout mice were bred and the progeny was genotyped by PCR for the presence of the homozygous knockout allele. Several lines failed to generate viable homozygous knockout offspring, indicating that the targeted deletion of the *acf* gene results in embryonic lethality. $ACF^{-/-}$ embryos were detectable in washings from uterine horns of timed matings up to day E 3.5, corresponding to the blastocyst stage. Analysis of uterine sections at E 6.5 and E 7.5 showed no putative resorbed or empty deciduae, suggesting that the absence of $ACF^{-/-}$ progeny was not due to premature death of the newly implanted embryo, but rather indicated an early implantation defect of the $ACF^{-/-}$ blastocysts. This hypothesis was experimentally confirmed by showing that only isolated wild-type and heterozygous ACF blastocysts could develop into blastocyst outgrowths. $ACF^{-/-}$ blastocysts, by contrast, failed to develop.

These results confirmed that $ACF^{-/-}$ blastocysts undergo cell death before the implantation step and strongly suggests a role of ACF at an early stage of development that is completely independent of apoB mRNA editing. Indeed the critical role of ACF in blastocyst development precedes the development of either the liver or small intestine, the primary sites of ACF expression in the adult mammal. Taken a step further, the data strongly suggest that complementing apobec-1 in apoB mRNA C-to-U editing is *not* the primary function of ACF. Along these lines, it is interesting to recall that targeted deletion of NAT1 (the novel target of apobec-1) is embryonic lethal, even though this lethality does not occur until E7.5 (91). It would be tempting to speculate again that the absence of ACF could lead to aberrant editing of NAT1, resulting in a loss of NAT1 and the concomitant embryonic lethality. However, nothing is known about the editing status of NAT1 mRNA in E 3.5 blastocysts.

Supporting the role of ACF in cell survival, ACF knockdown using siRNA-mediated gene silencing results in a greater number of apoptotic cells and increased abundance of the active form of caspase 3. These effects were observed in all cell lines studied, independent of the expression of apobec-1 (25). These findings suggest that cell death following knockdown of ACF is independent of apobec-1 expression, but the findings leave unanswered just what these ACF targets might be.

What might some of these other targets of ACF include? Like apobec-1, ACF exhibits a high-affinity binding for AU-rich mRNAs, with apoB RNA being its first identified target (3, 4). Furthermore, it is interesting to note that several AU-rich cytokines such as interleukin 6, interleukin 1β, COX-2, and γ-interferon are also expressed in the very early stages of development, including at the blastocyst stage (92). Although the interaction of ACF with such RNAs is yet to be demonstrated, it is reasonable to speculate that dysregulation of the binding of ACF to one or more of these RNAs could impair the implantation and development of $ACF^{-/-}$ blastocysts and early embryos. This speculation needs to be balanced against the observation that individual germ-line deletions of these cytokine and AU-rich mRNAs is well tolerated since IL-$6^{-/-}$ mice survive to adulthood as do COX-$2^{-/-}$ and INFγ $^{-/-}$ mice. However, the possibility exists that ACF deletion results in alterations in one or more of these targets and further studies will focus on the mechanisms underlying this unexplained early embryonic lethality in the hopes of identifying high-affinity targets.

10.4.2.3 Animal Models to Study ACF Function in Proliferation If indeed ACF participates in cell growth and/or cell survival, one might further speculate that modulation of ACF expression, or its subcellular redistribution, could exert an effect (positive or negative) on cancer development. Experimental testing of this speculation requires developing both conditional ACF knockout and transgenic lines overexpressing ACF in different tissues, tasks currently being undertaken in our laboratory. It would be of interest also to study the role of ACF in cell growth and proliferation in the adult animal. An obvious target organ in this regard would be the liver. In response to hepatic injury, the remaining hepatocytes leave their quiescent state to reenter the cell cycle and proliferate. The regenerative process persists until the liver mass, structure, and function are restored. At that point, the regenerating hepatocytes reenter their quiescent state. This event requires early expression of cytokines such as TNF-α and IL6 that activate, in turn, transcription factors such as NF-KB and STAT-3 (93). Again, many of the protein products involved in the regeneration process are encoded by AU-rich mRNAs. It would be of great interest to know if tissue-specific ACF deletor mice exhibit a reduced capacity for hepatic regeneration. Altered hepatic regeneration represents an attractive experimental target in which to examine the role of ACF in carcinogenesis, since several models exist in which altered proliferation is a requisite step in the initiation process (94). In addition, studies have demonstrated that chemical hepatocarcinogenesis is greatly accelerated in the setting of altered proliferation (95). In this regard, the development of liver-specific transgenic ACF expressor lines would be of great interest.

10.4.2.4 ACF: Putative Involvement in Cancer Development Analysis of the structural organization of ACF reveals homology with members of the embryonic lethal abnormal vision (ELAV)/Hu family, including HuR. The members of this family were initially identified as tumor inducers as inferred by the presence of antibodies in the sera of patients with cancers presenting with paraneoplastic neurological disorders (96, 97). These proteins, including the ubiquitously expressed HuR, were found to regulate the expression of AU-rich mRNAs by altering their stability and/or translation (98). HuR, like ACF, exhibits a predominantly nuclear localization and shuttles between nucleus and cytoplasm (99). Several reports indicate that HuR may be a major player in cancer transformation by binding and modulating the stability of RNAs encoding proteins involved in cell cycle regulation (100) and proliferation, including *c-foc* and *c-myc* (101) as well as factors that influence tumor formation such as COX-2, VEGF, TNF-α, and several interleukins (102, 103). A growing number of reports show a consistent correlation between elevated levels of HuR and advanced stages of malignancy in breast, colon, and lung cancers (104, 105). It is extremely tempting to speculate that there might exist parallels between these two homologous proteins, ACF and HuR. One might speculate for example that ACF and HuR recognize similar or overlapping mRNA targets, raising the possibility of functional cooperation or antagonism. But more importantly, these findings suggest a putative role for ACF in malignancy and underscore the usefulness of ACF as a potential target for molecular tumor therapy.

10.5 CONCLUSIONS

These collective studies and accompanying speculations suggest that important and yet-to-be-defined targets exist for both apobec-1 and for ACF. In addition, unanswered questions remain for these gene products in regard to the composition and functional organization of the C-to-U RNA editing machinery. The development of transgenic and conditional deletor lines in which both gain and loss of function of apobec-1 and/or ACF will hopefully provide important information regarding the longer-range functions of these proteins.

ACKNOWLEDGMENTS

The authors are grateful to all members of the Davidson laboratory, past and presents, for their stimulating discussions and insights. In particular, we would like to acknowledge the insights and support of Dr. Jeffrey Henderson, whose work on some of the murine cancer models is cited in the text. Work cited from the authors' laboratory was supported by the following grants from the National Institutes of Health: HL-38180, DK-56260, and DK-52574.

REFERENCES

1. Maas, S., and Rich, A. (2000) Changing genetic information through RNA editing. *Bioessays* **22**, 790–802.
2. Blanc, V., and Davidson, N. O. (2003) C-to-U RNA editing: Mechanisms leading to genetic diversity. *J Biol Chem* **278**, 1395–1398.
3. Mehta, A., Kinter, M. T., Sherman, N. E., and Driscoll, D. M. (2000) Molecular cloning of apobec-1 complementation factor, a novel RNA-binding protein involved in the editing of apolipoprotein B mRNA. *Mol Cell Biol* **20**, 1846–1854.
4. Lellek, H., Kirsten, R., Diehl, I., Apostel, F., Buck, F., and Greeve, J. (2000) Purification and molecular cloning of a novel essential component of the apolipoprotein B mRNA editing enzyme-complex. *J Biol Chem* **275**, 19848–19856.

5. Davidson, N. O., and Shelness, G. S. (2000) APOLIPOPROTEIN B: mRNA editing, lipoprotein assembly, and presecretory degradation. *Annu Rev Nutr* **20**, 169–193.
6. Greeve, J., Altkemper, I., Dieterich, J. H., Greten, H., and Windler, E. (1993) Apolipoprotein B mRNA editing in 12 different mammalian species: Hepatic expression is reflected in low concentrations of apoB-containing plasma lipoproteins. *J Lipid Res* **34**, 1367–1383.
7. Hadjiagapiou, C., Giannoni, F., Funahashi, T., Skarosi, S. F., and Davidson, N. O. (1994) Molecular cloning of a human small intestinal apolipoprotein B mRNA editing protein. *Nucleic Acids Res* **22**, 1874–1879.
8. Lau, P. P., Zhu, H. J., Baldini, A., Charnsangavej, C., and Chan, L. (1994) Dimeric structure of a human apolipoprotein B mRNA editing protein and cloning and chromosomal localization of its gene. *Proc Natl Acad Sci USA* **91**, 8522–8526.
9. Teng, B., Black, D. D., and Davidson, N. O. (1990) Apolipoprotein B messenger RNA editing is developmentally regulated in pig small intestine: nucleotide comparison of apolipoprotein B editing regions in five species. *Biochem Biophys Res Commun* **173**, 74–80.
10. Young, S. G. (1990) Recent progress in understanding apolipoprotein B. *Circulation* **82**, 1574–1594.
11. Shah, R. R., Knott, T. J., Legros, J. E., Navaratnam, N., Greeve, J. C., and Scott, J. (1991) Sequence requirements for the editing of apolipoprotein B mRNA. *J Biol Chem* **266**, 16301–16304.
12. Backus, J. W., and Smith, H. C. (1992) Three distinct RNA sequence elements are required for efficient apolipoprotein B (apoB) RNA editing in vitro. *Nucleic Acids Res* **20**, 6007–6014.
13. Hersberger, M., Patarroyo-White, S., Arnold, K. S., and Innerarity, T. L. (1999) Phylogenetic analysis of the apolipoprotein B mRNA-editing region. Evidence for a secondary structure between the mooring sequence and the 3′ efficiency element. *J Biol Chem* **274**, 34590–34597.
14. Richardson, N., Navaratnam, N., and Scott, J. (1998) Secondary structure for the apolipoprotein B mRNA editing site. AU-binding proteins interact with a stem loop. *J Biol Chem* **274**, 31707–31717.
15. Davidson, N. O. (2002) The challenge of target sequence specificity in C→U RNA editing. *J Clin Invest* **109**, 291–294.
16. Hersberger, M., and Innerarity, T. L. (1998) Two efficiency elements flanking the editing site of cytidine 6666 in the apolipoprotein B mRNA support mooring-dependent editing. *J Biol Chem* **273**, 9435–9442.
17. Teng, B., Burant, C. F., and Davidson, N. O. (1993) Molecular cloning of an apolipoprotein B messenger RNA editing protein. *Science* **260**, 1816–1819.
18. Giannoni, F., Bonen, D. K., Funahashi, T., Hadjiagapiou, C., Burant, C. F., and Davidson, N. O. (1994) Complementation of apolipoprotein B mRNA editing by human liver accompanied by secretion of apolipoprotein B48. *J Biol Chem* **269**, 5932–5936.
19. Navaratnam, N., Fujino, T., Bayliss, J., Jarmuz, A., How, A., Richardson, N., Somasekaram, A., Bhattacharya, S., Carter, C., and Scott, J. (1998) Escherichia coli cytidine deaminase provides a molecular model for ApoB RNA editing and a mechanism for RNA substrate recognition. *J Mol Biol* **275**, 695–714.
20. Teng, B., Ochsner, S., Zhang, Q., Soman, K. V., Lau, P. P., and Chan, L. (1999) Mutational analysis of apolipoprotein B mRNA editing enzyme (Apobec-1): Structure-function relationships of RNA editing and dimerization. *J Lipid Res* **40**, 623–635.
21. Oka, K., Kobayashi, K., Sullivan, M., Martinez, J., Teng, B. B., Ishimura-Oka, K., and Chan, L. (1997) Tissue-specific inhibition of apolipoprotein B mRNA editing in the liver by adenovirus-mediated transfer of a dominant negative mutant APOBEC-1 leads to increased low density lipoprotein in mice. *J Biol Chem* **272**, 1456–1460.
22. Navaratnam, N., Morrison, J. R., Bhattacharya, S., Patel, D., Funahashi, T., Giannoni, F., Teng, B. B., Davidson, N. O., and Scott, J. (1993) The p27 catalytic subunit of the apolipoprotein B mRNA editing enzyme is a cytidine deaminase. *J Biol Chem* **268**, 20709–20712.
23. Teng, B., and Davidson, N. O. (1992) Evolution of intestinal apolipoprotein B mRNA editing. Chicken apolipoprotein B mRNA is not edited, but chicken enterocytes contain *in vitro* editing enhancement factor(s). *J Biol Chem* **267**, 21265–21272.

24. Henderson, J. O., Blanc, V., and Davidson, N. O. (2001) Isolation, characterization and developmental regulation of the human apobec-1 complementation factor (ACF) gene. *Biochim Biophys Acta* **1522**, 22–30.
25. Blanc, V., Henderson, J. O., Newberry, E. P., Kennedy, S., Luo, J., and Davidson, N. O. (2005) Targeted deletion of the murine apobec-1 complementation factor (acf) gene results in embryonic lethality. *Mol Cell Biol* **25**, 7260–7269.
26. Harris, S. G., Sabio, I., Mayer, E., Steinberg, M. F., Backus, J. W., Sparks, J. D., Sparks, C. E., and Smith, H. C. (1993) Extract-specific heterogeneity in high-order complexes containing apolipoprotein B mRNA editing activity and RNA-binding proteins. *J Biol Chem* **268**, 7382– 7392.
27. Blanc, V., Henderson, J. O., Kennedy, S., and Davidson, N. O. (2001) Mutagenesis of apobec-1 complementation factor reveals distinct domains that modulate RNA binding, protein-protein interaction with apobec-1, and complementation of C to U RNA-editing activity. *J Biol Chem* **276**, 46386–46393.
28. Schock, D., Kuo, S. R., Steinburg, M. F., Bolognino, M., Sparks, J. D., Sparks, C. E., and Smith, H. C. (1996) An auxiliary factor containing a 240-kDa protein complex is involved in apolipoprotein B RNA editing. *Proc Natl Acad Sci USA* **93**, 1097–10102.
29. Markovtsov, V., Nikolic, J. M., Goldman, J. A., Turck, C. W., Chou, M. Y., and Black, D. L. (2000) Cooperative assembly of an hnRNP complex induced by a tissue-specific homolog of polypyrimidine tract binding protein. *Mol Cell Biol* **20**, 7463–7479.
30. Rossoll, W., Kroning, A. K., Ohndorf, U. M., Steegborn, C., Jablonka, S., and Sendtner, M. (2002) Spec- ific interaction of Smn, the spinal muscular atrophy determining gene product, with hnRNP-R and gry-rbp/hnRNP-Q: A role for Smn in RNA processing in motor axons? *Hum Mol Genet* **11**, 93–105.
31. Lau, P. P., Xiong, W. J., Zhu, H. J., Chen, S. H., and Chan, L. (1991) Apolipoprotein B mRNA editing is an intranuclear event that occurs posttranscriptionally coincident with splicing and polyadenylation. *J Biol Chem* **266**, 20550–20554.
32. Mourelatos, Z., Abel, L., Yong, J., Kataoka, N., and Dreyfuss, G. (2001) SMN interacts with a novel family of hnRNP and spliceosomal proteins. *Embo J* **20**, 5443–5452.
33. Anant, S., Henderson, J. O., Mukhopadhyay, D., Navaratnam, N., Kennedy, S., Min, J., and Davidson, N. O. (2001) Novel role for RNA-binding protein CUGBP2 in mammalian RNA editing. CUGBP2 modulates C to U editing of apolipoprotein B mRNA by interacting with apobec-1 and ACF, the apobec-1 complementation factor. *J Biol Chem* **276**, 47338–47351.
34. Jarmuz, A., Chester, A., Bayliss, J., Gisbourne, J., Dunham, I., Scott, J., and Navaratnam, N. (2002) An anthropoid-specific locus of orphan C to U RNA-editing enzymes on chromosome 22. *Genomics* **79**, 285–296.
35. Harris, R. S., and Liddament, M. T. (2004) Retroviral restriction by APOBEC proteins. *Nat Rev Immunol* **4**, 868–877.
36. Yang, Y., Yang, Y., and Smith, H. C. (1997) Multiple protein domains determine the cell type-specific nuclear distribution of the catalytic subunit required for apolipoprotein B mRNA editing. *Proc Natl Acad Sci USA* **94**, 13075–13080.
37. Chester, A., Somasekaram, A., Tzimina, M., Jarmuz, A., Gisbourne, J., O'Keefe, R., Scott, J., and Navaratnam, N. (2003) The apolipoprotein B mRNA editing complex performs a multifunctional cycle and suppresses nonsense-mediated decay. *EMBO J* **22**, 3971–3982.
38. Yang, Y., Sowden, M. P., Yang, Y., and Smith, H. C. (2001) Intracellular trafficking determinants in APOBEC-1, the catalytic subunit for cytidine to uridine editing of apolipoprotein B mRNA. *Exp Cell Res* **267**, 153–164.
39. Giannoni, F., Chou, S. C., Skarosi, S. F., Verp, M. S., Field, F. J., Coleman, R. A., and Davidson, N. O. (1995) Developmental regulation of the catalytic subunit of the apolipoprotein B mRNA editing enzyme (APOBEC-1) in human small intestine. *J Lipid Res* **36**, 1664–1675.
40. Navaratnam, N., Bhattacharya, S., Fujino, T., Patel, D., Jarmuz, A. L., and Scott, J. (1995) Evolutionary origins of apoB mRNA editing: Catalysis by a cytidine deaminase that has acquired a novel RNA-binding motif at its active site. *Cell* **81**, 187–195.
41. MacGinnitie, A. J., Anant, S., and Davidson, N. O. (1995) Mutagenesis of apobec-1, the catalytic subunit of the mammalian apolipoprotein B mRNA editing enzyme, reveals distinct domains that mediate cytosine nucleoside deaminase, RNA binding, and RNA editing activity. *J Biol Chem* **270**, 14768–14775.

42. Anant, S., MacGinnitie, A. J., and Davidson, N. O. (1995) apobec-1, the catalytic subunit of the mammalian apolipoprotein B mRNA editing enzyme, is a novel RNA-binding protein. *J Biol Chem* **270**, 14762–14767.
43. Xie, K., Sowden, M. P., Dance, G. S., Torelli, A. T., Smith, H. C., and Wedekind, J. E. (2004) The structure of a yeast RNA-editing deaminase provides insight into the fold and function of activation-induced deaminase and APOBEC-1. *Proc Natl Acad Sci USA* **101**, 8114–8119.
44. Maris, C., Masse, J., Chester, A., Navaratnam, N., and Allain, F. H. (2005) NMR structure of the apoB mRNA stem-loop and its interaction with the C-to-U editing APOBEC1 complementary factor. *RNA* **11**, 173–186.
45. Anant, S., and Davidson, N. O. (2000) An AU-rich sequence element (UUUN[A/U]U) downstream of the edited C in apolipoprotein B mRNA is a high-affinity binding site for Apobec-1: Binding of Apobec-1 to this motif in the 3′ untranslated region of c-myc increases mRNA stability. *Mol Cell Biol* **20**, 1982–1992.
46. Barreau, C., Paillard, L., and Osborne, H. B. (2005) AU-rich elements and associated factors: Are there unifying principles? *Nucleic Acids Res* **33**, 7138–7150.
47. Navaratnam, N., Patel, D., Shah, R. R., Greeve, J. C., Powell, L. M., Knott, T. J., and Scott, J. (1991) An additional editing site is present in apolipoprotein B mRNA. *Nucleic Acids Res* **19**, 1741–1744.
48. Skuse, G. R., Cappione, A. J., Sowden, M., Metheny, L. J., and Smith, H. C. (1996) The neurofibromatosis type I messenger RNA undergoes base-modification RNA editing. *Nucleic Acids Res* **24**, 478–485.
49. Mukhopadhyay, D., Anant, S., Lee, R. M., Kennedy, S., Viskochil, D., and Davidson, N. O. (2002) C→U editing of neurofibromatosis 1 mRNA occurs in tumors that express both the type II transcript and apobec-1, the catalytic subunit of the apolipoprotein B mRNA-editing enzyme. *Am J Hum Genet* **70**, 38–50.
50. Sowden, M. P., Ballatori, N., Jensen, K. L., Reed, L. H., and Smith, H. C. (2002) The editosome for cytidine to uridine mRNA editing has a native complexity of 27S: identification of intracellular domains containing active and inactive editing factors. *J Cell Sci* **115**, 1027–1039.
51. Baum, C. L., Teng, B. B., and Davidson, N. O. (1990) Apolipoprotein B messenger RNA editing in the rat liver. Modulation by fasting and refeeding a high carbohydrate diet. *J Biol Chem* **265**, 19263–19270.
52. Inui, Y., Giannoni, F., Funahashi, T., and Davidson, N. O. (1994) REPR and complementation factor(s) interact to modulate rat apolipoprotein B mRNA editing in response to alterations in cellular cholesterol flux. *J Lipid Res.* **35**, 1477–1489.
53. Cho, S. J., Blanc, V., and Davidson, N. O. (2007) Mouse models as tools to explore C to U RNA editing. *Meth Enzymol* **424**, 417–435.
54. Sowden, M., Hamm, J. K., and Smith, H. C. (1996) Overexpression of APOBEC-1 results in mooring sequence-dependent promiscuous RNA editing. *J Biol Chem* **271**, 3011–3017.
55. Hersberger, M., Patarroyo-White, S., Qian, X., Arnold, K. S., Rohrer, L., Balestra, M. E., and Innerarity, T. L. (2003) Regulatable liver expression of the rabbit apolipoprotein B mRNA-editing enzyme catalytic polypeptide 1 (APOBEC-1) in mice lacking endogenous APOBEC-1 leads to aberrant hyperediting. *Biochem J* **369**, 255–262.
56. Yamanaka, S., Poksay, K. S., Driscoll, D. M., and Innerarity, T. L. (1996) Hyperediting of multiple cytidines of apolipoprotein B mRNA by APOBEC-1 requires auxiliary protein(s) but not a mooring sequence motif. *J Biol Chem* **271**, 11506–11510.
57. Yamanaka, S., Balestra, M. E., Ferrell, L. D., Fan, J., Arnold, K. S., Taylor, S., Taylor, J. M., and Innerarity, T. L. (1995) Apolipoprotein B mRNA-editing protein induces hepatocellular carcinoma and dysplasia in transgenic animals. *Proc Natl Acad Sci USA* **92**, 8483–8487.
58. Yamanaka, S., Poksay, K. S., Arnold, K. S., and Innerarity, T. L. (1997) A novel translational repressor mRNA is edited extensively in livers containing tumors caused by the transgene expression of the apoB mRNA-editing enzyme. *Genes Dev* **11**, 321–333.
59. Fukuchi-Shimogori, T., Ishii, I., Kashiwagi, K., Mashiba, H., Ekimoto, H., and Igarashi, K. (1997) Malignant transformation by overproduction of translation initiation factor eIF4G. *Cancer Res* **57**, 5041–5044.

60. Hirano, K., Young, S. G., Farese, R. V., Jr., Ng, J., Sande, E., Warburton, C., Powell-Braxton, L. M., and Davidson, N. O. (1996) Targeted disruption of the mouse apobec-1 gene abolishes apolipoprotein B mRNA editing and eliminates apolipoprotein B48. *J Biol Chem* **271**, 9887–9890.
61. Morrison, J. R., Paszty, C., Stevens, M. E., Hughes, S. D., Forte, T., Scott, J., and Rubin, E. M. (1996) Apolipoprotein B RNA editing enzyme-deficient mice are viable despite alterations in lipoprotein metabolism. *Proc Natl Acad Sci USA* **93**, 7154–7159.
62. Kendrick, J. S., Chan, L., and Higgins, J. A. (2001) Superior role of apolipoprotein B48 over apolipoprotein B100 in chylomicron assembly and fat absorption: An investigation of apobec-1 knock-out and wild-type mice. *Biochem J* **356**, 821–827.
63. Xie, Y., Nassir, F., Luo, J., Buhman, K., and Davidson, N. O. (2003) Intestinal lipoprotein assembly in apobec-1^{-}/$^{-}$ mice reveals subtle alterations in triglyceride secretion coupled with a shift to larger lipoproteins. *Am J Physiol Gastrointest Liver Physiol* **285**, G735–G746.
64. Powell-Braxton, L., Veniant, M., Latvala, R. D., Hirano, K. I., Won, W. B., Ross, J., Dybdal, N., Zlot, C. H., Young, S. G., and Davidson, N. O. (1998) A mouse model of human familial hypercholesterolemia: Markedly elevated low density lipoprotein cholesterol levels and severe atherosclerosis on a low-fat chow diet. *Nat Med* **4**, 934–938.
65. Mehta, A., and Driscoll, D. M. (2002) Identification of domains in apobec-1 complementation factor required for RNA binding and apolipoprotein-B mRNA editing. *RNA* **8**, 69–82.
66. Burd, C. G., and Dreyfuss, G. (1994) Conserved structures and diversity of functions of RNA-binding proteins. *Science* **265**, 615–621.
67. Sowden, M. P., Lehmann, D. M., Lin, X., Smith, C. O., and Smith, H. C. (2004) Identification of novel alternative splice variants of APOBEC-1 complementation factor with different capacities to support apolipoprotein B mRNA editing. *J Biol Chem* **279**, 197–206.
68. Chen, Z., Eggerman, T. L., and Patterson, A. P. (2007) ApoB mRNA editing is mediated by a coordinated modulation of multiple apoB mRNA editing enzyme components. *Am J Physiol Gastrointest Liver Physiol* **292**, G53–G65.
69. Linden, D., Sjoberg, A., Asp, L., Carlsson, L., and Oscarsson, J. (2000) Direct effects of growth hormone on production and secretion of apolipoprotein B from rat hepatocytes. *Am J Physiol Endocrinol Metab* **279**, E1335–E1346.
70. Funahashi, T., Giannoni, F., DePaoli, A. M., Skarosi, S. F., and Davidson, N. O. (1995) Tissue-specific, developmental and nutritional regulation of the gene encoding the catalytic subunit of the rat apolipoprotein B mRNA editing enzyme: Functional role in the modulation of apoB mRNA editing. *J Lipid Res* **36**, 414–428.
71. Giangreco, A., Sowden, M. P., Mikityansky, I., and Smith, H. C. (2001) Ethanol stimulates apolipoprotein B mRNA editing in the absence of de novo RNA or protein synthesis. *Biochem Biophys Res Commun* **289**, 1162–1167.
72. Lehmann, D. M., Galloway, C. A., Sowden, M. P., and Smith, H. C. (2006) Metabolic regulation of apoB mRNA editing is associated with phosphorylation of APOBEC-1 complementation factor. *Nucleic Acids Res* **34**, 3299–3308.
73. Van Mater, D., Sowden, M. P., Cianci, J., Sparks, J. D., Sparks, C. E., Ballatori, N., and Smith, H. C. (1998) Ethanol increases apolipoprotein B mRNA editing in rat primary hepatocytes and McArdle cells. *Biochem Biophys Res Commun* **252**, 334–339.
74. Lehmann, D. M., Galloway, C. A., MacElrevey, C., Sowden, M. P., Wedekind, J. E., and Smith, H. C. (2007) Functional characterization of APOBEC-1 complementation factor phosphorylation sites. *Biochim Biophys Acta* **1773**, 408–418.
75. Mukhopadhyay, D., Houchen, C. W., Kennedy, S., Dieckgraefe, B. K., and Anant, S. (2003) Coupled mRNA stabilization and translational silencing of cyclooxygenase-2 by a novel RNA binding protein, CUGBP2. *Mol Cell* **11**, 113–126.
76. Lau, P. P., Cahill, D. J., Zhu, H. J., and Chan, L. (1995) Ethanol modulates apolipoprotein B mRNA editing in the rat. *J Lipid Res* **36**, 2069–2078.
77. Conne, B., Stutz, A., and Vassalli, J. D. (2000) The 3′ untranslated region of messenger RNA: A molecular 'hotspot' for pathology? *Nat Med* **6**, 637–641.
78. Chen, C. Y. and Shyu, A. B. (1995) AU-rich elements: Characterization and importance in mRNA degradation. *Trends Biochem Sci* **20**, 465–470.

79. Herschman, H. R. (1991) Primary response genes induced by growth factors and tumor promoters. *Annu Rev Biochem* **60**, 281–319.
80. Espel, E., Garcia-Sanz, J. A., Aubert, V., Menoud, V., Sperisen, P., Fernandez, N., and Spertini, F. (1996) Transcriptional and translational control of TNF-alpha gene expression in human monocytes by major histocompatibility complex class II ligands. *Eur J Immunol* **26**, 2417–2424.
81. Audic, Y., and Hartley, R. S. (2004) Post-transcriptional regulation in cancer. *Biol Cell* **96**, 479–498.
82. Anant, S., Murmu, N., Houchen, C. W., Mukhopadhyay, D., Riehl, T. E., Young, S. G., Morrison, A. R., Stenson, W. F., and Davidson, N. O. (2004) Apobec-1 protects intestine from radiation injury through posttranscriptional regulation of cyclooxygenase-2 expression. *Gastroenterology* **127**, 1139–1149.
83. Riehl, T., Cohn, S., Tessner, T., Schloemann, S., and Stenson, W. F. (2000) Lipopolysaccharide is radioprotective in the mouse intestine through a prostaglandin-mediated mechanism. *Gastroenterology* **118**, 1106–1116.
84. Gou, Q., Liu, C. H., Ben-Av, P., and Hla, T. (1998) Dissociation of basal turnover and cytokine-induced transcript stabilization of the human cyclooxygenase-2 mRNA by mutagenesis of the 3′-untranslated region. *Biochem Biophys Res Commun* **242**, 508–512.
85. Cok, S. J., and Morrison, A. R. (2001) The 3′-untranslated region of murine cyclooxygenase-2 contains multiple regulatory elements that alter message stability and translational efficiency. *J Biol Chem* **276**, 23179–23185.
86. Stenson, W. F. (2007) Prostaglandins and epithelial response to injury. *Curr Opin Gastroenterol* **23**, 107–110.
87. Eberhart, C. E., Coffey, R. J., Radhika, A., Giardiello, F. M., Ferrenbach, S., and DuBois, R. N. (1994) Up-regulation of cyclooxygenase 2 gene expression in human colorectal adenomas and adenocarcinomas. *Gastroenterology* **107**, 1183–1188.
88. DuBois, R. N., Radhika, A., Reddy, B. S., and Entingh, A. J. (1996) Increased cyclooxygenase-2 levels in carcinogen-induced rat colonic tumors. *Gastroenterology* **110**, 1259–1262.
89. Williams, C. S., Luongo, C., Radhika, A., Zhang, T., Lamps, L. W., Nanney, L. B., Beauchamp, R. D., and DuBois, R. N. (1996) Elevated cyclooxygenase-2 levels in Min mouse adenomas. *Gastroenterology* **111**, 1134–1140.
90. Blanc, V., Henderson, J. O., Cho, S., Newberry, R. D., Xie, Y., Newberry, E. P., Rubin, D. C., Wang, H. L., Luo, J., Kennedy, S., and Davidson, N. O. (2007) Deletion of the AU-rich RNA binding protein apobec-1 reduces intestinal tumor burden in APCmin mice. *Cancer Res.* **67**, 8565–8573.
91. Yamanaka, S., Zhang, X. Y., Maeda, M., Miura, K., Wang, S., Farese, R. V., Jr., Iwao, H., and Innerarity, T. L. (2000) Essential role of NAT1/p97/DAP5 in embryonic differentiation and the retinoic acid pathway. *EMBO J* **19**, 5533–5541.
92. Rothstein, J. L., Johnson, D., DeLoia, J. A., Skowronski, J., Solter, D., and Knowles, B. (1992) Gene expression during preimplantation mouse development. *Genes Dev* **6**, 1190–1201.
93. Taub, R. (2004) Liver regeneration: From myth to mechanism. *Nat Rev Mol Cell Biol* **5**, 836–847.
94. Verna, L., Whysner, J., and Williams, G. M. (1996) N-nitrosodiethylamine mechanistic data and risk assessment: Bioactivation, DNA-adduct formation, mutagenicity, and tumor initiation. *Pharmacol Ther* **71**, 57–81.
95. Eferl, R., Ricci, R., Kenner, L., Zenz, R., David, J. P., Rath, M., and Wagner, E. F. (2003) Liver tumor development. c-Jun antagonizes the proapoptotic activity of p53. *Cell* **112**, 181–192.
96. Dalmau, J., Furneaux, H. M., Gralla, R. J., Kris, M. G., and Posner, J. B. (1990) Detection of the anti-Hu antibody in the serum of patients with small cell lung cancer—a quantitative western blot analysis. *Ann Neurol* **27**, 544–552.
97. Szabo, A., Dalmau, J., Manley, G., Rosenfeld, M., Wong, E., Henson, J., Posner, J. B., and Furneaux, H. M. (1991) HuD, a paraneoplastic encephalomyelitis antigen, contains RNA-binding domains and is homologous to Elav and Sex-lethal. *Cell* **67**, 325–333.
98. Brennan, C. M., and Steitz, J. A. (2001) HuR and mRNA stability. *Cell Mol Life Sci* **58**, 266–277.
99. Fan, X. C., and Steitz, J. A. (1998) HNS, a nuclear-cytoplasmic shuttling sequence in HuR. *Proc Natl Acad Sci USA* **95**, 15293–15298.
100. Wang, W., Caldwell, M. C., Lin, S., Furneaux, H., and Gorospe, M. (2000) HuR regulates cyclin A and cyclin B1 mRNA stability during cell proliferation. *EMBO J* **19**, 2340–2350.

101. Wang, W., Yang, X., Cristofalo, V. J., Holbrook, N. J., and Gorospe, M. (2001) Loss of HuR is linked to reduced expression of proliferative genes during replicative senescence. *Mol Cell Biol* **21**, 5889–5898.
102. Dixon, D. A., Tolley, N. D., King, P. H., Nabors, L. B., McIntyre, T. M., Zimmerman, G. A., and Prescott, S. M. (2001) Altered expression of the mRNA stability factor HuR promotes cyclooxygenase-2 expression in colon cancer cells. *J Clin Invest* **108**, 1657–1665.
103. Raineri, I., Wegmueller, D., Gross, B., Certa, U., and Moroni, C. (2004) Roles of AUF1 isoforms, HuR and BRF1 in ARE-dependent mRNA turnover studied by RNA interference. *Nucleic Acids Res* **32**, 1279–1288.
104. Denkert, C., Weichert, W., Pest, S., Koch, I., Licht, D., Kobel, M., Reles, A., Sehouli, J., Dietel, M., and Hauptmann, S. (2004) Overexpression of the embryonic-lethal abnormal vision-like protein HuR in ovarian carcinoma is a prognostic factor and is associated with increased cyclooxygenase 2 expression. *Cancer Res* **64**, 189–195.
105. Blaxall, B. C., Dwyer-Nield, L. D., Bauer, A. K., Bohlmeyer, T. J., Malkinson, A. M. and Port, J. D. (2000) Differential expression and localization of the mRNA binding proteins, AU-rich element mRNA binding protein (AUF1) and Hu antigen R (HuR), in neoplastic lung tissue. *Mol Carcinog* **28**, 76–83.

CHAPTER 11

ANTIVIRAL FUNCTION OF APOBEC3 CYTIDINE DEAMINASES

Ann M. Sheehy

THE MOLECULAR battles waged between viruses and hosts are ancient. The identification of the APOBEC3 family of cytidine deaminases as innate defense ammunition against one of the most devastating viruses in recorded history has mobilized many talented individuals to fully understand the interplay between these host proteins and the viruses they can suppress. Understanding and harnessing the unique cellular power of the APOBEC3 family of proteins has the potential to open new avenues of antiviral therapies.

11.1 EXPLANATION OF VIF PHENOTYPE UNCOVERS A UNIQUE INNATE RESISTANCE TO HIV-1 INFECTION

With the recent identification of human APOBEC3G (*apo*lipoprotein *B* mRNA-editing *e*nzyme *c*atalytic polypeptide 1-like protein 3G; hA3G) as the intracellular defense protein that inhibits the replication of *vif*-deficient human immunodeficiency virus-1 (HIV-1) in T cells, a long-standing, enigmatic observation was explained, namely, the cellular activity that allowed both primary CD4+ T cells and certain T-cell lines (termed nonpermissive cells) to successfully suppress a *vif*-deficient HIV-1 infection (104). HIV-1, regardless of the presence of the *vif* gene, establishes a robust and productive infection in most transformed T-cell lines, spreading unchecked until most cells in the culture are killed. However, in the absence of Vif (viral infectivity factor), an auxiliary viral protein, HIV-1 replication is impaired in nonpermissive cells. Heterokaryon experiments provided the first tantalizing suggestion of the presence of an innate resistance factor expressed in nonpermissive cells that functioned to inhibit HIV-1 infection (68, 107). Using the information provided by these experiments, a complementary DNA subtraction strategy was adopted, and the search for gene

RNA and DNA Editing: Molecular Mechanisms and Their Integration into Biological Systems,
Edited by Harold C. Smith

expression differences between a permissive and a nonpermissive cell led to the identification of hA3G (originally named CEM15) (104). The initial characterizations of the antiviral capacity of hA3G revealed its mutagenic capabilities and delineated its anti-HIV-1 activity to an enzymatic cytidine deaminase activity (39, 60, 69, 136). With these observations came the optimistic speculation that the Vif: hA3G regulatory circuit could provide a unique and meaningful point of therapeutic intervention. Several years and hundreds of papers later, the investigation of hA3G as a potent antiviral continues; it is more fully appreciated that this protein can exert its antiviral effects against not only HIV-1, but also other retroviruses and retroelements, as well as other viruses. In addition, the innate cellular defense activity provided by hA3G extends beyond the oft-observed cytidine deamination, implicating an as-yet uncharacterized, novel antiviral activity. The only aspect of this field to remain unchanged over the course of the last several years is the continued optimism and hope that further understanding of this innate defense will lead to a potential therapeutic that successfully harnesses the cellular ability to resist an HIV-1 invasion.

11.2 ANTIVIRAL FUNCTIONALITY OF THE APOBEC3 FAMILY OF PROTEINS

11.2.1 Mechanism of Action

h*A3G* is just one gene of the seven members APOBEC3 family of genes that reside on chromosome 22 (Figure 11.1) (48). While intense scrutiny of the APOBEC3 family is a relatively new phenomenon, similarity of the APOBEC3 members to APOBEC1 (Chapter 10) and AID (activation-induced deaminase; Chapter 2), aided in the initial delineation of the antiviral function of the APOBEC3 family. Neither APOBEC1 nor AID exhibit any detectable antiviral capacity, yet, both proteins, along with APOBEC3C (hA3C) and hA3G, had been shown to display strong DNA mutational capacity in an *E. coli* assay (90, 40, 91). This mutational activity was exerted through an enzymatic cytidine deaminase function. This enzymatic activity of both APOBEC1 and AID had been shown to play important roles *in vivo*; APOBEC1 acts as a tissue-specific, site-specific RNA-editing enzyme resulting in the expression of a novel form of apolipoprotein B in the small intestine, while AID plays a central role as a DNA deamination enzyme in the somatic hypermutation and affinity maturation processes during antibody gene diversification (24, 44, 82, 83, 93, 120). The genetic similarity, along with the common mutational activity observed among APOBEC1, AID, and hA3G, prompted a number of groups to experimentally address whether hA3G's capacity to function as a cytidine deaminase correlated with its documented antiviral activity. Using a variety of techniques and a number of different experimental systems, several laboratories convincingly characterized the ability of hA3G to enzymatically function as a cytidine deaminase in virally infected human cells, resulting in the mutation of retroviral minus (first)-strand cDNAs produced during the critical step of reverse transcription (39, 60, 69, 131, 136). These cytidine-to-uracil deamination events were observed as accumulated guanine-to-adenine mutations implicating a DNA deamination event; this speculation was experimentally verified and DNA editing became the hallmark of APOBEC3 antiviral activity.

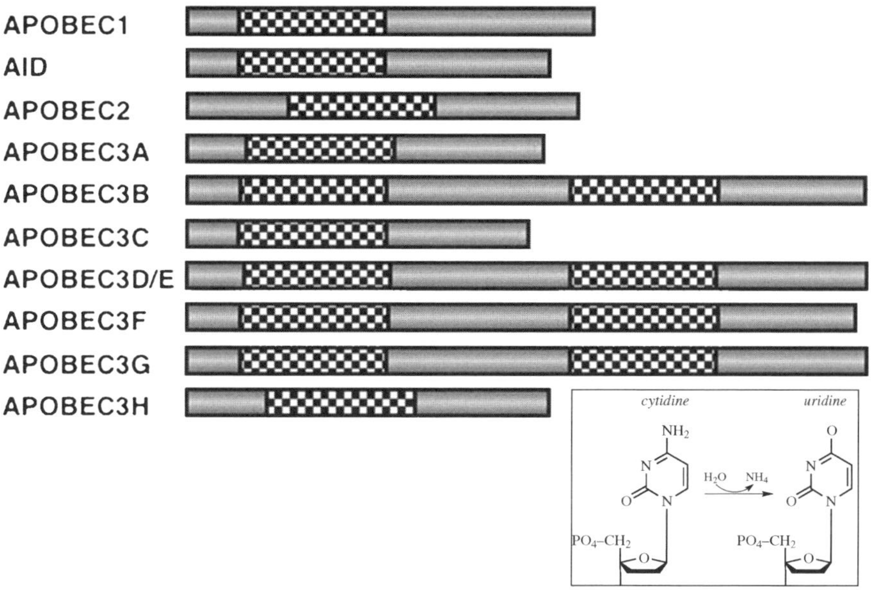

Figure 11.1 The human APOBEC3 family of cytidine deaminases. Alignment of the human APOBEC3 family members with founding member, APOBEC1, and B-cell-specific, AID. The checkered boxes indicate the delineated cytidine deaminase domains present in individual family members.

As the cytidine deamination function of APOBEC3 family members was being investigated, so were the differential abilities of the family members to suppress not only HIV-1 but also a range of viruses. While the focus remained on retroviruses, including HIV-2, SIV (simian immunodeficiency virus), MLV (murine leukemia virus), MPMV (Mason–Pfizer monkey virus), HTLV-1 (human T-cell leukemia virus 1), and the foamy viruses, investigation of the hepadnavirus, HBV (hepatitis B virus), was also intriguing. HBV's unique genome replication, which proceeds through an unusual single-stranded DNA step, suggested that it too might be vulnerable to APOBEC3 antiviral activity. Follow-up experiments convincingly verified this speculation (86, 87, 97, 116, 122). As shown in Table 11.1, six of the seven APOBEC3 family members have been shown to exhibit an impressively diverse range of antiviral activity; in addition to anti-retroviral function, activity against a hepadnavirus and a parvovirus has been observed. In many cases, original analyses of the antiviral activity of the APOBEC3 family members correlated with the measurable introduction of mutations, indicative of the editing capacity of the investigated APOBEC3, into viral DNA. However, there were several experimental systems that seemed to suggest that the mutational capacity of the APOBEC3 proteins was not necessarily required to confer the APOBEC proteins' viral suppressive effects, and, in fact, a number of investigators were unable to detect mutated viral sequences (8, 13, 18, 42, 85, 87, 97, 116, 122). The engineering, via site-directed mutagenesis, of a catalytically inactive hA3G protein revealed an unexpected observation (85). While this mutant protein had lost both its ability to mutate bacterial DNA in an *E. coli*-based mutator assay and its function as a cytidine deaminase in a eukaryotic system, it

TABLE 11.1 Suppression of Viral/Retroelement Replication via APOBEC3 Family of Proteins

APOBEC3 Family Member	Viral/Retroelement Suppression	Detection of DNA/ RNA Editing[a]	References
APOBEC3A	Adeno-associated	N	16
	virus	Y	89
	HIV-1	N	11, 16
	LTR-containing	N	16, 80
	retrotransposons:		
	IAP, MusD		
	L1 retrotransposon		
APOBEC3B	HIV-1	Y	9, 26
	SIVagm, SIVmac	Y, nr	130
	HBV	Y	13, 116
	LTR-containing	nr	12, 16
	retrotransposons:	nr, N	7, 12, 16, 111
	IAP, MusD		
	Alu, LI		
	retrotransposons		
APOBEC3C	SIVagm, SIVmac	Y, nr	130
	HBV	Y	4, 116
	Retroelements: IAP,	nr, N, nr, nr	7, 11, 12, 16, 30
	MusD, Ty1, Alu, L1		
APOBEC3DE	HIV-1	Y	22
	SIVagm, SIVmac	nr	22
APOBEC3F	HIV-1	Y	9, 62, 128, 137
	Foamy viruses	Y	64
	HBV	Y	13, 96, 116
	Retroelements: IAP,	N, Y, Y, N	11, 16, 30, 32, 80, 103, 111
	MusD, Ty1, L1		
APOBEC3G	HIV-1, HIV-2,	Y, nr, N	39, 60, 69, 94, 100, 104
	HTLV-1SIVagm,	nr	131, 136
	SIVmac	Y, nr, nr, nr,	10, 16, 71, 102
	MLV, MPMV, EIAV,	Y	9, 25, 39, 55, 64, 69, 70, 98
	Foamy viruses	(N/Y)	13, 61, 86, 96, 116, 119, 122
	HBV	Y,Y, Y, nr	11, 16, 19, 30–32, 45, 103
	Retroelements:		
	IAP, MusD, Ty1, Alu		
APOBEC3H	No documented inhibition	NA	88

[a]Nr, not recorded

nevertheless retained its anti-HIV character, suppressing the virus (exhibiting up to 90% of wild-type protein activity) in a single-round viral infectivity assay. Subsequent to this initial hA3G report, a number of other well-designed experimental systems revealed striking antiviral activity in the absence of detectable mutation: hA3G activity against HTLV-1 lacked obvious hypermutation, and the suppression of viral replication still occurred with a catalytically inactive hA3G protein; APOBEC3A (hA3A) antiviral activity against the adeno-associated virus (AAV) occurred without concomitant hypermutation; and the striking ability of resting CD4+ T cells (present in peripheral blood) to resist incoming HIV-1 virions was finally attributed to the expression of cellular hA3G, but this resistance was not accompanied by observable mutation (16, 18, 100). It is now clear that DNA editing and the APOBEC3 antiviral effect are not inextricably linked. While the suggestion of an antiviral function in the absence of cytidine deamination has been met with skepticism and the precise mechanism of this viral inhibition awaits complete elucidation, recent work is beginning to offer revealing insight into this alternative basis of the antiviral activity of the APOBEC3 family. Several groups working with HBV have noted hA3G's interference with viral cDNA synthesis and a hA3G-dependent perturbation of the normal association of viral RNA with viral cores (86, 97, 122). Other data probing hA3G activity against HIV-1 have implicated an antiviral function of hA3G that interferes with the critical steps of both reverse transcription and viral integration (8, 42, 66, 73, 86, 96, 108, 117, 122).

11.2.2 APOBEC3 Proteins and the Prevention of Zoonosis

With the expanding number of viruses that appear to be restricted by various APOBEC3 family members, it is interesting to speculate whether these proteins function not only as innate defense factors, but also to contribute to the limitation of zoonotic transmission of various viruses. As a specific example, consider the observation that there are a large number of SIVs that have been isolated from at least 30 different primate species and that only a small number of those SIVs (namely, viruses from either chimpanzees or sooty mangabeys) have been able to clear the hurdle of the species barrier and infect humans. In an attempt to address this observation in the laboratory, *vif* alleles from 10 different SIVs (representing six phylogenetic lineages of primate immunodeficiency lineages) were analyzed for their ability to function in nonpermissive cells to support HIV-1 infectivity (36). Only *vif* genes from SIVmac (macaque), SIVsmm (sooty mangabey), and SIVcpz (chimpanzee) could functionally substitute for HIV-1 *vif*. This observation suggested that the hA3G present in the nonpermissive cells was no longer inhibited by the vast majority of Vif proteins from different SIV strains and was fully capable of suppressing a viral infection.

In complementary studies, three groups conclusively showed that the species-specific Vif:A3G interaction could, in fact, play a critical role in limiting cross-species transmission (10, 70, 102). These groups delineated amino acid 128 of hA3G in governing its inhibition by HIV-1 Vif; in human A3G this residue is an aspartic acid, while in African green monkey (agm) A3G, the amino acid residue in the corresponding position is a lysine. By swapping the identity of the amino acids at this residue, human A3G could no longer suppress an SIVagm infection, mimicking the breakdown of a putative species restriction. While it is unlikely that the expression of hA3G is the sole barrier to zoonotic transmissions of all the viruses listed above that appear susceptible to hA3G activity, it is certainly clear that the Vif-hA3G protein interaction is a significant contributor to the block of cross-species transmission of lentiviruses. It will certainly be of interest to determine the extent of the role hA3G plays in limiting zoonosis.

11.2.3 *In Vivo* Correlations Between APOBEC3 Expression and Disease Course

One of the more intriguing lines of investigation that has stemmed from the identification of hA3G as an innate cellular defense has been the exploration of whether (and how) *hA3G* expression *in vivo* might impact HIV-1 disease. The observation of G $\rightarrow$ A hypermutated viral genomes present in HIV-1 infected individuals significantly predates the identification of hA3G; an editing mechanism was even implicated in the generation of these mutations (14, 47, 124). While these reportings were interesting at the time, their significance would become clearer as the function of hA3G in the suppression of HIV-1 replication emerged. A recent study comparing the local sequence context of the G $\rightarrow$ A transitions found in HIV-1-infected patients was nicely correlated with the preferred sequence target site selection of APOBEC3 family members hA3F and hA3G, providing experimental support for the supposition that these two proteins are in fact functioning in the context of a natural HIV-1 infection (5). Follow-up analyses in HIV-1 infected individuals on highly active antiretroviral therapy (whose viral loads are suppressed below current limits of detection) have revealed the presence of G $\rightarrow$ A hypermutated viral genomes detectable in resting CD4+ T cells (53). While it is not thought that these genomes contribute to continued viral replication, likely due to the suppressive presence of hA3G in these cells, the data do suggest that mutations arising in *vif*, due to the intrinsic error-prone nature of the HIV-1 reverse transcriptase, can lead to a dominant hA3G effect and result in a "dead end" viral infection.

Single-nucleotide polymorphisms (SNPs) of various genes involved in the Vif-hA3G protein complex have also been noted. In particular, a specific hA3G allele variant, as well as a specific polymorphism in the CUL5 gene (encoding the Cullin 5 protein; Cullin 5 plays a critical role in Vif-dependent regulation of hA3G and is discussed below), seen primarily in African-American populations, have been associated with a more rapid CD4+ T cell loss and subsequent more rapid disease progression (2, 3, 132). As interesting as these initial findings are, verification of the importance of these SNPs awaits a fundamental mechanistic explanation that to date remains uncharacterized. It is obvious that the Vif-hA3G interaction involves a number of cellular factors, and probing the intricacies of these complexes will be critical to precisely defining the contribution gene variants may make on the modulation of HIV-1 infection.

Perhaps the most relevant and interesting studies examining the potential attenuating effect of hA3G on HIV-1 infection comes from a pair of groups examining specific cohorts of HIV-1-infected individuals who are managing to maintain a dominance over the virus, namely long-term nonprogressors (LTNP), and HIV-exposed seronegative individuals (7, 49). The study examining LTNPs correlated elevated *hA3G* mRNA expression with a slower progression to disease providing an *in vivo* example of what had been previously shown in the laboratory-overexpression of *hA3G* can begin to circumvent Vif inhibition and suppress wild-type HIV-1 replication (132). Arguably, the most intriguing *in vivo* observation to date is a recent examination of 30 individuals who have been repeatedly exposed to the virus yet have successfully maintained a seronegative status. This study compared *hA3G* mRNA expression in HIV-exposed-seronegative individuals to the expression levels detected in HIV-infected and uninfected individuals. Remarkably, it was observed that hA3G mRNA expression was elevated in the unstimulated PBMCs of HIV-exposed-seronegative cohort. Upon further examination, it was shown that the subpopulation of CD14+, monocytic-lineage cells within the peripheral blood accounted for this increased expression. By specifically examining the ability of these CD14+ cells to resist a spreading HIV-1 infection in a tissue culture system, it was observed that the hA3G protein expression levels in these monocytic cells were also elevated and correspondingly able to partially resist an HIV-1 challenge. These studies all provide preliminary observations that support the premise that normal variants in cellular genes participating in the Vif-hA3G interaction may provide a genetic basis for both resistance to HIV-1 and the ability to modulate the virus once infected. The exciting question becomes whether chemotherapeutics may be developed to mimic these naturally occurring genetic differences.

11.3 THE BATTLE FOR CONTROL: VIRAL SUPPRESSION OF THE APOBEC3 PROTEINS

Since the earliest descriptions of the function of the HIV-1 *vif* gene (originally called *sor*, for short open reading frame), it was well-recognized and described that this viral gene was essential for the successful HIV-1 infection of primary CD4+ T cells and that this critical protein requirement extended to certain transformed T-lymphocyte lines (34, 35, 115). With the recent extensive investigation of the Vif-hA3G interaction, it is now fully appreciated how Vif functions as one of the most impressive viral counter-responses to a cellular innate defense arsenal.

11.3.1 The Hijacking of the Proteasomal Degradation Pathways

The packaging of hA3G into retroviral particles in virus-producing cells appears to be the critical step in the ability of the protein to exert its antiviral effect. These hA3G-containing particles subsequently enter target cells unimpeded. Once the viral process of reverse transcription begins in this target cell, hA3G, presumably as a component of the pre-integration complex (PIC), which is present in the virion, acts to block viral infection (8, 21, 42). In an effective viral counterattack, the HIV-1 Vif protein interferes with encapsidation of hA3G by binding directly to the cellular protein ultimately targeting it for ubiquitination and proteasomal degradation (20, 50, 63, 72, 76, 105,113). From this understanding of Vif suppression of hA3G came the question of whether Vif participated as a component of an E3 ligase complex. The field did not have to wait long for an answer to the query. Mere weeks after the elucidation of the Vif-dependent degradation of hA3G was reported, elegant experiments from the Yu group propelled the field forward again; they had determined that Vif functioned not as the E3 ligase, but as an E2-like substrate recognition adaptor protein molecule that binds hA3G and recruits the Cul5-ElonginB-ElonginC ubiquitin ligase (132). Further characterization of this critical complex revealed that Vif contained a SOCS (suppressor of cytokine signaling)-box motif that allowed it to selectively interact with Elongin C. The definitive delineation of this SOCS-box motif confirmed its identity as the E2/adaptor protein of the ubiquitin ligase known to degrade hA3G (133). The end result of this hA3G-imposed block is the effective short-circuiting of a viral infection. This fundamental understanding of the composition of the protein complex recruited by Vif, for the degradation of hA3G, will allow meaningful investigation of whether it will be possible to interfere with the assembly of this complex in some way, thereby perturbing the Vif-mediated destruction of hA3G.

11.3.2 Virion Exclusion of the APOBECs via a Viral-Dependent Mechanism

Proteasomal degradation of hA3G has not been the sole mechanism cited for the viral inhibition of the cellular defense protein. One group has noted Vif disruption of the translation of hA3G while others have speculated on the Vif-mediated interference with the molecules responsible for specifically packaging hA3G into budding virions (50, 71, 113). It is an attractive hypothesis that Vif functions in multiple ways to exclude hA3G from virions, and these experimental observations are intriguing; mechanistic detail awaits identification. It should also be noted that viruses lacking a Vif protein or an obvious Vif-like homolog are also capable of resisting the antiviral function of cognate APOBEC3 proteins. The effective exclusion of the APOBEC3 proteins from virions seems to have struck a common chord in the elucidation of these mechanisms.

MLV resists endogenous murine APOBEC3 (mA3) in the absence of an auxiliary protein (9, 27, 155). Two activities have been observed to contribute to this resistance: virion exclusion via disruption of the viral Gag protein-mA3 binding, a protein–protein interaction that has been shown to be essential for virion packaging, and proteolytic cleavage of mA3 via the virally encoded protease (1, 21, 27). Both MPMV resistance to rhesus APOBEC3G (rA3G) and HTLV-1 resistance to hA3G have also recently been attributed to the exclusion of the APOBEC3 proteins from virions via a Gag-mediated inhibitory mechanism (23, 25). Finally, foamy viruses, both feline and primate, appear to use the viral Bet protein to prevent packaging of the APOBEC3 family members (64, 98). Bet proteins directly interact with the APOBEC3 family members, and the speculation is that this interaction sequesters the APOBEC3 protein away from sites of virion assembly, thereby effectively inhibiting packaging of the cellular protein. The antiviral effects of the APOBEC3 family members have obviously forced vulnerable viruses to evolve a countermeasure to this intracellular defense, and the story behind the details of the mechanisms involved in combating the APOBEC3 proteins continues to unfold.

11.4 CELLULAR FUNCTION AND REGULATION OF THE APOBEC3 FAMILY

11.4.1 Guardians of the Genome: APOBEC3-Mediated Suppression of Cellular Retroelements

Until recently, investigation of the antiviral roles of the APOBEC3 family members has been the primary focus of laboratories characterizing these proteins while the defining of putative cellular functions of this family remained unaddressed. There are now a number of studies examining the role that the APOBEC3 family may play in genomic stability, namely, protecting the genome from the potentially dangerous, insertional mutagenesis activity of endogenous retroelements. These retroelements can be either endogenous retroviruses, such as human endogenous retroviruses (HERV), or retrotransposons, the latter of which may be further subdivided according to the presence of a long-terminal repeat (LTR) control region (21, 43, 51). One aspect of these studies that is becoming increasingly clear is that APOBEC3 proteins, particularly hA3A and hA3B, do function in an array of *in vitro* assays to inhibit a variety of mammalian retroelements (11, 12, 16, 19, 31, 32, 45,80, 111). An initial reporting that hA3G was unable to suppress LINEs (long interspersed nuclear elements) was perhaps not surprising in light of the fact that the reverse transcription associated with the replication of these elements occurs in the nucleus, and it had been shown that hA3G resides primarily in the cytoplasm; however, it does appear that multiple other APOBEC3 proteins—hA3A, hA3B, hA3C, and hA3F—do suppress the replication of these elements (6, 37, 57, 123, 127). In contrast to the non-LTR-containing retroelements (such as LINEs), the replicative cycle of retrotransposons with an LTR control region resemble retroviruses, assembling virus-like particles and undergoing a reverse transcription process in the cytoplasm. Perhaps due to this resemblance, both the murine intercisternal A particle (IAP) family and MusD family of LTR retrotransposons exhibit sensitivity to hA3G; hA3A, hA3B, hA3C, and hA3F also exhibit measurable activity suppressing the mobility of these elements.

One of the more intriguing questions revisited by these prolific studies is the mechanism of inhibition exerted by the APOBEC3 proteins. While the inhibition of endogenous retroelements by hA3F and hA3G occurs concomitantly with detectable cytidine deamination activity, suppression of the LTR-containing elements, IAP and MusD, via hA3A, occurs with a glaring absence of hypermutation of the retroelement, again providing suggestive evidence that while hA3G hypermutation of DNA sequences is a convenient laboratory marker used to track and establish the activity of this protein, editing is decidedly not the sole antiviral activity exerted by the APOBEC3 family (43).

11.4.2 Subcellular Localization (Sequestration)

As DNA mutators, the APOBEC3 family appears to exert remarkable restrictions on both viral invasion and mobility of retroelements, and both of these activities positively contribute to the overall health of the cell. Suppression of viral replication has its obvious benefits, while inhibition of endogenous mobile elements would appear to contribute to the prevention of deleterious insertional mutagenesis that could lead to either cell death, via disruption of an essential gene, or cellular transformation. However, as a family, the APOBEC3 proteins have the ability to mutate DNA, a function that in and of itself could have negative consequences for a cell expressing these factors; and as such, expression is likely to be effectively regulated. There are any number of processes by which this regulation could be exerted: Gene expression may be tightly controlled, coinciding with specific cellular needs that may occur during particular periods of cellular stress; viral invasion itself triggers a number of cellular response pathways; a regulatory posttranslational modification—for example, phosphorylation—may modify function; or subcellular localization may confine the APOBEC3 proteins to regions of the cell where these mutator proteins would not normally encounter their target DNA substrates.

The cytoplasmic location of both hA3F and hA3G was established early in the characterization of these proteins, and their subcellular address has been more precisely defined by recent experiments published by several groups (18, 37, 57, 69, 72, 113, 126, 127). Both immunohistochemistry and biochemical analysis have shown that hA3G is distributed throughout the cytoplasm, but also display a marked concentration in discrete cytoplasmic compartments. Both biochemical and immunohistological techniques were used to identify these foci as P-bodies; related structures, stress granules, that form when the cell is under environmental stress have also been shown to accumulate hA3G (37, 57, 126, 127). Foci that were later to become known as P-bodies were first described more than 10 years ago and are locations within eukaryotic cells that appear to be centrally involved in the regulation of cytoplasmic mRNA processing; they have also recently been implicated in the processes of both RNA interference and translation suppression via microRNA (33). Inactive mRNA accumulates within P-bodies and these stored mRNAs may then be degraded or, alternatively, may move to polysomes for translation. The signals controlling differential mRNA fate remain completely uncharacterized. The sequestration of hA3G within P-bodies may prove to be important for blunting the potentially damaging mutagenic properties of hA3 family members because there is experimental support for the inhibition of editing activity within these large RNP complexes (18). However, while these studies are spectacular in their photography and impressive in the obvious meticulous detail paid to defining a specific cellular localization of the hA3G protein, the question of why this protein localizes to P-bodies remains an open one. The explanation will likely be vigorously pursued.

In contrast to hA3G, which has not been found in the nucleus and, in fact, has recently been shown to be actively retained in the cytoplasm, hA3A and hA3B appear to reside primarily in the nucleus (6, 11, 16, 111). Both of these proteins exhibit an antiviral activity and are the APOBEC3 family's most potent suppressors of retrotransposons (11, 16, 43). The cytidine deaminase capacity of both proteins is detectable, but at least in the cases of retrotransposon inhibition, editing of the retrotransposed DNA is not observed. These data may suggest that posttranslational modifications of hA3A and hA3B proteins impair the cytidine deaminase capacity, but a putative "inhibitory domain" remains active; editing is inhibited but the antiviral/retroelement function is unaffected, thereby protecting cellular genomic DNA from inadvertent mutation.

11.4.3 Control of the Expression of the APOBEC3 Family is Exerted Transcriptionally

As a host restriction factor that can antagonize the replicative capacity of one of the deadliest viruses described in the human population, hA3G is an obvious, attractive target for therapeutic manipulation. However, a thorough understanding of the cellular regulation of the APOBEC3 genes is essential if the investigation of controlling their activities is to be intelligently and effectively examined and undertaken. The tissue distribution of several members of the APOBEC3 family suggests the following: hA3G is preferentially expressed in lymphocytic tissue; the distribution of hA3B expression appears strictly limited, although it is expressed in a number of tumor cell lines; and hA3C expression exhibits little specificity, with expression being observed in a wide array of tissues. In addition to the examination of tissue expression, the induction of hA3G and other members of the APOBEC3 family, both the mRNA and the resulting protein, have been probed. The emerging data addressing cellular control of *hA3G* gene suggests that this regulation exhibits both cell-type and development-stage specificity (13, 17, 48, 89, 90, 95, 99, 114, 119).

While it has been convincingly shown that IFNα and INFγ can control *hA3G* gene expression in hepatocytes and macrophages, transcriptional activation of *hA3G* in CD4+ T cells appears to be regulated by mitogens, such as phorbol myristate acetate (PMA) and phytohemagglutinin (PHA), and cytokines such as IL-2 (interleukin-2), IL-7, and IL-15 (58). However, in the discussion of control in CD4+ cells, it is interesting to note that the further characterization of these cells into resting and activated populations suggests differences in the regulation of *hA3G* expression: While INFα does not stimulate *hA3G* gene expression in activated CD4+ cells, it can dramatically increase gene expression in the resting CD4+ T cell population (18).

Recent work from two different groups, investigating both stimulated and basal transcription of the *hA3G* gene expression in hepatocytes and T cells, has begun to dissect the exact regulatory regions important for endogenous gene control (81, 90, 119). Investigation of the IFNα induction of *hA3G* expression in the HepG2 cell line identified a 128-bp fragment 5′ to the transcriptional start site (TSS) of the gene that functioned as the interferon-controlled promoter region in a luciferase reporter assay. A more extensive investigation was carried out in an effort to define the basal transcription of the *hA3G* gene. These investigators, using the same luciferase reporter assay but performing the experiments in the A3.01 T cell line, precisely defined the core promoter as a −114/+66-bp fragment (relative to the TSS). It should also be noted that this group also performed RACE (rapid amplification of cDNA ends) and redefined the major TSS as 66 bp upstream of the published mRNA sequence (GenBank accession number NM021822). This fragment was analyzed in the reporter assays in T cells and exhibited characteristic promoter activity. Sequence analysis and follow-up experiments identified the importance of the Sp1 and Sp3 transcription factor binding sites to the activity of this core promoter fragment.

In addition to the differences in hA3G transcriptional regulation observed among the various cell types that have been examined (macrophages, hepatocytes, and T lymphocytes), expression of the gene can vary dramatically across cellular differentiation transitions. Specifically, monocytes express significantly higher levels of enzymatically active hA3G than macrophages, suggesting that the protein is downregulated during this process of terminal differentiation (18, 89, 114). In contrast to this downregulation of hA3G, the maturation of dendritic cells results in the detectable upregulation of hA3G (114). These regulations of hA3G expression, across differentiation state transitions, likely contribute to the differential vulnerability of monocytes versus macrophages and immature versus mature dendritic cells to HIV-1 infection that has been observed and documented for many years in a variety of experimental systems (52, 59). It is also important to note that in almost all cases, stimulation of *hA3G* gene transcription resulted in a concomitant increase in protein expression, suggesting that the primary cellular control of regulation of this gene occurs at the level of transcriptional activation and does not seem to indicate a posttranscriptional effect, such as an increase in mRNA or protein stability. There may be posttranscriptional regulatory control, as has been suggested, but this does not seem to be the main mode of regulation (71, 113).

Finally, it would be remiss to discuss the cellular regulation of hA3G without including one of the more provocative observations recently reported concerning the cellular partitioning of this protein into high-molecular-mass (HMM) ribonucleoprotein complexes or low-molecular-mass complexes (LMM) (18). The initial biochemical dissection of this phenomenon explained observations that had been made by a number of groups citing that both circulating resting CD4+ T cells and monocytes were refractory to HIV infection; the differentiation of monocytes into macrophages or the activation of CD4+ T cells resulted in their vulnerability to HIV infection (56, 74, 84, 92, 112, 121, 134). The remarkable insight made was that the hA3G that was present in the LMM complexes was enzymatically active and functioned as a potent post-entry restriction factor in the resting CD4+ T cells and monocytes, protecting those cell types from viral invasion. Activation of the resting T cells or stimulation of the differentiation of the monocytes resulted in the migration of hA3G into HMM RNP complexes; and within these complexes, hA3G was inactive, unable to act against HIV and the cells became susceptible to infection (18). The distinguishing between active and inactive cellular hA3G also clarified the purported discordant result that had been made that resting CD4+ T cells present in tissues were unable to resist an HIV infection. Experiments examining the microenvironment of these cells showed that the local concentration of cytokines in the tissues, while not activating the cells themselves, stimulated the migration of hA3G from LMM to HMM complexes, thus explaining the apparent discrepancy (58). Subsequent characterization of the HMM RNP and LMM hA3G complexes complement other observations in the field: HMM RNP complexes localize to cytoplasmic P bodies (19). Offering hope on a therapeutic front, disassembly of the HMM RNP complex to the enzymatically active LMM hA3G is observable through the pointed degradation of RNA via RNase A and, in fact, "activation" of hA3G activity has already been shown to occur naturally via experiments examining the packaging of hA3G (110). It was shown that newly synthesized hA3G protein, which is seen LMM hA3G, is able to access the assembling virion and is actively packaged. Once within the virion, LMM hA3G assembles into RNA-containing high-molecular-weight complexes. The RNA within these complexes includes the viral genomic RNA, and these complexes may be analogous to the HMM RNP complexes seen intracellularly, in that the hA3G that is present appears to be enzymatically inactive. Only upon the degradation of the viral genomic RNA, which occurs naturally via virally encoded RNase H during the reverse transcription process, is the antiviral capacity of hA3G restored. While incredibly exciting, the understanding of this protein partitioning process remains rudimentary, but it is quite clear from the rapidly emerging details that thorough dissection of this sorting is essential when entering a discussion concerning the therapeutic manipulation of hA3G.

11.5 RESEARCH QUESTIONS AND THE HOPE OF THERAPEUTIC MANIPULATION OF THE APOBEC3 FAMILY

The interplay between host and viral factors during a pathogenic encounter has long been a source of intriguing and interesting investigation. In many cases an attack by one is met with a counterattack by the other. In the last several years, investigation of host restriction factors and HIV-1 has undergone an exciting renaissance with a significant leap forward in our current understanding of intracellular molecules that may be able to be utilized to shore up natural defenses against a viral assault. The APOBEC3 family of proteins certainly represents some of the more galvanizing celebrities of this group of recently identified innate defenses. Our knowledge of the details of the interplay between host hA3G and viral Vif is impressive, a testament to the hundreds of talented scientists who have pursued the questions with their own perspectives and expertise's (Figure 11.2). However, as much as we have learned in

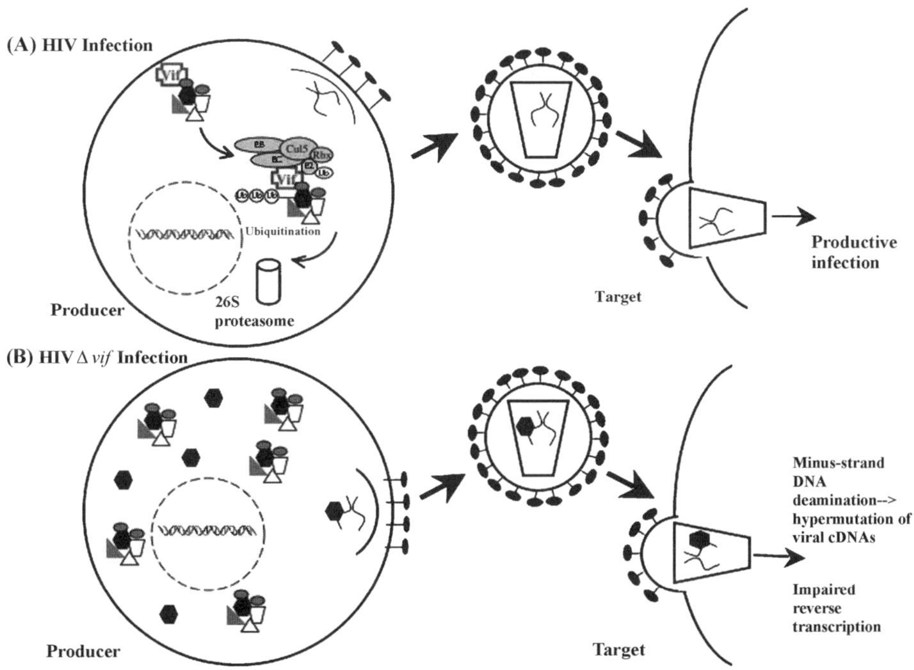

Figure 11.2 Current model of APOBEC3G function. (A) In the context of a cellular wild-type HIV infection, hA3G is primarily found in an HMM RNP complex via a specific interaction with the viral auxiliary protein, Vif. Within this complex, hA3G is subject to polyubiquitination by a well-defined E3 ubiquitin ligase and is subsequently degraded by the 26S proteasome. In the absence of hA3G packaging into budding virions, the produced particles are infectious and able to seed a productive infection in the next round of target cells. (B) In contrast to the context of a wild-type HIV infection, in the absence of Vif, intracellular hA3G can be found in both the HMM RNP complex and an LMM form. It is this LMM form that is efficiently and effectively packaged into budding virions. This packaged hA3G functions in a potent antiviral capacity, leading to a post-entry short-circuiting of viral replication in the target cells. ⬢ represents hA3G. (*See color insert.*)

regard to this host: viral protein interaction, there remain a significant number of unanswered questions and the solutions to these puzzles will be hard-won but, may reveal a path to harness the strength of the natural innate defense of hA3G (or other family member) and develop a therapeutic with potential to ultimately restrict HIV-1 replication.

11.5.1 The "Alternative Function"

Soon after the reporting of the cloning of hA3G, the capacity of this protein to lethally mutate nascent HIV-1 reverse transcripts was identified and the tacit, and altogether reasonable assumption was this mutational ability was the source of antiviral activity against HIV-1. That "dogma" stood for less than 2 years. While assessing mutational capacity is a convenient tool that can be used to track protein function, it is not the sole antiviral activity of this protein. There are numerous instances that have now been reported by multiple scientists who have observed antiviral effects exerted by APOBEC3 family members in the absence (or near absence) of DNA hypermutation (43). In fact, looking back through the literature, the instability of reverse transcripts synthesized after infection via Vif-deficient virus assembly in nonpermissive (hA3G-expressing) cells was reported more than 5 years before the identification of hA3G (108, 125). It is now understood that this instability results from an antiviral effect exerted via virion packaged hA3G; some investigators have begun to speculate that this instability is in fact as significant a marker of hA3G (and hA3F) antiviral function as the detectable hypermutation. It would be foolish to get bogged down in a debate of the relative importance of each of these functions (cytidine deamination versus the "alternative function"). It is not inconceivable that these family members, as critical as they appear to be in the restriction of retroviruses and retroelements, exert multiple functions in a continued effort to protect and defend the cell, and it may well turn out to be the case that editing continues to play a critical role in this alternative function; for example, just as APOBEC1 functions to deaminate a specific cytidine in a single mRNA transcript, perhaps some of the APOBEC3 family members do the same. These possible targets (be they viral or cellular in origin) await identification. Resources would be well spent in identifying and dissecting "secondary" functions. Fundamentally, what is the basis of the mechanism of action? Perhaps hA3G simply sterically interferes with the reverse transcription (RT) process or its inclusion in the pre-integration complex (PIC) disrupts appropriate protein complex formation necessary for the stable formation of products. Two recent reports implicate hA3G interference in the tRNA priming event of RT (38, 73).

A direct line of questioning that stems from the better delineation of the pleiotropic effects of hA3G is one that encompasses whether the need for virion packaging is absolute. To date, each antiviral effect of hA3G has been accompanied by a corresponding requirement for the presence of this protein in budding virions and no mutant has yet been identified in which virion packaging is inhibited yet antiviral function retained. While investigation has described the details of packaging, which includes an essential assembly between the nucleocapsid region of the Gag viral protein and hA3G that includes an RNA bridging molecule (it is of some debate as to whether the actual identity of this RNA bridge matters), and has delineated regions of hA3G that function in virion incorporation, the absolute necessity for its presence in the virion formally remains an open question (15, 46, 65, 117, 135). A broader understanding of the basis of the APOBEC3 family members antiviral and anti-retroelement functions will solidify and address these questions and will certainly lead to new and interesting avenues of study.

11.5.2 Protein Partitioning/Subcellular Localization

The description of HMM and LMM complexes containing inactive and active, respectively, hA3G, along with mass spectrometry analysis and immunohistochemistry localizing hA3G to P bodies, seems to suggest that the cellular localization is a critical parameter in the function and regulation of this protein. Characterization of the mechanisms and pathways important to the subcellular location of hA3G is certainly underway. Is the localization of hA3G to P bodies critical to its function either as an antiviral or as a suppressor of retroelements, and what is the significance of hA3G shuttling to stress granules under conditions of cellular stress? It is likely that specific cellular chaperones are involved in both the partitioning of hA3G into LMM versus HMM RNP complexes and subcellular localization. Identification of particular players important for hA3G localization may suggest unique pathways and targets in the quest to manipulate this protein to therapeutic benefit. For example, can the segregation of hA3G into HMM RNP complexes be avoided or remobilized to LMM hA3G, thereby possibly allowing the increased cytoplasmic concentration of enzymatically active hA3G that is capable of protecting the cell against an HIV invasion? The importance of the HMM RNP complexed hA3G is also unexplored. These large structures do localize to P-bodies, suggesting that they may play a critical role in the suppression of retroelement mobilization; a dramatic drop in the cellular concentration of these complexes may have unexpected consequences for genomic stability. If relocalization of hA3G from HMM RNP to LMM complexes is achievable, what are the consequences of elevated expression of a protein capable of mutating DNA?

In stark contrast to the amplified APOBEC3 locus in humans, the mouse genome contains a single APOBEC3 gene. The creation of an APOBEC3-deficient mouse revealed very little in the determination of either novel or essential roles for the APOBEC3 family (79). In the human system, each APOBEC3 family member does appear to have its own rather distinctive expression pattern and signature of suppression. Is the redundancy seen within human cells relevant and what are the important oligomerization partners (if any) for each protein's cellular role.

11.5.3 Protein Cofactors and Posttranslational Modifications

HIV-1 Vif function as the substrate-specific adaptor protein for recruitment of the E3 ligase that effectively degrades hA3G has been well established. This construction of the virally controlled complex is mediated by specific protein contacts between Vif and Elongin C and Vif and hA3G (54, 67, 75, 77, 106, 132, 133). These protein–protein interactions have been well characterized (10, 46, 102, 129). The molecular interactions, which lead to specific virion packaging of hA3G have also been precisely defined (15, 46, 65, 101, 117, 135). However, other proteins involved in these processes await full investigation. For instance, the defining cellular and/or viral proteins involved in the assembly of the HMM RNP hA3G versus LMM hA3G are not understood. It is likely that there are a number of unidentified hA3G-binding partners, some acting as chaperones, as suggested above, and others acting as more active cofactors. Both AID and APOBEC-1 require cofactors for enzymatic activity, and it not yet appreciated whether any of the APOBEC3 family of proteins require other cellular proteins for specificity or function (41, 78, 118). Each of these specifically defined protein–protein interactions that impact hA3G localization and function present opportunities for therapeutic manipulation and intervention. Small molecule inhibitors of either the Vif:hA3G or Vif:Elongin C interfaces present novel drug targets that could dramatically impact the ability of the cell to resist an HIV infection. To aid in this type of drug discovery, it will be important to invest in meticulous structural studies designed to reveal the important protein–protein contacts. To date, most of the structural information gathered on these protein interactions comes from a genetic approach involving the introduction of a series of point mutations. What are now needed to complement this data are biophysical examinations that allow characterization of individual molecules as well as the complexes.

Few confirmations of posttranslational modifications of hA3G have been reported, and these protein modifications that critically participate in the regulation of expression and function remain relatively uncharacterized for hA3G. Interestingly, one group has reported both (a) the positive role ubiquitination of hA3G via Nedd4-1 can play in viral encapsidation and (b) a putative enhancement of hA3G encapsidation via a tyrosine phosphorylation; corroboration of these modifications and additional proteins important in these putative pathways await characterization (28, 29).

11.5.4 Therapeutic Potential

With the revelations of the last several years on the dynamic interaction between the viral Vif and the cellular hA3G protein came the optimism that inevitably accompanies the uncovering of a novel anti-HIV-1 target. As fundamental knowledge has accumulated in the continued investigation of this protein–protein interplay, the hope of a viable therapeutic continues. hA3G represents a natural defense against HIV-1, and manipulating this defense may tip the balance in favor of the host. The investigation within cohorts looking for polymorphisms within *hA3G* and *hA3F* (and possibly *hA3B*) should continue, with close attention paid to both long-term nonprogressors and highly-exposed, seronegative individuals. Single reportings of discovered polymorphisms should be confirmed and expanded in order to support the contention that perturbation of hA3F/G expression can contribute positively to the ability to resist, even if only partially, an HIV-1 challenge.

Velcade® (Millennium products), a proteasome inhibitor, was approved for use by the FDA in 2003 and is currently used as a chemotherapeutic in the treatment of multiple myeloma and mantle cell lymphoma. While HIV-1 infection is obviously different from cancer, could a similar strategy be adopted to interfere with the Vif-dependent degradation of hA3G? If this perturbation is possible, what are the consequences of interfering with this delicate balance? While preliminary experiments suggest that the stabilization of hA3G can begin to suppress the infectivity of wild-type (Vif-containing) HIV-1 virions, a recent report would argue that *in vivo*, hA3G expression itself can contribute to the genetic variation of replicating HIV-1. What would be the ultimate consequence, then, of eliminating the viral curtailing of hA3G expression and elevating endogenous expression (109, 132)?

Finally, if the supposition is made that more is better, and elevation of hA3G expression can effectively combat an HIV-1 infection, how is that control to be achieved? [An obvious caveat to this assumption is the potential "mislocalization" of overexpressed protein; for instance, if overexpression results in an increase in HMM RNP-complexed hA3G, this would not be expected to concomitantly increase the innate resistance to viral infection. However, recent data would suggest that cellular overexpression results in increased virion packaging of hA3G and that this packaged hA3G is enzymatically active (110).] IFNα can stimulate the expression of hA3G in specific cell types, *in vitro*. How (if at all) can this information be translated into the setting of the clinic? And, most certainly, interferon stimulation is a crude and unrefined treatment, often accompanied with a number of unacceptable side effects. Can this manipulation of hA3G expression be more precisely accomplished? A better understanding of signal transduction pathways, the resulting transcription factors, and the critical regulatory sequences driving expression of hA3G is desperately needed. While initial studies delineating the core promoter have been offered, the understanding of how the expression of *hA3G* is achieved on the molecular is in its infancy. The Sp1 And Sp3 proteins likely do contribute to the core promoter activity of the gene, but these are ubiquitously expressed transcription factors and cannot themselves explain why *hA3G* is expressed in some cell types but not in others. And, while it appears a viable IFN-responsive element that participates in the induction of *hA3G* expression in macrophages and hepatocytes upon IFN-α stimulation has been identified, the delineation of elements important in the T-cell regulation of this gene has not been documented. In the same vein, can the expression of other family members in the appropriate and relevant cell types be accomplished? hA3B is a particularly tantalizing and interesting family member because it functions independently of Vif control and, as a consequence, can partially suppress a wild-type HIV-1 infection.

While we have uncovered so much in such a short period of time, the quest to utilize the information in a meaningful way continues. Boosting an innate intracellular defense in a concerted effort to combat a viral invasion has an intrinsic logic to it. However, it remains to be determined whether perturbing the natural expression of a protein with the potential to mutate DNA is sound reasoning. Currently though, enthusiasm and excitement propel the field forward in a desire to further understand the hA3G–Vif interaction and to discover whether the hA3G protein can, in fact, dramatically change the course of human history.

REFERENCES

1. Abudu, A., Takaori-Kondo, A., Izumi, T., Shirakawa, K., Kobayashi, M., Sasada, A., Fukunaga, K., and Uchiyama, T. (2006) Murine retrovirus escapes from murine APOBEC3 via two distinct novel mechanisms. *Curr Biol* **16**, 1565–1570.

2. An, P., Bleiber, G., Duggal, P., Nelson, G., May, M., Mangeat, B., Alobwede, I., Trono, D., Vlahov, D., Donfield, S., Goedert, J. J., Phair, J., Buchbinder, S., O'Brien, S. J., Telenti, A., and Winkler, C. A. (2004) APOBEC3G genetic variants and their influence on the progression to AIDS. *J Virol* **78**, 11070–11076.
3. An, P., Duggal, P., Wang, L. H., O'Brien, S. J., Donfield, S., Goedert, J. J., Phair, J., Buchbinder, S., Kirk, G. D., and Winkler, C. A. (2007) Polymorphisms of CUL5 are associated with CD4+ T cell loss in HIV-1 infected individuals. *PLoS Genet* **3**, e19.
4. Baumert, T. F., Rosler, C., Malim, M. H., and von Weizsacker, F. (2007) Hepatitis B virus DNA is subject to extensive editing by the human deaminase APOBEC3C. *Hepatology* **46**, 682–689.
5. Beale, R. C., Petersen-Mahrt, S. K., Watt, I. N., Harris, R. S., Rada, C., and Neuberger, M. S. (2004) Comparison of the differential context-dependence of DNA deamination by APOBEC enzymes: correlation with mutation spectra in vivo. *J Mol Biol* **337**, 585–596.
6. Bennett, R. P., Diner, E., Sowden, M. P., Lees, J. A., Wedekind, J. E., and Smith, H. C. (2006) APOBEC-1 and AID are nucleo-cytoplasmic trafficking proteins but APOBEC3G cannot traffic. *Biochem Biophys Res Commun* **350**, 214–219.
7. Biasin, M., Piacentini, L., Caputo, S. L., Kanari, Y., Magri, G., Trabattoni, D., Naddeo, V., Lopalco, L., Clivio, A., Cesana, E., Fasano, F., Bergamaschi, C., Mazzotta, F., Miyazawa, M., and Clerici, M. (2007) Apolipoprotein B mRNA-editing enzyme, catalytic polypeptide-like 3G: A possible role in the resistance to HIV of HIV-exposed seronegative individuals. *J Infect Dis* **195**, 960–964.
8. Bishop, K. N., Holmes, R. K., and Malim, M. H. (2006) Antiviral potency of APOBEC proteins does not correlate with cytidine deamination. *J Virol* **80**, 8450–8458.
9. Bishop, K. N., Holmes, R. K., Sheehy, A. M., Davidson, N. O., Cho, S. J., and Malim, M. H. (2004) Cytidine deamination of retroviral DNA by diverse APOBEC proteins. *Curr Biol* **14**, 1392–1396.
10. Bogerd, H. P., Doehle, B. P., Wiegand, H. L., and Cullen, B. R. (2004) A single amino acid difference in the host APOBEC3G protein controls the primate, species specificity of HIV type 1 virion infectivity factor. *Proc Natl Acad Sci USA* **101**, 3770–3774.
11. Bogerd, H. P., Wiegand, H. L., Doehle, B. P., Lueders, K. K., and Cullen, B. R. (2006) APOBEC3A and APOBEC3B are potent inhibitors of LTR-retrotransposon function in human cells. *Nucleic Acids Res* **34**, 89–95.
12. Bogerd, H. P., Wiegand, H. L., Hulme, A. E., Garcia-Perez, J. L., O'Shea, K. S., Moran, J. V., and Cullen, B. R. (2006) Cellular inhibitors of long interspersed element 1 and Alu retrotransposition. *Proc Natl, Acad Sci USA* **103**, 8780–8785.
13. Bonvin, M., Achermann, F., Greeve, I., Stroka, D., Keogh, A., Inderbitzin, D., Candinas, D., Sommer, P., Wain-Hobson, S., Vartanian, J. P., and Greeve, J. (2006) Interferon-inducible expression of APOBEC3 editing enzymes in human hepatocytes and inhibition of hepatitis B virus replication. *Hepatology* **43**, 1364–1374.
14. Bourara, K., Litvak, S., and Araya, A. (2000) Generation of G-to-A and C-to-U changes in HIV-1 transcripts by RNA editing. *Science* **289**, 1564–1566.
15. Cen, S., Guo, F., Niu, M., Saadatmand, J., Deflassieux, J., and Kleiman, L. (2004) The interaction between HIV-1 Gag and APOBEC3G. *J Biol Chem* **279**, 33177–33184.
16. Chen, H., Lilley, C. E., Yu, Q., Lee, D. V., Chou, J., Narvaiza, I., Landau, N. R., and Weitzman, M. D. (2006) APOBEC3A is a potent inhibitor of adeno-associated virus and retrotransposons. *Curr Biol* **16**, 480–485.
17. Chen, K., Huang, J., Zhang, C., Huang, S., Nunnari, G., Wang, F. X., Tong, X., Gao, L., Nikisher, K., and Zhang, H. (2006) Alpha interferon potently enhances the anti-human immunodeficiency virus type 1 activity of APOBEC3G in resting primary CD4 T cells. *J Virol* **80**, 7645–7657.
18. Chiu, Y. L., Soros, V. B., Kreisberg, J. F., Stopak, K., Yonemoto, W., and Greene, W. C. (2005) Cellular APOBEC3G restricts HIV-1 infection in resting CD4+ T cells. *Nature* **435**, 108–114.
19. Chiu, Y. L., Witkowska, H. E., Hall, S. C., Santiago, M., Soros, V. B., Esnault, C., Heidmann, T., and Greene, W. C. (2006) High-molecular-mass APOBEC3G complexes restrict Alu retrotransposition. *Proc Natl, Acad Sci USA* **103**, 15588–15593.
20. Conticello, S. G., Harris, R. S., and Neuberger, M. S. (2003) The Vif protein of HIV triggers degradation of the human antiretroviral DNA deaminase APOBEC3G. *Curr Biol* **13**, 2009–2013.
21. Cullen, B. R. (2006) Role and mechanism of action of the APOBEC3 family of antiretroviral resistance factors. *J Virol* **80**, 1067–1076.

22. Dang, Y., Wang, X., Esselman, W. J., and Zheng, Y. H. (2006) Identification of APOBEC3DE as another antiretroviral factor from the human APOBEC family. *J Virol* **80**, 10522–10533.
23. Derse, D., Hill, S. A., Princler, G., Lloyd, P., and Heidecker, G. (2007) Resistance of human T cell leukemia virus type 1 to APOBEC3G restriction is, mediated by elements in nucleocapsid. *Proc Natl Acad Sci USA* **104**, 2915–2920.
24. Di Noia, J. M., and Neuberger, M. S. (2007) Molecular mechanisms of antibody somatic hypermutation. *Annu Rev Biochem* **76**, 1–22.
25. Doehle, B. P., Bogerd, H. P., Wiegand, H. L., Jouvenet, N., Bieniasz, P. D., Hunter, E., and Cullen, B. R. (2006) The betaretrovirus Mason–Pfizer monkey virus selectively excludes simian APOBEC3G from virion particles. *J Virol* **80**, 12102–12108.
26. Doehle, B. P., Schafer, A., and Cullen, B. R. (2005) Human APOBEC3B is a potent inhibitor of HIV-1 infectivity and is resistant to HIV-1 Vif. *Virology* **339**, 281–288.
27. Doehle, B. P., Schafer, A., Wiegand, H. L., Bogerd, H. P., and Cullen, B. R. (2005) Differential sensitivity of murine leukemia virus to APOBEC3-mediated inhibition is governed by virion exclusion. *J Virol* **79**, 8201–8207.
28. Douaisi, M., Dussart, S., Courcoul, M., Bessou, G., Lerner, E. C., Decroly, E., and Vigne, R. (2005) The tyrosine kinases Fyn and Hck favor the recruitment of tyrosine-phosphorylated APOBEC3G into vif-defective HIV-1 particles. *Biochem Biophys Res Commun* **329**, 917–924.
29. Dussart, S., Douaisi, M., Courcoul, M., Bessou, G., Vigne, R., and Decroly, E. (2005) APOBEC3G ubiquitination by Nedd4-1 favors its packaging into HIV-1 particles. *J Mol Biol* **345**, 547–558.
30. Dutko, J. A., Schafer, A., Kenny, A. E., Cullen, B. R., and Curcio, M. J. (2005) Inhibition of a yeast LTR retrotransposon by human APOBEC3 cytidine deaminases. *Curr Biol* **15**, 661–666.
31. Esnault, C., Heidmann, O., Delebecque, F., Dewannieux, M., Ribet, D., Hance, A. J., Heidmann, T., and Schwartz, O. (2005) APOBEC3G cytidine deaminase inhibits retrotransposition of endogenous retroviruses. *Nature* **433**, 430–433.
32. Esnault, C., Millet, J., Schwartz, O., and Heidmann, T. (2006) Dual inhibitory effects of APOBEC family proteins on retrotransposition of mammalian endogenous retroviruses. *Nucleic Acids Res* **34**, 1522–1531.
33. Eulalio, A., Behm-Ansmant, I., and Izaurralde, E. (2007) P bodies: at the crossroads of post-transcriptional pathways. *Nat Rev Mol Cell Biol* **8**, 9–22.
34. Fisher, A. G., Ensoli, B., Ivanoff, L., Chamberlain, M., Petteway, S., Ratner, L., Gallo, R. C., and Wong-Staal, F. (1987) The sor gene of HIV-1 is required for efficient virus transmission in vitro. *Science* **237**, 888–893.
35. Gabuzda, D. H., Lawrence, K., Langhoff, E., Terwilliger, E., Dorfman, T., Haseltine, W. A., and Sodroski, J. (1992) Role of vif in replication of human immunodeficiency virus type 1 in CD4+ T lymphocytes. *J Virol* **66**, 6489–6495.
36. Gaddis, N. C., Sheehy, A. M., Ahmad, K. M., Swanson, C. M., Bishop, K. N., Beer, B. E., Marx, P. A., Gao, F., Bibollet-Ruche, F., Hahn, B. H., and Malim, M. H. (2004) Further investigation of simian immunodeficiency virus Vif function in human cells. *J Virol* **78**, 12041–12046.
37. Gallois-Montbrun, S., Kramer, B., Swanson, C. M., Byers, H., Lynham, S., Ward, M., and Malim, M. H. (2007) Antiviral protein APOBEC3G localizes to ribonucleoprotein complexes found in P bodies and stress granules. *J Virol* **81**, 2165–2178.
38. Guo, F., Cen, S., Niu, M., Saadatmand, J., and Kleiman, L. (2006) Inhibition of formula-primed reverse transcription by human APOBEC3G during human immunodeficiency virus type 1 replication. *J Virol* **80**, 11710–11722.
39. Harris, R. S., Bishop, K. N., Sheehy, A. M., Craig, H. M., Petersen-Mahrt, S. K., Watt, I. N., Neuberger, M. S., and Malim, M. H. (2003) DNA deamination mediates innate immunity to retroviral infection. *Cell* **113**, 803–809.
40. Harris, R. S., Petersen-Mahrt, S. K., and Neuberger, M. S. (2002) RNA editing enzyme APOBEC1 and some of its homologs can act as DNA mutators. *Mol Cell* **10**, 1247–1253.
41. Henderson, J. O., Blanc, V., and Davidson, N. O. (2001) Isolation, characterization and developmental regulation of the human apobec-1 complementation factor (ACF) gene. *Biochim Biophys Acta* **1522**, 22–30.

42. Holmes, R. K., Koning, F. A., Bishop, K. N., and Malim, M. H. (2007) APOBEC3F can inhibit the accumulation of HIV-1 reverse transcription products in the absence of hypermutation. Comparisons with APOBEC3G. *J Biol Chem* **282**, 2587–2595.
43. Holmes, R. K., Malim, M. H., and Bishop, K. N. (2007) APOBEC-mediated viral restriction: not simply editing? *Trends Biochem Sci* **33**, 118–118.
44. Honjo, T., Kinoshita, K., and Muramatsu, M. (2002) Molecular mechanism of class switch recombination: linkage with somatic hypermutation. *Annu Rev Immunol* **20**, 165–196.
45. Hulme, A. E., Bogerd, H. P., Cullen, B. R., and Moran, J. V. (2007) Selective inhibition of Alu retrotransposition by APOBEC3G. *Gene* **390**, 199–205.
46. Huthoff, H., and Malim, M. H. (2007) Identification of amino acid residues in APOBEC3G required for regulation by HIV-1 Vif and virion encapsidation. *J Virol* **81**, 3807–3815.
47. Janini, M., Rogers, M., Birx, D. R., and McCutchan, F. E. (2001) Human immunodeficiency virus type 1 DNA sequences genetically damaged by hypermutation are often abundant in patient peripheral blood mononuclear cells and may be generated during near-simultaneous infection and activation of CD4(+) T cells. *J Virol* **75**, 7973–7986.
48. Jarmuz, A., Chester, A., Bayliss, J., Gisbourne, J., Dunham, I., Scott, J., and Navaratnam, N. (2002) An anthropoid-specific locus of orphan C to U RNA-editing enzymes on chromosome 22. *Genomics* **79**, 285–296.
49. Jin, X., Brooks, A., Chen, H., Bennett, R., Reichman, R., and Smith, H. (2005) APOBEC3G/CEM15 (hA3G) mRNA levels associate inversely with human immunodeficiency virus viremia. *J Virol* **79**, 11513–11516.
50. Kao, S., Khan, M. A., Miyagi, E., Plishka, R., Buckler-White, A., and Strebel, K. (2003) The human immunodeficiency virus type 1 Vif protein reduces intracellular expression and inhibits packaging of APOBEC3G (CEM15), a cellular inhibitor of virus infectivity. *J Virol* **77**, 11398–11407.
51. Kazazian, H. H., Jr (2004) Mobile elements: drivers of genome evolution. *Science* **303**, 1626–1632.
52. Kedzierska, K., and Crowe, S. M. (2002) The role of monocytes and macrophages in the pathogenesis of HIV-1 infection. *Curr Med Chem* **9**, 1893–1903.
53. Kieffer, T. L., Kwon, P., Nettles, R. E., Han, Y., Ray, S. C., and Siliciano, R. F. (2005) G→A hypermutation in protease and reverse transcriptase regions of human immunodeficiency virus type 1 residing in resting CD4+ T cells in vivo. *J Virol* **79**, 1975–1980.
54. Kobayashi, M., Takaori-Kondo, A., Miyauchi, Y., Iwai, K., and Uchiyama, T. (2005) Ubiquitination of APOBEC3G by an HIV-1 Vif-Cullin5-Elongin B-Elongin C complex is essential for Vif function. *J Biol Chem* **280**, 18573–18578.
55. Kobayashi, M., Takaori-Kondo, A., Shindo, K., Abudu, A., Fukunaga, K., and Uchiyama, T. (2004) APOBEC3G targets specific virus species. *J Virol* **78**, 8238–8244.
56. Korin, Y. D., and Zack, J. A. (1998) Progression to the G1b phase of the cell cycle is required for completion of human immunodeficiency virus type 1 reverse transcription in T cells. *J Virol* **72**, 3161–3168.
57. Kozak, S. L., Marin, M., Rose, K. M., Bystrom, C., and Kabat, D. (2006) The anti-HIV-1 editing enzyme APOBEC3G binds HIV-1 RNA and messenger RNAs that shuttle between polysomes and stress granules. *J Biol Chem* **281**, 29105–29119.
58. Kreisberg, J. F., Yonemoto, W., and Greene, W. C. (2006) Endogenous factors enhance HIV infection of tissue naive CD4 T cells by stimulating high molecular mass APOBEC3G complex formation. *J Exp Med* **203**, 865–870.
59. Larsson, M. (2005) HIV-1 and the hijacking of dendritic cells: A tug of war. *Springer Semin Immunopathol* **26**, 309–328.
60. Lecossier, D., Bouchonnet, F., Clavel, F., and Hance, A. J. (2003) Hypermutation of HIV-1 DNA in the absence of the Vif protein. *Science* **300**, 1112.
61. Lei, Y. C., Hao, Y. H., Zhang, Z. M., Tian, Y. J., Wang, B. J., Yang, Y., Zhao, X. P., Lu, M. J., Gong, F. L., and Yang, D. L. (2006) Inhibition of hepatitis B virus replication by APOBEC3G in vitro and in vivo. *World J Gastroenterol* **12**, 4492–4497.
62. Liddament, M. T., Brown, W. L., Schumacher, A. J., and Harris, R. S. (2004) APOBEC3F properties and hypermutation preferences indicate activity against HIV-1 in vivo. *Curr Biol* **14**, 1385–1391.

63. Liu, B., Yu, X., Luo, K., Yu, Y., and Yu, X. F. (2004) Influence of primate lentiviral Vif and proteasome inhibitors on human immunodeficiency virus type 1 virion packaging of APOBEC3G. *J Virol* **78**, 2072–2081.
64. Lochelt, M., Romen, F., Bastone, P., Muckenfuss, H., Kirchner, N., Kim, Y. B., Truyen, U., Rosler, U., Battenberg, M., Saib, A., Flory, E., Cichutek, K., and Munk, C. (2005) The antiretroviral activity of APOBEC3 is inhibited by the foamy virus accessory Bet, protein. *Proc Natl Acad Sci USA* **102**, 7982–7987.
65. Luo, K., Liu, B., Xiao, Z., Yu, Y., Yu, X., Gorelick, R., and Yu, X. F. (2004) Amino-terminal region of the human immunodeficiency virus type 1 nucleocapsid is required for human APOBEC3G packaging. *J Virol* **78**, 11841–11852.
66. Luo, K., Wang, T., Liu, B., Tian, C., Xiao, Z., Kappes, J., and Yu, X. F. (2007) Cytidine deaminases APOBEC3G and APOBEC3F interact with human immunodeficiency virus type 1 integrase and inhibit proviral DNA formation. *J Virol* **81**, 7238–7248.
67. Luo, K., Xiao, Z., Ehrlich, E., Yu, Y., Liu, B., Zheng, S., and Yu, X. F. (2005) Primate lentiviral virion infectivity factors are substrate receptors that assemble with, cullin 5-E3 ligase through a HCCH motif to suppress APOBEC3G. *Proc Natl Acad Sci U S A* **102**, 11444–11449.
68. Madani, N., and Kabat, D. (1998) An endogenous inhibitor of human immunodeficiency virus in human lymphocytes is overcome by the viral Vif protein. *J Virol* **72**, 10251–10255.
69. Mangeat, B., Turelli, P., Caron, G., Friedli, M., Perrin, L., and Trono, D. (2003) Broad antiretroviral defence by human APOBEC3G through lethal editing of nascent reverse transcripts. *Nature* **424**, 99–103.
70. Mangeat, B., Turelli, P., Liao, S., and Trono, D. (2004) A single amino acid determinant governs the species-specific sensitivity of APOBEC3G to Vif action. *J Biol Chem* **279**, 14481–14483.
71. Mariani, R., Chen, D., Schrofelbauer, B., Navarro, F., Konig, R., Bollman, B., Munk, C., Nymark-McMahon, H., and Landau, N. R. (2003) Species-specific exclusion of APOBEC3G from HIV-1 virions by Vif. *Cell* **114**, 21–31.
72. Marin, M., Rose, K. M., Kozak, S. L., and Kabat, D. (2003) HIV-1 Vif protein binds the editing enzyme APOBEC3G and induces its degradation. *Nat Med* **9**, 1398–1403.
73. Mbisa, J. L., Barr, R., Thomas, J. A., Vandegraaff, N., Dorweiler, I. J., Svarovskaia, E. S., Brown, W. L., Mansky, L. M., Gorelick, R. J., Harris, R. S., Engelman, A., and Pathak, V. K. (2007) Human immunodeficiency virus type 1 cDNAs produced in the presence of APOBEC3G exhibit defects in plus-strand DNA transfer and integration. *J Virol* **81**, 7099–7110.
74. McElrath, M. J. Steinman, R. M., and Cohn, Z. A. (1991) Latent HIV-1 infection in enriched populations of blood monocytes and T cells from seropositive patients. *J Clin Invest* **87**, 27–30.
75. Mehle, A., Goncalves, J., Santa-Marta, M., McPike, M., and Gabuzda, D. (2004) Phosphorylation of a novel SOCS-box regulates assembly of the HIV-1 Vif-Cul5 complex that promotes APOBEC3G degradation. *Genes Dev* **18**, 2861–2866.
76. Mehle, A., Strack, B., Ancuta, P., Zhang, C., McPike, M., and Gabuzda, D. (2004) Vif overcomes the innate antiviral activity of APOBEC3G by promoting its degradation in the ubiquitin-proteasome pathway. *J Biol Chem* **279**, 7792–7798.
77. Mehle, A., Thomas, E. R., Rajendran, K. S., and Gabuzda, D. (2006) A zinc-binding region in Vif binds Cul5 and determines cullin selection. *J Biol Chem* **281**, 17259–17265.
78. Mehta, A., Kinter, M. T., Sherman, N. E., and Driscoll, D. M. (2000) Molecular cloning of apobec-1 complementation factor, a novel RNA-binding protein involved in the editing of apolipoprotein B mRNA. *Mol Cell Biol* **20**, 1846–1854.
79. Mikl, M. C., Watt, I. N., Lu, M., Reik, W., Davies, S. L., Neuberger, M. S., and Rada, C. (2005) Mice deficient in APOBEC2 and APOBEC3. *Mol Cell Biol* **25**, 7270–7277.
80. Muckenfuss, H., Hamdorf, M., Held, U., Perkovic, M., Lower, J., Cichutek, K., Flory, E., Schumann, G. G., and Munk, C. (2006) APOBEC3 proteins inhibit human LINE-1 retrotransposition. *J Biol Chem* **281**, 22161–22172.
81. Muckenfuss, H., Kaiser, J. K., Krebil, E., Battenberg, M., Schwer, C., Cichutek, K., Munk, C., and Flory, E. (2007) Sp1 and Sp3 regulate basal transcription of the human APOBEC3G gene. *Nucleic Acids Res* **35**, 3784–3796.

82. Muramatsu, M., Kinoshita, K., Fagarasan, S., Yamada, S., Shinkai, Y., and Honjo, T. (2000) Class switch recombination and hypermutation require activation-induced cytidine deaminase (AID), a potential RNA editing enzyme. *Cell* **102**, 553–563.
83. Muramatsu, M., Sankaranand, V. S., Anant, S., Sugai, M., Kinoshita, K., Davidson, N. O., and Honjo, T. (1999) Specific expression of activation-induced cytidine deaminase (AID), a novel member of the RNA-editing deaminase family in germinal center B cells. *J Biol Chem* **274**, 18470–18476.
84. Neil, S., Martin, F., Ikeda, Y., and Collins, M. (2001) Postentry restriction to human immunodeficiency virus-based vector transduction in human monocytes. *J Virol* **75**, 5448–5456.
85. Newman, E. N., Holmes, R. K., Craig, H. M., Klein, K. C., Lingappa, J. R., Malim, M. H., and Sheehy, A. M. (2005) Antiviral function of APOBEC3G can be dissociated from cytidine deaminase activity. *Curr Biol* **15**, 166–170.
86. Noguchi, C., Hiraga, N., Mori, N., Tsuge, M., Imamura, M., Takahashi, S., Fujimoto, Y., Ochi, H., Abe, H., Maekawa, T., Yatsuji, H., Shirakawa, K., Takaori-Kondo, A., and Chayama, K. (2007) Dual effect of APOBEC3G on hepatitis B virus. *J Gen Virol* **88**, 432–440.
87. Noguchi, C., Ishino, H., Tsuge, M., Fujimoto, Y., Imamura, M., Takahashi, S., and Chayama, K. (2005) G to A hypermutation of hepatitis B virus. *Hepatology* **41**, 626–633.
88. OhAinle, M., Kerns, J. A., Malik, H. S., and Emerman, M. (2006) Adaptive evolution and antiviral activity of the conserved mammalian cytidine deaminase APOBEC3H. *J Virol* **80**, 3853–3862.
89. Peng, G., Greenwell-Wild, T., Nares, S., Jin, W., Lei, K. J., Rangel, Z. G., Munson, P. J., and Wahl, S. M. (2007) Myeloid differentiation and susceptibility to HIV-1 are linked to APOBEC3 expression. *Blood* **110**, 393–400.
90. Peng, G., Lei, K. J., Jin, W., Greenwell-Wild, T., and Wahl, S. M. (2006) Induction of APOBEC3 family proteins, a defensive maneuver underlying interferon-induced anti-HIV-1 activity. *J Exp Med* **203**, 41–46.
91. Petersen-Mahrt, S. K., Harris, R. S., and Neuberger, M. S. (2002) AID mutates *E. coli* suggesting a DNA deamination mechanism for antibody diversification. *Nature* **418**, 99–103.
92. Pierson, T. C., Zhou, Y., Kieffer, T. L., Ruff, C. T., Buck, C., and Siliciano, R. F. (2002) Molecular characterization of preintegration latency in human immunodeficiency virus type 1 infection. *J Virol* **76**, 8518–8531.
93. Powell, L. M., Wallis, S. C., Pease, R. J., Edwards, Y. H., Knott, T. J., and Scott, J. (1987) A novel form of tissue-specific RNA processing produces apolipoprotein-B48 in intestine. *Cell* **50**, 831–840.
94. Ribeiro, A. C., Maia e Silva, A., Santa-Marta, M., Pombo, A., Moniz-Pereira, J., Goncalves, J., and Barahona, I. (2005) Functional analysis of Vif protein shows less restriction of human immunodeficiency virus type 2 by APOBEC3G. *J Virol* **79**, 823–833.
95. Rose, K. M., Marin, M., Kozak, S. L., and Kabat, D. (2004) Transcriptional regulation of APOBEC3G, a cytidine deaminase that hypermutates human immunodeficiency virus. *J Biol Chem* **279**, 41744–41749.
96. Rosler, C., Kock, J., Kann, M., Malim, M. H., Blum, H. E., Baumert, T. F., and von Weizsacker, F. (2005) APOBEC-mediated interference with hepadnavirus production. *Hepatology* **42**, 301–309.
97. Rosler, C., Kock, J., Malim, M. H., Blum, H. E., and von Weizsacker, F. (2004) Comment on "Inhibition of hepatitis B virus replication by APOBEC3G." *Science* **305**, 1403; author reply on, p1403.
98. Russell, R. A., Wiegand, H. L., Moore, M. D., Schafer, A., McClure, M. O., and Cullen, B. R. (2005) Foamy virus Bet proteins function as novel inhibitors of the APOBEC3 family of innate antiretroviral defense factors. *J Virol* **79**, 8724–8731.
99. Sarkis, P. T., Ying, S., Xu, R., and Yu, X. F. (2006) STAT1-independent cell type-specific regulation of antiviral APOBEC3G by IFN-alpha. *J Immunol* **177**, 4530–4540.
100. Sasada, A., Takaori-Kondo, A., Shirakawa, K., Kobayashi, M., Abudu, A., Hishizawa, M., Imada, K., Tanaka, Y., and Uchiyama, T. (2005) APOBEC3G targets human T-cell leukemia virus type 1. *Retrovirology* **2**, 32.
101. Schafer, A., Bogerd, H. P., and Cullen, B. R. (2004) Specific packaging of APOBEC3G into HIV-1 virions is mediated by the nucleocapsid domain of the gag polyprotein precursor. *Virology* **328**, 163–168.
102. Schrofelbauer, B., Chen, D., and Landau, N. R. (2004) A single amino acid of APOBEC3G controls its species-specific interaction with virion, infectivity factor (Vif). *Proc Natl Acad Sci USA* **101**, 3927–3932.

103. Schumacher, A. J., Nissley, D. V., and Harris, R. S. (2005) APOBEC3G hypermutates genomic DNA and inhibits Ty1 retrotransposition in yeast. *Proc Natl Acad Sci USA* **102**, 9854–9859.
104. Sheehy, A. M., Gaddis, N. C., Choi, J. D., and Malim, M. H. (2002) Isolation of a human gene that inhibits HIV-1 infection and is suppressed by the viral Vif protein. *Nature* **418**, 646–650.
105. Sheehy, A. M., Gaddis, N. C., and Malim, M. H. (2003) The antiretroviral enzyme APOBEC3G is degraded by the proteasome in response to HIV-1 Vif. *Nat Med* **9**, 1404–1407.
106. Shirakawa, K., Takaori-Kondo, A., Kobayashi, M., Tomonaga, M., Izumi, T., Fukunaga, K., Sasada, A., Abudu, A., Miyauchi, Y., Akari, H., Iwai, K., and Uchiyama, T. (2006) Ubiquitination of APOBEC3 proteins by the Vif-Cullin5-ElonginB-ElonginC complex. *Virology* **344**, 263–266.
107. Simon, J. H., Gaddis, N. C., Fouchier, R. A., and Malim, M. H. (1998) Evidence for a newly discovered cellular anti-HIV-1 phenotype. *Nat Med* **4**, 1397–1400.
108. Simon, J. H., and Malim, M. H. (1996) The human immunodeficiency virus type 1 Vif protein modulates the postpenetration stability of viral nucleoprotein complexes. *J Virol* **70**, 5297–5305.
109. Simon, V., Zennou, V., Murray, D., Huang, Y., Ho, D. D., and Bieniasz, P. D. (2005) Natural variation in Vif: differential impact on APOBEC3G/3F and a potential role in HIV-1 diversification. *PLoS Pathog* **1**, e6.
110. Soros, V. B., Yonemoto, W., and Greene, W. C. (2007) Newly synthesized APOBEC3G is incorporated into HIV virions, inhibited by HIV RNA, and subsequently activated by RNase H. *PLoS Pathog* **3**, e15.
111. Stenglein, M. D., and Harris, R. S. (2006) APOBEC3B and APOBEC3F inhibit L1 retrotransposition by a DNA deamination-independent mechanism. *J Biol Chem* **281**, 16837–16841.
112. Stevenson, M., Stanwick, T. L., Dempsey, M. P., and Lamonica, C. A. (1990) HIV-1 replication is controlled at the level of T cell activation and proviral integration. *Embo J* **9**, 1551–1560.
113. Stopak, K., de Noronha, C., Yonemoto, W., and Greene, W. C. (2003) HIV-1 Vif blocks the antiviral activity of APOBEC3G by impairing both its translation and intracellular stability. *Mol Cell* **12**, 591–601.
114. Stopak, K. S., Chiu, Y. L., Kropp, J., Grant, R. M., and Greene, W. C. (2007) Distinct patterns of cytokine regulation of APOBEC3G expression and activity in primary lymphocytes, macrophages, and dendritic cells. *J Biol Chem* **282**, 3539–3546.
115. Strebel, K., Daugherty, D., Clouse, K., Cohen, D., Folks, T., and Martin, M. A. (1987) The HIV 'A' (sor) gene product is essential for virus infectivity. *Nature* **328**, 728–730.
116. Suspene, R., Guetard, D., Henry, M., Sommer, P., Wain-Hobson, S., and Vartanian, J. P. (2005) Extensive editing of both hepatitis B virus DNA strands by APOBEC3 cytidine, deaminases in vitro and in vivo. *Proc Natl Acad Sci USA* **102**, 8321–8326.
117. Svarovskaia, E. S., Xu, H., Mbisa, J. L., Barr, R., Gorelick, R. J., Ono, A., Freed, E. O., Hu, W. S., and Pathak, V. K. (2004) Human apolipoprotein B mRNA-editing enzyme-catalytic polypeptide-like 3G (APOBEC3G) is incorporated into HIV-1 virions through interactions with viral and nonviral RNAs. *J Biol Chem* **279**, 35822–35828.
118. Ta, V. T., Nagaoka, H., Catalan, N., Durandy, A., Fischer, A., Imai, K., Nonoyama, S., Tashiro, J., Ikegawa, M., Ito, S., Kinoshita, K., Muramatsu, M., and Honjo, T. (2003) AID mutant analyses indicate requirement for class-switch-specific cofactors. *Nat Immunol* **4**, 843–848.
119. Tanaka, Y., Marusawa, H., Seno, H., Matsumoto, Y., Ueda, Y., Kodama, Y., Endo, Y., Yamauchi, J., Matsumoto, T., Takaori-Kondo, A., Ikai, I., and Chiba, T. (2006) Anti-viral protein APOBEC3G is induced by interferon-alpha stimulation in human hepatocytes. *Biochem Biophys Res Commun* **341**, 314–319.
120. Teng, B., Burant, C. F., and Davidson, N. O. (1993) Molecular cloning of an apolipoprotein B messenger RNA editing protein. *Science* **260**, 1816–1819.
121. Triques, K., and Stevenson, M. (2004) Characterization of restrictions to human immunodeficiency virus type 1 infection of monocytes. *J Virol* **78**, 5523–5527.
122. Turelli, P., Mangeat, B., Jost, S., Vianin, S., and Trono, D. (2004) Inhibition of hepatitis B virus replication by APOBEC3G. *Science* **303**, 1829.
123. Turelli, P., Vianin, S., and Trono, D. (2004) The innate antiretroviral factor APOBEC3G does not affect human LINE-1 retrotransposition in a cell culture assay. *J Biol Chem* **279**, 43371–43373.
124. Vartanian, J. P., Meyerhans, A., Asjo, B., and Wain-Hobson, S. (1991) Selection, recombination, and G—A hypermutation of human immunodeficiency virus type 1 genomes. *J Virol* **65**, 1779–1788.

125. von Schwedler, U., Song, J., Aiken, C., and Trono, D. (1993) Vif is crucial for human immunodeficiency virus type 1 proviral DNA synthesis in infected cells. *J Virol* **67**, 4945–4955.
126. Wichroski, M. J., Ichiyama, K., and Rana, T. M. (2005) Analysis of HIV-1 viral infectivity factor-mediated proteasome-dependent depletion of APOBEC3G: Correlating function and subcellular localization. *J Biol Chem* **280**, 8387–8396.
127. Wichroski, M. J., Robb, G. B., and Rana, T. M. (2006) Human retroviral host restriction factors APOBEC3G and APOBEC3F localize to mRNA processing bodies. *PLoS Pathog* **2**, e41.
128. Wiegand, H. L., Doehle, B. P., Bogerd, H. P., and Cullen, B. R. (2004) A second human antiretroviral factor, APOBEC3F, is suppressed by the HIV-1 and HIV-2 Vif proteins. *Embo J* **23**, 2451–2458.
129. Xu, H., Svarovskaia, E. S., Barr, R., Zhang, Y., Khan, M. A., Strebel, K., and Pathak, V. K. (2004) A single amino acid substitution in human APOBEC3G antiretroviral enzyme confers, resistance to HIV-1 virion infectivity factor-induced depletion. *Proc Natl Acad Sci USA* **101**, 5652–5657.
130. Yu, Q., Chen, D., Konig, R., Mariani, R., Unutmaz, D., and Landau, N. R. (2004) APOBEC3B and APOBEC3C are potent inhibitors of simian immunodeficiency virus replication. *J Biol Chem* **279**, 53379–53386.
131. Yu, Q., Konig, R., Pillai, S., Chiles, K., Kearney, M., Palmer, S., Richman, D., Coffin, J. M., and Landau, N. R. (2004) Single-strand specificity of APOBEC3G accounts for minus-strand deamination of the HIV genome. *Nat Struct Mol Biol* **11**, 435–442.
132. Yu, X., Yu, Y., Liu, B., Luo, K., Kong, W., Mao, P., and Yu, X. F. (2003) Induction of APOBEC3G ubiquitination and degradation by an HIV-1 Vif-Cul5-SCF complex. *Science* **302**, 1056–1060.
133. Yu, Y., Xiao, Z., Ehrlich, E. S. Yu, X. and Yu. X. F. (2004), Selective assembly of HIV-1 Vif-Cul5-ElonginB-ElonginC E3 ubiquitin ligase complex through a novel SOCS box and upstream cysteines. *Genes Dev* **18**, 2867–2872.
134. Zack, J. A., Arrigo, S. J., Weitsman, S. R., Go, A. S., Haislip, A., and Chen, I. S. (1990) HIV-1 entry into quiescent primary lymphocytes: molecular analysis reveals a labile, latent viral structure. *Cell* **61**, 213–222.
135. Zennou, V., Perez-Caballero, D., Gottlinger, H., and Bieniasz, P. D. (2004) APOBEC3G incorporation into human immunodeficiency virus type 1 particles. *J Virol* **78**, 12058–12061.
136. Zhang, H., Yang, B., Pomerantz, R. J., Zhang, C., Arunachalam, S. C., and Gao, L. (2003) The cytidine deaminase CEM15 induces hypermutation in newly synthesized HIV-1 DNA. *Nature* **424**, 94–98.
137. Zheng, Y. H., Irwin, D., Kurosu, T., Tokunaga, K., Sata, T., and Peterlin, B. M. (2004) Human APOBEC3F is another host factor that blocks human immunodeficiency virus type 1 replication. *J Virol* **78**, 6073–6076.

PART III

PREDICTIVE STRUCTURES

CHAPTER 12

A-TO-I EDITING OF Alu REPEATS

Eli Eisenberg
Erez Y. Levanon

A-TO-I RNA editing by members of the ADAR (adenosine deaminases acting on RNA) family is a posttranscriptional processes that alters the information encoded in gene transcripts. The phenomenon has been known for some time but was considered to be rare. In recent years, a number of computational approaches have been developed for detection and prediction of A-to-I RNA editing sites. These studies have revealed that in fact most human genes are subjected to A-to-I RNA editing, mainly within their untranslated regions and introns. The editing events occur primarily in *Alu* repeats, primate-specific repetitive elements that comprise ~10% of the human genome. The role of this abundant phenomenon is yet a mystery.

12.1 BACKGROUND

12.1.1 Indirect Evidence for Abundant A-to-I Editing

Following the discovery of the first A-to-I editing site in gluR (1), a handful of edited human genes were serendipitously identified. Some of these editing targets have been extensively studied, and their editing has been shown to be functionally crucial. Naturally, the possibility of many more such targets has attracted much interest. During the following decade, a number of tantalizing hints have been accumulating: Paul and Bass (2) have purified inosine monophosphate from cellular RNA and quantified its relative abundance in various tissues. They observed surprisingly high inosine abundance, up to 1 : 17,000 nt in rat's brain, certainly far more than can be accounted for by the few mammalian ADAR substrates known at the time. Moreover, the inosine abundance correlated well with the levels of ADAR1 and ADAR2 mRNA found in the various tissues, strongly suggesting the inosines to be the product of A-to-I editing events. The above result translates to at least one adenosine being modified by editing per 100 brain RNAs. The possibility of even a fraction of these modifications to

RNA and DNA Editing: Molecular Mechanisms and Their Integration into Biological Systems,
Edited by Harold C. Smith

result in functionally important genetic recoding event has raised hope that a systematic search could uncover many more edited targets, revealing a whole new world of RNA and protein diversity and regulation beyond genomic information.

12.1.2 Early Screens for A-to-I Editing Targets

In an effort to identify additional ADAR substrates, Morse *et al.* (3) have developed a method to identify inosine-containing RNAs using specific cleavage reaction and applied it to discover five novel human edited transcripts. Interestingly, all of these editing sites do not occur within codons, but rather in untranslated regions (UTRs) of coding RNAs or in noncoding RNAs. In a subsequent work (4), five more *C. elegans* substrates have been identified, including three mRNAs that encode proteins known to affect neuronal functions. Applying the same method to human brain resulted in 19 ADAR substrates (4). All substrates found exhibit a common structure: They are edited in long stem-loop structures located in noncoding regions, mostly in repetitive elements. Typically, these targets are not edited in a single location but rather undergo deamination of multiple adenosines. This was the first hint suggesting genetic recoding by A-to-I conversion in coding regions to be the exception rather than the rule.

Another indication for transcriptome-wide A-to-I RNA editing emerged in the same year from the Kazusa cDNA sequencing project (5), in which thousands of human novel large (>4 kb) cDNAs were sequenced. In the course of this project, tables of nucleotide substitutions and insertions/deletions found between each cDNA sequence and the corresponding genome sequence have been constructed. A prominent feature of the nucleotide substitutions between the cDNA and genomic sequences is the predominance of A-to-G changes (from genome to cDNA), which is difficult to explain by single nucleotide polymorphisms, reverse transcription errors, and/or sequencing errors. Most cDNAs of this project were derived from human brain, which was known to be rich in adenosine deaminase activity specific to double-stranded RNA (2). Thus, it was suggested that deamination by ADARs may be a plausible explanation for the predominance of clusters of A-to-G substitutions. Another sign for A-to-I editing being responsible for the overrepresentation of A-to-G mismatches came from the observation that these substitutions were frequently clustered in a relatively short region, and most of the clusters resided in *Alu* elements. In particular, clusters were frequently observed in regions including two *Alu* elements inversely oriented. These inversely oriented *Alu* elements are very likely to form a double-stranded intrastructure in the mRNA, providing the link to the dsRNA-specific ADARs.

Most of the editing sites found by Morse *et al.*, as well as almost all A-to-G mismatch clusters found by the Kazusa cDNA sequencing project, were located in noncoding regions. Thus, these observations suggest editing to occur mostly in nonconserved genomic regions. Paradoxically, the first few chance discoveries of editing targets in coding regions reside within ultra-conserved regions, even more conserved than a typical coding sequence. This higher-than-usual conservation can be explained by the evolutionary constraint of keeping the RNA secondary structure required for RNA editing in addition to conserving the protein structure that allows for synonymous changes. Therefore, in a complementary attempt to identify A-to-I editing sites, Hoopengardner *et al.* (6) have sequenced 41 highly conserved exons in *Drosophila* and found 16 of them to undergo A-to-I conversion. Applying the same strategy to the mammalian Shaker genes resulted in one additional, highly conserved, editing site within the coding region of the *KCNA1* gene.

12.1.3 The *Alu* Repeats

A significant number of the newly discovered ADAR substrates reside within *Alu* repeats. Almost half of the human genome comprises mobile repetitive elements that were introduced by retrotransposition through RNA intermediates. These repetitive elements encompass autonomous long terminal repeats (LTRs), long interspersed elements (LINEs), processed pseudo-genes, and the most abundant, nonautonomous short interspersed elements (SINEs). SINEs depend on the open reading frame of LINEs for amplification, because they have no protein coding ability (7). The most common SINE in humans is, by far, the primate-specific *Alu* sequence, with about 1.2 million copies comprising more than one-tenth of the human genome. *Alu* elements are short interspersed elements (SINEs), typically 280 nucleotides long. The prototype *Alu* structure is a tandem dimer in which two monomers are linked by an A-rich region. *Alu* elements are divided into several subfamilies whose relative ages are estimated based on sequence divergence from an *Alu* consensus sequence (8). During the early steps of primate evolution, the rate of *Alu* amplification was about one in every primate birth (9), but this rate has now decreased by two orders of magnitude (10). Indeed, several thousand *Alu* elements have integrated into the human genome since the divergence of mankind from the African apes (11), although very few are still potentially active (12).

The genomic distribution of *Alu* elements is nonuniform, with a strong bias toward GC-rich and gene-rich regions (13). *Alu* does not encode any functional protein, is transcribed by RNA polymerase III, and is thought to use the reverse transcriptase of another non-LTR retrotransposon, LINE-1, to make cDNA copies and retro-transpose back to the host genome (7). The vast majority of integrated *Alu*s seem to have no influence on the genome, and no obvious function has been assigned to them. Despite being considered genetically functionless, *Alu* elements were suggested to have broad evolutionary impacts. Being clustered in gene-rich regions, *Alu* repeats are capable of modifying genes through the insertion of mutations, gene conversion, and recombination (10). They can also disrupt promoter regions, change methylation status, insert new regulatory features, interfere with alternative splicing, and regulate the initiation of translation (10). *Alu* RNA levels have been reported to increase in response to cell stress (14). Finally, some *Alu* elements are incorporated into exons (15) and therefore directly influence the open reading frame of mature mRNAs.

Given the above characteristics of the *Alu* elements, their appearance in the few edited substrates identified by Morse *et al.* is well understood. The high abundance of the *Alu* repeats and their bias toward gene-rich regions make them a perfect target for ADARs. A transcript including two neighboring *Alu* repeats with inverted orientation might form a long (~280 nt), almost perfect, hairpin dsRNA structure, which is the ideal ADAR substrate (Figure 12.1) (16). As we shall see below, this mechanism is indeed highly abundant, and the results first observed by Morse *et al.* have merely revealed the tip of the iceberg.

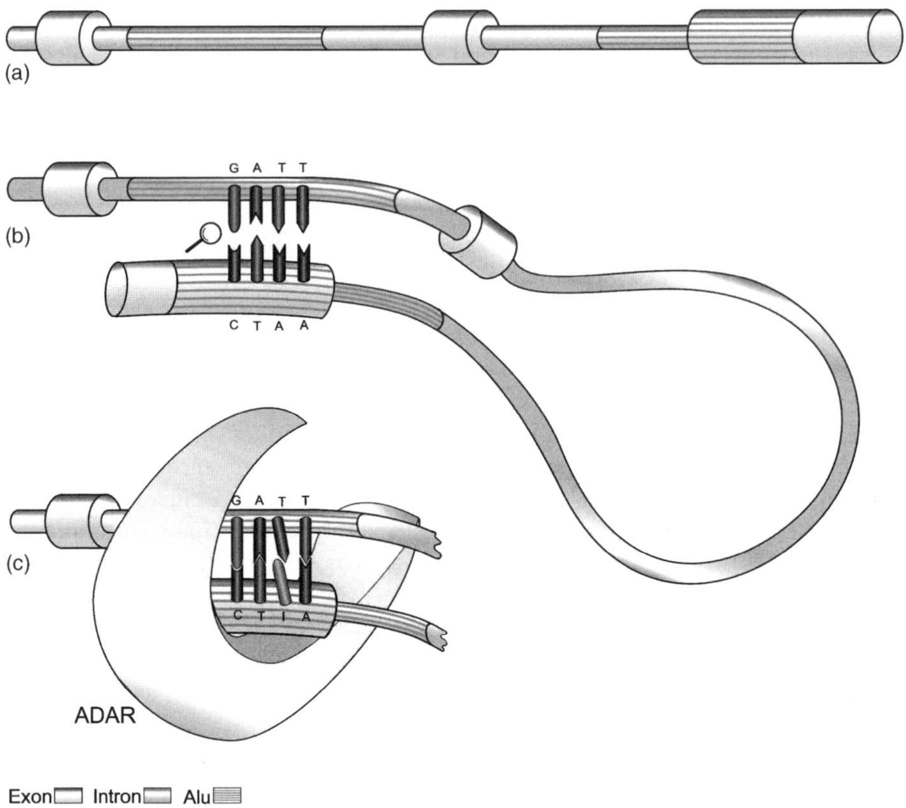

Figure 12.1 Editing in *Alu*. (**a**) Pre-mRNA as transcribed from DNA. The gene contains two *Alu* repeats with opposite orientations, one of which overlaps with an exon. (**b**) The two oppositely oriented *Alu* sequences form a dsRNA structure. (**c**) An enzyme of the ADAR family edits some of the adenosines in the dsRNA structure into inosines. [Used with permission from Levanon *et al.* 20.]

Alu elements are found in all primates but in no other organism. Therefore, they were suggested to have a role in primate evolution. However, the nature of this role is still under debate. The intriguing coupling between A-to-I editing and *Alu* repeats is a unique phenomenon to the primate genome, which implies that it is connected to the evolution of this mammalian branch.

12.2 COMPUTATIONAL DETECTION OF A-TO-I EDITING

12.2.1 Sifting through db EST

In principle, editing may be detected using the large-scale database of ESTs (expressed sequence tags) and RNAs (17), which currently holds over 8 million human records. Editing sites show up when a sequence is aligned with the genome: While the adenosine in the DNA reads A, sequencing identifies the inosine in the edited site

TCAAGATCAGCCTGACCAACATGGAGAAACCCTACTAAAAATACAAAGTTAGCCAGGCATAGTGG	Genome
TC***GG***GATCAGCCTGACCAACATGGAGAAACCCTACT***GGG***AATACA***G***AGTTAGCCAGGCAT***G***GTGG	AI093487
TC***G***AGATCAGCCTGACCAACATGGAGAAACCCTACT***G***AAAATACAAAGTTAGCCAGGCAT***G***GTGG	BF771639
TC***GG***GATCAGCCTGACCAACATGGAGAAACCCTACT***G***AAAAT***G***CAAAGTTAGCCAGGCAT***G***GTGG	BM681047
TCA***G***GATCAGCCTGACCAACATGGAGAAACCCTACT***G***N***G***AATACA***G***AGTTAGCCAGGCAT***G***GTGG	BQ307221
TCA***G***GATCAGCCTGACCAACATGGAGAAACCCTACT***GG***AAATACAAAGTTAGCCAGGCAT***G***GTGG	R01692
TCA***G***GATCAGCCTGACCAACATGGAGAAACCCTACT***GGG***AATACA***G***AGTTAGCCAGGCAT***G***GTGG	BQ305305
TCA***G***GATCAGCCTGACCAACATGGAGAAACCCTACT***G***NGAATACA***G***AGTTAGCCAGGCAT***G***GTGG	BQ305460
TCG***G***GATCAGCCTGACCAACATGGAGAAACCCTACT***GG***AAATACAAAGTTAGCCAGGCAT***G***GTGG	AA101562
TCA***G***GATCAGCCTGACCAACATGGAGAAACCCTACT***GG***AAATACAAAGTTAGCCAGGCAT***G***GTGG	AW190875
TC***G***AGATCAGCCTGACCAACATGGAGAAACCCT***G***CT***G***AAAATACAAAGTTAGCCAGGCAT***G***GTGG	BE350662
TC***GG***GATCAGCCTGACCAACATGGAGAAACCCTACT***GGG***AATACA***G***AGTT***G***GCCAGGCAT***G***GTGG	AA149993
TCAAGATCAGCCTGACCAACATGGAGAAACCCTACTAAAAATACAAAGTTAGCCAGGCATAGTGG	AI925871
TCA***G***GATCAGCCTGACCAACATGGAGAAACCCTACT***GGG***AATACAAAGTTAGCCAGGCAT***G***GTGG	AI333843
TC***GG***GATCAGCCTGACCAACATGGAGAAACCCTACT***GG***AAATACAAAGTTAGCC***G***GGCAT***G***GTGG	AW338261

Figure 12.2 Editing in the F11 receptor (JAM1) gene. Some of the publicly available expressed sequences covering this gene, together with the corresponding genomic sequence. The evidence for editing is highlighted.

as guanosine (G) (see some examples in Figure 12.2). However, the poor sequencing quality of the sequence database [up to 3% sequencing errors (18)] precludes a straightforward application of this approach. Moreover, millions of single-nucleotide polymorphisms (SNPs) and mutations are erroneously identified as editing events by this method. Thus, large-scale identification of editing substrates by bioinformatics tools was previously considered practically impossible (19).

In recent years, three independent groups have launched systematic searches using various computational algorithms to overcome these challenges. All of these methods are based on the alignment of RNA transcripts with the genome, the identification of mismatches, and the elimination of sequencing errors, single-nucleotide polymorphisms (SNPs), mutations, and other "contaminating mismatches." The main difference between the three approaches is their use of either the larger database of more than six million expressed sequence tags (ESTs), which is of lower quality, with up to 3% sequencing errors (18), or the more accurate, but smaller, set of longer mRNAs (currently holding about 200,000 human transcripts).

ADAR substrates are usually imperfect dsRNA stems formed by base pairing of an exon containing the adenosine to be edited with a complementary portion of the pre-mRNA (up to several thousand nucleotides apart). Levanon *et al.* (20) have used this characteristic of the editing phenomenon and restricted the search for A-to-G mismatches to genomic regions which were bioinformatically identified as potential double-stranded regions. This procedure eliminated most of the noise and facilitated the identification of true editing sites.

First, the human ESTs and cDNAs were aligned to the genome and assembled into clusters representing genes or partial genes. Then, the expressed part of each gene was aligned with the corresponding genomic region, looking for reverse complement alignments longer than 30 nt with identity levels higher than 85%. About 429,000 candidate dsRNAs were found in 14,512 different genes, mostly resulting from alignment of an exon to an intron. In addition, the following cleaning procedures were employed to the ESTs and cDNAs supporting the stem region in order to further decrease the number of random mismatches, SNPs, and mutations: Since sequencing errors tend to cluster in certain regions, especially in low-complexity areas and toward

TABLE 12.1 Number of Potential Editing Sites as a Function of the Background Level

AG	GA	Noise-to-Signal Ratio
12,220	216	0.018
17,011	2,324	0.137
24,116	1,129	0.047
31,600	5,825	0.184
76,457	28,882	0.378

Numbers of A-to-G mismatches that pass the cleaning procedures of Levanon *et al.* (20). Most of these are assumed to result from A-to-I editing, and the noise level is estimated from the number of G-to-A mismatches passing the same filters. Different rows in the table correspond to various sets of parameters. In particular, the last one corresponds to the most liberal set of parameters, providing the largest set with an estimated number of ~50,000 true editing sites.

sequence ends, all single-letter repeats longer than 4 nt, as well as 150 nt at both ends of each sequence, were discarded (note that this procedure has resulted in losing many true RNA editing sites which reside within the poly-A part of the *Alu* sequence). In addition, all 50-nt-wide windows in which the total number of mismatches is six or more (excluding four or more identical sequential mismatches, to allow for clusters of editing events) were considered as having low sequencing quality and were discarded. In addition, mismatches supported by less than 2% of available sequences were also discarded, and known SNPs of genomic origin were removed. Finally, employing these criteria yielded putative editing sites that tend to group together, a fact that is also supported by the few known cases that were available from previous studies (3). Thus, all mismatches that occur fewer than three times in an exon were ignored. These cleaning procedures yielded 12,723 A-to-I substitutions in 1637 human genes (for sensitivity to parameters, see Table 12.1). The false-positive rate was estimated from the number of mismatches of other types which passed the above filters, and it was reported to be <5% (see Figure 12.3). This method fails to detect the known editing sites within coding regions. This is because the associated dsRNA are relatively short and are not detected by the algorithm. Moreover, these sites are isolated and not part of a cluster.

12.2.2 Clusters of Mismatches in RNAs

Other works have taken a different route, using only the full-length cDNA transcripts (21–23). These sequences are of much higher quality compared to ESTs, and thus most of the sequencing error noise is suppressed and the need for complicated cleaning procedures as described above is avoided. Kim *et al.* (23) have calculated for each transcript the weighted mean of observed rates of all substitutions other than A-to-G to, which was then used to build a statistical model for all substitutions (including A-to-G) for that transcript. This model yielded the expected number of A-to-G mismatches, assuming that they are described by the model. Then, for each transcript, a reduced transcript consisting only of the genomic A bases was formed. This reduced transcript was scanned to find the maximum number of A-to-G substitutions in a sliding window of size 80 (size was

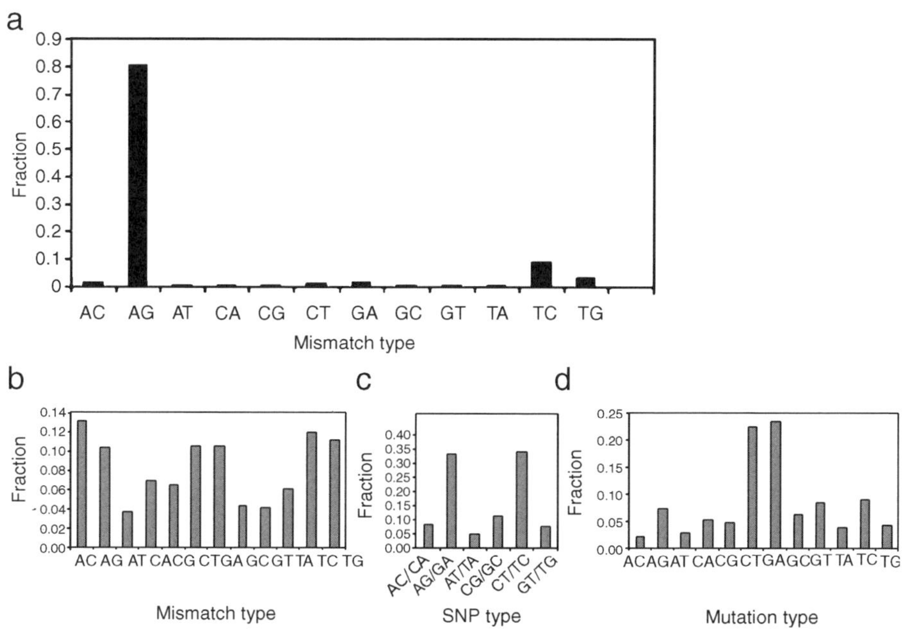

Figure 12.3 Distribution of mismatches between the DNA and the expressed RNA sequences that pass the cleaning algorithm. (**a**) Results of algorithm application to dsRNAs only. A-to-G mismatches clearly dominate the distribution. Notably, T-to-C mismatches are also overrepresented, likely being A-to-I editing events that were aligned to the opposite strand. (**b**) The distribution of mismatches resulting from applying the algorithm to random expressed sequences covering about 20% of the transcriptome. (**c**,**d**) The distributions for known SNP (77) (**c**) and mutations 78 (**d**), respectively. A-to-G mismatches do not stand out in the distributions b–d.

chosen because 80 As approximated the number of As within an *Alu* repeat). Using the statistical model, the probability for each transcript of the observed scan statistic was calculated, resulting in 2674 cDNA sequences with a significantly increased likelihood of A-to-G substitutions, with false-detection rate <0.01. These transcripts include 41,848 substitutions, most of which are A-to-G substitutions (35,085, 84%).

A similar approach was taken by Athanasiadis *et al.* (21): cDNAs derived from the Kazusa cDNA sequencing project (5) were searched for clusters of A-to-G mismatches. They have identified several examples of cDNA sequences harboring several such mismatches within a window of 200–300 nt. All clusters found, except one, were located within *Alu* repeat element in the cDNA.

Finally, Eisenberg *et al.* (22) have also looked at clusters of A-to-G mismatches within cDNA sequences, providing a second method to corroborate the earlier results in Levanon *et al.* UCSC alignments of human and mouse RNA sequences to their genomes were used (24), looking for clusters of mismatches of the same type. Eisenberg *et al.* scanned 128,068 human RNA sequences (total length 259 Mbp) and 102,895 mouse RNA sequences (total length 198 Mbp). Their distribution is presented

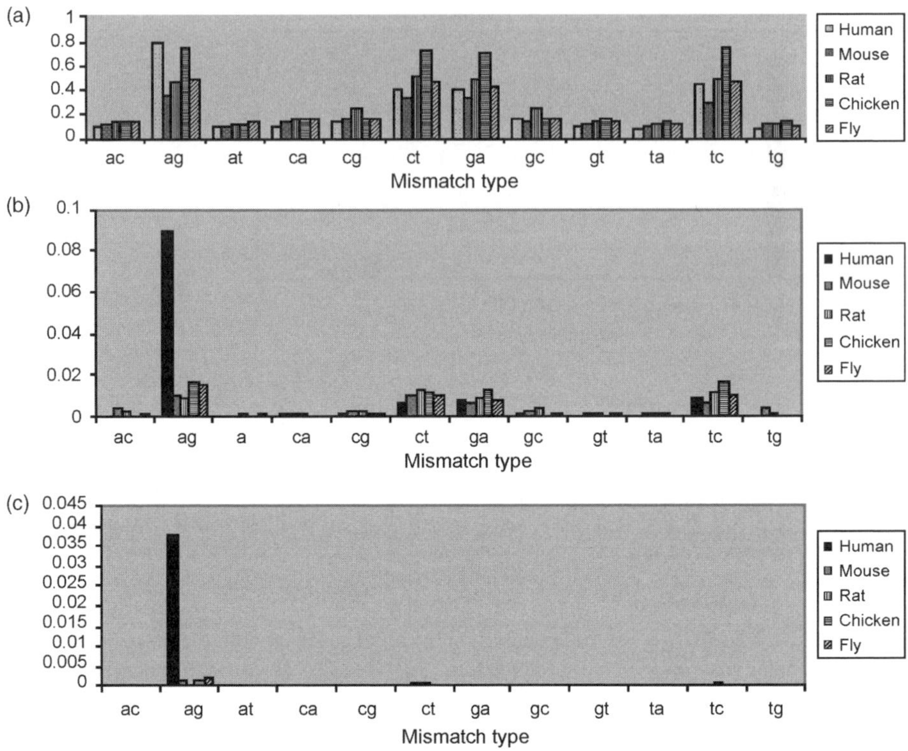

Figure 12.4 Mismatches distributions: a multi-species comparison. (**a**) Single mismatches; (**b**) instances of consecutive three identical mismatches per sequence; and (**c**) instances of consecutive five identical mismatches per sequence. A comparison of the average number of mismatches per sequence in human and mouse RNA sequences shows a significant overabundance of A-to-G mismatches, most probably representing A-to-I editing, in humans but not in the other four organisms that were examined. The distribution of other types of mismatches is similar in all of the organisms examined. Note that the background mismatch rate in chicken RNAs is significantly higher, presumably as a result of the lower quality of its current genome draft. *X*-axis: types of mismatch. The label xy refers to x in the DNA sequence and y in the expressed sequence [e.g., ac refers to genomic A(s), which reads as C(s) in the expressed data]. *Y*-axis: the average number of instances of consecutive one, three, or five mismatches per sequence.

in Figure 12.4. Even a simple count of all mismatches exhibits a vast overrepresentation of A-to-G mismatches in human sequences, suggesting ~50,000 inosines in these sequences (~1 : 5200 bp), as was found also by Athanasiadis *et al.* (21). Looking for sequences that exhibit three or more consecutive identical mismatches, a vast overrepresentation of A-to-G consecutive mismatches was observed in human, but not in mouse. A set of 4864 human RNA sequences with three or more consecutive A-to-G mismatches was found, with estimated accuracy of 80%, suggesting that at least ~4000 RNA sequences are actually multi-site edited (3% of the total number of RNA sequences scanned).

12.2.3 Numbers of Editing Sites Detected

In the work of Levanon *et al.*, 429,000 putative dsRNA regions from 14,512 different genes were scanned (20). Following the laborious cleaning procedures, 12,723 A-to-I substitutions in 1637 human genes were identified as editing events. The work by Kim *et al.* (23) has revealed 30,085 A-to-I substitutions in 2674 human transcripts, and Athanasiadis *et al.* have reported on 1445 human mRNAs subject to RNA editing at more than 14,500 sites (21). The variability in the numbers is largely due to differences in the fidelity between the methods. One study estimates the false-positive rate per predicted editing site to be about 2% (20), while the other two groups report on false positive rate per transcript less than 1% (23), comparable with a reported false positive rate per *Alu* of less than 0.6% (21), while the rate per specific editing sites predicted is presumably higher. In addition, another group (25) has conducted an experimental survey of RNA editing by comparing sequences of clones from a library of human brain cDNA that cover a 3-Mb genomic locus, with the reference human genome sequence and genomic DNA of the same individual, and found that the average editing rate within intronic and intergenic regions is approximately 1:1000 bp.

Despite the use of different computational algorithms and statistical methods, all works conclude that A-to-I substitutions within *Alu* repeats are widespread. Validation of the results by concomitant sequencing of both DNA and RNA of candidate genes predicted to have a cluster of editing sites adds to the degree of confidence in the computational results. We note that the actual prevalence of the editing phenomenon is still probably underestimated, owing to the use of strict "cleaning" procedures of the sequences (aimed at very low false-positive rates). This is supported by the only partial overlap between various editing datasets, the failure to predict already known editing sites (within coding regions), and the experimental observation of more editing sites than predicted in almost all highly edited sequences tested. Direct sequencing of 3 Mb of human brain cDNA has shown that the average editing rate within intronic and intergenic regions is approximately 1 : 1000 bp (25). Based on the above and the prevalence of *Alu* elements in the human genes, it is believed that the vast majority of human gene transcripts are subject to A-to-I RNA editing in either their UTRs or their introns (21).

12.2.4 Characterization of the Edited Transcripts

A-to-I substitutions formed more than 80% of the total number of all 12 possible types of mismatches in the selected set of transcripts. Virtually all editing events occurred within *Alu* repetitive elements, which formed only 20% of the total length of the tested transcripts. More than 50% of the substitutions occurred in 3′-UTR regions, 12% in 5′-UTRs, and 33% in introns. Many of the edited exons exhibited nonconformist characteristics such as intron retention, extension beyond the known UTR, and a marked deviation from those listed in RefSeq, all of which are related to defective splicing (23).

The characterization of a typical editing site is not straightforward, because there is no known consensus sequence to which the editing complex binds

TABLE 12.2 Nucleotides Distribution at the Editing Sites

	A	C	G	U
Upstream Site	29%	36%	6%	29%
Downstream Site	21%	22%	43%	14%
Same site, opposite strand	1.4%	**19.2%**	1.2%	**78.2%**

Distributions of the nucleotides in the upstream and downstream positions relative to the edited adenosine. The most pronounced effect is the preference for a downstream G, and the underrepresentation of upstream G. This is also the only pattern that can be easily observed for mouse editing sites (26) and therefore might represent ADARs preference rather than properties of the human *Alu* substrate. In addition, the distribution of the nucleotide that is located at the edited site on the opposite strand is given. For each predicted site, the best opposite-strand alignment within the genomic region covered by the same gene is used. Even though a strong dsRNA structure is a prerequisite for editing, a surprisingly high number of putative editing sites (22%) are located in a mismatch site, compared to ~10% mismatches in the surrounding nucleotides. Furthermore, virtually all editing events at a mismatch site occur at an A–C mismatch, where editing changes the less stable A–C pair into the more stable I–C pair. Changes that do not have a significant effect on the dsRNA stability—that is, change of A–A pairs into I–A pairs or change of A–G pairs into I–G pairs—are rare. This suggests that editing is directed at regulating the dsRNA stability.

(see Chapters 1, 4, 6, 10, and 13). Examining nucleotide triplets centered on the edited A, it is possible to identify more favorable sites (Table 12.2): C and T are significantly overrepresented one nucleotide 5′ to the edited A, whereas G is underrepresented; at the 3′ nucleotide, G is significantly overrepresented (20–23, 25, 26). This finding is consistent with the previously described pattern of action of both ADAR1 and ADAR2 (27). Focusing on the *Alu* sequence, there are several hotspots that are prone to editing. In particular, the A nucleotides at positions 27, 28, 136, and 162 of the consensus *Alu-J* sequence account for 25% of the editing events within *Alu* repeats (20). The internal poly-A region is also prone to editing.

The level of editing strongly depends on tissue type. Higher levels of editing were reported in different brain tissues and in the thymus, whereas lower editing levels were generally found in transformed cell lines (20–23).

An absolute prerequisite for the editing process is the temporary existence of a dsRNA. Indeed, a clear correlation between the proximity of an inverted *Alu* repeat with the rate of A-to-I substitution has been shown (21, 25, 26) The distance of an *Alu* sequence to an adjacent inverted *Alu* is indeed shorter for edited *Alu*s than that for unedited ones. The shorter the distance (up to 2 kb), the more probable the editing of the *Alu* element. Moreover, editing frequency correlates with the number of *Alu* sequences in opposite orientation and close proximity.

Nearly all A-to-I substitutions result in changing the stability of the dsRNAs formed by *Alu* sequences. As can be seen in Table 12.2, about 80% resulted in the destabilization of an A–U pair (changed to I–U) and 20% in the stabilization of an A–C pair (changed to I–C) (20, 25). Nevertheless, the proportion of editing events that "correct a mismatch" within dsRNA structures significantly outnumbers the prevalence of such mismatches, a fact that substantiates previous claims about the stabilizing effect of editing (28).

12.2.5 Looking for Conserved Polymorphism Sites

Another approach to identify mismatches due to editing events among the tens of millions of other mismatches has been motivated by the common characteristic of the known recoding editing sites, where editing affects the resulting protein (see also Chapters 1, 4, 10, and 13). In all these targets, the vicinity of the editing sites is highly conserved between species (6), due to the evolutionary constraint to keep the dsRNA structure intact in addition to maintaining the coding information. This constraint leads to conservation in the DNA level and has proven to be very useful for bioinformatic searches for more candidates (6, 29, 30), since it is used as a sieve through which one sifts the few editing recoding sites out of the tens of millions of mismatches between ESTs/RNAs and the genome. The main underlying observation is that sequencing errors and single-nucleotide polymorphisms (SNPs) are not evolutionarily conserved between species while editing recoding sites are. Five novel human recoding sites have been recently found using this approach (6, 29, 30). In the first study (6), only the conservation has been used. Clearly, conservation cannot be used as the single filter and must be accompanied by extensive sequencing effort. Indeed, 41 *Drosophila* genes (out of 914 examined) have been found to contain regions within coding sequences that display unusually high sequence conservation compared with surrounding sequences. Subsequent sequencing of these suspected regions resulted in 16 of them exhibiting editing in one or more sites. Applying the same method to the mammalian *Shaker* genes resulted in the discovery of one additional mammalian recoding site. Levanon *et al.* (29) have extended this strategy looking for conserved mismatches—that is, mismatches between the DNA and expressed data which occur in exactly the same position in human and mouse. The search resulted in four additional A-to-I editing substrates, all of which were confirmed to be edited in human and mouse. Finally, Clutterbuck *et al.* (30) have used a more sophisticated approach, looking also for conservation of the complementary region forming the dsRNA. Their search resulted in one novel target, which is one of the four identified by Levanon *et al.* (29). It thus turns out that one seems better off not using this additional filter, because the typical dsRNA structures of the few known targets are rather weak and hard to predict computationally (31).

It is interesting to note that the four editing sites identified by Levanon *et al.* all appear in dbSNP, since the variability of the expressed sequences in these sites was (erroneously) interpreted as a sign for genomic polymorphism (32). Hundreds more dbSNP records have been identified as editing sites, mostly within *Alu* repeats (32). The bright side of this is that one could actually use dbSNP as an alternative starting point when looking for additional editing targets.

12.2.6 Additional Potential Targets for Abundant A-to-I Editing

In an early attempt to solve the mystery of the unexplained high inosine levels, another potential source for abundant RNA editing was suggested (33–36). It has been pointed out that about 10% of all human transcripts have a cis antisense counterpart (37–39)—that is, another mRNA transcript containing sequences transcribed from the complementary strand. Similar results were reported for the mouse (40),

Drosophila (41), *Arabidopsis* (42), and rice genomes (43). Pioneering studies in several eukaryotic systems have identified several mechanisms by which antisense transcription can regulate gene expression, including transcriptional interference, RNA masking, RNA interference (RNAi), and RNA editing (reviewed in reference 44). Most of these mechanisms assume pairing of antisense transcripts to form double-stranded RNA (dsRNA) structures. Hypothetically, these long (average of about 300 bp) and perfect intermolecular duplexes can serve as editing substrates. Indeed, such long and perfectly matching dsRNAs are extensively edited following transfection to mammalian cells *in vitro* (45). Several studies have, therefore, suggested that naturally occurring antisense dsRNA duplexes are heavily edited *in vivo*, proposing a possible general regulatory role for antisense transcripts (34, 36, 46) (see Chapters 6 and 9). Indeed, two cases of editing in sense–antisense transcript pairs have been reported so far, none of them in mammals (47, 48).

However, only 1% of editing in *Alu* elements found by Athanasiadis *et al.* (21) could be attributed to intermolecular dsRNAs, with the rest probably guided by intramolecular *Alu–Alu* dsRNA. In addition, another computational search for editing within antisense genomic loci has found virtually all putative editing sites to reside within *Alu* repeats, comprising merely 7% of the total human antisense genomic area (~320,000 base pairs). Thus, one might conclude that intermolecular dsRNA formed by antisense transcripts are not extensively edited (49). The same picture emerges from a recent experimental work studying the *NFκB1* gene (50). These results suggest that intermolecular sense–antisense RNA pairings do not normally occur after transcription in the nucleus, or that these duplexes are retained in the nucleus or being degraded by RNAi or other mechanisms. These possibilities give a new perspective, and might shed some doubt, on possible roles of abundant antisense transcripts in gene expression regulation.

12.3 EDITING IN OTHER ORGANISMS

Naturally, following the discovery of abundant editing in the human transcriptome, efforts have been made to extend the search to other organisms (see Chapters 1 and 13). *Alu* repeats are primate specific (8), but other mammals have a similar number of different SINEs. For example, the number of rodent-specific SINEs in the mouse genome is larger than the number of *Alu* SINEs in humans, and they occupy a similar portion of the genome (7.6% in mouse, 10.7% in human) (51). It was therefore expected that similar levels of editing would be observed for other mammals. However, two studies estimating the total level of editing in mouse have found that this is not the case (22, 23). The computational algorithms that were used to explore A-to-I editing in humans were applied to the mouse, rat, chicken, and fly transcriptome databases to estimate the abundance of editing in *Alu*-deficient organisms. Indeed, there are about 40-fold less A-to-I substitutions in the mouse mRNA and ESTs data set (22). By comparing human and mouse mRNA sequences with their genomes, a 35-fold increase in editing events in humans was found. Furthermore, the number of multi-edited transcripts in humans is estimated to be about 30-fold higher than in mouse (22, 23). These results are consistent with prior experimental reports on higher editing levels in human as compared to *Caenorhabditis elegans* (4) and with recent experiments showing that up to 1 : 2000 nt of human brain RNA are A-to-I edited (1 : 1000 nt in intronic and intergenic regions), an order of magnitude more than the previous observation of 1 : 17,000 nt in rat brain (2). Combined together, the available data thus imply that abundant editing might be primate specific and strongly interlinked with *Alu* sequences.

12.3.1 Uniqueness of the *Alu* Repeat

This global difference between human and mouse came as a surprise, since it is generally believed that cellular mechanisms are generally conserved between human and mouse. A number of studies have attempted to pinpoint the reason underlying the striking difference in editing levels between human and mouse. Most vertebrates' genomes contain a high proportion of retroelements, although their character and number vary significantly. However, the dominance of *Alu* among the repetitive elements in the human genome is unique. The mouse genome, for instance, contains four types of SINE besides *Alu* repeats (Figure 12.5), and their total number is similar to that of *Alu* elements alone in humans (51). Thus, it was previously suggested (22, 23) that the dominance of one SINE in human makes it more likely to find two adjacent, identical, and inverted SINEs within a single transcript, and thus the formation of dsRNA by two is far more likely to occur in the human genome.

However, a detailed analysis of the repetitive elements repertoire (26) has revealed that this is probably not the case. A similar number of paired repeats (i.e, repetitive elements having an adjacent, inverted repeat of the same type within a single transcript) is found in human and mouse. While the mouse repeats are typically shorter, this leads to only a twofold difference in the total length of the paired repeats regions (4.84 M nt in human compared to 2.78 M nt in mouse). Therefore, having four

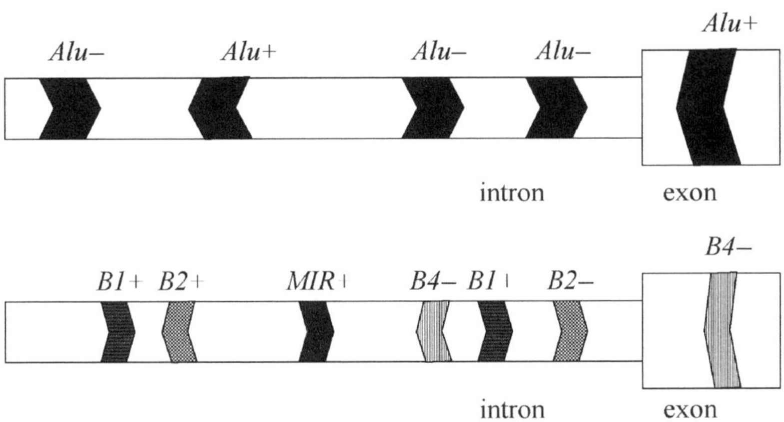

Figure 12.5 Schematic of the distribution of short interspersed elements in human and mouse genomes. (**A**) The *Alu* repetitive element is highly abundant in the human genome, and inverse copies tend to form transient double-stranded RNA (dsRNA) structures. (**B**) The mouse genome contains four main, divergent, shorter types of short interspersed element, named B1, B2, B4, and MIR.

dominant SINEs in the mouse compared to the dominant *Alu* in humans is most likely not the main factor responsible for the observed difference in editing levels between the species.

Instead, it seems that the divergence level of repeats can account for the lower editing level of the mouse repeats (26). The average divergence of the human *Alu*s is ~12%, much lower than even the least diverged mouse SINE. This is a result of a more than twofold higher substitution rate in mouse (51), which presumably reflects the lower number of human generations since the primate–rodent split. Thus, if one looks for paired repeats that are highly similar (thus potentially producing long and stable dsRNA suplexes), the picture changes dramatically. For example, even though the number of paired repeats is roughly the same for human and mouse, there are 6354 pairs producing dsRNA candidates in human, with total length of 1185 K nt, compared to only 960 such pairs in mouse, with a total length of 102 K nt. This 12-fold ratio is thus probably the main factor underlying the global difference between human and mouse editing levels. In addition, possible preferential targeting of *Alu* sequences by ADARs, or the preference of ADARs to longer and tighter dsRNAs (average length of human alignments, 187 nt; median, 181 nt; average length of mouse alignments, 107 nt; median, 91 nt), might contribute further to the elevated editing levels in humans.

The same line of analysis has been used to explain the large differences in editing frequencies between different repeat families within the mouse (26). Roughly speaking, the more divergent a certain repeat family is, the less likely it is to be edited. These findings support the validity of the above analysis and allow for high accuracy detection of editing sites in the mouse transcriptome. A database containing 833 A-to-I editing sites within mouse repeats was recently published (26).

Finally, we note that an exceptionally large number of potential A-to-I RNA editing sites were found in zebrafish (52). The explanation of this observation in terms of the underlying repeats is yet to be established. Note that even the zebrafish shows a considerably lower editing level as compared to that in humans. Thus, the abundant editing within *Alu* repeats might still be unique to primates. Given the observations of higher A-to-I editing in brain tissues (2) and the association of aberrant editing within a number of neurological disorders (53–57), some speculations have been made suggesting that widespread editing, as a result of the introduction of *Alu* elements, might have had a role in the evolution of primate (22, 58).

12.3.2 Predicting Editing Sites from Genomic Data

The analysis mentioned in the previous section opens a window for an additional computational strategy for predicting editing sites in other organisms, based solely on their genome sequence. Current computational detection methods are severely limited to organisms with a sizable amount of expressed sequences data, while the above-described model predicts which repeats are likely to be edited based on genomic sequence alone and may be applied to many more organisms.

Looking carefully at the properties of repeats, one might be able to predict a priori which repeats will contribute to editing, roughly estimate the overall editing level expected for this organism, and target the search for editing to these regions. The

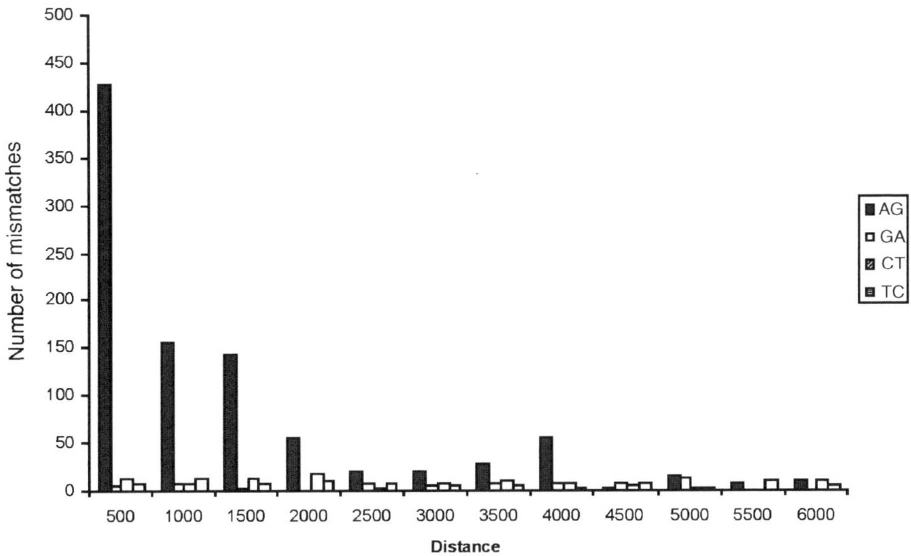

Figure 12.6 Mismatch distributions as a function of the distance to the closest inverted repeat of the same family in mouse. The overrepresentation of A-to-G mismatches is almost entirely due to repeats whose closest inverted repeat resides within 2000 nt.

main factor determining the relative level of editing in different repeat families within the same organism, and even in different species, is the number of close-by paired repeats that produce a strong putative dsRNA (Figure 12.6). Thus, enumerating the putative dsRNAs formed by paired repeats may serve as a tool for estimating the global editing level in a given organism.

Furthermore, this method can provide one with a list of specific genomic loci that are expected to be edited, to be experimentally verified. In order to estimate the efficiency of this approach, the database of editing sites in mouse was compared against the 960 paired repeats producing dsRNA in the mouse genome. Clear signal for editing was observed in 39 cases (4%). While the false-positive rate is very low, the model still gives a relatively small number of predicted loci, a reasonable starting point for an experimental validation study. In humans, editing was identified in 1968 out of the 6354 paired repeats producing a dsRNA candidate; that is, the positive predictive value is as high as 31%. It should be stressed that these estimates for the positive predictive value are based on computational analysis of RNA sequence passing through the suspected regions and are therefore limited by both (a) the availability of sequences supporting the given region in public databases and (b) the power of the computational algorithms that typically underestimate the level of editing (20). Moreover, weak editing sites, where much less than 50% of all transcripts are edited, do not usually show up in RNA sequence analysis. Thus, one should thus bear in mind that the above estimates are just lower bounds for the predictive value, and a higher success rate is to be expected in actual experimental validation studies.

12.4 BIOLOGICAL ROLE OF ALU EDITING

12.4.1 Possible Regulatory Roles

While the existence of abundant editing in the noncoding part of a gene is currently well established, little is known about its biological implications. However, a number of recent discoveries start to reveal possible regulatory roles.

One mechanism that may utilize the editing phenomenon for regulatory purposes is the nuclear retention of inosine-containing RNAs, through binding proteins such as $P54^{nrb}$ (59). This possibility was recently demonstrated for the *Cat2* gene in which several SINE sequences located in the 3′ UTR are edited, and the edited transcript is then retained in the nuclear speckles. Upon a stress signal, the edited part of the UTR is cleaved, followed by de novo polyadenylation, and the cleaved transcript can then be used to produce protein (60) (see Chapters 6 and 9). Most of the *Alu* sequences are located in introns, and thus the occurrence of two adjacent inverted repeats within a mature mRNA is rare. It is therefore not surprising to observe that massive editing in a mature RNA molecule is associated with defective splicing (23). This raises the possibility of editing being used as a marker to identify defectively spliced transcripts and mark them for degradation. Indeed, ribonuclease activity that specifically cleaves edited dsRNA region by Tudor-SN was described recently (61) (see Chapter 6). In addition, degradation of highly edited transcripts might be utilized as an additional expression level regulator.

12.4.2 *Alu* Editing and miRNA

The important role of micro-RNAs (miRNAs) in regulation of genes is increasingly recognized (see Chapter 9). miRNAs are an abundant class of endogenous ~22-nucleotide (nt) RNAs that specify posttranscriptional gene repression by base-pairing to the messages of protein-coding genes. miRNA usually have "seed sequences" located in their 5′, about 6–8 bp long, that are complementary matched to sequences in the 3′ UTR of their targeted genes. Given the sequence of the miRNA, its target sites can be detected by bioinformatics search of reverse complementary "seed" sequence in 3′ UTR of genes. Typically, such searches result in many targets for each miRNA.

Alu elements comprise a significant part of the 3′ UTR of many genes (62). The question thus arises whether the *Alu* elements harbor signature of miRNA targets. In order to answer this question, Smalheiser *et al.* (63) took all known miRNA seed sequences and search the *Alu* repeats located in 3′ UTR of all human genes for potential targets. Their results show that most miRNA have hits in only a few *Alu* elements, but two miRNAs seem to predominantly hit *Alu* sequences. These seeds (CAAAGUGC and AAGUGCUG) were highly overlapping and did not contain low-complexity sequence or unusual nucleotide composition. These seeds shared a common 6-mer core sequence (AAGUGC) that was also shared in the 5′ seeds of a set of 27 different human microRNAs (63).

In all, 28 microRNAs hit a 14-nucleotide region of *Alu* in sense orientation that is fully conserved across the consensus sequences for all 32 human *Alu* subfamilies described in Repbase (http://www.girinst.org/) (63). These results may suggest that these sequences in *Alu* might have a regulatory purpose. Consistent with these ideas, a significant fraction of the miRNAs that target the *Alu* is considered to be primate-specific miRNAs (e.g, the mi520 family). The coupled primate-specific development of these miRNA and their *Alu* targets raises some intriguing questions regarding the evolutionary route, whether the primate specific appearance of miRNA and its target evolved independently or not, and if they carry out primate specific regulation.

The basic mechanism through which miRNAs regulate gene expression is complimentary hybridization of RNA sequences. Therefore, RNA editing of either the miRNAs or their targets is an additional mechanism that might control or interfere with the miRNA regulation. This observation and the coupling of miRNAs to the *Alu* elements have triggered some research effort looking for the effect of RNA editing on miRNA regulation. A number of recent works have reported editing events in miRNA (64–67), and in one case the editing event was shown to alter the set of target of the miRNA (68).

An additional step toward establishing a connection between editing in *Alu* and miRNA regulation were demonstrated lately. Using the edited transcripts list from previous large-scale studies, Liang and Landweber (69) have studied 3236 editing sites located in 3′ UTRs, almost all of them located in *Alu* repeats. Among these sites, 292 editing positions reside in a miRNA's target site. This number was less than the expected by random, suggesting that miRNA targets sites are selected against editing events. Although most of the predicted sites are probably not true miRNA target, due to either lack of co-expression of the miRNA and its target in the same tissue or false-positive rate of current miRNA target prediction tools, it is reasonable to speculate that at least some of the targets are real miRNA targets, and editing serves as a fine-tuning mechanism for their regulation by miRNA. Since the database of editing sites used in this research is only a small portion of the actual editing repertoire in humans, the scope of regulation of miRNA by editing could be much larger.

Experimental evidence demonstrating that A-to-I editing does indeed take place in miRNA targets located in *Alu* is still lacking. Such a primate-specific mechanism allows for a highly complex gene expression regulation, allowing distinct levels of regulation for each transcript depending on its editing status. It thus may be speculated that such a mechanism, if one exists, may have played an important role in the development of the primates.

12.4.3 Alternative Splicing and *Alu* Editing

Another interesting potential effect of RNA editing is the modification of the splicing mechanism (see Chapter 6). The splicing machinery recognizes inosine as guanosine, and therefore A-to-I editing can change, for example, an intronic dinucleotide AA into the canonical 3′ splicing site AG. Such an editing-dependent splicing event was shown to occur in the auto editing of the ADAR2 gene (70).

Alu elements contain sequence motifs that are very similar to the consensus of the 5′ and 3′ potential splice sites. Indeed, hundreds of cases in which evolutionary changes have turned these motifs into functional splice sites were detected in the human genome (71). These events have led to the birth of new, *Alu*-based exons (15, 72). *Alu*-derived internal exons are almost always alternatively spliced (73), allowing the novel variant to be added to the old one, allowing for an additional splice-variant without erasing the former version of the gene.

The high abundance of editing within *Alu* repeats have thus raised the possibility of editing events creating functional splice sites in silent intronic *Alu* elements, thus promoting their exonization (21). Several example candidates for such exonization were proposed. In a recent study (74), all *Alu*-containing exons flanked by either AA at

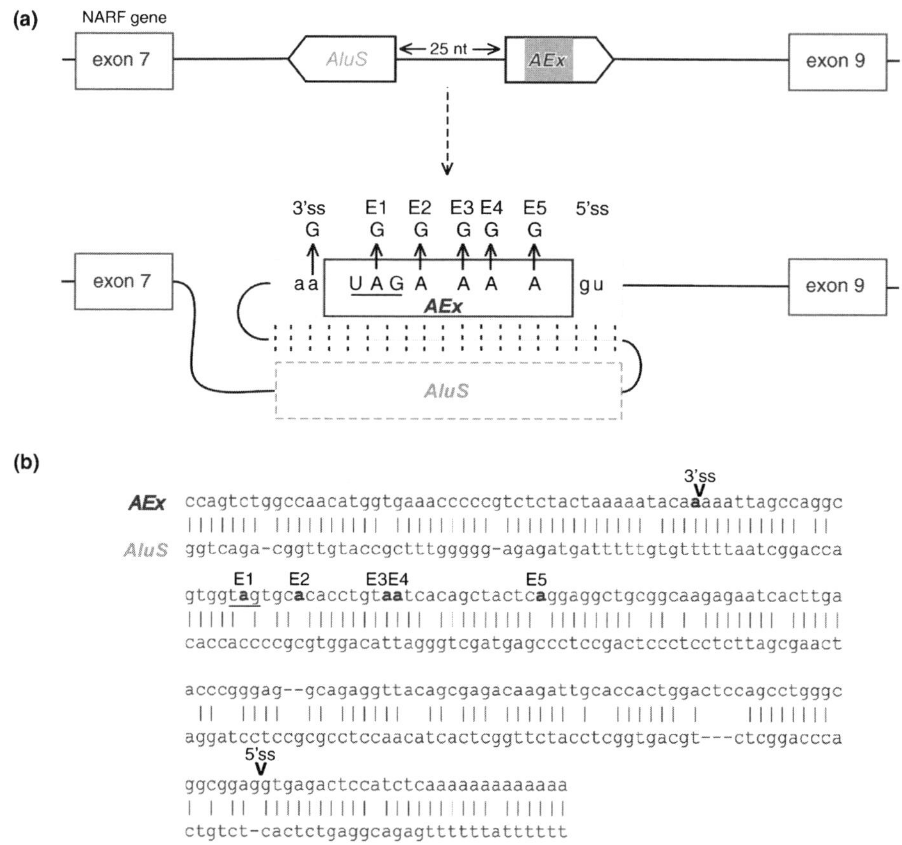

Figure 12.7 The birth of an *Alu*-exon through RNA editing. Editing prediction was inferred from alignment of cDNAs to human genomic DNA. (**a**) Schematic illustration of exons 7 to 9 of the NARF gene. Exons are depicted as rectangles; the *Alu*-exon, derived from *Alu*Sx (AEx), is in a sense orientation and is shown in the middle. The intronic, antisense-orientation *Alu* sequence (*Alu*S) is 25 base pairs upstream of the exonized *Alu*. Sense and antisense *Alu*s are expected to create a dsRNA secondary structure, thus allowing RNA editing. RNA editing changes an AA dinucleotide into a functional AG 3′ splice site (lower panel). RNA editing also occurs in five positions in the *Alu*-exon itself (E1, E2, E3, E4, and E5). In the first position (E1), editing changes a UAG stop codon into a UGG Trp codon. (**b**) Predicted folding between the sense and antisense *Alu* sequences (upper and lower lines, respectively). Adenosines that undergo editing are indicated in red. Splice sites utilized for *Alu* exonization are marked as 5′ss and 3′ss on the alignment. [Adapted from Lev-Maor et al. 74 with permission.]

the 3′ splice site or AT at the 5′ splice site were searched. These noncanonical splice sites will normally not be selected by the splicing machinery; however, each of these splice sites is theoretically capable of becoming a bona fide splice site through A-to-I RNA editing. For each such exon, the EST database was searched for evidence for editing of the splice site, in terms of an A-to-G mismatch. One example of editing-mediated *Alu* exonization has been found in the nuclear prelamin A recognition factor (NARF), a protein that interacts with the carboxyl-terminal tail of prelamin A and localizes to the nuclear lamina (75). The exon inserts 46 in-frame additional amino acids into the coding sequence of *NARF*. Lev-Maor *et al.* (74) have demonstrated experimentally that the exonization is regulated by RNA editing in a tissue-dependent manner, both through the creation of a functional AG 3′ splice site, and through alteration of functional exonic splicing enhancers within the exon (Figure 12.7). Furthermore, a premature stop codon within the *Alu*-exon is eliminated by another, exceptionally efficient RNA editing event.

12.5 CONCLUDING REMARKS

In the past few years, the bioinformatic approach of analyzing DNA–RNA mismatches has proven itself very effective in detecting A-to-I editing sites. In particular, it has revealed the unexpected scope of the phenomenon within noncoding RNA. However, many challenges are still open.

1. First, a full account for the sequence and structure characteristics determining the level of editing of specific repetitive elements is still lacking. Some advancement in this direction have been described above, but we are still far from being able to predict which specific elements are likely to be edited, what would be the relative level of editing, and what is the expected level of editing in a given organism. Given the large amount of editing sites available, such a task is not beyond possibility.
2. More importantly, the intriguing question of functional role of the abundant editing of noncoding RNA is yet to be answered. Obviously, much experimental work is needed in order to attempt solving this puzzle, but there is a room for computational contributions here as well. In particular, comparative analysis of the tissue origin of the edited sequences might provide us with hints to direct future experiments, looking for differential editing levels in different pathologies. In addition, extending the bioinformatic analysis to more organisms could facilitate experimental studies of convenient model organisms, looking at various states of development and disease.
3. While the regulatory importance of editing in noncoding regions is yet to be established, nonsynonymous editing of the coding sequences is clearly of high functional impact. Some algorithms attempting to identify such sites have been reviewed above, but current approaches are yet to be improved. In particular, better understanding of the dsRNA structure prerequisite for editing might enable more efficient targeted search.

4. Finally, computational strategies for searching for other types of RNA editing are yet to be developed yet. In particular, the C-to-U editing phenomenon, with very few examples known in mammals (76), is anticipated to have a much broader impact than what is currently known. Here too, better characterization of the sequence properties in the vicinity of the editing site is required for devising a successful computational search. A combined approach, using expressed data and evolutionary conservation accompanied by additional unique features of these types of editing, will be of use to reveal the full spectrum of the transcriptome.

REFERENCES

1. Sommer, B., Kohler, M., Sprengel, R., and Seeburg, P. H. (1991) RNA editing in brain controls a determinant of ion flow in glutamate-gated channels. *Cell* **67** (1), 11–19.
2. Paul, M. S., and Bass, B. L. (1998) Inosine exists in mRNA at tissue-specific levels and is most abundant in brain mRNA. *EMBO J* **17** (4), 1120–1127.
3. Morse, D. P., and Bass, B. L. (1999) Long RNA hairpins that contain inosine are present in *Caenorhabditis elegans* poly(A)+ RNA. *Proc Natl Acad Sci USA* **96** (11), 6048–6053.
4. Morse, D. P., Aruscavage, P. J., and Bass, B. L. (2002) RNA hairpins in noncoding regions of human brain and *Caenorhabditis elegans* mRNA are edited by adenosine deaminases that act on RNA. *Proc Natl Acad Sci USA* **99** (12), 7906–7911.
5. Kikuno, R., Nagase, T., Waki, M., and Ohara, O. (2002) HUGE: A database for human large proteins identified in the Kazusa cDNA sequencing project. *Nucleic Acids Res* **30** (1), 166–168.
6. Hoopengardner, B., Bhalla, T., Staber, C., and Reenan, R. (2003) Nervous system targets of RNA editing identified by comparative genomics. *Science* **301** (5634), 832–836.
7. Dewannieux, M., Esnault, C., and Heidmann, T. (2003) LINE-mediated retrotransposition of marked Alu sequences. *Nat Genet* **35** (1), 41–48.
8. Batzer, M. A., and Deininger, P. L. (2002) Alu repeats and human genomic diversity. *Nat Rev Genet* **3** (5), 370–379.
9. Shen, M. R., Batzer, M. A., and Deininger, P. L. (1991) Evolution of the master Alu gene(s). *J Mol Evol* **33** (4), 311–320.
10. Deininger, P. L., and Batzer, M. A. (1999) Alu repeats and human disease. *Mol Genet Metab* **67** (3), 183–193.
11. Roy, A. M., Carroll, M. L., Kass, D. H., Nguyen, S. V., Salem, A. H., Batzer, M. A., and Deininger, P. L. (1999) Recently integrated human Alu repeats: Finding needles in the haystack. *Genetica* **107** (1–3), 149–161.
12. Deininger, P. L., Batzer, M. A., Hutchison, C. A., 3rd, and Edgell, M. H. (1992) Master genes in mammalian repetitive DNA amplification. *Trends Genet* **8** (9), 307–311.
13. Versteeg, R., van Schaik, B. D., van Batenburg, M. F., Roos, M., Monajemi, R., Caron, H., Bussemaker, H. J., and van Kampen, A. H. (2003) The human transcriptome map reveals extremes in gene density, intron length, GC content, and repeat pattern for domains of highly and weakly expressed genes. *Genome research* **13** (9), 1998–2004.
14. Chu, W. M., Ballard, R., Carpick, B. W., Williams, B. R., and Schmid, C. W. (1998) Potential Alu function: Regulation of the activity of double-stranded RNA-activated kinase PKR. *Mol Cell Biol* **18** (1), 58–68.
15. Lev-Maor, G., Sorek, R., Shomron, N., and Ast, G. (2003) The birth of an alternatively spliced exon: 3′ splice-site selection in Alu exons. *Science* **300** (5623), 1288–1291.
16. Lehmann, K. A., and Bass, B. L. (1999) The importance of internal loops within RNA substrates of AD AR1. *J Mol Biol* **291** (1), 1–13.
17. Boguski, M. S., Lowe, T. M., and Tolstoshev, C. M. (1993) dbEST—database for "expressed sequence tags." *Nat Genet* **4** (4), 332–333.

18. Hillier, L. D., Lennon, G., Becker, M., Bonaldo, M. F., Chiapelli, B., Chissoe, S., Dietrich, N., DuBuque, T., Favello, A., Gish, W., Hawkins, M., Hultman, M., Kucaba, T., Lacy, M., Le, M., Le, N., Mardis, E., Moore, B., Morris, M., Parsons, J., Prange, C., Rifkin, L., Rohlfing, T., Schellenberg, K., Marra, M., and *et al.* (1996) Generation and analysis of 280, 000 human expressed sequence tags. *Genome Res* **6** (9), 807–828.
19. Seeburg, P. H. (2002) A-to-I editing: New and old sites, functions and speculations. *Neuron* **35** (1), 17–20.
20. Levanon, E. Y., Eisenberg, E., Yelin, R., Nemzer, S., Hallegger, M., Shemesh, R., Fligelman, Z. Y., Shoshan, A., Pollock, S. R., Sztybel, D., Olshansky, M., Rechavi, G., and Jantsch, M. F. (2004) Systematic identification of abundant A-to-I editing sites in the human transcriptome. *Nat Biotechnol* **22** (8), 1001–1005.
21. Athanasiadis, A., Rich, A., and Maas, S. (2004) Widespread A-to-I RNA editing of Alu-containing mRNAs in the human transcriptome. *PLoS Biol* **2** (12), e391.
22. Eisenberg, E., Nemzer, S., Kinar, Y., Sorek, R., Rechavi, G., and Levanon, E. Y. (2005) Is abundant A-to-I RNA editing primate-specific? *Trends Genet* **21** (2), 77–81.
23. Kim, D. D., Kim, T. T., Walsh, T., Kobayashi, Y., Matise, T. C., Buyske, S., and Gabriel, A. (2004) Widespread RNA editing of embedded alu elements in the human transcriptome. *Genome Res* **14** (9), 1719–1725.
24. Karolchik, D., Baertsch, R., Diekhans, M., Furey, T. S., Hinrichs, A., Lu, Y. T., Roskin, K. M., Schwartz, M., Sugnet, C. W., Thomas, D. J., Weber, R. J., Haussler, D., and Kent, W. J. (2003) The UCSC genome browser database. *Nucleic Acids Res* **31** (1), 51–54.
25. Blow, M., Futreal, P. A., Wooster, R., and Stratton, M. R. (2004) A survey of RNA editing in human brain. *Genome Res* **14** (12), 2379–2387.
26. Neeman, Y., Levanon, E. Y., Jantsch, M. F., and Eisenberg, E. (2006) RNA editing level in the mouse is determined by the genomic repeat repertoire. *RNA* **12** (10), 1802–1809.
27. Lehmann, K. A., and Bass, B. L. (2000) Double-stranded RNA adenosine deaminases ADAR1 and ADAR2 have overlapping specificities. *Biochemistry* **39** (42), 12875–12884.
28. Wong, S. K., Sato, S., and Lazinski, D. W. (2001) Substrate recognition by ADAR1 and AD AR2. *RNA* **7** (6), 846–858.
29. Levanon, E. Y., Hallegger, M., Kinar, Y., Shemesh, R., Djinovic-Carugo, K., Rechavi, G., Jantsch, M. F., and Eisenberg, E. (2005) Evolutionarily conserved human targets of adenosine to inosine RNA editing. *Nucleic Acids Res* **33** (4), 1162–1168.
30. Clutterbuck, D. R., Leroy, A., O'Connell, M. A., and Semple, C. A. (2005) A bioinformatic screen for novel A-I RNA editing sites reveals recoding editing in BC10. *Bioinformatics* **21** (11), 2590–2595.
31. Bhalla, T., Rosenthal, J. J., Holmgren, M., and Reenan, R. (2004) Control of human potassium channel inactivation by editing of a small mRNA hairpin. *Nat Struct Mol Biol* **11** (10), 950–956.
32. Eisenberg, E., Adamsky, K., Cohen, L., Amariglio, N., Hirshberg, A., Rechavi, G., and Levanon, E. Y. (2005) Identification of RNA editing sites in the SNP database. *Nucleic Acids Res* **33** (14), 4612–4617.
33. Wang, Q., Khillan, J., Gadue, P., and Nishikura, K. (2000) Requirement of the RNA editing deaminase ADAR1 gene for embryonic erythropoiesis. *Science* **290** (5497), 1765–1768.
34. Tonkin, L. A., and Bass, B. L. (2003) Mutations in RNAi rescue aberrant chemotaxis of ADAR mutants. *Science* **302** (5651), 1725.
35. Kumar, M., and Carmichael, G. G. (1998) Antisense RNA: Function and fate of duplex RNA in cells of higher eukaryotes. *Microbiol Mol Biol Rev* **62** (4), 1415–1434.
36. Carmichael, G. G. (2003) Antisense starts making more sense. *Nat Biotechnol* **21** (4), 371–372.
37. Lehner, B., Williams, G., Campbell, R. D., and Sanderson, C. M. (2002) Antisense transcripts in the human genome. *Trends Genet* **18** (2), 63–65.
38. Yelin, R., Dahary, D., Sorek, R., Levanon, E. Y., Goldstein, O., Shoshan, A., Diber, A., Biton, S., Tamir, Y., Khosravi, R., Nemzer, S., Pinner, E., Walach, S., Bernstein, J., Savitsky, K., and Rotman, G. (2003) Widespread occurrence of antisense transcription in the human genome. *Nat Biotechnol* **21** (4), 379–386.
39. Shendure, J., and Church, G. M. (2002) Computational discovery of sense–antisense transcription in the human and mouse genomes. *Genome Biol* **3** (9), 1–14.

40. Kiyosawa, H., Yamanaka, I., Osato, N., Kondo, S., and Hayashizaki, Y. (2003) Antisense transcripts with FANTOM2 clone set and their implications for gene regulation. *Genome Res* **13** (6B), 1324–1334.
41. Misra, S., Crosby, M. A., Mungall, C. J., Matthews, B. B., Campbell, K. S., Hradecky, P., Huang, Y., Kaminker, J. S., Millburn, G. H., Prochnik, S. E., Smith, C. D., Tupy, J. L., Whitfied, E. J., Bayraktaroglu, L., Berman, B. P., Bettencourt, B. R., Celniker, S. E., de Grey, A. D., Drysdale, R. A., Harris, N. L., Richter, J., Russo, S., Schroeder, A. J., Shu, S. Q., Stapleton, M., Yamada, C., Ashburner, M., Gelbart, W. M., Rubin, G. M., and Lewis, S. E. (2002) Annotation of the *Drosophila melanogaster* euchromatic genome: A systematic review. *Genome Biol* **3** (12), 000–000.
42. Meyers, B. C., Vu, T. H., Tej, S. S., Ghazal, H., Matvienko, M., Agrawal, V., Ning, J., and Haudenschild, C. D. (2004) Analysis of the transcriptional complexity of *Arabidopsis thaliana* by massively parallel signature sequencing. *Nat Biotechnol* **22** (8), 1006–1011.
43. Osato, N., Yamada, H., Satoh, K., Ooka, H., Yamamoto, M., Suzuki, K., Kawai, J., Carninci, P., Ohtomo, Y., Murakami, K., Matsubara, K., Kikuchi, S., and Hayashizaki, Y. (2003) Antisense transcripts with rice full-length cDNAs. *Genome Biol* **5** (1), R5.
44. Lavorgna, G., Dahary, D., Lehner, B., Sorek, R., Sanderson, C. M., and Casari, G. (2004) In search of antisense. *Trends Biochem Sci* **29** (2), 88–94.
45. Polson, A. G., and Bass, B. L. (1994) Preferential selection of adenosines for modification by double-stranded RNA adenosine deaminase. *EMBO J* **13** (23), 5701–5711.
46. Bass, B. L. (2002) RNA editing by adenosine deaminases that act on RNA. *Annu Rev Biochem* **71**, 817–846.
47. Peters, N. T., Rohrbach, J. A., Zalewski, B. A., Byrkett, C. M., and Vaughn, J. C. (2003) RNA editing and regulation of *Drosophila* 4f-rnp expression by sas-10 antisense readthrough mRNA transcripts. *RNA* **9** (6), 698–710.
48. Kimelman, D., and Kirschner, M. W. (1989) An antisense mRNA directs the covalent modification of the transcript encoding fibroblast growth factor in *Xenopus* oocytes. *Cell* **59** (4), 687–696.
49. Neeman, Y., Dahary, D., Levanon, E. Y., Sorek, R., and Eisenberg, E. (2005) Is there any sense in antisense editing? *Trends Genet* **21** (10), 544–547.
50. Kawahara, Y., and Nishikura, K. (2006) Extensive adenosine-to-inosine editing detected in Alu repeats of antisense RNAs reveals scarcity of sense–antisense duplex formation. *FEBS Lett* **580** (9), 2301–2305.
51. Waterston, R. H., *et al.* (2002) Initial sequencing and comparative analysis of the mouse genome. *Nature* **420** (6915), 520–562.
52. Guryev, V., Koudijs, M. J., Berezikov, E., Johnson, S. L., Plasterk, R. H., van Eeden, F. J., and Cuppen, E. (2006) Genetic variation in the zebrafish. *Genome Res* **16** (4), 491–497.
53. Palladino, M. J., Keegan, L. P., O'Connell, M. A., and Reenan, R. A. (2000) A-to-I pre-mRNA editing in *Drosophila* is primarily involved in adult nervous system function and integrity. *Cell* **102** (4), 437–449.
54. Maas, S., Patt, S., Schrey, M., and Rich, A. (2001) Underediting of glutamate receptor GluR-B mRNA in malignant gliomas. *Proc Natl Acad Sci USA* **98** (25), 14687–14692.
55. Kawahara, Y., Ito, K., Sun, H., Aizawa, H., Kanazawa, I., and Kwak, S. (2004) Glutamate receptors: RNA editing and death of motor neurons. *Nature* **427** (6977), 801.
56. Gurevich, I., Tamir, H., Arango, V., Dwork, A. J., Mann, J. J., and Schmauss, C. (2002) Altered editing of serotonin 2C receptor pre-mRNA in the prefrontal cortex of depressed suicide victims. *Neuron* **34** (3), 349–356.
57. Brusa, R., Zimmermann, F., Koh, D. S., Feldmeyer, D., Gass, P., Seeburg, P. H., and Sprengel, R. (1995) Early-onset epilepsy and postnatal lethality associated with an editing-deficient GluR-B allele in mice. *Science* **270** (5242), 1677–1680.
58. Mehler, M. F., and Mattick, J. S. (2006) Non-coding RNAs in the nervous system. *The Journal of Physiol* **575** (Pt 2), 333–341.
59. Zhang, Z., and Carmichael, G. G. (2001) The fate of dsRNA in the nucleus: A p54(nrb)-containing complex mediates the nuclear retention of promiscuously A-to-I edited RNAs. *Cell* **106** (4), 465–475.
60. Prasanth, K. V., Prasanth, S. G., Xuan, Z., Hearn, S., Freier, S. M., Bennett, C. F., Zhang, M. Q., and Spector, D. L. (2005) Regulating gene expression through RNA nuclear retention. *Cell* **123** (2), 249–263.
61. Scadden, A. D., and Smith, C. W. (2001) Specific cleavage of hyper-edited dsRNAs. *EMBO J* **20** (15), 4243–4252.

62. Yulug, I. G., Yulug, A., and Fisher, E. M. (1995) The frequency and position of Alu repeats in cDNAs, as determined by database searching. *Genomics* **27** (3), 544–548.
63. Smalheiser, N. R., and Torvik, V. I. (2006) Alu elements within human mRNAs are probable microRNA targets. *Trends Genet* **22** (10), 532–536.
64. Blow, M. J., Grocock, R. J., van Dongen, S., Enright, A. J., Dicks, E., Futreal, P. A., Wooster, R., and Stratton, M. R. (2006) RNA editing of human microRNAs. *Genome Biol* **7** (4), R27.
65. Luciano, D. J., Mirsky, H., Vendetti, N. J., and Maas, S. (2004) RNA editing of a miRNA precursor. *RNA* **10** (8), 1174–1177.
66. Pfeffer, S., Sewer, A., Lagos-Quintana, M., Sheridan, R., Sander, C., Grasser, F. A., van Dyk, L. F., Ho, C. K., Shuman, S., Chien, M., Russo, J. J., Ju, J., Randall, G., Lindenbach, B. D., Rice, C. M., Simon, V., Ho, D. D., Zavolan, M., and Tuschl, T. (2005) Identification of microRNAs of the herpesvirus family. *Nature Methods* **2** (4), 269–276.
67. Yang, W., Chendrimada, T. P., Wang, Q., Higuchi, M., Seeburg, P. H., Shiekhattar, R., and Nishikura, K. (2006) Modulation of microRNA processing and expression through RNA editing by ADAR deaminases. *Nature Struct Mol Biol* **13** (1), 13–21.
68. Kawahara, Y., Zinshteyn, B., Sethupathy, P., Iizasa, H., Hatzigeorgiou, A. G., and Nishikura, K. (2007) Redirection of silencing targets by adenosine-to-inosine editing of miRNAs. *Science* **315** (5815), 1137–1140.
69. Liang, H., and Landweber, L. F. (2007) Hypothesis: RNA editing of microRNA target sites in humans? *RNA* **13** (4), 463–467.
70. Rueter, S. M., Dawson, T. R., and Emeson, R. B. (1999) Regulation of alternative splicing by RNA editing. *Nature* **399** (6731), 75–80.
71. Dagan, T., Sorek, R., Sharon, E., Ast, G., and Graur, D. (2004) AluGene: A database of Alu elements incorporated within protein-coding genes. *Nucleic Acids Res* **32** (database issue), D489–D492.
72. Sorek, R., Lev-Maor, G., Reznik, M., Dagan, T., Belinky, F., Graur, D., and Ast, G. (2004) Minimal conditions for exonization of intronic sequences: 5′ splice site formation in alu exons. *Molecular cell* **14** (2), 221–231.
73. Sorek, R., Ast, G., and Graur, D. (2002) Alu-containing exons are alternatively spliced. *Genome Res* **12** (7), 1060–1067.
74. Lev-Maor, G., Sorek, R., Levanon, E. Y., Paz, N., Eizenberg, E., and Ast, G. (2007) RNA-editing-mediated exon evolution. *Genome Biol* **8** (2), R29.
75. Barton, R. M., and Worman, H. J. (1999) Prenylated prelamin A interacts with Narf, a novel nuclear protein. *J Biol Chem* **274** (42), 30008–30018.
76. Wedekind, J. E., Dance, G. S., Sowden, M. P., and Smith, H. C. (2003) Messenger RNA editing in mammals: New members of the APOBEC family seeking roles in the family business. *Trends Genet* **19** (4), 207–216.
77. Jiang, R., Duan, J., Windemuth, A., Stephens, J. C., Judson, R., and Xu, C. (2003) Genome-wide evaluation of the public SNP databases. *Pharmacogenomics* **4** (6), 779–789.
78. Antonarakis, S. E., Krawczak, M., and Cooper, D. C. (2002) The nature and mechanisms of human gene mutation. In: Vogelstein, B., and Kinzler, K. W., (eds.). *The Genetic Basis of Human Cancer*, 2nd ed., McGraw-Hill, New York.

CHAPTER 13

RNA EDITING IN DINOFLAGELLATES AND ITS IMPLICATIONS FOR THE EVOLUTIONARY HISTORY OF THE EDITING MACHINERY

Senjie Lin
Huan Zhang
Michael W. Gray

IT IS well accepted that RNA editing emerged independently in individual phyla of organisms. It is unclear, however, what evolutionary trends there are with regard to acquisition and loss of this phenomenon and diversification of the editing machinery. Insights into this question are provided from work on dinoflagellates, a branch of Alveolata that also comprises ciliates and apicomplexans. While not detected in ciliates, apicomplexans, and basal lineages of dinoflagellates, RNA editing is widespread in the diverse, later-diverging dinoflagellate lineages. This chapter will discuss an apparent trend in dinoflagellate evolution, from A → G RNA editing that would require relatively simple enzymatic machinery (e.g., deaminase-mediated transitional modification) to more complex editing types needed to account for the presence of both transitions and transversions evident in later-branching dinoflagellate lineages.

13.1 INTRODUCTION

RNA editing is a nucleotide re-tailoring process, occurring co- or posttranscriptionally, that can be recognized by changes in an RNA (or the corresponding cDNA) sequence compared to that of its encoding DNA. The term "RNA editing" was initially

RNA and DNA Editing: Molecular Mechanisms and Their Integration into Biological Systems,
Edited by Harold C. Smith

coined to describe the insertion and deletion of uridine (U) residues in mitochondrial transcripts in the kinetoplastid protozoans *Crithidia fasciculata*, *Trypanosoma brucei*, and *Leishmania tarentolae* (1, 2); however, this term has now come to encompass a wide range of RNA re-tailoring processes that occur in organisms as disparate as animals (mammals, including marsupials, as well as flies, frogs, worms, and squid), fungi, plants, slime molds, dinoflagellates, and other unicellular eukaryotes (for review, see reference 3). Some eukaryotic viral RNAs are also edited. Based on the phylogenetic distribution of homologs of known editing enzymes, editing likely occurs in many other species in which it has not yet been formally described, including all metazoans (4). Known RNA editing systems, which appear mechanistically and evolutionarily unrelated (5), are of two general types: insertional/deletional, in which nucleotides are added to (and in some cases also removed from) a transcript (1, 6, 7); and substitutional, where one kind of nucleotide (nt) is replaced by another at a given position (8–12). The substitution type of editing occurs in a wider array of organisms than the insertion/deletion type (Table 13.1).

The diverse mechanisms of editing make it challenging to reconstruct the evolutionary history of RNA editing in general, although there is little doubt that most RNA editing systems represent early emerging traits within the specific lineages in which they occur. In the case of particular types of editing (e.g., C → U substitution) that are found in phylogenetically disparate lineages, it is unclear whether a general RNA editing machinery existed in a remote common ancestor of these lineages, with part or all of the machinery being lost subsequently in many taxa, or whether similar RNA editing machinery arose independently (convergently) in individual lineages. To address this issue, it is necessary to systematically investigate the nature and phylogenetic distribution of editing.

Dinoflagellates constitute a morphologically and ecologically diverse phylum consisting of both photosynthetic and nonphotosynthetic members. These marine-dominant protists constitute a sister group to apicomplexans (13), a phylum of obligate parasites [Sporozoa (14)] that includes *Plasmodium*, the causative agent of malaria. Together with ciliates, which are important micrograzers in the marine ecosystem, dinoflagellates and apicomplexans comprise the crown group Alveolata. Dinoflagellate fossils date back to the Silurian Period, with a major radiation in the Jurassic (15). Geochemical evidence pushes the emergence of dinoflagellates back to the early Cambrian (16). In today's oceans, dinoflagellates are among the most important marine primary producers, and some lineages (crudely classified as *Symbiodinium*) are essential for the growth of coral reefs. Many dinoflagellates form red tides, some of which produce a suite of toxins.

Despite their ecological and economic significance, our understanding of these organisms at the cytological and molecular levels is still limited. Dinoflagellate nuclear genomes are enormous in size and highly variable in amount of DNA (3–200 pg DNA/cell) and yet lack the typical histones that function in maintenance of chromatin structure and in its compaction. Mitosis occurs via mitotic spindles localized outside the nucleus, with the nuclear envelope remaining intact. Chromosomes are permanently condensed. Although 50% of dinoflagellate lineages do not possess a chloroplast, the photosynthetic lineages harbor a diversity of plastids (18). In "typical" dinoflagellate phototrophs, the chloroplast [acquired via secondary endosymbiosis, (19)] is

TABLE 13.1 Taxonomic Distribution of Substitution-type RNA Editing

Organism	Editing Type	Examples	References
Mammals	$C \rightarrow U$	Apolipoprotein B mRNA	56, 57
	$A \rightarrow I$	Serotonin receptor and ion channel mRNAs, microRNA	4, 44, 45, 49, 58–60
Marsupials	$C \rightarrow U$	Mitochondrial tRNAs	61
Plants	$C \rightarrow U$, $U \rightarrow C$	Mitochondrial mRNAs, chloroplast mRNAs, mitochondrial tRNAs	9, 62–65
Hepatitis delta virus (HDV)	$A \rightarrow I$	HDV antigenome	66
Human immunodeficiency virus (HIV)	$G \rightarrow A$, $C \rightarrow T$	HIV-1 mRNAs	67
Drosophila	$A \rightarrow I$, $U \rightarrow C$	Ion channel mRNA	44, 47, 68–70
Squid	$A \rightarrow I$	Ion channel mRNA	71
Caenorhabditis elegans	$A \rightarrow I$	5′ and 3′ UTRs	4
Physarum polycephalum	$C \rightarrow U$	*cox1* mRNA	54
Trypanosomes	$C \rightarrow U$	7 SL RNA, mitochondrial tRNA	53, 72
Dinoflagellates	$A \rightarrow G$	cob and *cox1*mRNA, chloroplast mRNAs and rRNA	31–36
	$G \rightarrow A$	cob and *cox1* mRNA, chloroplast mRNAs and rRNA	
	$C \rightarrow U$	*cob* and *cox1* mRNA, chloroplast mRNAs and rRNA	
	$U \rightarrow C$	*cob* and *cox1* mRNA, chloroplast mRNAs and rRNA	
	$U \rightarrow G$	*cob* mRNA, chloroplast rRNA	
	$G \rightarrow U$	*cob* mRNA	
	$U \rightarrow A$	*cob* and chloroplast mRNA	
	$A \rightarrow C$	*cox1* mRNA, chloroplast mRNAs	
	$G \rightarrow C$	*cob* and *cox1* mRNA, chloroplast mRNAs and rRNA	

Source: Modified from Gott (44).

characterized by having peridinin as the major accessory pigment, three layers of membranes, and a proteobacteria-like form II Rubisco, likely originating by way of horizontal gene transfer. Other dinoflagellate chloroplasts are of haptophyte, diatom, cyanophyte, or prasinophyte ancestry. Excepting the prasinophyte type, which is believed to have arisen through serial secondary endosymbiosis (20), all of the latter types of chloroplast were presumably acquired via tertiary replacement.

There is a dearth of information on gene regulation and genome function in dinoflagellates. Sporadic investigations have shown that clock-related (21) and some other genes are under transcriptional control (22, 23). Introns are not common (26, 27), and those that exist do not always have the canonical GT/AG boundaries. The few comprehensive studies of dinoflagellate gene structure reveal high gene copy number and arrangement in polycistronic or otherwise tandem arrays (e.g., references 26–29). Recently, it was found that dinoflagellate nucleus-encoded mRNAs undergo spliced leader RNA *trans*-splicing (30). These cytological, physiological, and evolutionary features, rather unusual for eukaryotes, make dinoflagellates a highly interesting system for studies on evolution of RNA editing.

We first reported editing in mitochondrial cytochrome *b* (*cob*) and cytochrome *c* oxidase I (*cox1*) mRNAs in three dinoflagellate species (31). Later, analysis of *cob* mRNA editing was extended to six additional species (32). Recently, RNA editing was documented for two dinoflagellate chloroplasts as well (33, 34). To gain a more comprehensive picture of mRNA editing in dinoflagellates, we have now surveyed a number of other taxa for *cob* and *cox1* mRNA editing, including more basal lineages (32, 33). Together, these data provide an opportunity to discuss the nature and evolution of RNA editing in this important phylum of protists and to speculate on the implications of this information for the evolution of RNA editing machinery in general.

13.2 INFERRED RNA EDITING IN DINOFLAGELLATES

13.2.1 *cob* and *cox1* mRNA

Most of the dinoflagellate genomic and cDNA sequences of mitochondrial genes retrieved so far (Table 13.2) allow us to infer editing within almost entire *cob* and *cox1* coding regions. No insertions or deletions have to be assumed in aligning each pair of genomic DNA and corresponding cDNA sequences except at their ends (to facilitate cross-taxa comparisons, we have excluded the variable/extended 5′ and 3′ portions of the sequences). The resulting alignments range from 893 to 1039 nt for *cob* and from 1132 to 1320 nt for *cox1*. In some cases, *cob* and *cox1* appear to have unusual structures. In *Coolia monotis*, *Lingulodinium polyedrum*, *Gyrodinium uncatenatum*, *G. dorsum*, and *Karenia mikimotoi*, for instance, *cob* and *cox1* diverge radically from typical dinoflagellate sequences and in some cases are even interrupted by partial sequences of each other, *cox3*, or unknown sequences (Zhang and Lin, unpublished data). We consider these partial gene sequences to be pseudogenes and on that basis have excluded them from further consideration. In some other cases, PCR-based isolation of *cob* and *cox1* genes was unsuccessful. In all remaining cases (Table 13.2), gene and corresponding cDNA sequences are completely colinear.

TABLE 13.2 Dinoflagellate and *Plasmodium* Species Included in the Phylogenetic and Mitochondrial *cob* and *cox1* Editing Analyses

Taxa	Species Names Abbreviated	Strain and Source	SSU rDNA	*cob* cDNA	*cob* gDNA	*cox1* cDNA	*cox1* gDNA
Non-dinoflagellate							
Plasmodium berghei			AJ243513		AAD01526		AAD01525
P. yoelii			AF180727		AAC25924		AAC25923
Dinoflagellates							
Adenoides eludens		CCMP1891[a]	AF274249	EF036541		EF036565	
Akashiwo sanguinea	Asa	LIS1[b]	AY456106	EF036542	AY456105	EF036566	EU126138
Alexandrium affine		CCMP112[a]	AY831409	EF036543		EF377324	
Alexandrium tamarense	Ata	CB307; D. M. Anderson	AF022191[c]	DQ082987	AY456116	EF036567	EU126139
Amphidinium carterae	Aca	CCMP1314[a]	AF274251	EF036544	EU126130	EF036568	EU126140
A. operculatum	Aop	CCMP123[a]	AY443011[c]	EF036545	EU126131	EF036569	EU126141
Ceratium longipes		CCMP1770[a]	DQ388462	EF036546		EF036570	
Ceratocorys horrida		CCMP157[a]	DQ388456	EF036547		EF036571	
Coolia monotis		CCMP304[a]	AJ415509[c]			EF036572	
Crypthecodinium cohnii	Cco	WHd; M. Gray	M64245	AF403221	AF403220	AF487783	AF186994
Dinophysis acuminata	Dac	Narragansett Bay sample	EU130569	EU130567	EU130568	EU130565	EU130566
Gambierdiscus toxicus		CCMP401[a]	DQ388463	EF036550		EF036575	
Gonyaulax cochlea		CCMP1592[a]	DQ388465	EF036551		EF036576	
Gymnodinium catenatum		CCMP1937[a]	AF022193[c]	EF036552			
G. simplex		CCMP419[a]	DQ388466	EF036553		EF036577	
Heterocapsa triquetra	Htr	CCMP449[a]	AJ415514[c]	EF036554	EU126132	EF036578	EU126142
H. rotundata (=*Katodinium rotundatum*)	Hro	CCMP1542[a]	DQ388464	EF036556	EU126133	EF036582	EU126143
Karenia brevis	Kbr	CCMP2229[a]	AF274259[c]	EF036555	AY456104	EF036580	EU126144

K. mikimotoi		CCMP429[a]	AF009131[c]			EF036581	
Karlodinium veneficum (formerly K. micrum)	Kve	CCMP1975[a]	EF036540	DQ082989	AY345908	EF036579	AF463416
Noctiluca scintillans	Nsc	NS3; E. J. Buskey	DQ388461			EF036583	EU126145
Oxyrrhis marina	Oma	CCMP1795[a]	AF482425[c]	EF036557	EU126134	EF036584	EU126146
Peridinium aciculiferum		PAER1	AY970653	DQ094825			
Pfiesteria piscicida	Ppi	CCMP1831[a,d]	AF330620	AF357518	AF357519	AF463413	AF463412
Pfiesteria-like	Ppi-like	CCMP1828[a]	AY590476	EF036558	AY456119	EF036585	EU126147
Prorocentrum cassubicum (=*Exuviaella cassubica*)	Pca	LB1596[d]	DQ388460	EF036548	EU126135	EF036573	
P. dentatum	Pde	CCMP1517[a]	DQ336057	DQ336059	DQ336058		
P. donghaiense	Pdo	S. Lü	DQ336054	DQ336056	DQ336055	EF036587	
P. lima (=*E. lima*)		CCMP1966[a]	EF377326	EF036559		EF377325	
P. micans	Pmic	CCMP1589[a]	AY585526	AY745238	AY585525	EF036588	EU126148
P. minimum strain 696	Pmin	CCMP696[a]	DQ336072	AY030285	AY030286	AF463415	AF463414
P. minimum strain EXUV	PmEXUV	CCMP1329[a]	DQ336060,	DQ336062	DQ336061		
P. minimum strain JA9801	Pm9801	JA9801; P. Glibert	DQ336063	DQ336065	DQ336064		
P. minimum strain PTPM	PmPTPM	PTPM; P. Tester	DQ336069	DQ336071	DQ336070		
P. nanum (=*E. pusilla*)	Pna	LB1008[d]	DQ388459	EF036549	EU126136	EF036574	
Protoceratium reticulatum	Pre	CCMP1721[a]	AF274273[c]	EF036560	EU126137	EF036589	EU126149
Pseudopfiesteria shumwayae	Psh	T4; P. Tester,	AF218805[c]	DQ082988	AF502593	EF036586	EU130570
Pyrodinium bahamense	Pba	Azanza-Corrales	AY456115		AY456114		
Pyrocystis lunula		J.W. Hastings	AF274274[c]	EF036561		EF036590	
P. noctiluca		CCMP732[a]	AF022156	EF036562		EF036591	
Scrippsiella sp.	Scrsp	LIS[b]	AY743960	AY743962	AY743961	EF036592	EU130571
S. sweeneyae		CCCM280[e]	AF274276	EF036563		EF036593	

(continued)

TABLE 13.2 (*Continued*)

Taxa	Species Names Abbreviated	Strain and Source	SSU rDNA	*cob* cDNA	*cob* gDNA	*cox1* cDNA	*cox1* gDNA
Symbiodinium goreaui	Sgo	CCMP2466[a]	EF036539	EF036564	EU130574		
S. microadriaticum	Smi	CCMP830[a]	AY456111	DQ082985	AY456110	EF036594	EU130573
Symbiodinium sp.	Symsp	CCMP832[a]	AY456113	DQ082986	AY456112	EF036595	EU130572

[a]Provasoli–Guillard National Center for Culture of Marine Phytoplankton, West Boothbay Harbor, Maine.

[b]Isolated from Long Island Sound in April 2003, morphologically similar to *Akashiwo sanguinea* based on the description in Steidinger and Tengen (73) and SSU rDNA sequence (identical to AF276818). The LIS isolate of *Scrippsiella* sp. is most closely related to *S. trochoidea* based on morphology (73) and SSU rDNA identity (1738 bp out of 1754bp identical to AY421792).

[c]Sequences from GenBank originally obtained from other strains of the same species as used in this study.

[d]The Culture Collection of Algae (UTEX), The University of Texas at Austin, Texas.

[e]Canadian Center for the Culture of Microorganisms, University Boulevard, Vancouver, British Columbia, Canada.

As reported previously (31), when *Pflesteria piscicida* cDNA and genomic DNA are used as templates with *P. piscicida* cDNA-specific primers (which generated no DNA product from the genomic template), PCR analysis confirms that at least in this species, and likely in other dinoflagellates as well (supported by the clean direct sequencing result of genomic DNA-templated PCR products), the *cob* cDNA originates from transcribed sequences and does not represent a second (mitochondrial or nuclear) copy of *cob* that might have been derived evolutionarily from an edited mRNA, as in the case of the nucleus-encoded *cox2* gene in legumes (37).

In all species examined, base substitutions constitute the only differences observed between gene and corresponding cDNA sequences, with such comparisons revealing numerous reproducible single-nt changes (both transitions and transversions), indicative of RNA editing. All of the dinoflagellate taxa examined earlier (31, 32), ranging from relatively basal (e.g., *Symbiodinium*) to later branching lineages (e.g., *Alexandrium*), showed evidence of RNA editing, leading us to believe that

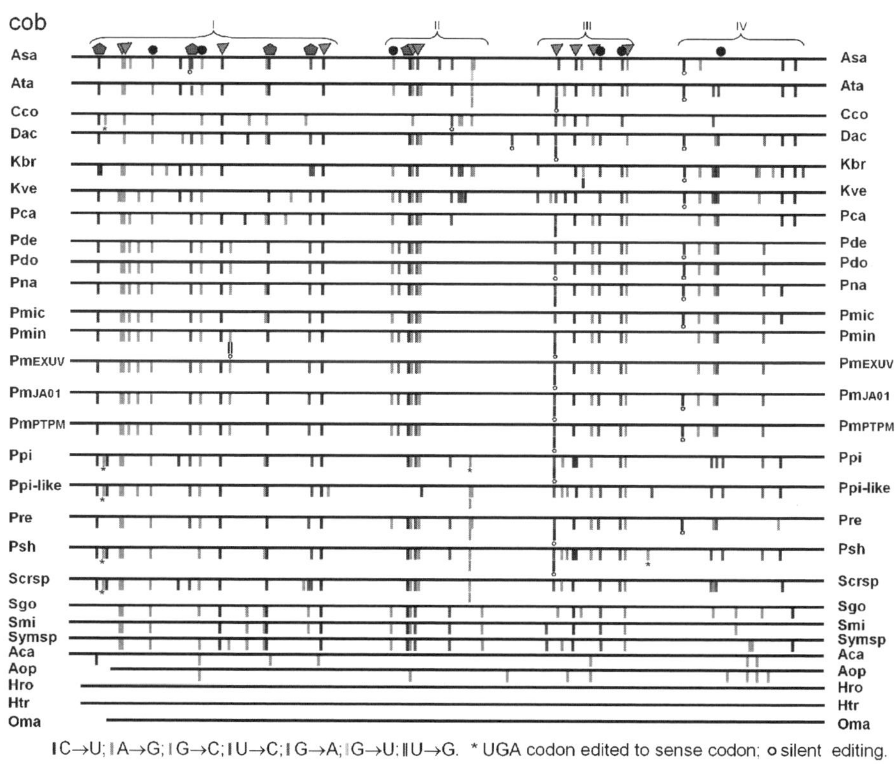

Figure 13.1 Schematic diagram of distribution of editing events in mitochondrial cytochrome *b* (*cob*) mRNA in dinoflagellates. Sites at which editing increases similarity between dinoflagellate and non-dinoflagellate species are designated by either a closed circle (editing to an amino acid identical to consensus) or a pentagon (editing to an amino acid chemically similar to consensus). A triangle denotes dinoflagellate-specific editing. Brackets with Roman numerals indicate discrete clusters of editing events. Species abbreviations are as listed in Table 13.2. [Data from Lin et al. (31), Zhang and Lin (31, 35), and Zhang et al. 36.] (*See color insert.*)

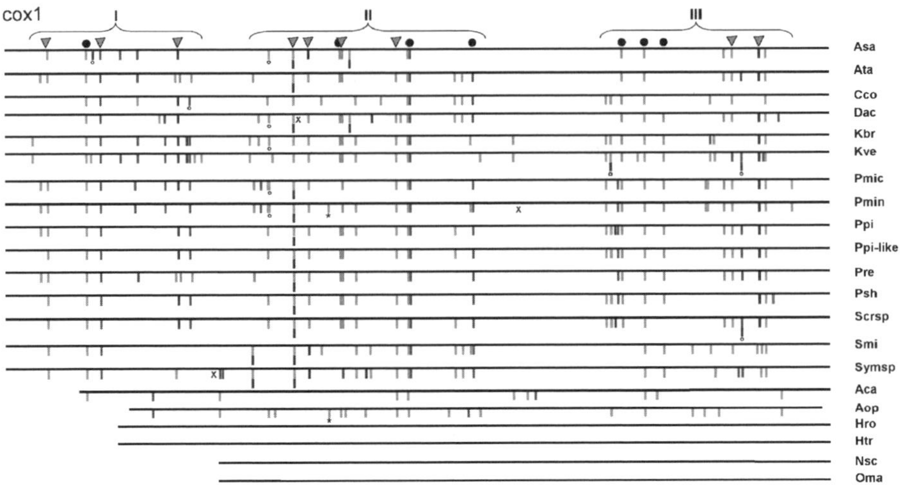

Figure 13.2 Schematic diagram of distribution of editing events in mitochondrial cytochrome *c* oxidase subunit I (*cox1*) mRNA in dinoflagellates. Symbols are as denoted in Figure 13.1. Brackets with Roman numerals indicate discrete clusters of editing events. Species abbreviations are as listed in Table 13.2. [Data from Lin et al. (31), Zhang and Lin (32, 35), and Zhang et al. (36).] (*See color insert.*)

mitochondrial mRNA editing might be universal within Dinoflagellata. However, further analyses involving additional taxa reveal that *cob* and *cox1* mRNA editing appears to be absent in the most basal lineages (Figures 13.1 and 13.2).

13.2.2 Chloroplast Transcripts

To date, editing of chloroplast RNA has been reported for two species. In *Ceratium horridum*, *psaA*, *psbA*, and *psbE* transcripts and 16S rRNA all undergo editing (33). In *L. polyedrum*, RNA editing has been detected for *petB* and *psbA* transcripts and 16S rRNA. Both species belong to the order of Gonyaulacales, whose members are considered to be typical dinoflagellates because their chloroplasts contain peridinin as the major accessory pigment and form II Rubisco as the CO_2-fixing enzyme. In typical photosynthetic dinoflagellates, the chloroplast genome is drastically reduced, with the majority of its genes transferred to the nucleus (17, 38). The remaining plastid genome evidently exists as a collection of single-gene minicircles. Surprisingly, the minicircle DNA carrying the photosynthetic genes in *C. horridum* apparently has been transferred to the nucleus (39). This observation raises the question of whether these nucleus-localized genes are the result of retroposition involving reverse transcription of post-edited RNA, and whether the chloroplast still contains a set of the original genes. If the latter possibility is excluded, it would be of great interest to know whether RNA editing alters the transcripts of other plastid genes that are known to have been transferred to the nucleus [e.g., in *Alexandrium* (17) and *Heterocapsa* (38)].

13.3 BIOCHEMICAL CHARACTERISTICS OF EDITING

13.3.1 Unusually Diverse Types of Editing

In total, nine types of substitutional editing have been recognized to date for all life forms, with A → G (operationally A → I) and C → U edits being most common (Table 13.1). Dinoflagellates share A → G, C → U, and U → C editing types with other organisms, and in addition display six unique types that so far have not been recognized elsewhere: G → A, U → G, G → U, U → A, A → C, and G → C (Table 13.1). As is evident from Tables 13.3, 13.4 and 13.5, five of the nine types of editing (A → G, G → A, C → U, U → C, and G → C) are common among mitochondrial *cob*, *cox1*, and chloroplast sequences. Eight types of editing occur in *cob* mRNA (a U → A edit detected only in *P. piscicida* is outside the region included in the present analysis, hence not shown in Figure 13.1), with only A → C absent, whereas *cox1* mRNA exhibits six types, with U → G, G → U, and U → A absent. In the two dinoflagellate species examined so far, chloroplast gene transcripts exhibit eight editing types, with G → U absent. Overall, of the 12 possible nt transitions and transversions, only A → U, C → G, and C → A substitutions have not been detected in dinoflagellate RNA editing so far.

In *cob*, A → G substitution accounts for 22–60% of all editing events in most species. In *Amphidinium carterae* and *A. operculatum*, this particular substitution amounts to 86% and 100%, respectively, of the editing observed, a proportion considerably higher than in any other dinoflagellate species analyzed so far (Table 13.3). The *Amphidinium* genus has been assigned as an early branch of dinoflagellates in some phylogenetic analyses (e.g., reference 40) and as a late branching taxon in others (e.g., reference 41). Recently, a three-gene phylogeny (*cob*, *cox1*, 18S rDNA) suggested that this genus is most likely a basal lineage (42), diverging prior to any other lineages analyzed in this study. The C → U edit comprises 23–36% of the observed editing events across all the taxa except *Amphidinium*, in which this edit occurs only once (in *A. carterae*) or not at all (*A. operculatum*) (Table 13.3). The U → C change occurs at a frequency of 0% in *Amphidinium* and 14–31% in other species. Minor edits include G → C (0% in *Amphidinium* and 5–10% in other lineages) and G → A (0% in *Amphidinium* and 0–7% in other lineages). G → U, U → G, and U → A edits are rare, each occurring only once in one of the three species affected (*Akashiwo sanguinea*, *Prorocentrum minimum*, and *P. piscicida*, respectively). In the case of *cox1* mRNA, A → G edits are even more prominent: 82% in *A. carterae*, 90% in *A. operculatum*, and 41–73% in other lineages, followed by U → C (0–31%), G → C (0–22%), and C → U (0–18%). Compared to *cob*, *cox1* mRNA editing exhibits a slightly higher frequency of U → C than C → U edits. Only one G → A edit has been observed, in *Pseudopfiesteria shumwayae* (Table 13.4). Edits G → U, U → G, and U → A do not seem to occur in *cox1* mRNA in any dinoflagellate species examined.

Although only two dinoflagellates have been examined thus far with respect to editing of chloroplast transcripts, inspection of editing events in these cases reveals some interesting characteristics (Table 13.5). Of the eight types of editing we observed, A → G and C → U occur in transcripts of all seven genes examined in both species, with A → G being overwhelmingly dominant in the 16S rRNA (77% in *C. horridum* and 84% in *L. polyedrum*). G → A, U → C, and G → C edits occur in most

TABLE 13.3 Distribution of Number and Type of Editing in *cob* mRNA in Dinoflagellates[a]

	Aca	Aop	Asa	Ata	Cco	Dac	Kbr	Kve	Pca	Pde	Pdo	Pna	Pmic	Pmin	Pm EXUV	Pm JA01	Pm PTPM	Ppi	Ppi-like	Pre	Psh	Scrsp	Sgo	Smi	Symsp sp.
No. of edits	7	9	33	33	20	31	40	39	32	30	30	33	30	31	30	31	31	33	32	30	34	31	25	20	24
(%)[b]	(0.75)	(1.01)	(3.60)	(3.54)	(2.15)	(3.35)	(4.30)	(4.19)	(3.45)	(3.22)	(3.22)	(3.54)	(3.24)	(3.33)	(3.22)	(3.33)	(3.33)	(3.57)	(3.52)	(3.22)	(3.68)	(3.35)	(2.67)	(2.13)	(2.56)
A → G	6	9	10	10	12	7	11	11	12	12	12	13	12	12	12	12	12	9	11	11	11	10	14	9	12
(%)	(85.7)	(100)	(30.3)	(30.3)	(60.0)	(22.6)	(27.5)	(28.2)	(37.5)	(40.0)	(40.0)	(39.4)	(40.0)	(38.7)	(40.0)	(38.7)	(38.7)	(27.3)	(34.4)	(36.7)	(32.4)	(32.3)	(56.0)	(45.0)	(50.0)
G → A	0	0	0	0	0	1	1	1	0	2	2	2	1	2	2	2	2	1	1	0	1	0	0	0	0
(%)	(0)	(0)	(0)	(0)	(0)	(3.2)	(2.5)	(2.6)	(0)	(6.7)	(6.7)	(6.1)	(3.3)	(6.5)	(6.7)	(6.5)	(6.5)	(3.0)	(3.2)	(0)	(2.9)	(0)	(0)	(0)	(0)
C → U	1	0	10	11	3	11	11	12	10	7	7	9	8	7	7	8	8	11	8	8	10	9	6	5	6
(%)	(14.3)	(0)	(30.3)	(33.3)	(15.0)	(35.5)	(27.5)	(30.8)	(31.3)	(23.3)	(23.3)	(27.3)	(26.7)	(22.6)	(23.3)	(25.8)	(25.8)	(33.3)	(25.0)	(26.7)	(29.4)	(29.0)	(24.0)	(25.0)	(25.0)
U → C	0	0	9	9	4	9	14	12	8	7	7	7	7	7	7	7	7	9	8	8	8	9	3	4	4
(%)	(0)	(0)	(27.3)	(27.3)	(20.0)	(29.0)	(35.0)	(30.8)	(25.0)	(23.3)	(23.3)	(21.2)	(23.3)	(22.6)	(23.3)	(22.6)	(22.6)	(27.3)	(25.0)	(26.7)	(23.5)	(29.0)	(12.0)	(20.0)	(16.7)
U → G	0	0	0	0	0	0	0	0	0	0	0	0	0	1	0	0	0	0	0	0	0	0	0	0	0
(%)	(0)	(0)	(0)	(0)	(0)	(0)	(0)	(0)	(0)	(0)	(0)	(0)	(0)	(3.2)	(0)	(0)	(0)	(0)	(0)	(0)	(0)	(0)	(0)	(0)	(0)
G → U	0	0	1	0	0	0	0	0	0	0	0	0	0	0	0	0	0	0	0	0	0	0	0	0	0
(%)	(0)	(0)	(3.0)	(0)	(0)	(0)	(0)	(0)	(0)	(0)	(0)	(0)	(0)	(0)	(0)	(0)	(0)	(0)	(0)	(0)	(0)	(0)	(0)	(0)	(0)
U → A[c]	0	0	0	0	0	0	0	0	0	0	0	0	0	0	0	0	0	0	0	0	0	0	0	0	0
(%)	(0)	(0)	(0)	(0)	(0)	(0)	(0)	(0)	(0)	(0)	(0)	(0)	(0)	(0)	(0)	(0)	(0)	(0)	(0)	(0)	(0)	(0)	(0)	(0)	(0)
A → C	**0**	**0**	**0**	**0**	**0**	**0**	**0**	**0**	**0**	**0**	**0**	**0**	**0**	**0**	**0**	**0**	**0**	**0**	**0**	**0**	**0**	**0**	**0**	**0**	**0**
(%)	**(0)**	**(0)**	**(0)**	**(0)**	**(0)**	**(0)**	**(0)**	**(0)**	**(0)**	**(0)**	**(0)**	**(0)**	**(0)**	**(0)**	**(0)**	**(0)**	**(0)**	**(0)**	**(0)**	**(0)**	**(0)**	**(0)**	**(0)**	**(0)**	**(0)**
G → C	0	0	3	3	1	3	3	3	2	2	2	2	2	2	2	2	2	3	4	3	4	3	2	2	2
(%)	(0)	(0)	(9.1)	(9.1)	(5.0)	(9.7)	(7.5)	(7.7)	(6.3)	(6.7)	(6.7)	(6.1)	(6.7)	(6.5)	(6.7)	(6.5)	(6.5)	(9.1)	(12.5)	(10.0)	(11.8)	(9.7)	(8.0)	(10)	(8.3)
No. of A, U → G, C	6	9	19	19	16	16	25	23	20	19	19	20	19	20	19	19	19	18	19	19	19	19	17	13	16
% of A, U → G, C	*85.7*	*100*	*57.6*	*57.6*	*80.0*	*51.6*	*62.5*	*59.0*	*62.5*	*63.3*	*63.3*	*60.6*	*63.3*	*64.5*	*63.3*	*61.3*	*61.3*	*54.6*	*59.4*	*63.3*	*55.9*	*61.3*	*68.0*	*65.0*	*66.7*

[a]Types of editing that are not detected in any species for this mRNA are shown in boldface. Abbreviations of species names are as listed in Table 13.2. Data from Lin *et al.* (31), Zhang and Lin (32, 35), and Zhang et al. (36).

[b]Editing density.

[c]One edit of this type has been documented in Ppi (31), but it is outside the region of sequence considered in this study [i.e., nt 88–1012 of *Ppi cob* mRNA, GenBank accession number AF357518 (for Aop, the corresponding region is nt 126–1012)].

TABLE 13.4 Distribution of Number and Type of Editing in *cox1* mRNA in Dinoflagellates[a]

	Aca	Aop	Asa	Ata	Cco	Dac	Kbr	Kve	Pmic	Pmin	Ppi	Ppi-like	Pre	Psh	Scrsp	Smi	Sysp.
No. of edits (%)[b]	11 (0.92)[c]	19 (1.68)	23 (1.75)	27 (2.05)	22 (1.67)	30 (2.27)	30 (2.27)	39 (2.95)	32 (2.42)	34 (2.58)	30 (2.27)	27 (2.05)	27 (2.05)	25 (1.89)	28 (2.12)	26 (1.97)	28 (2.12)
A → G (%)	9 (81.8)	17 (89.5)	9 (39.1)	14 (51.9)	15 (68.2)	15 (50.0)	16 (53.3)	20 (51.3)	15 (46.9)	18 (52.9)	15 (50.0)	13 (48.1)	12 (44.4)	12 (48.0)	14 (50.0)	18 (69.2)	15 (53.9)
G → A (%)	0 (0)	0 (0)	0 (0)	0 (0)	0 (0)	0 (0)	0 (0)	0 (0)	0 (0)	0 (0)	0 (0)	0 (0)	0 (0)	1 (4.0)	0 (0)	0 (0)	0 (0)
C → U (%)	0 (0)	0 (0)	5 (21.7)	2 (7.4)	1 (4.5)	5 (16.7)	3 (10.0)	5 (12.8)	3 (9.4)	2 (5.9)	4 (13.3)	3 (11.1)	1 (3.7)	2 (8.0)	4 (14.3)	2 (7.7)	3 (10.7)
U → C (%)	2 (18.2)	2 (10.5)	4 (17.4)	5 (18.5)	5 (22.7)	5 (16.7)	6 (20.0)	9 (23.1)	8 (25.0)	8 (23.5)	6 (20.0)	6 (22.2)	7 (25.9)	6 (24.0)	6 (21.4)	3 (11.5)	9 (28.6)
U → G (%)	**0 (0)**	**0 (0)**	**0 (0)**	**0 (0)**	**0 (0)**	**0 (0)**	**0 (0)**	**0 (0)**	**0 (0)**	**0 (0)**	**0 (0)**	**0 (0)**	**0 (0)**	**0 (0)**	**0 (0)**	**0 (0)**	**0 (0)**
G → U (%)	**0 (0)**	**0 (0)**	**0 (0)**	**0 (0)**	**0 (0)**	**0 (0)**	**0 (0)**	**0 (0)**	**0 (0)**	**0 (0)**	**0 (0)**	**0 (0)**	**0 (0)**	**0 (0)**	**0 (0)**	**0 (0)**	**0 (0)**
U → A (%)	**0 (0)**	**0 (0)**	**0 (0)**	**0 (0)**	**0 (0)**	**0 (0)**	**0 (0)**	**0 (0)**	**0 (0)**	**0 (0)**	**0 (0)**	**0 (0)**	**0 (0)**	**0 (0)**	**0 (0)**	**0 (0)**	**0 (0)**
A → C (%)	0 (0)	0 (0)	1 (4.3)	0 (0)	0 (0)	0 (0)	0 (0)	0 (0)	0 (0)	0 (0)	0 (0)	0 (0)	0 (0)	0 (0)	0 (0)	1 (3.8)	1 (3.6)
G → C (%)	0 (0)	0 (0)	4 (17.4)	6 (22.2)	1 (4.5)	5 (16.7)	5 (16.7)	5 (12.8)	6 (18.8)	6 (17.6)	5 (16.7)	5 (18.5)	7 (25.9)	4 (16.0)	4 (14.3)	2 (7.7)	1 (3.6)
No. of A, U → G, C	11	19	13	19	20	20	22	29	23	26	21	19	19	18	20	21	24
% of A, U → G, C	100	100	56.5	70.4	90.9	66.7	73.3	74.4	71.9	76.5	70.0	70.4	70.4	72.0	71.4	80.8	82.8

[a]Types of editing that are not detected in any species for this mRNA are shown in boldface. Abbreviations of species names are as listed in Table 13.2. Data from Lin et al. (31), Zhang and Lin (32, 36), and Zhang et al. (36).

[b]Editing density.

[c]The region of sequence considered in this study is equivalent to Cco *cox1* genomic clone, coding region nt 85–1404, GenBank accession number AF186994 (for Aca and Aop, the corresponding regions are nt 215–1404 and nt 274–1404, respectively).

TABLE 13.5 Distribution of Types of Editing in Dinoflagellate Chloroplasts[a]

	Ceratium horridum				*Lingulodinium polyedrum*		
	psaA	*psbB*	*psbE*	16S rRNA	*petB*	*psbA*	16S rRNA
No. of edits (%)[b]	105 (8.1)	86 (5.5)	13 (7.6)	35 (3.6)	15 (1.8)	10 (0.9)	19 (1.6)
A → G (%)	41 (39.4)	32 (37.2)	2 (15.4)	27 (77.1)	10 (66.7)	6 (60.0)	16 (84.2)
G → A (%)	11 (10.6)	10 (11.6)	3 (23.1)	0 (0.0)	2 (13.3)	0 (0.0)	1 (5.3)
C → U (%)	12 (11.5)	9 (10.5)	2 (15.4)	6 (17.1)	2 (13.3)	1 (10.0)	1 (5.3)
U → C (%)	28 (26.9)	29 (33.7)	5 (38.5)	1 (2.9)	1 (6.7)	0 (0.0)	0 (0.0)
U → G (%)	0 (0.0)	0 (0.0)	0 (0.0)	1 (2.9)	0 (0.0)	0 (0.0)	1 (5.3)
G → U (%)	**0 (0.0)**	**0 (0.0)**	**0 (0.0)**	**0 (0.0)**	**0 (0.0)**	**0 (0.0)**	**0 (0.0)**
U → A (%)	1 (0.0)[c]	0 (0.0)	0 (0.0)	0 (0.0)	0 (0.0)	0 (0.0)	0 (0.0)
A → C (%)	3 (2.9)	1 (1.2)	0 (0.0)	0 (0.0)	0 (0.0)	0 (0.0)	0 (0.0)
G → C (%)	9 (8.7)	5 (5.8)	1 (7.7)	0 (0.0)	0 (0.0)	3 (30.0)	0 (0.0)
No. of A, U → G, C	72	62	7	29	11	6	17
% of A, U → G, C	69.2	72.1	53.8	82.9	73.3	60.0	89.5

[a]Types of editing that are not detected in any species for mRNA or rRNA in question are shown in boldface. Data from Zauner et al. (33) and Wang and Morse (34).

[b]Editing density.

[c]Denotes an unverified edit (this position was found to be edited in some PCR clones but not in others).

of the seven cases. U → G is only found in the 16S rRNA in both species, whereas A → C is found in *psaA* and *psbB* transcripts only in *C. horridum*. In the three protein-coding transcripts, the U → C edit appears to be as frequent (27–39%) as does A → G, making it the most prominent representative in transcripts of all of the dinoflagellate chloroplast genes studied so far.

Of the nine reported types of substitutional editing, U → C and C → U transitions have previously been documented in other cases of mitochondrial mRNA editing (7, 9); however, A → I editing (operationally equivalent to an A → G change) has previously been described only in nucleus-encoded mRNAs (11, 12) (see Chapters 1, 6, and 12) or tRNAs (see Chapter 5). In dinoflagellates, A → I editing occurs in transcripts of both mitochondrial and chloroplast genes (see also Chapter 5). Of the other six types of editing exclusively occurring in dinoflagellates, the G → C transversion is found in almost all taxa that carry out editing (*Amphidinium* being the only exception), both in mitochondrial and chloroplast genes (Tables 13.3–13.5). U → G and G → U transversions are detected only in a few cases and only one edit in each. The G → A transition appears to be widespread in *cob* and the chloroplast genes, albeit at low levels, but in *cox1* mRNA this edit only occurs in *P. shumwayae* as mentioned above. The U → A transversion appears to be unique to *P. piscicida*, where it changes <u>U</u>UG(Leu) to <u>A</u>UG(Met) at the 5′ end of the *cob* mRNA (31). This edit creates what may be the initiation codon for this particular mRNA. Another U → A edit possibly occurs in the third position of a CUU codon in the *C. horridum psaA* mRNA, but this putative edit needs to be verified because in some transcripts this position is found to be unchanged. In any event, this edit would be silent because both CUU and CUA code for leucine. Incomplete, silent editing has previously been described at codon third-position sites in plant mitochondria (43).

13.3.2 Varying Editing Density and Discrete Distribution of Editing Events in Coding Sequences

Editing density (percentage of nucleotides that are edited) in transcripts of each affected gene varies from species to species and apparently from gene to gene as well (31, 32, 35, 36). For *cob* (893–1039 bp), editing density ranges between 0.8% and 4.3% in species in which editing occurs (Table 13.3). For *cox1* (1132–1320 bp), the proportion is 0.9–3.0% (Table 13.4). In transcripts of *L. polyedrum* plastid genes such as *petB* (842 bp), *psbA* (1,074 bp), and 16S rDNA (1219 bp), editing density is at the lower end of the range observed in the case of mitochondrial genes (Table 13.5). In *Ceratium horridum*, editing density for *psaA* (1331 bp), *psbB* (1571 bp), *psbE* (1959 bp), and 16S rRNA (1024 bp) is notably higher, in the range 3.6–8.1% (Table 13.5 and reference 33). Whether this elevated editing density is related to the nuclear location of the corresponding genes is unknown.

The physical distribution of editing events along each impacted gene sequence is decidedly nonrandom. Rather, editing events occur in discrete clusters [(31, 32); also see Figures 13.1–13.3]. In *cob*, four main islands of dense editing events are separated by long stretches of sequence that are devoid of editing events (Figure 13.1). These "barren intervals" are 81 nt between clusters I and II, 73 nt between clusters II and III (except for one edit within this interval in *Dinophysis acuminata* and another in *A. operculatum*), and 72 nt between clusters III and IV (except for the one edit within this interval in *P. shumwayae* and another in the *Pfiesteria*-like taxon). In *cox1*, editing events can be grouped into three major clusters. The intervals between these clusters are 77 nt (except for the interruption in *Amphidinium* spp. and *Symbiodinium* sp.; Figure 13.2) and 211 nt

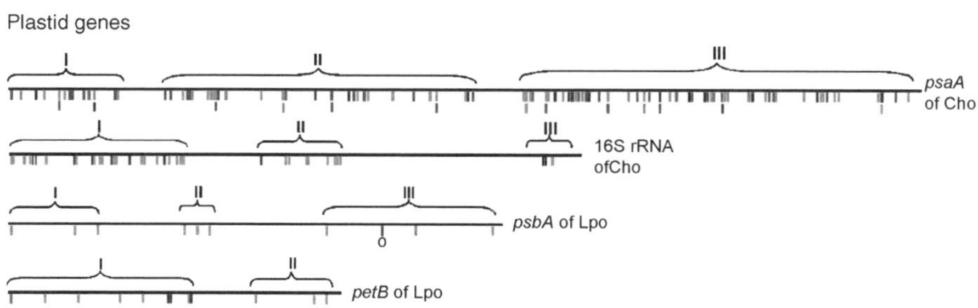

IC→U; I A→G; I G→C; I U→C; I G→A; A→C; II U→G; U→A; *, UGA codon edited to sense codon;
O, silent third-position editing; X, UGA codon that did not edit to sense condon.

Figure 13.3 Schematic diagram of distribution of editing events in plastid genes in dinoflagellates. Cho, *Ceratium horridum*; Lpo, *Lingulodinium polyedrum*. Symbols are as denoted in Figure 13.1. Brackets with Roman numbers indicate discrete clusters of editing events [Data from Zauner et al. (33) and wang and Morse (34)]. (*See color insert.*)

(except for five edits in *Karenia brevis*, *Karlodinium veneficum*, and *A. carterae*). Similarly, clustering of edits is observed in plastid gene sequences. For instance, *psaA* and 16S rRNA sequences in *Ceratium horridum* both contain three clusters of edits, 83 and 86 nt apart in the former and 135 and 364 nt apart in the latter (Figure 13.3). A similarly patchy distribution of editing events with comparable intervals is also evident in *L. polyedrum*.

13.3.3 Markedly Nonrandom Distribution in the Type of Codon Edited and in the Position of the Editing Site Within Codons

Although the range of codon and resultant amino acid changes is broad for the *cob* and *cox1* mRNAs (30 replacements for both), changes are highly skewed toward a few codons in each case. In *cob*, the most frequent changes are AUN(Ile) → GUN(Val) N = A, U, C), and UUY(Phe) → CUY(Leu), each accounting for 19% of all amino acid changes across all 25 taxa investigated. Following these in frequency are UCU (Ser) → UUU(Phe) and CUU(Leu) → UUU(Phe). Together, these seven codons account for 55% ± 7% (range 33–63% in different taxa) of the total inferred editing for *cob* mRNA. Because some of the codons are changed multidirectionally, editing results in a net increase or decrease of only 20 amino acid residues, and the most dramatic changes involve only six residues (Table 13.6). The overall content of Val, Ala, and Leu increases by 110, 103, and 65 residues, respectively, for the 25 taxa examined, whereas the content of Ile, Gly, and Thr decreases by 125, 58, and 55 residues, respectively. In *cox1*, the most frequent change (at 36%) is AUN(Ile) → GUN (Val) (N = A, U, C); the next most frequently observed changes are UUY(Phe) → CUY(Leu) (15%) and GGA(Gly) → GCA(Ala) (12%). Together, these six codons account for 65% ± 7% (range 44–71%) of all the inferred editing events for *cox1* mRNA. As in the case of *cob*, editing in *cox1* affects 20 amino acid residues. The net change of encoded amino acid residues predominantly involves an increase of Val, Leu, and Ala (respectively by 158, 69, and 64 residues for all the 17 taxa analyzed) and a decrease of Ile, Gly, and Phe (respectively by 174, 51, and 50 residues for these taxa). In chloroplast genes, codons affected by editing are even more diverse. For the three protein genes whose sequences are available to us, we find a total of 44 codon changes associated with editing, two of which convert a conventional stop codon (UGA or UAG) to a sense codon. The majority of the diversity is found in *psaA* of *C. horridum*, in which Ile → Val, Thr → Ala, Val → Ile, and Phe→Leu dominate, together accounting for 27.6% of the 105 edits. However, the net effect of editing in these three genes only involves 18 amino acid residues, with the most dramatic increase in Val (+22) and Ala (+14) and decrease in Phe (−10) and Ile (−16). Common to all the genes analyzed is a marked increase in the content of Val and Ala and decrease in Ile content.

As is also the case with the mRNA editing that occurs in plant organelles (9), substitutions occur almost exclusively at first or second positions of affected codons in dinoflagellate mitochondrial and chloroplast mRNA (Table 13.7), resulting in all of these cases in an amino acid change relative to the gene prediction at the position in question, with the exception of a silent UUA(Leu) → CUA(Leu) change in *C. cohnii cox1*. In the case of dinoflagellate *cob* mRNA, 88–100% (average 92% ± 4%) of the edits are located in the first or second position of the affected codons (Table 13.7). Of

TABLE 13.6 Editing-Mediated Increase (+) or Decrease (−) in the Content of the 20 Amino Acid Substituents of Cob[a]

AA	Aca	Aop	Asa	Ata	Cco	Dac	Kbr	Kve	Pca	Pde	Pdo	Pna	Pmic	Pmin	PmEVXU	PmJA01	PmPTPM	Ppi	Ppi-like	Pre	Psh	Scrsp	Sgo	Smi	Symsp	Total
V	3	3	5	4	6	2	6	4	4	4	4	6	5	4	4	4	4	3	5	4	3	3	8	5	7	110
A	0	1	4	5	2	4	4	5	5	6	6	6	4	4	4	4	4	4	5	6	5	6	3	3	3	103
L	0	0	2	4	2	4	6	7	2	3	3	3	3	3	3	3	3	3	3	4	2	3	−1	1	0	66
C	2	3	1	1	0	0	2	2	2	2	2	2	2	2	2	2	2	1	−1	1	0	1	1	1	1	34
M	0	0	1	1	1	1	0	0	2	1	1	2	2	0	0	0	0	1	2	0	1	0	2	1	1	20
W	0	0	0	0	1	0	0	0	0	0	0	0	0	0	0	0	0	1	1	0	2	1	0	0	0	6
D	0	0	0	0	0	0	1	0	0	0	0	0	0	0	0	0	0	0	0	0	0	0	0	0	0	1
Q	0	0	1	0	0	0	0	0	0	0	0	0	0	0	0	0	0	0	0	0	0	0	0	0	0	1
P	0	0	0	0	0	0	0	0	0	0	0	0	0	0	0	0	0	0	−1	0	0	0	0	0	0	−1
E	0	0	−1	0	0	0	0	0	0	0	0	0	0	0	0	0	0	0	0	0	0	0	0	0	0	−1
K	−1	−1	0	0	0	0	0	0	0	0	0	0	0	0	0	0	0	0	0	0	0	0	0	0	0	−2
H	0	0	−1	0	0	0	0	0	0	0	0	0	0	0	0	0	0	0	0	0	0	0	0	0	0	−1
F	1	0	3	0	−1	0	−4	−2	1	−1	−1	0	0	−1	−1	−1	−1	1	0	−2	1	0	2	0	1	−5
R	1	1	−1	−1	0	0	−1	0	0	0	0	0	0	0	0	0	0	0	−1	−1	−1	−1	0	0	0	−5
N	0	−1	0	−1	−1	−1	−1	−2	−1	0	0	0	0	0	0	0	0	−1	0	−2	−1	−1	−2	−1	−2	−18
S	−1	1	−3	−3	1	−1	−1	−3	−2	−1	−1	−2	−3	−1	−1	−1	−1	−1	−1	0	0	−2	1	0	1	−25
Y	−2	−3	0	−1	−1	−1	−2	−2	−2	−2	−2	−2	−2	−2	−2	−2	−2	−1	0	−1	−1	−1	−1	−1	−1	−37
T	0	−1	0	−2	−1	−3	−3	−3	−3	−4	−4	−4	−2	−2	−2	−2	−2	−2	−1	−3	−2	−3	−2	−2	−2	−55
G	0	0	−3	−2	0	−3	−3	−3	−2	−3	−3	−3	−2	−3	−3	−3	−3	−3	−3	−2	−3	−2	−2	−2	−2	−58
I	−3	−3	−7	−5	−8	−2	−4	−3	−6	−5	−5	−8	−7	−4	−4	−4	−4	−4	−7	−4	−4	−3	−9	−5	−7	−125

[a]Taxon names are abbreviated as in Table 13.2. Amino acids (AA) are denoted by their standard one-letter abbreviations.

TABLE 13.7 Codon Positions of Editing Events in Dinoflagellate Mitochondrial and Plastid Genes[a]

	Aca	Aop	Asa	Ata	Cco	Dac	Kbr	Kve	Pca	Pde	Pdo	Pna	Pmic	Pmin	Pm EXUV	Pm JA01	Pm PTPM	Ppi	Ppi- like	Pre	Psh	Scrsp	Sgo	Smi	Symsp.
cob/total editing sites	7	9	33	33	20	31	40	39	32	30	30	33	30	31	30	31	31	33	32	30	34	31	25	20	24
1st (%)	3 (42.9)	4 (44.4)	17 (51.5)	19 (57.6)	13 (65.0)	16 (51.6)	24 (60.0)	23 (59.0)	19 (59.4)	21 (70.0)	21 (70.0)	20 (60.6)	18 (60.0)	20 (64.5)	20 (66.7)	20 (64.5)	20 (64.5)	19 (57.6)	20 (62.5)	19 (63.3)	19 (55.9)	20 (64.5)	16 (64.0)	13 (65.0)	16 (66.7)
2nd (%)	4 (57.1)	5 (55.6)	12 (36.4)	11 (33.3)	5 (25.0)	11 (35.5)	14 (35.0)	15 (38.5)	10 (31.3)	7 (23.3)	7 (23.3)	9 (27.3)	9 (30.0)	8 (25.8)	8 (26.7)	8 (25.8)	8 (25.8)	11 (33.3)	9 (28.1)	9 (30.0)	11 (32.4)	10 (32.3)	7 (28.0)	6 (30.0)	7 (29.2)
3rd (%)	0 (0.0)	0 (0.0)	4 (12.1)	3 (9.1)	2 (10.0)	4 (12.9)	2 (5.0)	1 (2.6)	3 (9.4)	2 (6.7)	2 (6.7)	4 (12.1)	3 (10.0)	3 (9.7)	2 (6.7)	3 (9.7)	3 (9.7)	3 (9.1)	3 (9.4)	2 (6.7)	4 (11.8)	1 (3.2)	2 (8.0)	1 (5.0)	1 (4.2)
1st or 2nd (%)	7 (100)	9 (100)	29 (87.9)	30 (90.9)	18 (90.0)	27 (87.1)	38 (95.0)	38 (97.4)	29 (90.6)	28 (93.3)	28 (93.3)	29 (87.9)	27 (90.0)	28 (90.3)	28 (93.3)	28 (90.3)	28 (90.3)	30 (90.9)	29 (90.6)	28 (93.3)	30 (88.2)	30 (96.8)	23 (92.0)	19 (95.0)	23 (95.8)
% of editing that changes aa	7 (100)	9 (100)	31 (93.9)	31 (93.9)	19 (95.0)	29 (93.5)	39 (97.5)	38 (97.4)	31 (96.9)	29 (96.7)	29 (96.7)	31 (93.9)	29 (96.7)	29 (93.5)	29 (96.7)	29 (93.5)	29 (93.5)	32 (97.0)	31 (96.9)	28 (93.3)	33 (97.1)	31 (100)	25 (100)	20 (100)	24 (100)
cox1/total editing site	11	19	23	27	22	30	30	39	N/A	N/A	N/A	N/A	32	34	N/A	N/A	N/A	30	27	27	25	28	N/A	26	28
1st (%)	8 (72.7)	12 (63.2)	10 (43.5)	16 (59.3)	14 (63.6)	16 (53.3)	18 (60.0)	24 (61.5)	N/A	N/A	N/A	N/A	18 (56.3)	20 (58.8)	N/A	N/A	N/A	18 (60.0)	15 (55.6)	14 (51.9)	11 (44.0)	14 (50.0)	N/A	19 (73.1)	22 (78.6)
2nd (%)	1 (9.1)	0 (0.0)	10 (43.5)	10 (37.0)	7 (31.8)	13 (43.3)	11 (36.7)	11 (28.2)	N/A	N/A	N/A	N/A	12 (37.5)	11 (32.4)	N/A	N/A	N/A	11 (36.7)	11 (40.7)	12 (44.4)	12 (48.0)	12 (42.9)	N/A	6 (23.1)	5 (17.9)
3rd (%)	2 (18.2)	7 (36.8)	3 (13.0)	1 (3.7)	1 (4.5)	1 (3.3)	1 (3.3)	4 (10.3)	N/A	N/A	N/A	N/A	2 (6.3)	3 (8.8)	N/A	N/A	N/A	1 (3.3)	1 (3.7)	1 (3.7)	2 (8.0)	2 (7.1)	N/A	1 (3.8)	1 (3.6)
1st or 2nd (%)	9 (81.8)	12 (63.2)	20 (87.0)	26 (96.3)	21 (95.5)	29 (96.7)	30 (96.7)	35 (89.7)	N/A	N/A	N/A	N/A	30 (93.8)	31 (91.2)	N/A	N/A	N/A	29 (96.7)	26 (96.3)	26 (96.3)	23 (92.0)	26 (92.9)	N/A	25 (96.2)	27 (96.4)
Plastid	*Ceratium horridum*														*Lingulodinium polyedrum*										
1st (%)	61.1														66.7										
2nd (%)	30.0														28.6										
3rd (%)	8.9														4.7										
1st or 2nd (%)	91.1														95.3										

[a]Abbreviations of species names for *cob* and *cox1* are as listed in Table 13.2.

the small number of third-position edits, some also predict an amino acid change: AUA (Ile)→AUG(Met) and UGA(stop) → UGG(Trp). The latter edit takes place in *C. cohnii* and several Peridiniale lineages (*Pseudopfiesteria*, *Pfiesteria* and related species, and *Scrippsiella*), converting a standard termination codon to UGG at the position of a highly conserved Trp. In total, there are only five alignable codons where editing at the third position does not lead to a change in the predicted amino acid (Figure 13.1). Only two of these edits are shared by multiple species, one of which is accompanied by a first-position edit within the same codon that does result in an amino acid change.

Similarly, high proportions of *cox1* edits, 63–97% (average 92% ± 8%) occur in the first and second positions, generating amino acid changes (Table 13.7). Again, of the small number of third-position edits, some also predict an amino acid change: AUA (Ile) → AUG(Met) and UGA(stop) → UGG(Trp). At only four aligned codons does third-position editing not lead to substitution of the encoded amino acid (Figure 13.2). One of these edits is an A → G change shared by five species [*A. sanguinea*, *Karenia brevis*, *Karlodinium veneficum* (formerly *K. micrum*), *P. minimum*, and *P. micans*], while another is a C → U edit shared by two species (*K. veneficum* and *Scrippsiella* sp.) and accompanied by a first-position U → C edit. In the other two cases, the edit in question occurs in only one species.

First- and second-position editing is also prominent in transcripts of chloroplast genes. For *C. horridum psaA*, *psbB*, and *psbE* data pooled together, 91% of the edits occur in the first and second positions, whereas in the combined *L. polyedrum petB* and *psbA* data, the proportion is as high as 95%. Of the 105 editing events in *C. horridum psaA*, only seven occur in the codon third position. Two of these edits involve a Met codon, one changing the AUG (Met) to AUC (Ile) and the other AUC (Ile) to AUG (Met). Four others are found in codons whose first (one case) or second (three cases) position is also edited; one of these double edits changes UAG (stop codon) to UGC (Cys).

In a number of cases, two positions in the same codon are edited. In *cob* and *cox1*, this "double editing" affects only one or two codons (Figures 13.1 and 13.2), whereas in chloroplast sequences, this figure can be much higher. In the *psaA* sequence of *C. horridum*, for instance, a total of 13 codons of this type account for 12% of all editing events. Such a high number of doubly edited codons is not observed for transcripts of other dinoflagellate chloroplast genes that have been examined so far (Figure 13.3).

13.4 CONSEQUENCES OF EDITING

The clustering of editing sites observed in all the dinoflagellate species suggests that specific regions in the transcripts are selectively targeted for editing. However, alignment of dinoflagellate *cob* and *cox1* sequences with their counterparts from other organisms provides little insight into the function of editing, because none of the editing events seems to impact on amino acid residues crucial for protein function (31, 32). Only about 10% of the editing sites predict an obvious increase in amino acid sequence similarity between the dinoflagellate Cob and Cox1 proteins and their non-dinoflagellate homologs. About 15% of the edited sites display dinoflagellate-specific editing. In some cases, editing actually results in a decrease in similarity compared to the highly conserved consensus. Other cases exhibit more or less random amino acid changes that do not affect sequence similarity. No significant net effect can be predicted from the most dominant amino acid changes caused by editing (Tables 13.6, 13.8 and 13.9). The only effect observed is a small increase in the average hydrophobicity for both Cob and Cox1 (from 8.4 to 9.1 and from 9.0 to 9.3, respectively, as estimated using PepTool [BioTools, Inc., Alberta, Canada]). These results are at odds with the general characteristics of mRNA editing in other organisms, where such editing seems to serve an obvious role in genetic information transfer: generation of continuous, translatable ORFs; creation of translation initiation and/or termination codons; and/or restoration of critical amino acid residues at functional sites within the protein (9). In the nervous systems of fruit fly and mammals, editing of various gene transcripts (targets of deaminases) has been shown to be critical for normal signal generation and transmission, and mutations leading to a deficiency in editing are associated with abnormal development or fatal disease (44, 45). Editing diversifies protein repertoire in some organisms (46) and hence, in the case of insect sodium channel protein, fine-tunes neuronal activities and regulates resistance to insecticides (47). More subtly, editing in siRNA targets or microRNA may play a role in regulating gene expression (48, 49).

Although no effect of mRNA editing is apparent in dinoflagellates on the structure and function of proteins in question, a notable asymmetry is observed in the nature of the nucleosides generated by editing (Tables 13.3–13.5). In the case of *cob* mRNA, 52–100% of editing events in each species are from A or U to a G or C. Over the 26 taxa analyzed, A or U to G or C edits account for an average of 64 $\pm$ 10%. The figure is even higher in the case of *cox1* mRNA (63–100%, average 76 $\pm$ 11%), due to a notably higher proportion of A $\rightarrow$ G and lower proportion of C $\rightarrow$ U editing in transcripts of this gene compared to those of *cob*. Thus, editing has a pronounced tendency to increase the G+C content of edited codons in these dinoflagellate mRNAs. If, as in other alveolates, mitochondria in dinoflagellates need to import most or even all of the tRNAs required for translation in the organelle, elevation of GC content (or decrease in AU) might be expected to enhance the efficiency of translation of the edited mitochondrial mRNAs by nucleus-encoded tRNAs. This re-tailoring might involve a reduction in the number of certain AU-rich codons whose presence would be rate limiting for mitochondrial translation (31). Notably, editing in transcripts of chloroplast genes, including the 16S rRNA, also raises the GC content of the transcripts (Table 13.5). Edits of A or U to G or C account for 60–90% of all the chloroplast editing events, and those in the two 16S rRNA transcripts account for 83% and 90%, respectively, in *C. horridum* and *L. polyedrum*. In this regard, it would be of interest to determine whether the chloroplast protein synthesis apparatus also relies on imported tRNAs.

Editing clearly diversifies the function of the UGA codon. As mentioned above, an A $\rightarrow$ G transition at the third position of a UGA codon in the *cob* mRNA of *C. cohnii* and several Peridiniale lineages (*Pseudopfiesteria*, *Pfiesteria* and related species, and *Scrippsiella*) converts a standard termination codon to UGG at the position of a highly conserved Trp. Similarly, at the third position of an internal UGA codon in *Pseudopfiesteria shumwayae cob*, an A $\rightarrow$ G transition converts the codon to UGG (Trp). In addition, a G $\rightarrow$ C edit occurs in the second position of the

TABLE 13.8 Editing-Mediated Increase (+) or Decrease (−) in the Content of the 20 Amino Acid Substituents of Cox1[a]

AA	Aca	Aop	Asa	Ata	Cco	Dac	Kbr	Kve	Pmic	Pmin	Ppi	Ppi-like	Pre	Psh	Scrsp	Smi	Symsp	Total
V	6	8	4	8	11	10	10	13	9	11	9	8	6	8	8	16	13	158
L	1	0	5	3	1	4	4	6	8	8	5	6	5	4	4	1	4	69
A	0	3	4	5	1	4	7	7	5	5	4	4	6	4	3	1	1	64
M	1	4	1	1	1	0	0	2	0	0	1	1	1	1	1	1	1	17
C	0	0	0	1	1	1	1	1	1	1	1	1	1	1	1	0	0	12
R	0	0	0	0	1	0	1	1	1	1	1	1	0	1	1	0	0	9
S	1	0	0	1	2	−1	−3	−4	0	1	1	1	0	0	1	3	2	5
P	0	0	0	1	0	1	0	1	0	0	0	0	1	0	0	0	1	5
W	1	2	0	0	0	0	0	0	0	1	0	0	0	0	0	0	0	4
Q	1	1	0	0	0	0	0	0	0	0	0	0	0	0	0	0	0	2
H	0	0	0	0	0	0	0	0	0	0	0	0	0	0	0	1	0	1
D	0	0	0	0	0	0	0	0	0	0	0	0	0	0	0	−1	0	−1
E	0	0	0	0	0	−1	0	0	0	0	0	0	0	0	0	0	0	−1
K	0	0	0	0	−1	0	−1	−1	−1	−1	−1	−1	0	−1	−1	0	0	−9
Y	0	0	−1	−1	−1	−1	−1	−1	−1	−1	−1	−1	−1	−1	−1	−1	−1	−15
N	−1	0	−1	−1	−2	−1	0	0	−1	−1	−2	−2	−1	−1	−2	−1	−1	−18
T	0	−3	−2	−1	0	−3	0	0	−1	−1	−1	−1	−2	−1	−1	−2	−2	−21
F	−1	0	−2	−2	−1	−2	−2	−4	−6	−7	−3	−5	−3	−3	−2	−2	−5	−50
G	0	0	−3	−4	−1	−3	−4	−4	−5	−5	−4	−4	−5	−4	−3	−1	−1	−51
I	−7	−12	−5	−11	−12	−8	−12	−17	−9	−11	−10	−8	−8	−8	−9	−15	−12	−174

[a]Taxon names are abbreviated as in Table 13.2. Amino acids (AA) are denoted by their standard one-letter abbreviations.

TABLE 13.9 Editing-Mediated Increase (+) or Decrease (−) in the Content of the 18 Amino Acid Substituents of Chloroplast PsaA of *Ceratium horridum* (Cho) and PsbA and PetB of *Lingulodinium polyedrum* (Lpo)[a]

		Lpo		
AA	Cho, psaA	psbA	petB	Total
V	11	6	5	22
A	11	2	1	14
G	7	−2	0	5
L	3	0	0	3
P	3	0	0	3
H	2	0	0	2
E	2	0	0	2
S	2	−1	0	1
C	0	0	0	0
R	−3	0	2	−1
M	−2	0	0	−2
Q	−1	0	−1	−2
K	−3	0	0	−3
Y	−4	0	0	−4
T	−4	1	−1	−4
N	−2	0	−2	−4
F	−10	0	0	−10
I	−7	−6	−3	−16

[a]Amino acids (AA) are denoted by their standard one-letter abbreviations.

UGA codon in *P. piscicida cob*, which converts this stop codon to a Ser codon. In *cox1*, a similar third-position A→G transition is detected in the UGA codon in *Amphidinium* spp. and *P. minimum*, converting the standard termination codon to a Trp codon. All these cases seem to suggest that UGA probably does not code for Trp in dinoflagellates as it does in mitochondrial genes in apicomplexans (50, 51) and ciliates (52). However, we also observe nonedited UGA codons in *cox1* mRNA in *Dinophysis acuminatum*, *P. minimum*, and *Symbiodinium* sp. (Figure 13.2). In the later two cases the UGA codon is present at the position of a highly conserved UGG (Trp) in other dinoflagellate species. In dinoflagellate mitochondria, the $tRNA^{Trp}$ (presumably imported from the cytosol and operating in a nucleocytoplasmic translation system that uses the standard genetic code) would normally be expected to have a 5′-CCA-3′ anticodon capable of decoding only UGG codons. The observation of nonedited UGA codons suggests that in at least some cases, the 5′-CCA-3′ anticodon might be edited to 5′-UCA-3′, able to decode both UGG and UGA, as in the case of *Leishmania tarentolae* mitochondria (53). Among the chloroplast gene transcripts analyzed, there is one case (mentioned earlier) in which a stop codon is converted to a sense codon by editing: In *psaA* mRNA of *C. horridum*, UAG is edited to UGC (Cys) via substitution at two positions.

13.5 PHYLOGENETIC TREND

Initial investigation had revealed a broad phylogenetic distribution of mRNA editing in *cob* and *cox1* transcripts (31, 32). Among the lineages analyzed, *C. cohnii* and *Symbiodinium* spp. represent relatively early branches although their exact phylogenetic positions have yet to be resolved. Later branching lineages include *Alexandrium*, *Karlodinium*, and others. Based on these analyses, it has become apparent that editing density increases from basal to late branching lineages (32). Editing density is relatively low in *C. cohnii* and *Symbiodinium* but considerably higher in later branching lineages. *C. cohnii* is a benthic species and *Symbiodinium* an endosymbiotic lineage, and presumably their natural habitats are relatively stable (or protected). It follows that the lower editing density in these species might be related to a reduced mutation pressure associated with their more stable (or protected) habitats, or it might simply reflect their ancestral status (32). More recent analyses provide some insights into this question. First, we detect no editing in *cob* or *cox1* mRNA sequences in *Oxyrrhis marina* 35, a lineage basal to any described dinoflagellates, which as plankton inhabits as highly variable an environment as other planktonic species that display high editing density (e.g., *Karlodinium*). Similarly, no editing is apparent in the *cox1* mRNA of *Noctiluca* (attempts to amplify *cob* sequences in this case were unsuccessful), another early diverging lineage despite some uncertainty in its affiliation. Second, a three-gene phylogeny (*cob*, *cox1*, 18S rDNA) reconstructed recently placed *Heterocapsa* and *Amphidinium* at positions basal to *C. cohnii* and *Symbiodinium* spp., albeit later than *Oxyrrhis* (Figure 13.4A), quite a surprising result. Intriguingly, we observe that editing densities in *Amphidinium* spp. are lower than in *C. cohnii* and *Symbiodinium* spp. No editing is found in *cob* or *cox1* sequences of *Heterocapsa* spp. (Figures 13.1 and 13.2). Evidently, differences in editing density among different taxa are correlated with the phylogenetic positions of the taxa: Basal lineages have no or few editing events whereas later branching lineages exhibit higher levels of editing.

Does editing density continue to increase in later branching lineages? Based on earlier, taxon-limited analysis (32), the answer could have been positive. However, when additional lineages are analyzed, the picture becomes less clear. As shown in Figure 13.4, following the initial dramatic increase in editing density from basal to later branching lineages, the overall variation among the latter groups becomes rather small. For instance, between the clades of *Pfiesteria*, *Prorocentrum*, and *Alexandrium*, editing density is very similar. When examined more closely, however, the correlation holds up better in local regions of the phylogenetic tree. For example, in the clade comprising *Akashiwo*, *Karenia*, and *Karlodinium*, editing density is lower in *Akashiwo* than in the other two taxa, consistent with their relative branching orders. *Akashiwo* is presumably basal to the other two lineages because it contains a typical dinoflagellate plastid whereas the other two taxa harbor a haptophyte-type tertiary-replacement plastid. Absence of a correlation between editing density and phylogenetic position in the later branching lineages possibly reflects the often ambiguous evolutionary relationships among these lineages, especially the Gonyaulales, Prorocentrales, and Peridiniales.

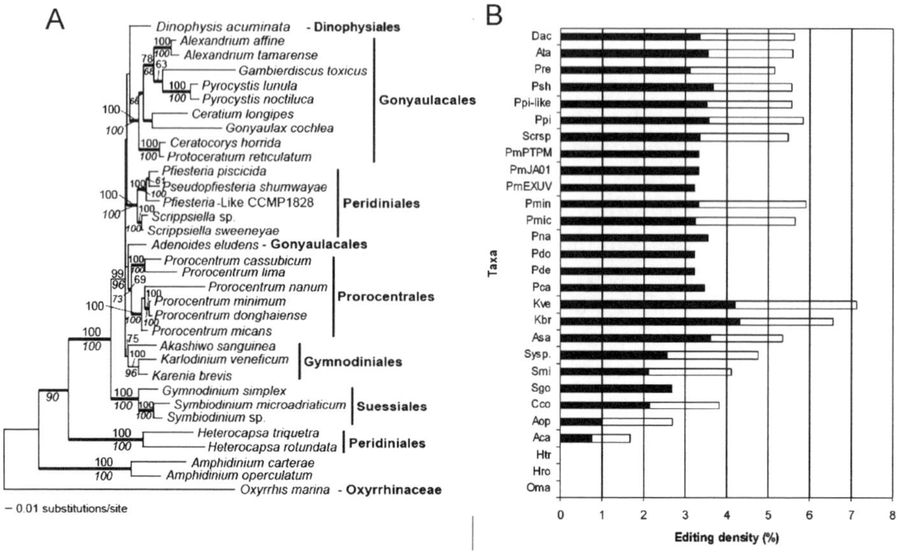

Figure 13.4 Phylogenetic mapping of editing density based on concatenated mitochondrial *cob* and *cox1* and 18S rDNA sequences. (**A**) The three-gene maximum likelihood phylogenetic tree was inferred using PAUP with ssGTR model (four categories of nucleotide substitution rates). Bootstrap support was assessed using PHYML with a GTR + I + G model (numbers above branches) and maximum parsimony analysis (numbers below branches). Posterior probabilities from Bayesian analysis greater than 0.95 are indicated as thick branches. The tree is rooted with *Oxyrrhis*, the lineage that is consistently placed at the base of all dinoflagellates when trees are rooted with the apicomplexan *Plasmodium*. [Adapted from Zhang et al. (36) with permission.] (**B**) Editing density is plotted against taxa (abbreviations of species names are as in Table 13.2), with filled bars of *cob* extended to the right by open bars of *cox1*. Note that in the phylogenetic tree, there is a well-supported early-to-later branching order from *Oxyrrhis*, *Amphidinium*, *Heterocapsa*, and *Symbiodinium*, to the clade of *Akashiwo/Karlodinium/Karenia*. Accordingly, editing density increases in this order. For the remaining lineages, phylogenetic resolution is less clear, and editing densities among those lineages are similar.

To further examine whether features of RNA editing other than editing density have any correlation with evolutionary position, we performed phylogenetic reconstructions based on the location (LOE) and type (TOE) of editing. As reported in Zhang and Lin (32), for analysis of the LOE-based tree, each location of editing ("editable site") is designated a character, whereas for the TOE-based analysis, each type of editing (e.g., A → G, C → U) is treated as a character. These reconstructions show that both the LOE and TOE trees (Figure 13.5) are highly similar to the gene sequence trees (Figure 13.4A). In the *Oxyrrhis*-rooted trees, *Heterocapsa*, *Noctiluca*, and *Amphidinium* occupy basal positions, followed by *C. cohnii* and *Symbiodinium* spp. In these trees, *Pfiesteria* and *Pseudopfiesteria* cluster with *Scrippsiella*, in full agreement with our current understanding that *Pfiesteria* and related lineages are peridinioid. In addition, the fucoxanthin-containing lineages *Karenia* and *Karlodinium* cluster together in all of the trees, also consistent with our current understanding of their phylogenetic positions.

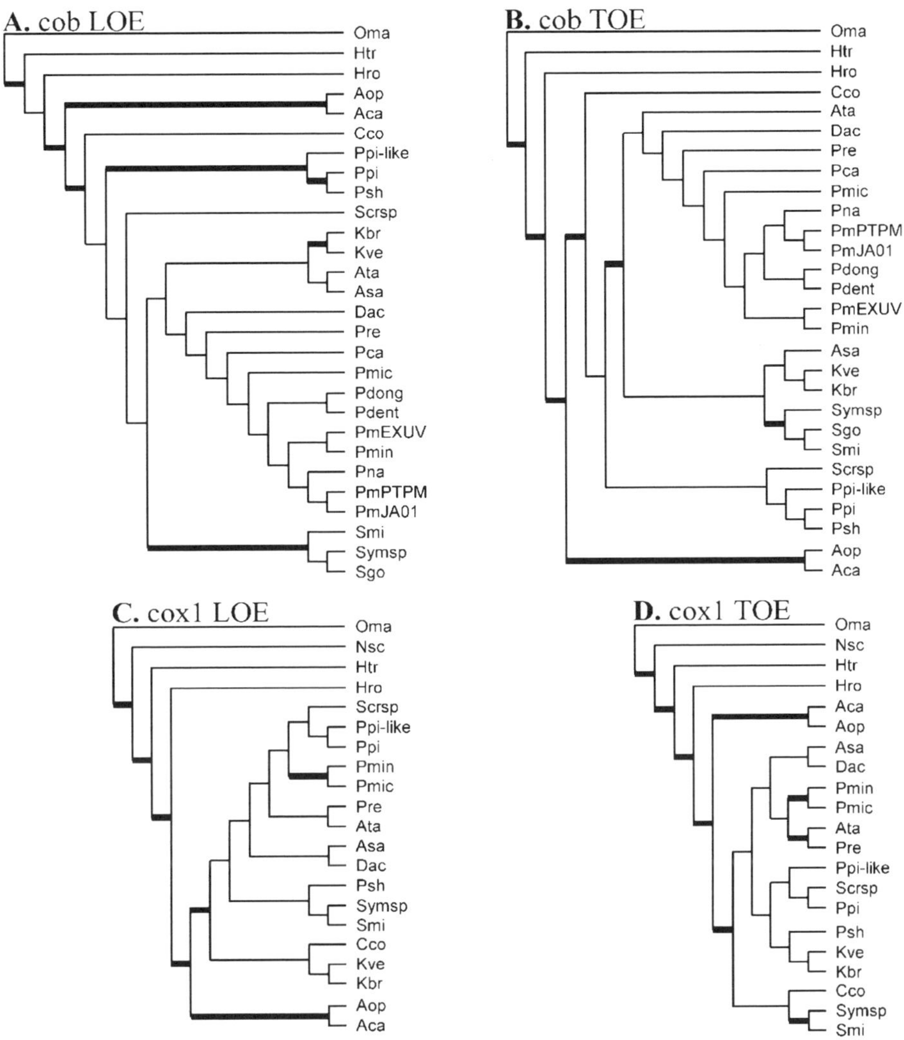

Figure 13.5 Maximum parsimony tree inferred on the basis of (**A**, **C**) location (LOE) and (**B**, **D**) type (TOE) of editing. For analysis of the LOE-based tree, each location of editing ("editable site") was designated a character, whereas for the TOE-based analysis, each type of editing (e.g., A $\rightarrow$ G, C $\rightarrow$ U) was treated as a character. Presence (1) or absence (0) of editing at each editable site (LOE) or of each editing type (TOE) was set as the state of the character. Maximum parsimony analysis was performed using the PENNY program in the Phylip v3.65 package with a Camin–Sokal model. Confidence in branching order was evaluated by bootstrap analysis (1000 resamplings); branches with bootstrap values higher than 70% are indicated by thick lines. Note that in all these trees, there is a consistent branching order from *Oxyrrhis*, *Heterocapsa*, *Amphidinium*, and *Crypthecodinium*. The major difference between the *cob* and *cox1* editing-based trees lies in the relative positions of Gymnodiniales, Suessiales, and Periniales (*Pfiesteria* and related lineages).

13.6 IMPLICATIONS FOR THE ORIGIN AND EVOLUTION OF THE RNA EDITING MACHINERY

The fact that RNA editing comprises a wide variety of biochemically distinct types that are found in phylogenetically disparate groups of organisms argues strongly that the various enzymatic systems that underlie this editing emerged independently in individual lineages within the eukaryotic domain (3, 5). No mitochondrial RNA editing has been reported in the specific relatives of dinoflagellates (i.e., apicomplexans and ciliates), suggesting that this process in Alveolata did not emerge until after the split of dinoflagellates from these groups. The apparent absence of mitochondrial RNA editing in *Oxyrrhis* and other basal lineages, if confirmed for all genes, suggests that the emergence of RNA editing in Dinoflagellata likely occurred later than the origin of this phylum per se (i.e., after the separation of *O. marina* and other nonediting basal lineages). We cannot completely exclude the possibility that editing emerged earlier than this but was lost subsequently in the basal lineages, as a result of retroposition of cDNA copies of post-edited mRNA, with replacement of the original gene whose transcript would have required editing. If the former scenario is correct, then the higher editing density in late branching lineages of dinoflagellates indicates an expansion of editing during dinoflagellate evolution. On the contrary, if the latter conjecture were true, then the lower editing density in basal lineages could be attributed to a loss of editing sites through the progressive fixation of these edited nucleotides in the genome.

The "editing expansion" hypothesis proposed above would benefit from an analysis of how different types of editing have emerged. In this regard, the observation of almost exclusive A → G edits in the basal dinoflagellate lineages such as *Amphidinium* may be revealing. In the nucleus of organisms in which posttranscriptional A → I changes occur, this transition type of substitution editing operates through an adenosine deaminase acting on RNA (ADAR), a mechanistically simple biochemical reaction. Following A → G is the appearance of U → C and C → U (in *Amphidinium* and *Symbiodinium*; Figures 13.1 and 13.2), which are also widespread in other organisms (Table 13.1). The C → U transition is believed to be carried out by a system similar to that for A → G (i.e., a cytidine deaminase acting on RNA, a "simple" edit as well). Subsequently, and concurrent with the relatively rapid and extensive dinoflagellate radiation evident from the fossil record (15), editing types diversify to include transversions as well as transitions (Figures 13.1 and 13.2). When all of the dinoflagellate data are considered together, a temporal sequence emerges for the appearance of these additional types of editing (Figure 13.6). Both G → C and A → C first appeared in *Symbiodinium*, a lineage diverging after *Amphidinium* according to the latest three-gene phylogeny (Figure 13.4) and the *cox1* TOE-based tree. Then, G → A and U → A emerged in *Pfiesteria* (Peridiniales) and *Ceratium* (Gonyaulacales) and U → G in *Ceratium* and *Prorocentrum* (Prorocentrales). A single G → U edit is detected in *Akashiwo sanguinea*. This species, along with *Karenia* and *Karlodinium*, belongs to Gymnodiniales, whose phylogenetic status cannot be clearly resolved relative to Prorocentrales and Gonyaulacales (Figure 13.4A). Due to the unresolved phylogenetic positions of these latter lineages, the evolutionary order of editing types found in these lineages is not clear.

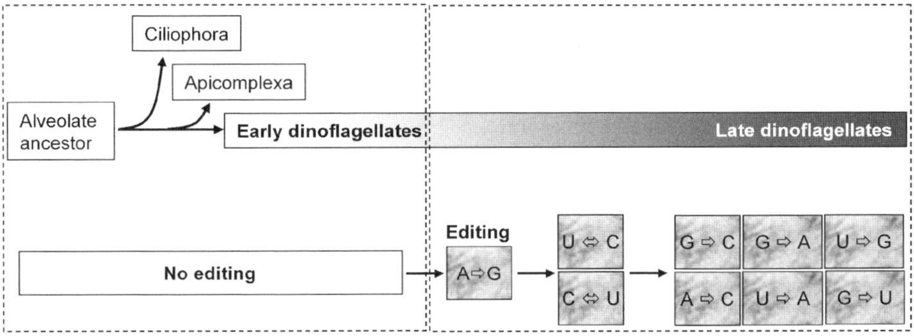

Figure 13.6 Hypothetical evolutionary history of editing in dinoflagellates. From the common ancestor of Alveolata, the ciliates, apicomplexans, and dinoflagellates emerged. Indicated in the dashed box to the left are lineages (Ciliophora, Apicomplexa, and early diverging lineages of dinoflagellates) without RNA editing. In the dashed box to the right are late diverging dinoflagellates with editing. The filled squares with arrows indicate a temporal sequence of emergence of different types of editing. The last six types are placed together on the right because the order of emergence of lineages carrying these types of editing is currently unclear.

Two possibilities can be suggested to account for the emergence of the molecular machinery that mediates RNA editing in dinoflagellates: (1) a single system responsible for every type of editing (transversions as well as transitions) or (2) multiple systems carrying out different types of editing (31). The data presented here might suggest the existence of separate systems in dinoflagellates, each emerging at different times in evolution: for example, transition-type A → G and C → U systems based on deaminases first, then tranversion-type systems presumably operating through some sort of excision-replacement mechanism. Data from the slime mold, *Physarum polycephalum*, indicate that insertion and substitution types of editing can coexist in the mitochondria of the same organism (54). Molecular components for each type of editing that exists (e.g., homologs of ADAR, guide RNAs) will have to be identified in dinoflagellates before we can begin to understand the biochemical mechanism(s) of these editing events.

In this context, it is noteworthy that the dinoflagellate genetic system shares a number of characteristics with its kinetoplastid counterpart. In both systems, genes appear to have few introns; many genes are arranged in polycistronic or otherwise tandem arrays; and nuclear pre-mRNA transcripts frequently acquire a common 5′-end spliced leader via *trans*-splicing (30, 55). In parallel, editing of mitochondrial transcripts in these two lineages of protists is extensive. Two questions that remain to be investigated are (a) whether similar evolutionary pressures are driving the appearance and spread of editing in these two phyla and (b) whether the enzymatic machinery and mechanism of mitochondrial RNA editing in the two systems exhibits any biochemical commonality.

The analysis presented here prompts a number of other questions concerning evolution of RNA editing within dinoflagellates. For example, do the remaining three possible types of editing (A → U, C → G and C → A) that have not yet been identified occur in dinoflagellates? What drives dinoflagellates to adopt such diverse types of

RNA editing? It is intriguing that in *C. horridum*, chloroplast *psaA* transcripts exhibit both the highest frequency of within-codon double edits and the highest editing density among the transcripts examined. Does this observation imply that as the number of editing events increases, the likelihood of within-codon double edits also increases? Furthermore, does the high editing density in *psaA* transcripts reflect selection pressure to increase G + C content, as we postulate for dinoflagellate mitochondrial mRNA editing? Finally, does the similarity of editing that is inferred for dinoflagellate mitochondrial and chloroplast transcripts mean that a single nucleus-encoded system is responsible for editing in both organelles? This is a question that also applies to the substitutional editing that occurs in plant mitochondria and chloroplasts, so it will be of great interest to see whether a similar evolutionary strategy has been adopted (presumably via a convergent evolutionary process) in these two unrelated eukaryotic groups.

ACKNOWLEDGMENT

We thank Dr. david Morse for providing *lingulodinium polyedrum* chloroplast RNA editing data.
This work is supported by the National Atmospheric and Oceanographic Administration ECOHAB grant #NA05NOS4781196 and National Science Foundation grants #DEB-0344186 and #OCE0452780.

REFERENCES

1. Stuart, K.D., Panigrahi, A.K., and Salavati, R. (2001) RNA editing in kinetoplastid mitochondria. In: Bass, B.L. (ed.), *RNA Editing*, Oxford University Press, Oxford, pp. 1- 19.
2. Horton, T. L., and Landweber, L. F. (2002) Rewriting the information in DNA: RNA editing in kinetoplastids and myxomycetes. *Curr Opin Microbiol* **5**, 620–626.
3. Gray, M. W. (2003) Diversity and evolution of mitochondrial RNA editing systems. *IUBMB Life* **55**, 227–233.
4. Bass, B. L. (2002) RNA editing by adenosine deaminases that act on RNA. *Annu Rev Biochem* **71**, 817–846.
5. Gray, M. W. (2001) Speculations on the origin and evolution of editing. In: Bass, B. L. (ed.), *RNA Editing (Frontiers in Molecular Biology)*, Oxford University Press, Oxford, pp. 160–184.
6. Vanfleteren, J. R., and Vierstraete, A. R. (1999) Insertional RNA editing in metazoan mitochondria: The cytochrome b gene in the nematode *Teratocephalus lirellus*. *RNA* **5**, 622–624.
7. Gott, J. M. (2001) RNA editing in *Physarum polycephalum*. In: Bass, B. L. (ed.), *RNA Editing (Frontiers in Molecular Biology)*, Oxford University Press, Oxford, pp. 20–37.
8. Gott, J. M., and Emeson, R. B. (2000) Functions and mechanisms of RNA editing. *Annu Rev Genet* **34**, 499–531.
9. Bock, R. (2001) RNA editing in plant mitochondria and chloroplasts. In: Bass, B. L. (ed.), *RNA Editing (Frontiers in Molecular Biology)*, Oxford University Press, Oxford, pp. 38–60.
10. Driscoll, D. M., and Innerarity, T. L. (2001) RNA editing by cytidine deamination in mammals. In: Bass, B. L. (ed.), *RNA Editing (Frontiers in Molecular Biology)*, Oxford University Press, , pp. 61–76.
11. Emeson, R. B., and Singh, M. (2001) Adenosine-to-inosine RNA editing: Substrates and consequences. In: Bass, B. L. (ed.), *RNA Editing (Frontiers in Molecular Biology)*, Oxford University Press, Oxford, pp. 109–138.
12. Hough, R. F., and Bass, B. L. (2001) Adenosine deaminases that act on RNA. In: Bass, B. L. (ed.), *RNA Editing (Frontiers in Molecular Biology)*, Oxford University Press, Oxford, pp. 77–108.

13. Gajadhar, A. A., Marquardt, W. C., Hall, R., Gunderson, J., Ariztia-Carmona, E. V., and Sogin, M. L. (1991) Ribosomal RNA sequences of *Sarcocystis muris*, *Theileria annulata* and *Crypthecodinium cohnii* reveal evolutionary relationships among apicomplexans, dinoflagellates, and ciliates. *Mol Biochem Parasitol* **45**, 147–154.
14. Cavalier-Smith, T. (1998) A revised six-kingdom system of life. *Biol Rev* **73**, 203–266.
15. Fensome, R. A., Saldarriaga, J. F., and Taylor, F. J.R. (1999) Dinoflagellate phylogeny revisited: reconciling morphological and molecular based phylogenies. *Grana* **38**, 66–80.
16. Moldowan, J. M., and Talyzina, N. M. (1998) Biogeochemical evidence for dinoflagellate ancestors in the early Cambrian. *Science* **281**, 1168–1170.
17. Hackett, J. D., Yoon, H. S., Soares, M. B., Bonaldo, M. F., Cassavant, T. L., Scheetz, T. E., Nosenko, T., and Bhattacharya, D. (2004) Migration of the plastid genome to the nucleus in a peridinin dinoflagella *Curr. Biol.* **14**, 213–218.
18. Schnepf, E., and Elbrachter, M. (1999) Dinophyte chloroplasts and phylogeny—A review. *Grana* **38**, 81–97.
19. Archibald, J. M., and Keeling, P. J. (2002) Recycled plastids: A "green movement" in eukaryotic evolution. *Trends Genet* **18**, 577-584.
20. Keeling, P. J., Archibald, J. M., Fast, N. M., and Palmer, J. D. (2004) Comment on "The evolution of modern eukaryotic phytoplankton." *Science* **306**, 2191.
21. Okamoto, O. K., and Hastings, J. W. (2003) Novel dinoflagellate clock-related genes identified through microarray analysis. *J Phycol* **39**, 519–526.
22. Taroncher-Oldenburg, G., and Anderson, D. M. (2000) Identification and characterization of three differentially expressed genes, encoding S-adenosylhomocysteine hydrolase, methionine aminopeptidase, and a histone-like protein, in the toxic dinoflagellate *Alexandrium fundyense*. *Appl Environ Microbiol* **66**, 2105–2112.
23. Okamoto, O. K., Robertson, D. L., Fagan, T. F., Hastings, J. W., and Colepicolo, P. (2001) Different regulatory mechanisms modulate the expression of a dinoflagellate iron-superoxide dismutase. *J Biol Chem* **276**, 19989–19993.
24. Erdner, D. and Anderson, D. (2006) Global Transcriptional profiling of the toxic dinoflagellate *Alexandrium Fundyense*. using massively parallel signature sequencing, *BMC Genomics* **7**, 88.
25. Van Dolah, F. M., Lide, K. B., Morey, J. S., Brunelle, S. A., Ryan, J. C., Monroe, E. A., and Haynes, B. L. (2007) Microarray analysis of diurnal- and circadian-regulated genes in the Florida red-tide dinoflagellate *Karenia brevis* (Dinophyceae) *J. Phycol.* **43**, 741–752.
26. Rowan, R., Whitney, S. M., Fowler, A., and Yellowlees, D. (1996) Rubisco in marine symbiotic dinoflagellates: Form II enzymes in eukaryotic oxygenic phototrophs encoded by a nuclear multigene family. *Plant Cell* **8**, 539–553.
27. Okamoto, O. K., Liu, L., Robertson, D. L., and Hastings, J. W. (2001) Members of a dinoflagellate luciferase gene family differ in synonymous substitution rates. *Biochemistry* **40**, 15862–15868.
28. Lin, S., and Zhang, H. (2003) Mitogen-activated protein kinase (MAPK) in *Pfiesteria piscicida* and its growth rate-related expression. *Appl Environ Microbiol* **69**, 343–349.
29. Zhang, H., Hou, Y., and Lin, S. (2006) Isolation and characterization of PCNA in the dinoflagellate *Pfiesteria piscicida*. *J Euk Microbiol* **53**, 142–150.
30. Zhang, H., Hou, Y., Miranda, L., Campbell, D. A., Sturm, N. R., Gaasterland, T., and Lin, S. (2007) Spliced leader RNA *trans*-splicing in dinoflagellates. *Proc Natl Acad Sci USA* **104**, 4618–4623.
31. Lin, S., Zhang, H., Spencer, D. F., Norman, J. E., and Gray, M. W. (2002) Widespread and extensive editing of mitochondrial mRNAs in dinoflagellates. *J Mol Biol* **320**, 727–739.
32. Zhang, H., and Lin, S. (2005) Mitochondrial cytochrome b mRNA editing in dinoflagellates: Possible ecological and evolutionary associations? *J Euk Microbiol* **52**, 538–545.
33. Zauner, S., Greilinger, D., Laatsch, T., Kowallik, K. V., and Maier, U. -G. (2004) Substitutional editing of transcripts from genes of cyanobacterial origin in the dinoflagellate *Ceratium horridum*. *FEBS Lett* **577**, 535–538.
34. Wang, Y., and Morse, D. (2006) Rampant polyuridylylation of plastid gene transcripts in the dinoflagellate *Lingulodinium*. *Nucleic Acids Res* **34**, 613–619.
35. Zhang, H., and Lin, S. (2007) Status of mRNA editing and SL RNA trans-splicing groups *Oxyrrhis*, *Noctiluca*, *Heterocapsa*, and *Amphidinium* as basal lineages of true dinoflagellates. Submitted

36. Zhang, H., Bhattacharya, D., Maranda, L., and Lin, S. (2007) Mitochondrial *cob* and *cox1* and their mRNA editing in dinoflagellates: With reference to phylogenetic position of *Dinophysis*. Submitted
37. Adams, K. L., Song, K., Roessler, P. G., Nugent, J. M., Doyle, J. L., Doyle, J. J., and Palmer, J. D. (1999) Intracellular gene transfer in action: Dual transcription and multiple silencings of nuclear and mitochondrial *cox2* genes in legumes. *Proc Natl Acad Sci USA* **96**, 13863–13868.
38. Waller, R. F., Patron, N. J., and Keeling, P. J. (2006) Phylogenetic history of plastid-targeted proteins in the peridinin-containing dinoflagellate *Heterocapsa triquetra. Int J Syst Evol Microbiol* **56**, 1439–1447.
39. Laatsch, T., Zauner, S., Stoebe-Maier, B., Kowallik, K. V., and Maier, U. -G. (2000) Plastid-derived single gene minicircles of the dinoflagellate *Ceratium horridum* are localized in the nucleus. *Mol Biol Evol* **21**, 1318–1322.
40. Flø-Jorgensen, M. F., Murray, S., and Daugbjerg, N. (2004) *Amphidinium* revisited. I. Redefinition of *Amphidinium* (Dinophyceae) based on cladistic and molecular phylogenetic analysis. *J. Phycol.* **40**, 351–365.
41. Murray, S., Flø-Jorgensen, M. F., Ho, S. Y., Patterson, D. J., and Jermiin, L. S. (2005) Improving the analysis of dinoflagellate phylogeny based on rDNA. *Protist* **156**, 269–286.
42. Zhang, H, Bhattacharya, D., and Lin, S. (2007) A three-gene dinoflagellate phylogeny suggests monophyly of Prorocentrales and a basal position for *Amphidinium* and *Heterocapsa. J Mol Evol* **65**, 463–474.
43. Covello, P. S., and Gray, M. W. (1990) Differences in editing at homologous sites in messenger RNAs from angiosperm mitochondria. *Nucleic Acids Res* **18**, 5189–5196.
44. Hoopengardner, B., Bhalla, T., Staber, C., and Reenan, R. (2003) Nervous system targets of RNA editing identified by comparative genomics. *Science* **301**, 832–836.
45. Kawahara, Y., Ito, K., Sun, H., Aizawa, H., Kanazawa, I., and Kwak, S. (2004) RNA editing and death of motor neurons. *Nature* **427**, 801.
46. Gott, J. M. (2003) Expanding genome capacity via RNA editing. *C R Biol* **326**, 901–908.
47. Dong, K. (2007) Insect sodium channels and insecticide resistance. *Invert Neurosci* **7**, 17–30.
48. Kawahara, Y., Zinshteyn, B., Sethupathy, P., Iizasa, H., Hatzigeorgiou, A. G., and Nishikura, K. (2007) Redirection of silencing targets by adenosine-to-inosine editing of miRNAs. *Science* **315**, 1137–1140.
49. Nishikura, K. (2006) Editor meets silencer: crosstalk between RNA editing and RNA interference. *Nat Rev Mol Cell Biol* **7**, 919–931.
50. Wilson, R. J.M., and Williamson, D. H. (1997) Extrachromosomal DNA in the Apicomplexa. *Microbiol Mol Biol Rev* **61**, 1–16.
51. Feagin, J. E. (2000) Mitochondrial genome diversity in parasites. *Int J Parasitol* **30**, 371–390.
52. Cummings, D. J. (1992) Mitochondrial genomes of the ciliates. *Int Rev Cytol* **141**, 1–64.
53. Alfonzo, J. D., Blanc, V., Estévez, A. M., Rubio, M. A., and Simpson, L. (1999) C to U editing of the anticodon of imported mitochondrial $tRNA^{Trp}$ allows decoding of the UGA stop codon in *Leishmania tarentolae. EMBO J* **18**, 7056–7062.
54. Gott, J. M., Visomirski, L. M., and Hunter, J. L. (1993) Substitutional and insertional RNA editing of the cytochrome *c* oxidase subunit 1 mRNA of *Physarum polycephalum. J Biol Chem* **268**, 25483–25486.
55. Mayer, M. M. and Floeter-Winter, L. M. (2005) Pre-mRNA *trans*-splicing: from kinetoplastids to mammals, an easy language for life diversity. Mem Inst Oswaldo Cruz, *Rio de Janeiro* **100**, 501–513.
56. Chen, S. H., Habib, G., Yang, C. Y., Gu, Z. W., Lee, B. R., Weng, S. A., Silberman, S. R., Cai, S. J., Deslypere, J. P., Rosseneu, M., *et al.* (1987) Apolipoprotein B-48 is the product of a messenger RNA with an organ-specific in-flame stop codon. *Science* **238**, 363–366.
57. Powell, L. M., Wallis, S. C., Pease, R. J., Edwards, Y. H., Knott, T. J., and Scott, J. (1987) A novel form of tissue-specific RNA processing produces apolipoprotein-B48 in intestine. *Cell* **50**, 831–840.
58. Keegan, L. P., Gallo, A., and O'Donnell, M. A. (2001) The many roles of an RNA editor *Nat Rev Genet* **2**, 869–878.
59. Burns, C. M., Chu, H., Rueter, S. M., Hutchinson, L. K., Canton, H., Sanders-Bush, E., and Emeson, R. B. (1997) Regulation of serotonin-2C receptor G-protein coupling by RNA editing. *Nature* **387**, 303–308.
60. Sommer, B., Kohler, M., Sprengel, R., and Seeburg, P. H. (1991) RNA editing in brain controls a determinant of ion flow in glutamate-gated channels. *Cell* **67**, 11–19.
61. Janke, A., and Päbo, S. (1993) Editing of a tRNA anticodon in marsupial mitochondrial changes its codon recognition. *Nucleic Acids Res* **21**, 1523–1525.

62. Covello, P. S., and Gray, M. W. (1989) RNA editing in plant mitochondria. *Nature* **341**, 662–666.
63. Gualberto, J. M., Lamattina, L., Bonnard, G., Weil, J. –H., and Grienenberger, J. –M. (1989) RNA editing in wheat mitochondria results in the conservation of protein sequences. *Nature* **341**, 660–662.
64. Hiesel, R., Wissinger, B., Schuster, W., and Brennicke, A. (1989) RNA editing in plant mitochondria. *Science* **246**, 1632–1634.
65. Maréchal-Drouard, L., Ramamonjisoa, D., Cosset, A., Weil, J. H., and Dietrich, A. (1993) Editing corrects mispairing in the acceptor stem of bean and potato mitochondrial phenylalanine transfer RNAs. *Nucleic Acids Res* **21**, 4909–4914.
66. Casey, J. L., and Gerin, J. L. (1995) Hepatitis D virus RNA editing: Specific modification of adenosine in the antigenomic RNA. *J Virol* **69**, 7593–7600.
67. Bourara, K., Litvak, S., and Araya, A. (2000) Generation of G-to-AS and C-to-U changes in HIV-1 transcripts by RNA editing. *Science* **289**, 1564–1566.
68. Smith, L. A., Wang, X., Peixoto, A. A., Neumann, E. K., Hall, L. M., and Hall, J. C. (1996) A Drosophila calcium channel alphal subunit gene maps to a genetic locus associated with behavioral and visual defects. *J Neurosci* **16**, 7868–7879.
69. Hanrahan, C. J. Palladino, M. J., Ganetzky, B., and Reenan, R. A. (2000) RNA editing of the *Drosophila para* Na (+) channel transcript, Evolutionary conservation and developmental regulation. *Genetics* **155**, 1149–1160.
70. Song, W., Liu, Z., Tan, J., Nomura, Y., and Dong K. (2004) RNA editing generates tissue-specific sodium channels with distinct gating properties. *J Biol Chem* **279**, 2554–2561.
71. Patton, D. E., Silva, T., and Bezanilla, F. (1997) RNA editing generates a diverse array of transcripts encoding squid kv2k+ channels with altered functional properties. *Neuron* **19**, 711–722.
72. Ben-Shlomo, H., Levitan, A., Shay, N. E., Goncharov, I., and Michaeli, S. (1999) RNA editing associated with the generation of two distinct conformations of the trypanosomatid *Leptomonas collosoma* 7SL RNA. *J Biol Chem* **274**, 25642–25650.
73. Steidinger, K. A., and Tengen, K. (1997) Dinoflagellates. In: Tomas, C. R. (ed) *Identifying Marine Phytoplankton*, Academic Press, New York, pp. 387–594.

PART IV

STRUCTURAL APPROACHES

CHAPTER 14

THE BOX C/D RNPs: EVOLUTIONARILY ANCIENT NUCLEOTIDE MODIFICATION COMPLEXES

Keith Gagnon
Xinxin Zhang
E. Stuart Maxwell

THE 1990s witnessed an explosion of small RNAs found localized to the eukaryotic nucleolus. Included in these small noncoding RNAs were the box C/D small nucleolar RNAs (snoRNAs), which base-pair with ribosomal RNA to direct site-specific, nucleotide 2′-*O*-methylation. Continuing investigations rapidly expanded the box C/D snoRNA population, identified new functions, defined the limited set of core proteins that bind the box C/D RNAs to assemble ribonucleoprotein (RNP) complexes, and even discovered box C/D RNAs and core proteins in Archaea. *In vitro* ribonucleoprotein assembly systems along with emerging crystal structures are now providing insight into the molecular details of both box C/D RNP structure and methylation function. This chapter reviews recent advances in box C/D RNA and RNP biology, concluding with a discussion of future directions and the experimental challenges that face investigators in this exciting and rapidly advancing field of RNA-guided nucleotide modification.

14.1 INTRODUCTION

Small RNAs were first identified in eukaryotic nuclei in the 1960s when a handful of abundant species were phenol-extracted from isolated nuclei and characterized on sucrose gradients and polyacrylamide gels (reviewed in reference 1). Subsequent nucleotide analysis revealed a base composition distinct from GC-rich ribosomal RNA leading to their designation as U-rich small nuclear RNAs or the U snRNAs

RNA and DNA Editing: Molecular Mechanisms and Their Integration into Biological Systems,
Edited by Harold C. Smith

(2, 3). It quickly became apparent that the U snRNAs are partitioned in the nucleoplasmic and nucleolar compartments (4–8), and sequencing in the 1970s defined U3 as the first nucleolar box C/D RNA (9–12). The box C/D snoRNA population remained small until the early 1990s with the discovery of numerous intronic snoRNAs encoded within pre-messenger RNA introns (reviewed in reference 13). Discovery of box C/D RNAs in archaeal organisms at the end of the decade indicated an evolutionarily ancient origin (14–16). The populations of eukaryotic and archaeal box C/D RNAs has continued to grow and now numbers in the hundreds.

Nucleolar localization of the box C/D snoRNAs implied a role in ribosome biogenesis, and experiments in *Saccharomyces cerevisiae* as well as *Xenopus laevis* demonstrated the essentiality of the major U3 box C/D snoRNA for ribosomal RNA processing (7, 17–19). Several additional snoRNAs were identified as crucial for pre-rRNA processing, although most tested snoRNA species revealed no growth phenotype when deleted in yeast (20–22). Critical to understanding the primary function of box C/D snoRNAs in nucleotide modification was the observation that they possessed regions of complementarity with rRNA and that these complementary regions corresponded to sites of nucleotide 2′-*O*-methylation (23). The list of target RNAs modified by the box C/D RNAs has now expanded to include the eukaryotic splicing snRNAs, some pre-messenger RNAs, and archaeal tRNAs (15, 24–26).

In the last decade, investigations have focused upon identifying box C/D RNA-binding proteins. Highly homologous eukaryotic and archaeal core proteins have been defined, and the identification of eukaryotic accessory proteins suggests that snoRNP assembly may be a complex and regulated process (27–33). *In vitro* assembly systems for the archaeal box C/D sRNP complex have been established and used to define box C/D RNP assembly pathways as well as reveal RNP structure and examine the mechanisms of box C/D RNA-guided nucleotide methylation function (34–36). These studies, coupled with emerging crystal structures of the core proteins and RNA–protein complexes, are now leading to a more detailed understanding of this RNA–protein enzyme and its function as an RNA-guided nucleotide modification complex.

14.2 DIVERSITY OF BOX C/D RNA POPULATIONS

14.2.1 Box C/D RNA Nomenclature

Box C/D RNAs are abundant in both Eukarya and Archaea, numbering in the hundreds. Following the original nomenclature, most animal and plant homologs are designated U(n) with the species number, reflecting the approximate order of discovery. Homologs identified in other organisms usually adopt the same designation. However, there are organisms that use a nonconventional nomenclature. Yeast snoRNAs are designated by a number preceded by the prefix snR(n) (37). In Archaea, the snoRNA-like small RNAs or sRNAs are numbered and preceded by sR(n) (14). Other examples of nonstandard nomenclature include the protozoa *Trypanosoma brucei* (TB*x*Cs*y*C*z*, x = chromosome number, y = cluster number, z = RNA number) and *Euglena gracilis* (Eg-m*x*, x = RNA number) (38, 39). Different naming schemes have also been employed for tissue-specific or disease-related box C/D snoRNAs, such as the mouse-brain-specific box C/D snoRNA II-52 (MBII-52) or the box C/D snoRNA linked to neonatal lethality in Prader–Willi syndrome (Pwcr1/MBII-85) (40, 41).

14.2.2 Box C/D RNA Structure

Box C/D RNAs are defined by their highly conserved nucleotide box C (RUGAUGA) and box D (CUGA) sequences positioned near the 5′ and 3′ termini, respectively (Figure 14.1), and frequently possess internal C′ and D′ boxes (42–44). The C′ and D′ nucleotide boxes are typically present and well-conserved in the archaeal sRNAs but often degenerate and difficult to identify, or even missing, in the eukaryotic snoRNAs (14–16, 45). Individual box C/D RNAs are defined by their unique guide sequences, which are located immediately adjacent and upstream of boxes D and D′. These guide sequences base-pair to the different target RNAs and determine the site of box C/D RNA action on the target RNA (44, 46–48). Archaeal box C/D sRNAs are typically smaller (50–70 nucleotides) than the eukaryotic box C/D snoRNAs (75–150 nucleotides) (15, 16, 37, 49–51). The highly conserved archaeal guide regions of 12 nucleotides most often constitute the entire spacer regions between the box C/D and C′/D′ motifs and account for the smaller size of the archaeal sRNAs (52). The well-defined box C/D and C′/D′ motifs fold to establish kink-turn or “K-turn” and

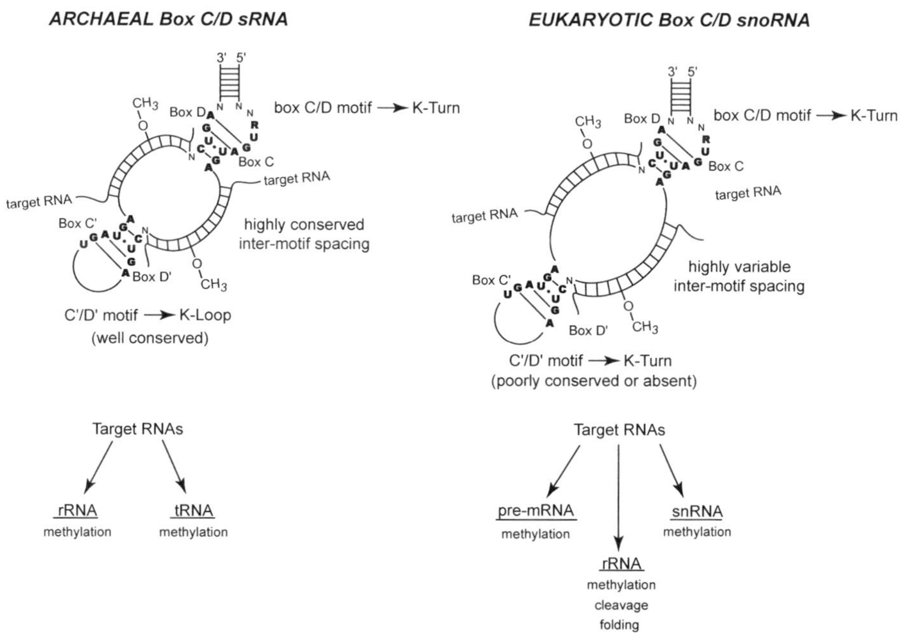

Figure 14.1 Archaeal and eukaryotic box C/D RNAs and their target RNAs. The secondary structures of archaeal and eukaryotic box C/D RNAs are shown with box C, D, C′, and D′ nucleotide sequences designated in bold. Representative target RNAs are base paired with the D and D′ guide sequences and the nucleotides targeted for methylation indicated. Box C/D and C′/D′ motifs are designated as well as those structural differences characteristic of the archaeal and eukaryotic box C/D RNAs. Known target RNAs of the archaeal and eukaryotic box C/D RNAs are shown below.

"K-loop" elements, respectively. The K-turn consists of canonical stem I and internal stem II separated by an asymmetric bulge that possesses tandem sheared G–A pairs essential for K-turn stability (30, 53, 54). The internal K-loop lacks stem I and is typically replaced with a small loop. Ultimately, both the K-turn and K-loop serve as core protein binding sites to assemble the box C/D and C′/D′ RNP complexes (34–36). The recent discovery of circularized box C/D sRNAs in *Pyrococcus furiosus* suggests a novel processing pathway that may augment box C/D sRNA stability in hyperthermophiles (55).

14.2.3 Diversity of Box C/D RNA Populations

Appreciation of box C/D RNA diversity in Ukarya and Archaea has come with the characterization of RNA populations in a variety of organisms in both kingdoms. In eukaryotes, approximately two-thirds of the yeast snoRNAs involved in rRNA maturation are conserved and found in plants and humans (56). The yeast box C/D snoRNA population has been well-defined and currently numbers 46 RNAs (57). This compares with more than double that number of human box C/D RNAs involved in rRNA maturation, but does not include additional RNAs modifying other cellular target RNAs (58). Thus, the box C/D snoRNA populations of metazoan organisms are characteristically more complex with many of these additional RNAs unique to specific organisms. Box C/D homologs in different organisms may possess unique guide sequences reflecting organism-specific rRNA sequences yet modify corresponding nucleotides in the respective ribosomal RNAs. However, many box C/D snoRNAs modify rRNA nucleotides or nucleotides of other target RNAs that are unique to that organism (38). Thus, different eukaryotes typically possess a unique set of snoRNAs characteristic of that organism. While the majority of box C/D snoRNAs are involved in nucleotide methylation of rRNA nucleotides, conserved box C/D snoRNAs such as U3, U8, and U14 function in animals, plants, and fungi in nonmodification roles such facilitating pre-rRNA folding or cleavage during precursor processing (13). To date, there appears to be no shared box C/D RNA homologs between Archaea and Eukarya. Similarly, box C/D sRNA populations among Crenarchaeota and Euryarchaeota are typically distinct with respect to the specific organism examined (15, 16, 59).

14.2.4 Box C/D RNA Identification

Hundreds of box C/D RNAs have been defined in numerous eukaryotic and archaeal organisms. Techniques to identify and define box C/D RNAs include both biochemical and computational approaches. Biochemical approaches include isolation of small RNAs fractionated on gradients or immunoprecipitated with antibodies against box C/D RNP proteins, followed by cDNA cloning and sequencing (60–62). Immunoprecipitation of box C/D RNPs recently identified 66 previously unknown box C/D snoRNAs in the protist *Euglena gracilis* (39). Computational approaches have utilized a variety of search engines and algorithms to analyze available databases using box C and D consensus sequences (Table 14.1), as well as appropriately positioned guide sequences that exhibit complementarity to cellular target RNAs (63, 64). Other sequence analysis-based searches have used snoRNA or intronic sequence databases to search for homologs

TABLE 14.1 Box C/D RNA Databases and Search Engines

Database	Website	Comments/Utility	References
snoRNA Database	http://lowelab.ucsc.edu/snoRNAd b/	Box C/D snoRNAs from *S. cerevisiae*, *A. thaliana*, and currently 8 Archaea are sorted by name, target, and genome locus.	14, 49, 173
Yeast snoRNA Database	http://people.biochem.umass.edu/f ournierlab/snornadb/main.php	Box C/D and H/ACA snoRNA database specific for yeast. Provides interactive 3-D view of rRNA modifications.	56, 57
snoRNA-LBME-db	http://www-snorna.biotoul.fr/	Human box C/D and H/ACA snoRNA database. Find snoRNAs by name, target sequence or modification, or genomic location.	8383
Plant snoRNA Database	http://bioinf.scri.sari.ac.uk/cgi-bin/ plant_snorna/home	Database of currently 18 plant species in a tabulated and downloadable sequence format.	83
The snoRNP Database	http://www.mbio.ncsu.edu:8001/h ome.html	Collection of 6349 small nucleolar RNAs and 302 associated proteins. Includes genus and species, accession numbers, loci, molecular family/class, and citations.	
Human snoRNA Database	http://www.trex.uqam.ca/∼snorna/	Contains the sequences of currently 463 human snoRNAs. Provides 2-D structure and energy information imported from mFold.	
sno/scaRNAbase	http://gene.fudan.sh.cn/snoRNAba se.nsf	Database consisting of 1979 sno/scaRNAs from 85 organisms. Contains sequence, target site, accession number, references and allows users to perform BLAST searches.	174

TABLE 14.1 (*Continued*)

Database	Website	Comments/Utility	References
RNAdb 2.0	http://research.imb.uq.edu.au/RN Adb	Comprehensive database of mammalian non-protein-coding RNAs (ncRNAs). Provides nucleotide sequences and annotations for tens of thousands of ncRNAs.	175, 176
Algorithm/Search Engine	Website	Function/Utility	References
snoSeeker	http://genelab.zsu.edu.cn/snoseeke r/	Searches for box C/D snoRNAs in the genomic alignment of two or more organisms.	64
snoScan	http://lowelab.ucsc.edu/snoscan/	Searches mammalian, yeast, or archaeal genomic sequences for box C/D snoRNA genes with a probabilistic model starting from a query or target RNA sequence.	63, 173
SNO.pl	http://hsc.utoledo.edu/bioinfo/eid/	Searches for conserved structures characteristic of C/D box snoRNAs within the Mammalian Orthologous Intron Database (MOID).	65

in particular organisms (50, 65). Computational approaches can identify potential box C/D RNAs that are expressed at low levels or in a tissue-specific manner whose detection by biochemical approaches could prove difficult. The utility of computational approaches was recently demonstrated by the identification of 62 novel box C/D RNAs from the trypanosome *Leishmania major* and 50 novel snoRNAs in *Caenorhabditis elegans* (66, 67). Both candidate and experimentally confirmed RNAs are found in an array of databases that report ever-increasing populations of archaeal and eukaryotic box C/D RNAs (Table 14.1). Both biochemical approaches have identified "orphan" box C/D RNAs where the associated guide sequence is not complementary to any known cellular RNA sequence, thus raising the question as to not only their potential target RNAs but also their possible function(s).

14.3 BOX C/D RNA FUNCTIONS AND TARGET RNAs

14.3.1 Folding and Cleavage of Pre-rRNA

The base-pairing of box C/D RNA guide sequences to their respective target RNAs determines the site of RNA function. The first identified eukaryotic snoRNAs were localized in the nucleolus, thus implying a role in pre-rRNA processing and/or ribosome biogenesis (7, 18, 68, 69). Early experiments indeed demonstrated the importance of U3 (snR17), U8, U14, snR10, and snR30 box C/D snoRNAs for pre-rRNA cleavage (20, 22, 70–73). Those snoRNAs required for rRNA processing are typically essential RNAs, whereas those snoRNAs guiding nucleotide modification are not. The box C/D snoRNAs required for pre-rRNA cleavage at specific sites in yeast is based upon snoRNA gene disruption experiments. Similar observations have been made for *Xenopus laevis* rRNA processing when these same essential snoRNA species are depleted in oligonucleotide knockout experiments (13, 17, 74, 75).

Some box C/D snoRNAs such as U3, U8, and U14 have been shown to be important for pre-rRNA folding and cleavage. These snoRNAs are considered to have "chaperone" function as their base-pairing to target rRNAs facilitates correct pre-RNA folding required for precursor maturation. U3 hydrogen-bonds with the 5′ ETS of the rRNA precursor and U14 base-pairs with 18S rRNA (76–78). Both snoRNAs are essential for pre-rRNA cleavage and production of mature 18S rRNA. U8 is required for proper folding of 5.8S and 28S rRNAs within the precursor transcript and is essential for 28 S rRNA production (74, 79, 80). Interestingly, U14 not only guides 18S rRNA processing using its C′/D′ guide sequence, but also guides nucleotide methylation using its box C/D guide sequence (75). Recent work utilizing computer simulations has suggested that snoRNAs may play important roles in long-range rRNA folding (81). To date, eukaryotic box C/D RNAs have not been implicated in chaperone folding or precursor processing functions for any RNA other than pre-ribosomal RNA.

14.3.2 2′-*O*-Methylation of Diverse RNA Targets

The primary role of most box C/D RNAs is to guide the modification of a targeted nucleotide's ribose sugar, catalyzing addition of a methyl group at the 2′-hydroxyl position (Figure 14.1). The nucleotide targeted for methylation is positioned within the RNA duplex formed by the hydrogen-bonded target:guide RNA sequences and base-paired to the fifth nucleotide upstream from the snoRNA's D or D′ box (44, 46, 47). The number of nucleotides modified by the box C/D RNAs in Eukaryotes and Archaea numbers in the hundreds and thus explains the large box C/D populations found in eukaryotic and archaeal organisms. Approximately 100 ribose methylations are found in human rRNA, about 120 in plant *Arabidopsis thaliana* rRNA, and 55 in yeast rRNA (82–84). The occurrence of methylated nucleotides in archaeal rRNA is also considerable with 67 identified in *Sulfolobus solfataricus* and at least 26 and 93 predicted in *Sulfolobus acidocaldarius* and *Pyrococcus abyssi*, respectively (15, 16, 85). The specific rRNA nucleotides methylated vary from organism to organism, but these modifications tend to be clustered in conserved core and functional regions of the rRNA (84). Disruption of one or a few modifications at specific nucleotides often does not result in an obvious phenotype (84), although complete disruption of nucleotide methylation in yeast has severe effects upon rRNA maturation and ribosome function (86).

The types of RNA targeted for methylation by box C/D RNAs extends beyond the ribosomal RNAs to include archaeal tRNAs and the eukaryotic splicing snRNAs. Many of the modifications in archaeal rRNA and tRNAs are predicted by newly defined box C/D sRNAs and their associated D and D′ guide sequences (15). Also found in Archaea and Eukarya are "orphan" RNAs where the guide sequence(s) of an identified C/D RNA is not complementary to any sequence in a known cellular RNA (48, 64). In eukaryotes, specific box C/D RNAs guide nucleotide methylation of U1, U2, U4, U5, U6, and U12 snRNAs (24, 87, 88) (see Chapter 8). Specific members of the box C/D RNA family are also found localized to the nuclear Cajal bodies and designated small Cajal RNAs or scaRNAs. These scaRNAs are unique in that they possess both box C/D and box H/ACA RNA motifs and direct the 2′-*O*-methylation and pseudouridylation of the RNA Pol II-transcribed snRNAs U1, U2, U4, and U5 within the Cajal bodies (24) (see Chapter 8).

14.3.3 Additional Roles and Targets for Box C/D RNAs

Investigation of specific tissues has suggested a larger population of box C/D RNAs than originally anticipated. Analysis of small RNAs in mouse, rat, and human brain has identified many brain-specific box C/D snoRNAs (40, 89, 90). Only the mouse/human-brain-specific box C/D snoRNA MBII-52/HBII-52 has been mapped to a putative target. This box C/D snoRNA appears to target an A-to-I editing site in the serotonin receptor 2C (5-HT_{2C}) pre-mRNA. Methylation of this site potentially modulates pre-mRNA editing and alternative splicing, thereby resulting in changes to serotonin signaling capacity (25). Alternatively, it has been suggested that the HBII-52 snoRNA uses its guide region sequences to affect alternative splicing of the 5-HT_{2C} pre-mRNA by masking splicing silencers, which then leads to the inclusion of an exon in the final spliced serotonin receptor mRNA (26). The loss of MBII-85/HBII-85 is implicated in Prader–Willi syndrome (91). Furthermore, regulation of MBII-52 and MBII-48 snoRNA levels has been noted in the early phase of memory consolidation during learning, thus suggesting a possible role for this snoRNA in higher brain function (92).

14.4 BOX C/D RNP STRUCTURE AND NUCLEOTIDE METHYLATION FUNCTION

14.4.1 Eukaryotic Box C/D Core Proteins and snoRNP Structure

Box C/D snoRNAs bind a small set of core proteins to establish ribonucleoprotein (RNP) complexes (Figure 14.2). Terminal box C/D and internal C′/D′ snoRNA motifs serve as protein binding sites for RNP assembly, and it is the bound core proteins that carry out the nucleotide 2′-*O*-methytransferase reaction. Four well-conserved core proteins bind eukaryotic box C/D snoRNAs. They are the 15.5kD protein (Snu13p in yeast), nucleolar proteins 56 (Nop56) and 58 (Nop58), and the methyltransferase fibrillarin. The 15.5kD protein directly binds the box C/D K-turn, but not box C′/D′ motifs, initiating assembly of box C/D snoRNPs (30, 93–95). The use of nucleotide analog interference mapping (NAIM) has suggested that 15.5kD's inability to bind C′/D′ motifs may lie in slight structural differences that arise when a K-turn possesses a terminal loop instead of a stem II structure (96).

The Nop56 and Nop58 core proteins are highly homologous and essential for nucleotide modification and ribosome biogenesis (27, 97). Crosslinking experiments have shown that Nop58 interacts with box C whereas Nop56 interacts with box C′ (94). This suggests that Nop58 and Nop56 core proteins bind the K-turn or K-loops, respectively, although their asymmetric distribution in the snoRNP complex has not yet been confirmed in RNP assembly experiments. Binding of both Nop56 and Nop58 requires stem II of the box C/D motif (98). These core proteins are likely to be RNP structural proteins, although their role in the methylation reaction or its regulation cannot be ruled out at this time.

The methyltransferase fibrillarin interacts with both Nop56 and Nop58 and can be crosslinked to both the box C/D and C′/D′ motifs (27, 94). Evidence establishing

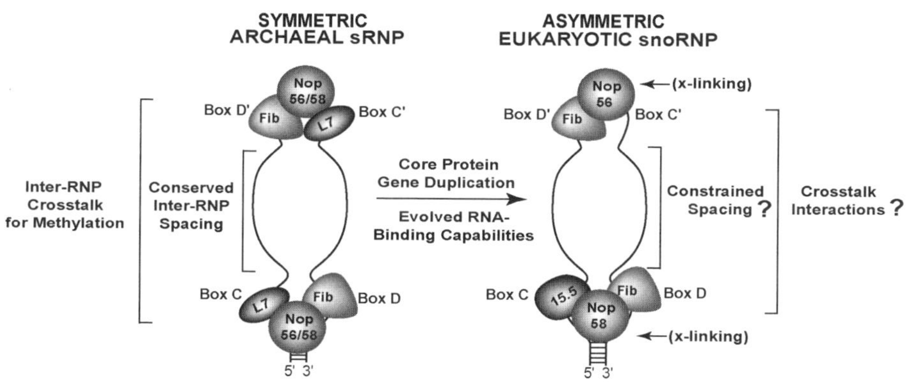

Figure 14.2 Archaeal and eukaryotic box C/D ribonucleoprotein complexes. Archaeal and eukaryotic box C/D RNAs are shown with the RNP core proteins bound to the box C/D and C′/D′ motifs. The binding of all three archaeal core proteins to both the box C/D and C′/D′ motifs assembles a symmetric sRNP. The differential distribution of the eukaryotic 15.5kD, Nop56, and Nop58 core proteins binding the box C/D and C′/D′ motifs establishes an asymmetric snoRNP. [Crosslinking experiments have implied the differential distribution of Nop58 and Nop56 on the eukaryotic box C/D and C′/D′ motifs, respectively (94).]

fibrillarin as the methyltransferase enzyme includes the presence of an *S*-adenosylmethionine (AdoMet)-dependent methyltransferase-like domain that is required for methylation activity (99, 100) and the observation that disruption of the yeast fibrillarin gene results in the loss of pre-rRNA methylation (86).

14.4.2 Archaeal Box C/D Core Proteins and *In Vitro* sRNP Assembly

Archaeal box C/D sRNAs bind three core proteins that are highly homologous to the eukaryotic core proteins (29, 34, 35). Ribosomal protein L7, homologous to eukaryotic 15.5kD, recognizes both box C/D and C′/D′ motifs to initiate box C/D and C′/D′ RNP assembly. A single Nop56/58 core protein, highly homologous to the eukaryotic Nop56 and Nop58 pair, also binds both box C/D and C′/D′ motifs. The third core protein fibrillarin binds both complexes to complete box C/D and C′/D′ RNP assembly. Much of what is presently known about box C/D RNP structure and assembly comes from examination of *in vitro* assembly of the archaeal sRNP using *in vitro* synthesized box C/D sRNAs and recombinant sRNP core proteins. These *in vitro* assembly studies revealed an order of assembly with L7 initiating RNP formation followed by Nop56/58 and then fibrillarin binding (34–36). Binding of the L7 core protein induces structural changes in both box C/D and C′/D′ motifs (101–103). These structural changes are required for subsequent Nop56/58, but not fibrillarin, binding. Nop56/58 and fibrillarin form a stable dimer in the absence of the sRNA and/or L7, and evidence suggests that it is the dimer that binds the assembling box C/D and C′/D′ RNP complexes (35, 103–105). In contrast to the “asymmetric” eukaryotic snoRNP complex, the binding of all three core proteins to the box C/D and C′/D′ motifs assembles a “symmetric” sRNP (Figure 14.2).

14.4.3 Emerging Core Protein and RNP Crystal Structures

Emerging crystal structures of core proteins, the K-turn, and sRNA–core protein subcomplexes are beginning to reveal the detailed architecture of the box C/D RNP. The ease of expressing soluble, recombinant archaeal core proteins has greatly facilitated these studies, and all three archaeal core proteins have yielded crystal structures at detailed resolution. Crystal structures of free L7 core protein, the K-turn, and the L7–K-turn RNP have shown that L7 binding induces a conformational change or “remodels” the sRNA (54, 95, 101, 102, 106). This induced fit has been confirmed in fluorescence resonance energy transfer (FRET) experiments (102). Both the L7 and 15.5kD core proteins lock the K-turn into a tightly kinked structure characterized by stacked stem I and stem II helices bent at approximately 60°, hence the designation of kink-turn for this RNA element (53, 107). Particularly important for binding is the requirement of a pyrimidine nucleotide from box C (UGAUGA) which is extended and inserted into a cleft of the L7 protein. A co-crystal structure of *Archaeoglobus fulgidus* Nop56/58 and fibrillarin has provided the molecular details of this core protein dimer (104). *S*-adenosyl-methionine is bound to the methyltransferase domain of fibrillarin and biophysical studies have indicated that Nop56/58 binding to fibrillarin helps stabilize Ado-Met in its binding pocket (108). A recent crystal structure of the Nop56/58 homolog hPrp31 bound to the 15.5kD–U4 K-turn RNP has revealed that Nop56/58 and hPrp31 bind the K-turn–L7/15.5kD protein complex through their Nop domains and require interaction with both the K-turn and the L7/15.5kD protein (109). Despite crystallization of all the individual components of the complex, a crystal structure of the fully assembled box C/D sRNP has thus far proved elusive.

14.4.4 Investigating Methylation Function Using *In Vitro* Assembled Archaeal Box C/D sRNP

The archaeal *in vitro* assembly system has enabled investigations of box C/D sRNP methylation function with respect to sRNP structure and the role of the sRNA and individual core proteins. The assembled box C/D sRNP is catalytically active and guides site-specific methylation of target RNAs using both D and D′ guide sequences. Target RNA methylation is nucleotide-specific and dependent upon fibrillarin binding and the presence of *S*-adenosyl-methionine (34). Further studies have shown that efficient methylation of D and D′ guide RNAs requires both box C/D and C′/D′ RNPs be juxtaposed in a fully assembled sRNP (35). The juxtaposed box C/D and C′/D′ RNPs are dependent upon the highly conserved spatial positioning of each RNP separated by 12 nucleotides (52). This suggests molecular interactions between the two complexes and/or induced sRNA remodeling is important for methylation function. This is not likely the case for the eukaryotic box C/D snoRNPs that guide the methylation using both D and D′ guide sequences where the spatial positioning of box C/D and C′/D′ RNPs is not conserved and is often quite distant. Finally, both *in vivo* and *in vitro* studies have examined the base- pairing interactions of the target RNA with the sRNA guide sequence (110, 111). A minimum RNA:RNA duplex of 9–11 nucleotides is required and Watson–Crick pairing is essential. Interestingly, the ability of the *in vitro* complex to methylate target nucleotides positioned within thermally stable double-stranded secondary structures may suggest an ability of the core complex to facilitate target RNA melting necessary for base-pairing with the sRNA guide sequence.

14.5 BOX C/D RNP BIOGENESIS

14.5.1 Genomic Organization of Eukaryotic Box C/D snoRNA Genes

The genomic organization of eukaryotic box C/D RNAs falls into two major categories. Some box C/D snoRNA genes are independently transcribed from snoRNA-specific promoters with the primary transcript possessing single or multiple snoRNA species. However, the majority of box C/D snoRNAs are encoded within introns of host genes and transcribed by RNA polymerase II as part of the host precursor transcript. Fungi, plants, trypanosomes, and unicellular organisms possess primarily independently transcribed snoRNA genes, although snoRNAs encoded within host gene introns are present. In metazoan organisms, snoRNA coding sequences are overwhelmingly positioned within host gene introns (reviewed in references 13 and 48). For both snoRNA categories, the primary transcripts undergo posttranscriptional processing to produce mature snoRNA species.

14.5.2 Independently Transcribed and Intronic Eukaryotic Box C/D snoRNA Genes

Box C/D snoRNA genes that are independently transcribed using an snoRNA-specific promoter may be found as single or clustered species. U3 is the best-characterized box C/D snoRNA that is independently transcribed as a single snoRNA. U3 genes have been identified and characterized in fungi, protists, vertebrates, insects, and plants (reviewed in reference 13). They are most often found in multiple copies and dispersed throughout the genome, although occurrence of pseudogenes in vertebrates can make determination of gene copy number difficult (112–114). Box C/D snoRNA genes in yeast, plants, and trypanosomes are often organized into clusters and are transcribed as a polycistronic precursor using an snoRNA-specific promoter (83, 115–118). These polycistronic transcripts are subsequently processed to produce the mature individual box C/D snoRNAs.

The predominant box C/D snoRNA gene organization in metazoans, and particularly vertebrates, is that of encoding within introns of host RNA Pol II transcripts. Typically, the intronic snoRNAs are located in protein-coding pre-mRNA introns. Intronic box C/D snoRNAs are frequently found in multiple introns of common host genes with each intron limited to a single-encoded snoRNA. Intronic box C/D snoRNAs are often found in the same host gene in different organisms, but this organization is not universal, and frequently an snoRNA species is found in the same host gene but in different introns (reviewed in reference 13). Host genes often possess isomers of the same intronic snoRNA species (119–121). Host genes also encode different snoRNA species in different introns (122,123; reviewed in reference 13). The intronic snoRNAs are characteristically found within host genes that encode proteins involved in ribosome biogenesis or protein synthesis. Ribosomal protein genes very often contain intronic snoRNAs (13, 48). The positioning of intronic snoRNAs within protein coding genes important for ribosome biogenesis and protein synthesis suggests a possible coordination of the snoRNA and ribosome biogenesis pathways. Of particular interest are the intronic box C/D snoRNAs encoded with pre-mRNAs that do not encode a protein such as the UHG genes of human and *Drosophila melanogaster* (124–128). In these cases, the host precursor transcript simply serves as a carrier for the encoded intronic snoRNAs.

14.5.3 Archaeal Box C/D sRNA Genes

Most of the limited information concerning archaeal box C/D sRNA genes comes from an analysis of several hyperthermophile *Pyrococcus* genomes (16). Box C/D sRNA genes are primarily intergenic, although some do overlap upstream and/or downstream open reading frames on the same DNA strand. They typically exhibit very little clustering, although in *S. solfataricus* and *P. fulgidus* two sRNA genes (sR10/sR11 and sR26/sR60, respectively) are separated by only a few nucleotides and appear to be co-transcribed (15). In *P. abyssi*, the box C/D sRNA genes are preferentially located in noncoding regions of the genome, again with little clustering. Interestingly, the box C/D sR40 of *Pyroccocus* and *A. fulgidus* is found within the intron of the $tRNA^{Trp}$ gene, and it is this sRNA that is responsible for 2′-*O*-methylation of two nucleotides in the $tRNA^{Trp}$ itself (129–131). Nothing is presently known about the RNA polymerase promoters responsible for sRNA transcription, because obvious promoter elements are not observed and detailed analysis of possible sRNA precursor transcription has not been performed.

14.5.4 Transcription and Processing of Independently Transcribed Box C/D snoRNAs

Independently transcribed box C/D snoRNA genes possess their own promoter and terminator regions as well as enhancer elements flanking the coding sequence. They typically possess promoters similar to protein coding genes and are typically transcribed by RNA Polymerase II (13, 132). Transcription of U3 is driven by a Pol II promoter in vertebrates and possesses several conserved sequence elements. These include TATA-like boxes, proximal and distal sequence elements (PSE and DSE), and a "U3 box" specific for U3 at the DSE (113, 133–135). Plant U3 genes also possess TATA-like boxes but are transcribed by RNA Pol III due to the shorter spacing between the TATA box and upstream sequence elements (USE) (136, 137). The U3 transcripts of vertebrates are capped at the 5′ terminus with trimethyl-guanosine (TMG), whereas those of plants are *O*-methyl capped (138). Other Pol III transcribed box C/D snoRNA genes include snR52 in yeast (139) and several plant box C/D snoRNAs clustered with tRNA genes. The plant snoRNA genes are co-transcribed with tRNA as a tRNA– snoRNA precursor using the tRNA gene's RNA Pol III promoter (140). Yeast promoters of box C/D snoRNA genes that are independently transcribed by Pol II often contain A/T-rich stretches and a TATA box. Upstream Rap1p or Abf1p binding sites that are typical for yeast ribosomal protein genes are also present sometimes (115). Box C/D snoRNA promoters for independent transcripts in plants are not well-characterized but do contain putative TATA boxes (116). Independently transcribed box C/D snoRNAs are often polycistronic. The nascent transcripts are capped with trimethylguanosine (TMG) but undergo 5′ end processing involving endonucleases to produce processing intermediates that are matured by trimming exonucleases. In yeast, the ortholog of bacterial endonuclease RNase III, Rnt1p, and the exonucleases Rat1p and Xrn1p are responsible for snoRNA processing and trimming (115, 141–143).

Termination of Pol II-transcribed genes can result in either polyadenylated or non-polyadenylated transcripts. The box C/D snoRNAs are not polyadenylated, but current understanding of 3′ end formation is complicated by the overlapping machinery involved in these two pathways and shared components which include the core cleavage and polyadenylation factor (CPF) complex and the exosome (144). 3′ End formation of independently transcribed box C/D snoRNA precursors involves factors Nrd1p, Nab3p, and Sen1p, which appear to be specific for non-poly (A) termination and 3′ end pre-snoRNA processing (145). Nrd1p and Nab3p recognize and bind specific sequence elements upstream of the termination signal of box C/D snoRNA transcripts, although their recognition of these sequences alone is not sufficient to prevent polyadenylation and direct proper 3′ end formation (146). Nrd1p interacts with the exosome, cap-binding complex, and Pol II and is therefore implicated in coordinating transcription and 3′ end formation with exosome processing (147). 3′ End formation is also dependent upon the co-transcriptional assembly of the box C/D snoRNP (148, 149). Thus, termination and 3′ end formation may rely on a monitoring mechanism whereby Nrd1p, Nab3p, and Sen1p are involved in detecting the assembled snoRNP upstream of the termination signal and directing a bypass of transcript polyadenylation while activating exosome-mediated 3′ end processing.

14.5.5 Transcription and Processing of Intronic Box C/D snoRNAs

Box C/D snoRNA genes, particularly those of metazoan organisms, are positioned within introns of host genes that may be protein or non-protein-coding. Intronic snoRNAs are excised from the intron during host precursor splicing (reviewed in reference 13). The intronic snoRNA host genes are primarily driven by promoters containing terminal oligo*p*yrimidine tracts, or TOP promoters, which are characteristic of a broad family of protein-coding genes involved in ribosome biogenesis and protein synthesis (128, 150–152). Processing of intronic box C/D snoRNAs requires specific positioning of the snoRNA within the host intron, approximately 50 nucleotides upstream of the branchpoint (153, 154). Recently, a splicing factor designated intron binding protein 160 (IBP160) has been identified that defines the snoRNA distance from the branchpoint (155). IBP160 binds the intron approximately 35–40 nucleotides upstream of the branch point in a sequence-independent manner. Intronic snoRNA processing also requires snoRNP assembly. Immunoprecipitation experiments with fibrillarin antibodies have shown that the snoRNP complex is assembled while the snoRNA is still part of the unspliced, host pre-mRNA transcript. Accordingly, the box C/D snoRNP was found to assemble in the C1 splicing complex where IBP160 exerts its function (156). Thus, it appears that IBP160 may be a key factor linking intronic box C/D snoRNP assembly and intronic snoRNA processing with host pre-mRNA splicing events.

Upon pre-mRNA splicing, box C/D snoRNAs are excised from the intron and trimmed to a mature snoRNA using exonucleases. In yeast, an RNase III enzyme Rnt1p debranches the lariat intron. Rnt1p cleavage requires interaction with Nop1p (fibrillarin) of the assembled box C/D snoRNP (157). This cleavage provides entry sites for exonucleolytic trimming by the 5′–3′ exonucleases Rat1p and Xrn1p and the 3′–5′ trimming activities of Rrp6p and the exosome (158, 159). Using a minor pathway, intron-encoded snoRNAs can be directly excised from unspliced pre-mRNA using an endonuclease followed by exonucleolytic trimming (160, 161). Box C/D snoRNAs that are processed directly from unspliced pre-mRNAs appear to be released by a mechanism similar to that observed for independently transcribed snoRNAs (157).

14.5.6 Box C/D snoRNP Transport

Box C/D snoRNA transcription, processing, and snoRNP assembly occurs in the nucleoplasm before snoRNP transport into the nucleolus. Transport to the nucleolus requires the box C/D core motif and all four core proteins for nucleolar localization (162–166). Experiments indicate that in higher eukaryotes, box C/D snoRNPs require transit through Cajal bodies for maturation before entering the nucleolus (165). In HeLa cells, independently transcribed U3 snoRNA precursors are found in the nucleoplasm as large multiprotein processing complexes containing factors linked to RNA processing (hRrp46, LSm 2–8, La protein), snoRNP assembly (Tip48, Tip49, Nopp140), and RNA transport (PHAX, CRM1) (167). The snRNA export factor PHAX is required for U3 transport to the Cajal body, where it is m^3G-capped and undergoes CRM1-dependent transport to the nucleolus (168, 169). Intronic and uncapped box C/D snoRNAs may use a variation of this nucleolar transport pathway. U14 nucleolar localization is impaired when PHAX is depleted and some intronic and uncapped box C/D snoRNAs can bind PHAX in the absence of the cap-binding complex (167, 169).

14.6 FUTURE DIRECTIONS AND EXPERIMENTAL CHALLENGES

14.6.1 Box C/D RNA Diversity, Targets, and Functions

Computational approaches will continue to define box C/D RNA populations in both archaeal and eukaryotic organisms. Improvements in bioinformatics approaches coupled with increasing genome databases will identify new box C/D RNAs to reveal new populations and how these populations vary between organisms. These analyses may reveal conserved box C/D RNAs implying conserved functions in different organisms. More certain is the identification of new RNAs unique to given organisms. Defining new RNAs and their potential targets may well suggest novel functions. Comparison of archaeal and eukaryotic organism-specific populations should also provide insight into the rate of evolution for this class of small noncoding RNAs and how rapidly box C/D RNA diversity (and function?) is evolving.

Biochemical approaches will be required to identify box C/D RNAs that do not exhibit well-conserved box sequence elements. Defining these RNAs will likely require either biochemical fractionation of small noncoding RNA populations or immunoprecipitation of these RNAs using antibodies against the RNP core proteins. The observation of myriad RNAs bound to the L7 core protein in Archaea may suggest a greater diversity of C/D box RNAs than originally anticipated (170). Nonconventional box C/D RNAs may be associated with novel box C/D RNA-binding proteins implying novel RNP structures or perhaps functions. Indeed, it is a formal possibility that the protein(s) of these RNPs is the functional component of the complex working independently of the RNA's guide sequence. Certainly, this is the case for the archaeal H/ACA core proteins that function as a pseudouridylate synthase for a tRNA nucleotide in the absence of bound H/ACA RNA (171). Evidence does not indicate that the box C/D core proteins can methylate nucleotides in the absence of the RNA, but what other "moonlighting" functions might the box C/D core or accessory proteins carry out? Certainly, this is the case for the L7/15.5kD core protein where it functions in Archaea as a ribosomal protein and in eukaryotes as a U4 snRNP protein (29, 30). Are there other functions for the core proteins?

Both computational and biochemical approaches are certain to identify not only new box C/D RNAs but also new orphan RNAs with unknown targets. The occurrence of orphan box C/D RNAs raises the formal possibility that the function of these orphan RNAs is not determined by their guide sequence. Thus, novel functions could be expected for these box C/D RNAs and RNPs. A particular challenge will be defining tissue-specific and development-specific box C/D RNAs. It is these unique and perhaps rare box C/D RNAs that are most likely to exhibit novel functions as do the human/mouse-brain-specific box C/D RNAs. Identification of tissue-specific RNAs will require biochemical approaches isolating box C/D RNAs specifically expressed in a given tissue or developmental stage and identification through cloning and sequencing approaches. It will be interesting to determine how widespread is the tissue-specific expression of the box C/D RNAs and ultimately if these same RNAs are specifically expressed in the same tissue of different eukaryotes or within different tissues of a given eukaryote. These uniquely expressed RNAs may well modify nucleotides that, in turn, regulate other RNA metabolic processes such as observed for the brain-specific box C/D RNAs. Thus, the effect of box C/D RNA-guided modification may extend well beyond methylation of a given nucleotide to simply affect target RNA structure or function.

14.6.2 Box C/D RNP Structure and Methylation Function

Archaeal *in vitro* assembly systems have facilitated detailed examinations of box C/D sRNP structure and methylation function. However, important questions concerning this minimal RNA:protein enzyme remain. What are the inter-RNP interactions of juxtaposed box C/D and C′/D′ RNPs (protein–protein contacts?) which are required for efficient nucleotide modification and how do they affect RNP structure and, in turn, methylation function? As with eukaryotic snoRNPs, do the archaeal complexes function in more than nucleotide methylation? One would predict that, at the very least, chaperone activity to assist in RNA folding should be anticipated.

Most archaeal box C/D sRNPs guide the methylation of two target nucleotides, sometimes positioned within different target RNAs. What is the mechanism of target substrate interaction with the respective guide sequences? Can a box C/D sRNP bind two target RNAs simultaneously to carry out guided methylation from both complexes or is there interplay between the two RNPs such that only one is catalytically active at any given time? Recent work indicates the importance of the RNP core proteins for target RNA binding (110). What are these target RNA–protein interactions and how do they enhance substrate turnover? Do the core proteins assist in unfolding highly structured RNAs, and how is this accomplished as none of the core proteins appear to possess helicase activity as evidenced by the lack of known helicase domains? What is the topology of the target RNA in an RNA duplex when bound to the guide sequence? A target RNA:guide sequence of 12 nucleotides is equivalent to a full turn of an A-form RNA. This would suggest that this RNA duplex is distorted or untwisted to accommodate target RNA binding to the sRNP. What is the function of the known RNA-binding motifs on both fibrillarin and Nop56/58, and do these function in target RNA binding and positioning of the target nucleotide at the catalytic site? Many questions remain as to the function of specific domains of each core protein and their roles in sRNP assembly and methylation function. The structural details obtained from a crystal structure of this sRNP both free and bound with target RNA substrate will help to address some of these questions. More challenging will be enzymatic analyses of the methylation mechanism(s). Assembled sRNPs have thus far exhibited low efficiencies of methylation. Perhaps additional accessory proteins will be required for full activity of *in vitro* assembled sRNPs. Certainly, *in vivo* genetic systems will ultimately be important to confirm *in vitro* results.

The archaeal sRNP is certain to serve as a minimal model complex for examination of the more structurally and functionally diverse eukaryotic snoRNP. The diversity of the eukaryotic complex poses interesting questions as well as specific challenges. How similar is the eukaryotic box C/D snoRNP both structurally and functionally with respect to the archaeal sRNP? Although there are only a few double-guide eukaryotic box C/D snoRNP, are they also constrained by spatial positioning of the box C/D and C′/D′ RNP for methylation activity? What is the stoichiometry of the snoRNP core proteins on each motif; and specifically, are Nop56 and Nop58 differentially distributed on the box C/D and C′/D′ RNPs as indicated by crosslinking experiments? If so, then how have these core proteins evolved their RNA binding domains to recognize the two very similar but different RNA motifs? A similar question can be asked for the L7/15.5kD homologs, where the archaeal protein recognizes K-turn and K-loop elements, whereas the eukaryotic homolog binds only the K-turn. Has this evolution of core protein binding capability allowed more evolutionary drift in the C′/D′ motifs that are presently observed, perhaps at the same time promoting greater diversity in eukaryotic box C/D snoRNA structure and function? Because of the diversity in size and folded structures of the eukaryotic box C/D snoRNAs, do individual snoRNPs possess unique proteins distinct for a given RNA species? Are there specific snoRNP proteins (either core or accessory) that are characteristic of the nonmethylation functions? These snoRNA-specific proteins might include helicases for chaperone or RNA folding activity as has been demonstrated for the Dbp4 protein of the yeast U14 snoRNP (172).

A particular challenge will be the establishment of *in vitro* RNP assembly systems for the eukaryotic box C/D snoRNP. Such systems can directly address specific structural and functional questions. However, the eukaryotic complex is likely to be difficult to assemble *in vitro* because it appears to require an array of accessory proteins or assembly factors for snoRNP formation. Of course, crystal structures of the assembled eukaryotic snoRNP will be valuable but are unlikely in the near future. In contrast to Archaea, genetic systems are now available to study snoRNP structure and function *in vivo*. Both genetic manipulations in yeast as well as transfection approaches in cell culture presently provide tools to address many questions. Of particular interest will be comparison of archaeal and eukaryotic complexes to reveal the evolution of this RNA–protein enzyme and perhaps suggest how these RNPs acquired new functions.

14.6.3 Box C/D RNP Biogenesis

Both intronic and independently transcribed eukaryotic snoRNA genes have been characterized in a number of eukaryotic organisms for a fundamental understanding of snoRNA transcription. Still lacking, however, is an understanding of snoRNA transcriptional regulation and how snoRNA biogenesis may be coordinated with other biosynthetic pathways, most specifically ribosome biogenesis. Is snoRNA transcription and processing coordinately regulated with pre-rRNA transcription and/or ribosome assembly? If so, are there snoRNA transcription or processing factors involved in this coordination? Are the clustered snoRNA genes, either intronic snoRNAs in common host genes or polycistronic transcripts, organized for coordinated function? In contrast, transcription of the archaeal snoRNA genes has not been characterized and much remains to be learned. What are the sRNA promoters, and which elements are shared with other sRNA genes or are unique to a given sRNA species? Are the sRNA genes coordinately expressed with overlapping genes or independently transcribed, and what are the precursor sRNA transcripts, if any?

With respect to box C/D RNP assembly, what are the eukaryotic proteins needed to assemble the complex and are there corresponding assembly factors in Archaea? For the eukaryotic snoRNP, how do these assembly factors interact, if at all, with the spliceosome and/or the transcription complex for coordinate regulation? Thus far, only IBP60 has been shown to coordinate splicing with snoRNP synthesis (155). What are the accessory proteins, and do they remain bound to the complex to affect RNP structure or function? Is snoRNP assembly completed in the nucleoplasm or are there additional assembly steps in the Cajal bodies, or even the cell cytoplasm? Do all snoRNP transit through the Cajal bodies, or are there several transport pathways to the nucleolus? Does this transport utilize specific trafficking proteins? Do independently transcribed and intronic snoRNAs utilize different assembly and trafficking pathways? Finally, how are the snoRNAs and snoRNPs turned over? While it is apparent that box C/D RNAs are relatively stable small non-coding RNAs, nothing is known about their degradation pathway. What are the nucleases involved in snoRNA turnover, and are they common to other RNA degradation pathways? Furthermore, is snoRNA/snoRNP turnover coordinated with repression of ribosome synthesis or other metabolic pathways? These and many more questions await further investigation. Results from these investigations are certain to provide new and unexpected insights into the assembly, structure, and function of this evolutionarily ancient RNA–protein complex.

ACKNOWLEDGMENTS

This work was supported by NSF Grant MCB 0543741 to ESM.

REFERENCES

1. Busch, H., Reddy, R., Rothblum, L., and Choi, Y. C. (1982) snRNAs, snRNPs, and RNA Processing. *Annu Rev Biochem* **51**, 617–654.
2. Muramatsu, M., Hodnett, J. L., and Busch, H. (1966) Base composition of fractions of nuclear and nucleolar ribonucleic acid obtained by sedimentation and chromatography. *J Biol Chem* **241**, 1544–1550.
3. Hodnett, J. L., and Busch, H. (1968) Isolation and characterization of uridylic acid-rich 7 S ribonucleic acid of rat liver nuclei. *J Biol Chem* **243**, 1368–1375.
4. Nakamura, T., Prestayko, A. W., and Busch, H. (1968) Studies on nucleolar 4 to 6 S ribonucleic acid of Novikoff hepatoma cells. *J Biol Chem* **243**, 1368–1375.
5. Wienberg, R. A., and Penman, S. (1969) Small molecular weight monodisperse nuclear RNA. *J Mol Biol* **38**, 289–304.
6. Zapisek, W. F., Saponara, A. G., and Enger, M. D. (1969) Low molecular weight methylated ribonucleic acid species from Chinese hamster ovary cells. I. Isolation and characterization. *Biochemistry* **8**, 1170–1181.
7. Prestayko, A. W., Tonato, M., and Busch, H. (1970) Low molecular weight RNA associated with 28 S nucleolar RNA. *J Mol Biol* **47**, 505–515.
8. Prestayko, A. W., Tonato, M., Lewis, B. C., and Busch, H. (1971) Heterogeniety of nucleolar U3 ribonucleic acid of the Novikoff hepatoma. *J Biol Chem* **246**, 182–187.
9. Reddy, R., Ro-Choi, T. S., Henning, D., and Busch, H. (1974) Primary sequence of U1 nuclear ribonucleic acid of Novikoff hepatoma ascites cells. *J Biol Chem* **249**, 6486–6494.
10. Reddy, R., Henning, D., and Busch, H. (1979) Nucleotide sequence of nucleolar U3B RNA. *J Biol Chem* **253**, 11097–11105.
11. Shibata, H., Reddy, R., Henning, D., Ro-Choi, T. S., and Busch, H. (1974) Low molecular weight nuclear RNA. The 3′-terminal sequence of the U2 RNA. *Mol Cell Biochem* **4**, 3–19.
12. Shibata, H., Ro-Choi, T. S., Reddy, R., Choi, Y. C., Henning, D., and Busch, H. (1975) The primary nucleotide sequence of nuclear U2 ribonucleic acid. The 5′-terminal portion of the molecule. *J Biol Chem* **250**, 3909–3920.
13. Maxwell, E. S., and Fournier, M. J. (1995) The small nucleloar RNAs. *Annu Rev Biochem* **64**, 897–934.
14. Omer, A. D., Lowe, T. M., Russell, A. G., Ebhardt, H., Eddy, S. R., and Dennis, P. P. (2000) Homologs of small nucleolar RNAs in *Archaea*. *Science* **288**, 517–522.
15. Dennis, P. P., Omer, A., and Lowe, T. (2001) A guided tour: Small RNA function in *Archaea*. *Mol Microbiol* **40**, 509–519.
16. Gaspin, C., Cavaille, J., Erauso, G., and Bachellerie, J. P. (2000) Archaeal homologs of eukaryotic methylation guide small nucleolar RNAs: Lessons from the *Pyrococcus* genomes. *J Mol Biol* **297**, 895–906.
17. Borovjagin, A. V., and Gerbi, S. A. (1999) U3 small nucleolar RNA is essential for cleavage at sites 1, 2 and 3 in pre-rRNA and determines which rRNA processing pathway is taken in *Xenopus* oocytes. *J Mol Biol* **286**, 1347–1363.
18. Zieve, G., and Penman, S. (1976) Small RNA species of the HeLa cell: Metabolism and subcellular localization. *Cell* **8**, 19–31.
19. Savino, R., and Gerbi, S. A. (1990) *In vivo* disruption of *Xenopus* U3 snRNA affects ribosomal RNA processing. *EMBO J* **9**, 2299–2308.
20. Tollervey, D. (1987) A yeast small nuclear RNA is required for normal processing of pre-ribosomal RNA. *EMBO J* **6**, 4169–4175.

21. Zagorski, J., Tollervey, D., and Fournier, M. J. (1988) Characterization of an SNR gene locus in *Saccharomyces cerevisiae* that specifies both dispensable and essential small nuclear RNAs. *Mol Cell Biol* **8**, 3282–3290.
22. Morrissey, J. P., and Tollervey, D. (1993) Yeast snR30 is a small nucleolar RNA required for 18S rRNA synthesis. *Mol Cell Biol* **13**, 2469–2477.
23. Bachellerie, J., Michot, B., Nicoloso, M., Balakin, A., Ni, J., and Fournier, M. (1995) Antisense snoRNAs: A family of nucleolar RNAs with long complementarities to rRNA. *Trends Biochem Sci* **20**, 261–264.
24. Darzacq, X., Jady, B. E., Verheggen, C., Kiss, A. M., Bertrand, E., and Kiss, T. (2002) Cajal body-specific small nuclear RNAs: A novel class of 2′-*O*-methylation and pseudouridylation guide RNAs. *EMBO J* **21**, 2746–2756.
25. Vitali, P., Basyuk, E., Le Meur, E., Bertrand, E., Muscatelli, F., Cavaille, J., and Huttenhofer, A. (2005) ADAR2-mediated editing of RNA substrates in the nucleolus is inhibited by C/D small nucleolar RNAs. *J Cell Biol* **169**, 745–753.
26. Kishore, S., and Stamm, S. (2006) The snoRNA HBII-52 regulates alternative splicing of the serotonin receptor 2C. *Science* **311**, 230–232.
27. Gautier, T., Berges, T., Tollervey, D., and Hurt, E. (1997) Nucleolar KKE/D repeat proteins Nop56p and Nop58p interact with Nop1p and are required for ribosome biogenesis. *Mol Cell Biol* **17**, 7088–7098.
28. Newman, D., Kuhn, J., Shanab, G., and Maxwell, E. (2000) Box C/D snoRNA-associated proteins: Two pairs of evolutionarily ancient proteins and possible links to replication and transcription. *RNA* **6**, 861–879.
29. Kuhn, J., Tran, E., and Maxwell, E. S. (2002) Archaeal ribosomal protein L7 is a functional homolog of the eukaryotic 15. 5 kD/Snu13p snoRNP core protein. *Nucleic Acids Res* **30**, 931–941.
30. Watkins, N. J., Segault, V., Charpentier, B., Nottrott, S., Fabrizio, P., Bachi, A., Wilm, M., Rosbash, M., Branlant, C., and Luhrmann, R. (2000) A common core RNP structure shared between the small nucleolar box C/D RNPs and the spliceosomal U4 snRNP. *Cell* **103**, 457–466.
31. Schimmang, T., Tollervey, D., Kern, H., Frank, R., and Hurt, E. (1989) A yeast nucleolar protein related to mammalian fibrillarin is associated with small nucleolar RNAs and is essential for viability. *EMBO J* **8**, 4015–4024.
32. Amiri, K.(1994) Firillarin-like proteins occur in the domain Archaea. *J Bacteriology* **176**, 2124–2127.
33. King, T., Decatur, W., Bertrand, E., Maxwell, E., and Fournier, M. (2001) A well-connected and conserved nucleoplasmic helicase is required for the production of box C/D and H/ACA snoRNAs and localization of snoRNP proteins. *Mol Cell Biol* **21**, 7731–7746.
34. Omer, A., Ziesche, S., Ebhardt, H., and Dennis, P. P. (2002) *In vitro* reconstitution and activity of a C/D box methylation guide ribonucleoprotein complex. *Proc Natl Acad Sci USA* **99**, 5289–5294.
35. Tran, E. J., Zhang, X., and Maxwell, E. S. (2003) Efficient RNA 2′-*O*-methylation requires juxtaposed and symmetrically assembled archaeal box C/D and C′/D′ RNPs. *EMBO J* **22**, 3930–3940.
36. Rashid, R., Aittaleb, M., Chen, Q., Spiegel, K., Demeler, B., and Li, H. (2003) Functional requirement for symmetric assembly of archaeal box C/D small ribonucleoprotein particles. *J Mol Biol* **33**, 295–306.
37. Riedel, N., Wise, J. A., Swerdlow, H., Mak, A., and Guthrie, C. (1986) Small nuclear RNAs from *Saccharomyces cerevisiae*: Unexpected diversity in abundance, size, and molecular complexity. *Proc Natl Acad Sci USA* **83**, 8097–8101.
38. Liang, X. H., Uliel, S., Hury, A., Barth, S., Doniger, T., Unger, R., and Michaeli, S. (2005) A genome-wide analysis of C/D and H/ACA-like small nucleolar RNAs in *Trypanosome brucei* reveals a trypanosome-specific pattern of rRNA modification. *RNA* **11**, 619–645.
39. Russell, A. G., Schnare, M. N., and Gray, M. W. (2006) A large collection of compact box C/D snoRNAs and their isoforms in *Euglena gracilis*: Structural, functional and evolutionary insights. *J Mol Biol* **357**, 1548–1565.

40. Cavaille, J., Buiting, K., Keifmann, M., Lalande, M., Brannan, C. I., Horsthemke, B., Bachellerie, J. P., Brosius, J., and Huttenhofer, A. (2000) Identification of brain-specific and imprinted small nucleolar RNA genes exhibiting an unusual genomic organization. *Proc Natl Acad Sci USA* **97**, 14311–14316.
41. De los Santos, T., Schweizer, J., Rees, C. A., and Francke, U. (2000) Small evolutionarily conserved RNA, resembling C/D box small nucleolar RNA, is transcribed from PWCR1, a novel imprinted gene in Prader–Willi deletion region, which is highly expressed in brain. *Am J Hum Genet* **67**, 1067–1082.
42. Balakin, A. G., Smith, L., and Fournier, M. J. (1996) The RNA world of the nucleolus: Two major families of small RNAs defined by different box elements with related functions. *Cell* **86**, 823–834.
43. Tycowski, K. T., Shu, M. D., and Steitz, J. A. (1996) A mammalian gene with introns instead of exons generating stable RNA products. *Nature* **379**, 464–466.
44. Kiss-Laszlo, Z., Henry, Y., Bachellerie, M., Caizergues-Ferrer, M., and Kiss, T. (1996) Site-specific ribose methylation of preribosomal RNA: A novel function for small nucleolar RNAs. *Cell* **85**, 1077–1088.
45. Kiss-Laszlo, Z., Henry, Y., and Kiss, T. (1998) Sequence and structural elements of methylation guide snoRNAs essential for site-specific ribose methylation of pre-rRNA. *EMBO J* **17**, 797–807.
46. Nicoloso, M., Qu, L. H., Michot, B., and Bachellerie, J. P. (1996) Intron-encoded, anti-sense small nucleolar RNAs: the characterization of nine novel species points to their direct role as guides for the 2′-*O*-ribose methylation of rRNAs. *J Mol Biol* **260**, 178–195.
47. Tollervey, D. (1996) Small nucleolar RNAs guide ribosomal RNA methylation. *Science* **273**, 1056–1057.
48. Bachellerie, J. P., Cavaille, J., and Huttenhofer, A. (2002) The expanding snoRNA world. *Biochimie* **84**, 775–790.
49. Brown, J. W., Clark, G. P., Leader, D. J., Simpson, C. G., and Lowe, T. (2001) Multiple snoRNA gene clusters from *Arabidopsis*. *RNA* **7**, 1817–1832.
50. Chen, C. L., Liang, D., Zhou, H., Zhuo, M., Chen, Y. Q., and Qu, L. H. (2003) The high diversity of snoRNAs in plants: identification and comparative study of 120 snoRNA genes from *Oryza sativa*. *Nucleic Acids Res* **31**, 2601–2613.
51. Huttenhofer, A., Kiefmann, M., Meier-Ewert, S., O'Brien, J., Lehrach, H., Bachellerie, J. P., and Brosius, J. (2001) RNomics: An experimental approach that identifies 201 candidates for novel, small, non-messenger RNAs in mouse. *EMBO J* **20**, 2943–2953.
52. Tran, E. J., Zhang, X., Lackey, L., and Maxwell, E. S. (2005) Conserved spacing between the box C/D and C′/D′ RNPs of the archaeal box C/D, sRNP complex is required for efficient 2′-*O*-methylation of target RNAs. *RNA* **11**, 285–293.
53. Klein, D. J., Schmeing, T. M., Moore, P. B., and Steitz, T. A. (2001) The kink-turn: a new RNA secondary structure motif. *EMBO J* **20**, 4214–4221.
54. Moore, T., Zhang, Y., Fenley, M. O., and Li, H. (2004) Molecular basis of box C/D RNA–protein interactions: Cocrystal structure of archaeal L7Ae and a box C/D RNA. *Structure* **12**, 807–818.
55. Starostina, N. G., Marshburn, S., Johnson, L. S., Eddy, S. R., Terns, R. M., and Terns, M. P. (2004) Circular box C/D RNAs in *Pyrococcus furiosus*. *Proc Natl Acad Sci USA* **101**, 14097–14101.
56. Samarsky, D. A., and Fournier, M. J. (1999) A comprehensive database for the small nucleolar RNAs from *Saccharomyces cerevisiae*. *Nucleic Acids Res* **27**, 161–164.
57. Piekna-Przybylska, D., Decatur, W. A., and Fournier, M. J. (2007) New bioinformatic tools for analysis of nucleotide modifications in eukaryotic rRNA. *RNA* **13**, 305–312.
58. Lestrade, L., and Weber, M. J. (2006) snoRNA-LBME-db, a comprehensive database of human H/ACA and C/D box snoRNAs. *Nucleic Acids Res* **34**, D158–D162.
59. Omer, A. D., Ziesche, S., Decatur, W. A., Fournier, M. J., and Dennis, P. P. (2003) RNA-modifying machines in archaea. *Mol Microbiol* **48**, 617–629.
60. Huttenhofer, A., Cavaille, J., and Bachellerie, J. P. (2004) Experimental RNomics: A global approach to identifying small nuclear RNAs and their targets in different model organisms. *Methods Mol Biol* **265**, 409–428.
61. Aspegren, A., Hinas, A., Larsson, P., Larsson, A., and Soderbom, F. (2004) Novel non-coding RNAs in *Dictyostelium discoideum* and their expression during development. *Nucleic Acids Res* **32**, 4646–4656.

62. Yang, C., Zhou, H., Luo, J., and Qu, L. (2005) Identification of 20 snoRNA-like RNAs from the primitive eukaryote *Giardia lamblia*. *Biochem Biophys Res Commun* **328**, 1224–1231.
63. Schattner, P., Brooks, A. N., and Lowe, T. M. (2005) The tRNAscan-SE, snoscan and snoGPS web servers for the detection of tRNAs and snoRNAs. *Nucleic Acids Res* **33**, W686–W689.
64. Yang, J., Zhang, X., Huang, Z., Zhou, H., Huang, M., Zhang, S., Chen, Y., and Qu, L. (2006) snoSeeker: An advanced computational package for screening of guide and orphan snoRNA genes in the human genome. *Nucleic Acids Res* **34**, 5112–5123.
65. Fedorov, A., Stombaugh, J., Harr, M. W., Yu, S., Nasalean, L., and Shepelev, V. (2005) Computer identification of snoRNA genes using a Mammalian Orthologous Intron Database. *Nucleic Acids Res* **33**, 4578–4583.
66. Liang, X. H., Hury, A., Hoze, E., Uliel, S., Myslyuk, I., Apatoff, A., Unger, R., and Michaeli, S. (2007) Genome-wide analysis of C/D and H/ACA-like small nucleolar RNAs in *Leishmania major* indicates conservation among trypanosomatids in their repertoire and in their rRNA targets. *Eukaryot Cell* **6**, 361–377.
67. Huang, Z. P., Chen, C. J., Zhou, H., Li, B. B., and Qu, L. H. (2007) A combined computational and experimental analysis of two families of snoRNA genes from *Caenorhabditis elegans*, revealing the expression and evolution pattern of snoRNAs in nematodes. *Genomics* **89**, 490–501.
68. Crouch, R. J., Kanaya, S., and Earl, P. L. (1983) A model for the involvement of the small nucleolar RNA (U3) in processing eukaryotic ribosomal RNA. *Mol Biol Rep* **9**, 75–78.
69. Bachellerie, J. P., Michot, B., and Raynal, F. (1983) Recognition signals for mouse pre-rRNA processing: A potential role for U3 nucleolar RNA. *Mol Biol Rep* **9**, 79–86.
70. Kass, S., Tyc, K., Steitz, J. A., and Sollner-Webb, J. (1990) The U3 small nucleolar ribonucleoprotein functions in the first step of preribosomal RNA processing. *Cell* **60**, 897–908.
71. Hughes, J. M., and Ares, M., Jr. (1991) Depletion of U3 small nucleolar RNA inhibits cleavage in the 5′ external transcribed spacer of yeast pre-ribosomal RNA and impairs formation of 18S ribosomal RNA. *EMBO J* **10**, 4231–4239.
72. Lempicki, R. A., Jarmolowski, A., Huang, G. Y., Li, H. V., and Fournier, M. J. (1990) Mutations in conserved domains of U14 RNA impair 18S ribosomal RNA production in *Saccharomyces cerevisiae*. *Mol Biol Rep* **14**, 119–120.
73. Li, H. D., Zagorski, J., and Fournier, M. J. (1990) Depletion of U14 small nuclear RNA (snR128) disrupts production of 18S rRNA in *Saccharomyces cerevisiae*. *Mol Cell Biol* **10**, 1145–1152.
74. Peculis, B., and Steitz, J. (1993) Disruption of U8 nucleolar snRNA inhibits 5.8S and 28S rRNA processing in the *Xenopus* oocyte. *Cell* **73**, 1233–1245.
75. Dunbar, D. A., and Baserga, S. J. (1998) The U14 snoRNA is required for 2′-O-methylation of the pre-18S rRNA in *Xenopus* oocytes. *RNA* **4**, 195–204.
76. Beltrame, M., and Tollervey, D. (1995) Base pairing between U3 and the pre-ribosomal RNA is required for 18S rRNA synthesis. *EMBO J* **14**, 4350–4356.
77. Borovjagin, A. V., and Gerbi, S. A. (2005) An evolutionary intra-molecular shift in the preferred U3 snoRNA binding site on pre-ribosomal RNA. *Nucleic Acids Res* **33**, 4995–5005.
78. Liang, W. Q., and Fournier, M. J. (1995) U14 base-pairs with 18S rRNA: A novel snoRNA interaction required for rRNA processing. *Genes Dev* **9**, 2433–2443.
79. Peculis, B. A. (1997) The sequence of the 5′ end of the U8 small nucleolar RNA is critical for 5. 8S and 28S rRNA maturation. *Mol Cell Biol* **17**, 3702–3713.
80. Michot, B., Joseph, N., Mazan, S., and Bachellerie, J. P. (1999) Evolutionarily conserved structural features in the ITS2 of mammalian pre-rRNAs and potential interactions with the snoRNA U8 detected by comparative analysis of new mouse sequences. *Nucleic Acids Res* **27**, 2271–2282.
81. Schoemaker, R. J., and Gultyaev, A. P. (2006) Computer simulation of chaperone effects of archaeal C/D box sRNA binding on rRNA folding. *Nucleic Acids Res* **34**, 2015–2026.
82. Maden, B. E. (1990) The numerous modified nucleotides in eukaryotic ribosomal RNA. *Prog Nucleic Acid Res Mol Biol* **39**, 241–303.
83. Brown, J. W., Echeverria, M., Qu, L. H., Lowe, T. M., Bachellerie, J. P., Huttenhofer, A., Kastenmayer, J. P., Green, P. J., Shaw, P., and Marshall, D. F. (2003) Plant snoRNA database. *Nucleic Acids Res* **31**, 432–435.
84. Decatur, W. A., and Fournier, M. J. (2002) rRNA modifications and ribosome function. *Trends Biochem Sci* **27**, 344–351.

85. Noon, K. R., Bruenger, E., and McCloskey, J. A. (1998) Posttranscriptional modifications in 16S and 23S rRNAs of the archaeal hyperthermophile *Sulfolobus solfataricus*. *J Bacteriol* **180**, 2883–2888.
86. Tollervey, D., Lehtonen, H., Jansen, R., Kern, H., and Hurt, E. C. (1993) Temperature-sensitive mutations demonstrate roles for yeast fibrillarin in pre-rRNA processing, pre-rRNA methylation, and ribosome assembly. *Cell* **72**, 443–457.
87. Tycowski, K. T., Aab, A., and Steitz, J. A. (2004) Guide RNAs with 5′ caps and novel box C/D snoRNA-like domains for modification of snRNAs in metazoans. *Curr Biol* **14**, 1985–1995.
88. Tycowski, K., You, Z., Graham, P., and Steitz, J. (1998) Modification of U6 spliceosomal RNA is guided by other small RNAs. *Mol Cell* **2**, 629–638.
89. Cavaille, J., Vitali, P., Basyuk, E., Huttenhofer, A., and Bachellerie, J. P. (2001) A novel brain-specific box C/D small nucleolar RNA processed from tandemly repeated introns of a noncoding RNA gene in rats. *J Biol Chem* **276**, 26374–26383.
90. Cavaille, J., Seitz, H., Paulsen, M., Ferguson-Smith, A. C., and Bachellerie, J. P. (2002) Identification of tandemly-repeated C/D snoRNA genes at the imprinted human 14q32 domain reminiscent of those at the Prader-Willi/Angelman syndrome region. *Hum Mol Genet* **11**, 1527–1538.
91. Ding, F., Prints, Y., Dhar, M. S., Johnson, D. K., Garnacho-Montero, C., Nicholls, R. D., and Francke, U. (2005) Lack of Pwcr1/MBII-85 snoRNA is critical for neonatal lethality in Prader–Willi syndrome mouse models. *Mamm Genome* **16**, 424–431.
92. Rogelj, B., Hartmann, C. E., Yeo, C. H., Hunt, S. P., and Giese, K. P. (2003) Contextual fear conditioning regulates the expression of brain-specific small nucleolar RNAs in hippocampus. *Eur J Neurosci* **18**, 3089–3096.
93. Szewczak, L. B., DeGregorio, S. J., Strobel, S. A., and Steitz, J. A. (2002) Exclusive interaction of the 15.5 kD protein with the terminal box C/D motif of a methylation guide snoRNP. *Chem Biol* **9**, 1095–1107.
94. Cahill, N., Friend, K., Speckman, W., Li, Z., Terns, R., Terns, M., and Steitz, J. A. (2002) Site-specific cross-linking analyses reveal an asymmetric distribution for a box C/D snoRNP. *EMBO J* **21**, 3816–3828.
95. Charron, C., Manival, X., Clery, A., Senty-Segault, V., Charpentier, B., Marmier-Gourrier, N., Branlant, C., and Aubry, A. (2004) The archaeal sRNA binding protein L7Ae has a 3D structure very similar to that of its eukaryal counterpart while having a broader RNA-binding specificity. *J Mol Biol* **342**, 757–773.
96. Szewczak, L. B., Gabrielsen, J. S., DeGregorio, S. J., Strobel, S. A., and Steitz, J. A. (2005) Molecular basis for RNA kink-turn recognition by the h15.5K small RNP protein. *RNA* **11**, 1407–1419.
97. Newman, D. R., Kuhn, J. F., Shanab, G. M., and Maxwell, E. S. (2000) Box C/D snoRNA-associated proteins: Two pairs of evolutionarily ancient proteins and possible links to replication and transcription. *RNA* **6**, 861–879.
98. Watkins, N. J., Dickmanns, A., and Luhrmann, R. (2002) Conserved stem II of the box C/D motif is essential for nucleolar localization and is required, along with the 15.5K protein, for the hierarchical assembly of the box C/D snoRNP. *Mol Cell Biol* **22**, 8342–8352.
99. Wang, H., Boisvert, D., Kim, K., Kim, R., and Kim, S. H. (2000) Crystal structure of a fibrillarin homologue from *Methanococcus jannaschii*, a hyperthermophile, at 1.6 Å resolution. *EMBO J* **19**, 317–323.
100. Galardi, S., Fatica, A., Bachi, A., Scaloni, A., Presutti, C., and Bozzoni, I. (2002) Purified box C/D snoRNPs are able to reproduce site-specific 2′-*O*-methylation of target RNA *in vitro*. *Mol Cell Biol* **22**, 6663–6668.
101. Suryadi, J., Tran, E. J., Maxwell, E. S., and Brown, B. A., II (2005) The crystal structure of *Methanocaldococcus jannaschii* multifunctional L7Ae RNA-binding protein reveals an induced-fit interaction with the box C/D RNAs. *Biochemistry* **44**, 9657–9672.
102. Turner, B., Melcher, S. A., Wilson, T. J., Norman, D. G., and Lilley, D. M. J. (2005) Induced fit of RNA on binding the L7Ae protein to the kink-turn motif. *RNA* **11**, 1192–1200.
103. Gagnon, K. T., Zhang, X., Agris, P. F., and Maxwell, E. S. (2006) Assembly of the archaeal box C/D sRNP can occur via alternative pathways and requires temperature-facilitated sRNA remodeling. *J Mol Biol* **362**, 1025–1042.
104. Aittaleb, M., Rashid, R., Chen, Q., Palmer, J., Daniels, C., and Li, H. (2003) Structure and function of archaeal box C/D sRNP core proteins. *Nat Struct Biol* **10**, 256–263.

105. Zhang, X., Champion, E. A., Tran, E., Brown, B. A.II, Baserga, S. J., and Maxwell, E. S. (2006) The coiled-coil domain of the Nop56/58 core protein is dispensable for sRNP assembly, but is critical for archaeal box C/D sRNP-guided nucleotide methylation. *RNA* **12**, 1092–1103.
106. Hama, T., and Ferre-D'Amare, A. R. (2004) Structure of protein L7Ae bound to a k-turn derived from an archaeal box H/ACA sRNA at 1.8 Å resolution. *Structure* **12**, 893–903.
107. Vidovic, I., Nottrot, S., Hartmuth, K., Luhrmann, R., and Ficner, R. (2000) Crystal structure of the spliceosomal 15.5 kD protein bound to a U4 snRNA fragment. *Mol Cell* **6**, 1331–1342.
108. Aittaleb, M., Visone, T., Fenley, M. O., and Li, H. (2004) Structural and thermodynamic evidence for a stabilizing role of Nop5p in *S*-adenosyl-L-methionine binding to fibrillarin. *J Biol Chem* **279**, 41822–41829.
109. Liu, S., Li, P., Dybkov, O., Nottrott, S., Hartmuth, K., Luhrmann, R., Carlomagno, T., and Wahl, M. C. (2007) Binding of the human Prp31 Nop domain to a composite RNA–protein platform in U4 snRNP. *Science* **316**, 115–120.
110. Appel, C. D., and Maxwell, E. S. (2007)*Structural features of the guide:target RNA duplex required for archaeal box C/D, sRNA-guided nucleotide* 2′-*O*-methylation.*RNA* 13, 899–911.
111. Cavaille, J., Nicoloso, M., and Bachellerie, J. P. (1996) Targeted ribose methylation of RNA in vivo directed by tailored antisense RNA guides. *Nature* **383**, 732–735.
112. Reddy, R., Henning, D., Chirala, S., Rothblum, L., Wright, D., and Busch, H. (1985) Isolation and Characterization of three rat U3 RNA pseudogenes collinear with U3 RNA. *J Biol Chem* 260, 5715–5719.
113. Stroke, I. L., and Weiner, A. M. (1985) Genes and pseudogenes for rat U3A and U3B small nuclear RNA. *J Mol Biol* **184**, 183–193.
114. Mazan, S., Michot, B., and Bachellerie, J. P. (1989) Mouse U3-RNA-processed pseudogenes are nonrandomly integrated into genomic DNA. Implications for the process of retrogene formation. *Eur J Biochem* **181**, 599–605.
115. Qu, L. H., Henras, A., Lu, Y. J., Zhou, H., Zhou, W. X., Zhu, Y. Q., Zhao, J., Henry, Y., Caizergues-Ferrer, M., and Bachellerie, J. P. (1999) Seven novel methylation guide small nucleolar RNAs are processed from a common polycistronic transcript by Rat1p and RNase III in yeast. *Mol Cell Biol* **19**, 1144–1158.
116. Leader, D. J., Clark, G. P., Watters, J., Beven, A. F., Shaw, P. J., and Brown, J. W. (1997) Clusters of multiple different small nucleolar RNA genes in plants are expressed as and processed from polycistronic pre-snoRNAs. *EMBO J* **16**, 5742–5751.
117. Dunbar, D. A., Chen, A. A., Wormsley, S., and Baserga, S. J. (2000) The genes for small nucleolar RNAs in *Trypanosoma brucei* are organized in clusters and are transcribed as a polycistronic RNA. *Nucleic Acids Res* **28**, 2855–2861.
118. Uliel, S., Liang, X. H., Unger, R., and Michaeli, S. (2004) Small nucleolar RNAs that guide modification in trypanosomatids: Repertoire, targets, genome organization, and unique features. *Int J Parasitol* **34**, 445–454.
119. Cecconi, F., Mariottini, P., Loreni, F., Pierandrei-Amaldi, P., Campioni, N., and Amaldi, F. (1994) U17XS8, a small nucleolar RNA with a 12 nt complementarity to 18S rRNA and coded by a sequence repeated in the six introns of *Xenopus laevis* ribosomal protein S8 gene. *Nucleic Acids Res* **22**, 732–741.
120. Liu, J., and Maxwell, E. S. (1990) Mouse U14 snRNA is encoded in an intron of the mouse cognate hsc70 heat shock gene. *Nucleic Acids Res* **18**, 6565–6571.
121. Pellizzoni, L., Crosio, C., Campioni, N., Loreni, F., and Pierandrei-Amaldi, P. (1994) Different forms of U15 snoRNA are encoded in the introns of the ribosomal protein S1 gene of *Xenopus laevis*. *Nucleic Acids Res* **22**, 4607–4613.
122. Prislei, S., Michienzi, A., Presutti, C., Fragapane, P., and Bozzoni, I. (1993) Two different snoRNAs are encoded in introns of amphibian and human L1 ribosomal protein genes. *Nucleic Acids Res* **21**, 5824–5830.
123. Xia, L., Lui, J., Sage, C., Trexler, B., Andrews, M., and Maxwell, E. (1995) Intronic U14 snoRNAs of *Xenopus laevis* are located in two different parent genes and can be processed from their introns during early oogenesis. *Nucleic Acids Res* **23**, 4844–4849.

124. Tycowski, K. T., and Steitz, J. A. (2001) Non-coding snoRNA host genes in *Drosophila:* expression strategies for modification guide snoRNAs. *Eur J Cell Biol* **80**, 119–125.
125. Huang, Z., Zhou, H., He, H., Chen, C., Liang, D., and Qu, L. (2005) Genome-wide analyses of two families of snoRNA genes from *Drosophila melanogaster*, demonstrating the extensive utilization of introns for coding of snoRNAs. *RNA* **11**, 1303–1316.
126. Tycowski, K., Shu, M., and Steitz, J. (1996) A mammalian gene with introns instead of exons generating stable RNA products. *Nature* **379**, 464–466.
127. Bortolin, M., and Kiss, T. (1998) Human U19 intron-encoded snoRNA is processed from a long primary transcript that possesses little potential for protein coding. *RNA* **4**, 445–454.
128. Pelczar, P., and Filipowicz, W. (1998) The host gene for intronic U17 small nucleolar RNAs in mammals has no protein-coding potential and is a member of the 5′-terminal oligopyrimidine gene family. *Mol Cell Biol* **18**, 4509–4518.
129. Singh, S. K., Gurha, P., Tran, E. J., Maxwell, E. S., and Gupta, R. (2004) Sequential 2′-*O*-methylation of archaeal pre-tRNA^{Trp} nucleotides is guided by the intron-encoded but trans-acting box C/D ribonucleoprotein of pre-tRNA. *J Biol Chem* **279**, 47661–47671.
130. Bortolin, M., Bachellerie, J., Clouet-d'Orval, B. (2003) *In vitro* RNP assembly and methylation guide activity of an unusual box C/D RNA, cis-acting archaeal pre-tRNATrp. *Nucleic Acids Res* **31**, 6524–6535.
131. Clouet-d'Orval, B., Bortolin, M., Gaspin, C., and Bachellerie, J. (2001) Box C/D RNA guides for the ribose methylation of archaeal tRNAs. The tRNATrp intron guides the formation of two ribose-methylated nucleosides in the mature tRNATrp. *Nucleic Acids Res* **29**, 4518–4529.
132. Steinmetz, E. J., Warren, C. L., Kuehner, J. N., Panbehi, B., Ansari, A. Z., and Brow, D. A. (2006) Genome-wide distribution of yeast RNA polymerase II and its control by Sen1 helicase. *Mol Cell* **24**, 735–746.
133. Suh, D., Busch, H., and Reddy, R. (1983) Isolation and characterization of a human U3 small nucleolar RNA gene. *Biochem Biophys Res Commun* **137**, 1133–1140.
134. Ach, R. A., and Weiner, A. M. (1991) Cooperation between CCAAT and octamer motifs in the distal sequence element of the rat U3 small nucleolar RNA promoter. *Nucleic Acids Res* **19**, 4209–4218.
135. Savino, R., Hitti, Y., and Gerbi, S. A. (1992) Genes for *Xenopus laevis* U3 small nuclear RNA. *Nucleic Acids Res* **20**, 5435–5442.
136. Kiss, T., and Solymosy, F. (1990) Molecular analysis of a U3 RNA gene locus in tomato: Transcription signals, the coding region, expression in transgenic tobacco plants and tandemly repeated pseudogenes. *Nucleic Acids Res* **18**, 1941–1949.
137. Connelly, S., Marshallsay, C., Leader, D., Brown, J. W., and and Filipowicz, W. (1994) Small nucleolar RNA genes transcribed by either RNA polymerase II or RNA polymerase III in monocot plants share three promoter elements and use a strategy to regulate gene expression different from that used by their dicot plant counterparts. *Mol Cell Biol* **14**, 5910–5919.
138. Shimba, S., Buckley, B., Reddy, R., Kiss, T., and Filipowicz, W. (1992) Cap structure of U3 small nucleolar RNA in animal and plant cells is different. Gamma-monomethyl phosphate cap structure in plant RNA. *J Biol Chem* **267**, 13772–13777.
139. Roberts, D. N., Stewart, A. J., Huff, J. T., and Cairns, B. R. (2003) The RNA polymerase III transcriptome revealed by genome-wide localization and activity–occupancy relationships. *Proc Natl Acad Sci USA* **100**, 14695–14700.
140. Kruszka, K., Barneche, F., Guyot, R., Ailhas, J., Meneau, I., Schiffer, S., Marchfelder, A., and Echeverria, M. (2003) Plant dicistronic tRNA-snoRNA genes: A new mode of expression of the small nucleolar RNAs processed by RNase Z. *EMBO J* **22**, 621–632.
141. Petfalski, E., Dandekar, T., Henry, Y., and Tollervey, D. (1998) Processing of the precursors to small nucleolar RNAs and rRNAs requires common components. *Mol Cell Biol* **18**, 1181–1189.
142. Chanfreau, G., Rotondo, G., Legrain, P., and Jacquier, A. (1998) Processing of a dicistronic small nucleolar RNA precursor by the RNA endonuclease Rnt1. *EMBO J* **17**, 3726–3737.
143. Chanfreau, G., Legrain, P., and Jacquier, A. (1998) Yeast RNase III as a key processing enzyme in small nucleolar RNA metabolism. *J Mol Biol* **284**, 975–988.
144. Kim, M., Vasiljeva, L., Rando, O. J., Zhelkovsky, A., Moore, C., and Buratowski, S. (2006) Distinct pathways for snoRNA and mRNA termination. *Mol Cell* **24**, 723–734.

145. Steinmetz, E. J., Conrad, N. K., Brow, D. A., and Corden, J. L. (2001) RNA-binding protein Nrd1 directs poly(A)-independent 3′-end formation of RNA polymerase II transcripts. *Nature* **413**, 327–331.
146. Carroll, K. L., Pradhan, D. A., Granek, J. A., Clarke, N. D., and Corden, J. L. (2004) Identification of cis elements directing termination of yeast nonpolyadenylated snoRNA transcripts. *Mol Cell Biol* **24**, 6241–6252.
147. Vasiljeva, L., and Buratowski, S. (2006) Nrd1 interacts with the nuclear exosome for 3′ processing of RNA polymerase II transcripts. *Mol Cell* **21**, 239–248.
148. Morlando, M., Greco, P., Dichtl, B., Fatica, A., Keller, W., and Bozzoni, I. (2002) Functional analysis of yeast snoRNA and snRNA 3′-end formation mediated by uncoupling of cleavage and polyadenylation. *Mol Cell Biol* **22**, 1379–1389.
149. Morlando, M., Ballarino, M., Greco, P., Caffarelli, E., Dichtl, B., and Bozzoni, I. (2004) Coupling between snoRNP assembly and 3′ processing controls box C/D snoRNA biosynthesis in yeast. *EMBO J* **23**, 2392–2401.
150. Smith, C. M., and Steitz, J. A. (1998) Classification of gas5 as a multi-small-nucleolar-RNA (snoRNA) host gene and a member of the 5′-terminal oligopyrimidine gene family reveals common features of snoRNA host genes. *Mol Cell Biol* **18**, 6897–6909.
151. Levy, S., Avni, D., Hariharan, N., Perry, R. P., and Meyuhas, O. (1991) Oligopyrimidine tract at the 5′ end of mammalian ribosomal protein mRNAs is required for their translational control. *Proc Natl Acad Sci USA* **88**, 3319–3323.
152. Amaldi, F., and Pierandrea-Amaldi, P. (1997) *Progress in Molecular and Subcellular Biology* (ed. Jeanteur P.) Vol. 18, PN Springer-Verlag, Berlin, 1–17.
153. Hirose, T., and Steitz, J. A. (2001) Position within the host intron is critical for efficient processing of box C/D snoRNAs in mammalian cells. *Proc Natl Acad Sci USA* **98**, 12914–12919.
154. Vincenti, S., De Chiara, V., and Bozzoni, I. (2007) The position of yeast snoRNA-coding regions within host introns is essential for their biosynthesis and for efficient splicing of the host pre-mRNA. *RNA* **13**, 138–150.
155. Hirose, T., Ideue, T., Nagai, M., Hagiwara, M., Shu, M. D., and Steitz, J. A. (2006) A spliceosomal intron binding protein, IBP160, links position-dependent assembly of intron-encoded box C/D snoRNP to pre-mRNA splicing. *Mol Cell* **23**, 673–684.
156. Hirose, T., Shu, M. D., and Steitz, J. A. (2003) Splicing-dependent and -independent modes of assembly for intron-encoded box C/D snoRNPs in mammalian cells. *Mol Cell* **12**, 113–123.
157. Giorgi, C., Fatica, A., Nagel, R., and Bozzoni, I. (2001) Release of U18 snoRNA from its host intron requires interaction of Nop1p with the Rnt1p endonuclease. *EMBO J* **20**, 6856–6865.
158. Kiss, T., and Filipowicz, W. (1995) Exonucleolytic processing of small nucleolar RNAs from pre-mRNA introns. *Genes Dev* **9**, 1411–1424.
159. Allmang, C., Kufel, J., Chanfreau, G., Mitchell, P., Petfalski, E., and Tollervey, D. (1999) Functions of the exosome in rRNA, snoRNA and snRNA synthesis. *EMBO J* **18**, 5399–5410.
160. Caffarelli, E., Fatica, A., Prislei, S., De Gregorio, E., Fragapane, P., and Bozzoni, I. (1996) Processing of the intron-encoded U16 and U18 snoRNAs: The conserved C and D boxes control both the processing reaction and the stability of the mature snoRNA. *EMBO J* **15**, 1121–1131.
161. Villa, T., Ceradini, F., Presutti, C., and Bozzoni, I. (1998) Processing of the intron-encoded U18 small nucleolar RNA in the yeast *Saccharomyces cerevisiae* relies on both exo- and endo-nucleolytic activities. *Mol Cell Biol* **18**, 3376–3383.
162. Samarsky, D., Fournier, M., Singer, R., and Bertrand, E. (1998) The snoRNA box C/D motif directs nucleolar targeting and also couples snoRNA synthesis and localization. *EMBO J* **17**, 3747–3757.
163. Lange, T. S., Borovjagin, A., Maxwell, E., and Gerbi, S. (1998) Conserved boxes C and D are essential nucleolar localization elements of U14 and U8 snoRNAs. *EMBO J* **17**, 3176–3187.
164. Lange, T. S., Ezrokhi, M., Borovjagin, A. V., Rivera-Leon, R., North, M. T., and Gerbi, S. A. (1998) Nucleolar localization elements of *Xenopus laevis* U3 small nucleolar RNA. *Mol Cell Biol* **9**, 2973–2985.
165. Narayanan, A., Speckman, W., Terns, R., and Terns, M. (1999) Role of the box C/D motif in localization of small nucleolar RNAs to coiled bodies and nucleoli. *Mol Cell Biol* **10**, 2131–2147.

166. Verheggen, C., Mouaikel, J., Thiry, M., Blanchard, J. M., Tollervey, D., Bordonne, R., Lafontaine, D. L., and Bertrand, E. (2001) Box C/D small nucleolar RNA trafficking involves small nucleolar RNP proteins, nucleolar factors and a novel nuclear domain. *EMBO J* **20**, 5480–5490.
167. Watkins, N. J., Lemm, I., Ingelfinger, D., Schneider, C., Hobbach, M., Urlaub, H., and Luhrmann, R. (2004) Assembly and maturation of the U3 snoRNP in the nucleoplasm in a large dynamic multiprotein complex. *Mol Cell* **16**, 789–798.
168. Verheggen, C., Lafontaine, D., Samarsky, D., Mouaikel, J., Blanchard, J. M., Bordonne, R., and Bertrand, E. (2002) Mammalian and yeast U3 snoRNPs are matured in specific and related nuclear compartments. *EMBO J* **21**, 2736–2745.
169. Boulon, S., Verheggen, C., Jady, B. E., Girard, C., Pescia, C., Paul, C., Ospina, J. K., Kiss, T., Matera, A. G. Bordonne, R. *et al.* (2004) PHAX and CRM1 are required sequentially to transport U3 snoRNA to nucleoli. *Mol Cell* **16**, 777–787.
170. Dennis, P. P., and Omer, A. (2005) Small non-coding RNAs in Archaea. *Curr Opin Microbiol* **8**, 685–694.
171. Roovers, M., Hale, C., Tricot, C., Terns, M., Terns, R., Grosjean, H., and Droogmans, L. (2006) Formation of the conserved pseudouridine at position 55 in archaeal tRNA. *Nucleic Acids Res* **34**, 4293–4301.
172. Liang, W., Clark, J., and Fournier, M. J. (1997) The rRNA-processing function of the yeast U14 small nucleolar RNA can be rescued by a conserved RNA helicase-like protein. *Mol Cell Biol* **17**, 4124–4132.
173. Lowe, T.M., and Eddy, S.R. (1999) A Computational screen for methylation guide snoRNAs in yeast *Science* **283**, 1168–1171.
174. Xie, J., Zhang, M., Zhou, T., Hua, x., Tang, L., and Wu, W. (2007) Sno/ScaRNA base: a curated database for small nucleolar RNAs and Cajal body-specific RNAs. *Nucleic Acids Res* **35**, D183–D187.
175. Pang, K.C., Stephen, S., Engstrom, P.G., Tajul-Arifin, K. Chen, W., Wahlestedt, C., Lenhard, B., Hayashizaki, Y. Mattick, J.S.(2005) RNAdb—a comprehensive mammalian noncoding RNA database. *Nucleic Acids Res* **33**, D125–D130.
176. Pang, K.C., Stephen, S., Dinger, M.E., Engstrom, P.G., Lenhard, B., and Mattick, J.S. (2007) RNAdb2.0—an expanded database of mammalian non-coding RNAs. Nucleic Acids Res **35**, D178–D182.

CHAPTER 15

STRUCTURAL FEATURES OF THE ADAR FAMILY OF ENZYMES AND THEIR SUBSTRATES

*M. Wadud Bhuiya**
Evan M. Ritter†
Bernard A. Brown II‡

THE ADAR family of enzymes recognizes a variety of nucleic acid targets for functional roles in site-specific RNA editing, hyper-editing, and perhaps modulation of the myriad small cellular RNAs. Specific recognition is achieved by a combination of ADAR binding domain features and fine structure of the nucleic acid targets. A structure of a complete ADAR enzyme alone or bound to a target RNA does not currently exist. However, by examining the structures of isolated domains and protein–nucleic acid co-complexes, a coarse view of ADAR-substrate recognition is beginning to emerge. This chapter will explore what is known about the structures of ADAR enzymes and their nucleic acid targets.

15.1 ADAR ENZYMES

RNA editing by adenosine deamination is catalyzed by members of an enzyme family known as adenosine deaminases active on RNA (ADAR) (1). ADARs catalyze the deamination of adenosine to inosine within double-stranded RNAs (Figure 15.1). Inosine mimics guanosine in its base-pairing properties, and the translational machinery interprets I as G. In this manner, A-to-I conversions in mRNA can alter the genetic information, resulting in changes in mRNA splicing and stability, as well as

**Present address:* Department of Biology, Brookhaven National Laboratory, Upton, NY, 11973.
†*Present address:* 504 Murdocksville Road, West End, NC, 27376.
‡*Present address:* Department of Molecular and Structural Biochemistry, North Carolina State University, Raleigh, NC 27695.

RNA and DNA Editing: Molecular Mechanisms and Their Integration into Biological Systems,
Edited by Harold C. Smith

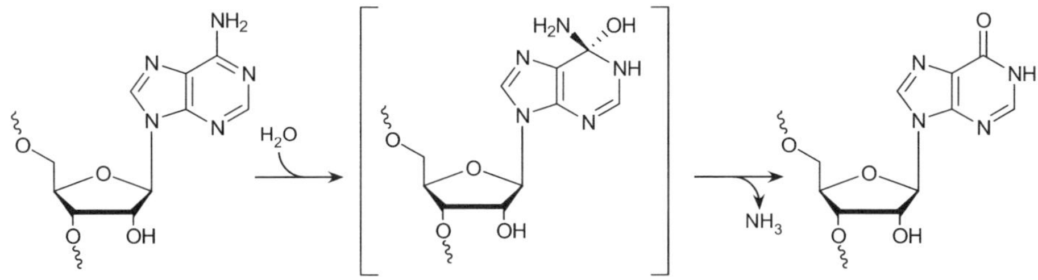

Adenosine Inosine

Figure 15.1 Conversion of adenosine to inosine. Hydrolytic deamination at C6 of the adenine base leads to inosine formation. The reaction is thought to proceed through a tetravalent intermediate and result in the release of ammonia.

altering protein functions. ADARs are essential in all animals examined to date (2). In humans, three ADARs and their corresponding genes have been identified—ADAR1, ADAR2, and ADAR3 (3–5)—whereas in lower metazoan organisms, such as *Drosophila melanogaster* (6) or *Caenorhabditis elegans* (7), only one ADAR gene has been identified. ADARs have not been identified in fungi or plants.

ADAR1 and ADAR2 are expressed in almost all cell types (8) and are able to convert A to I in long dsRNA duplexes, pre-mRNAs, and viral transcripts, either promiscuously or site-specifically (3, 4) (see also Chapters 1, 6, and 9). ADAR3 is exclusively expressed in the brain and is catalytically inactive on synthetic dsRNA as well as known pre-mRNA substrates *in vitro*, even though it can bind either double-stranded or single-stranded RNA (9, 10). Thus, it has been suggested that ADAR3 could act on target RNAs that have not yet been identified, or alternatively it could modulate the activity of ADAR1 and ADAR2 by competing for their RNA-target binding sites (10).

15.2 OVERVIEW AND FUNCTIONS OF ADARs

ADARs are modular enzymes consisting of a single polypeptide. The human ADAR3 protein is most similar to ADAR2, with 72% sequence similarity—compared to ADAR1, for which there is 53% sequence similarity. Moreover, all ADARs share a common modular domain organization (Figure 15.2). All identified ADARs contain an N-terminal region, which varies in length and sequence, followed by a variable number of double-stranded RNA-binding domains (dsRBDs, also known as double-stranded RNA-binding motif or DRBM) and the catalytic deaminase domain (DM) (see also Chapter 1). ADAR1 is the largest of the three family members and has an extended N-terminus that varies in length, depending on expression conditions. The constitutively expressed form of ADAR1 is shorter than the interferon-induced form and contains a Z-DNA/RNA binding motif, Zβ (11). The interferon-induced form of ADAR1 is approximately 300 amino acids longer at the N-terminus than the constitutively expressed form and contains an arginine–glycine repeat region, a Z-DNA/RNA-binding domain called Zα, and a tandem-repeated sequence adjacent to the Zβ domain of the constitutive form (12) (see also Chapter 6).

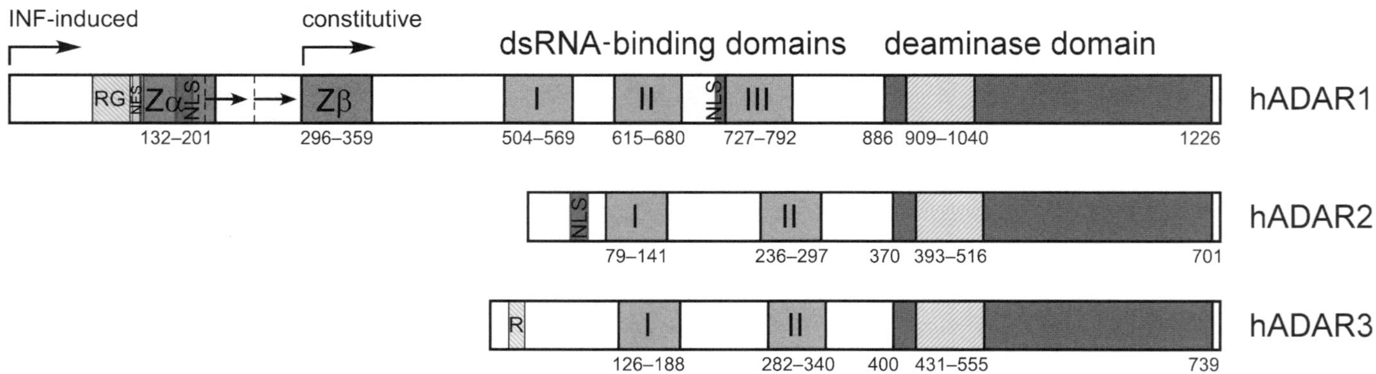

Figure 15.2 Human adenosine deaminases acting on RNA (ADAR) proteins and their important structural elements. Diagrams are to scale.

A nuclear localization signal (NLS) located near the N-terminus directs nuclear import of human ADAR1 (see also Chapters 1, 6, and 9). Human ADAR1 also has an additional NLS within dsRBD-III that gives it characteristics similar to a shuttling protein and suggests that RNA editing occurs during transcription or during RNA processing in the nucleolus (13). Nuclear export appears to be controlled by a nuclear export signal (NES) located close to the N-terminus. ADAR2 also contains an NLS in its N-terminal region, but does not contain an NES. ADAR3 has no NLS or NES, but does contain a unique N-terminal region containing an arginine-rich R domain (9, 10). ADAR1 has a region in the N-terminus that is rich in arginine and glycine residues. ADAR2 and ADAR3 each contain two dsRNA-binding motifs, in contrast to three dsRNA-binding motifs in ADAR1. Following the double-stranded binding region, all ADAR family members have a conserved deaminase domain containing the enzymatic active site.

The vertebrate ADAR1 was the first ADAR gene identified (14), as well as the first ADAR purified (15). The cloning and sequencing of ADAR1 led to the discovery of ADAR2 (9, 16, 17). A third ADAR gene, similar to ADAR2 but with unknown function, ADAR3, was discovered soon after the sequence of ADAR1 was determined (9, 10). It has the amino acid residues necessary for catalysis, but neither site-specific nor nonspecific dsRNA adenosine deaminase activity has been observed.

The sequences of ADAR1 from human, rat, and mouse are all highly conserved. Human ADAR1 is about 140 kD (1226 amino acids) and is composed of three major domains. The region closest to the C-terminus contains the adenosine deaminase domain, while the central region consists of three dsRNA-binding domains. The N-terminal region contains a Z-DNA-binding domain (Zab), which is composed of two homologous subdomains (Zα and Zβ) separated by a tandem-repeated linker.

The human ADAR1 Zα domain, known to bind left-handed Z-DNA, will also bind Z-RNA (18) (see also Chapter 9). Z-DNA and Z-RNA have remarkable structural similarities that do not appear to be discriminated by the Zα domain (19). Consequently, the actual biological targets (i.e., Z-DNA or Z-RNA) and the circumstances affording recognition by the ADAR1 Zα domain are presently unclear. The Zα domain has been identified in several other proteins involved in viral infection (E3L) and the antiviral interferon-response (DLM, PKR) (20–23). This circumstantial evidence suggests that the ADAR1 Zα domain may be involved in target recognition in antiviral hyper-editing.

ADAR2 contains only one NLS and is primarily localized in the nucleolus and nucleoplasm. NLS signals are important since both ADAR1 and ADAR2 are constantly being shuttled in and out of the nucleolus. In one experiment comparing functionality, ADAR2-mediated RNA editing in the nucleolus was observed, but the expected ADAR1-like editing was not. This suggests important differences between it and ADAR1.

ADAR3 is distinctively different from the other proteins in its family. Human ADAR3 is only found in certain regions of the brain such as the amygdala and the thalamus. Purified ADAR3 proteins have not been shown to interact with previously known RNA deamination sites, nor are they capable of deaminating adenosine residues of a long synthetic dsRNA substrate. Like ADAR2, ADAR3 does not have a Z-DNA binding domain. ADAR3 has been shown to bind dsRNA via its dsRBDs as well as single-stranded RNA via an R-domain that is lysine-rich and located near the N-terminus. This domain is unique to ADAR3, which may in fact play a role in inhibition. It has been shown to inhibit other ADAR proteins *in vitro*, which suggests a potential role in RNA editing regulation (10).

15.2.1 Double-Stranded RNA Binding Domains (dsRBDs)

ADARs from all organisms have a common domain structure consisting of one to three dsRNA binding domains (dsRBDs) near the N terminus (24, 25). The double-stranded RNA-binding domain (dsRBD) is a 65 to 70-amino-acid motif that mediates dsRNA interactions in a large variety of proteins with diverse functions found in organisms ranging from *Escherichia coli* to humans (26). The domain was first identified in the dsRNA-activated kinase PKR (27) and in the *Drosophila* Staufen protein (26). The dsRBD is highly specific for dsRNA, with little or no observable binding to ssRNA, dsDNA, or ssDNA (26, 28, 29). The functions of dsRBD-bearing proteins range from RNA transport to RNA processing to RNA chaperonin activity (30). Most dsRBD structures have an α–β–β–β–α topology in which the N- and C-terminal α-helices pack against one face of a three-stranded antiparallel β-sheet (Figure 15.3) (31, 32). A

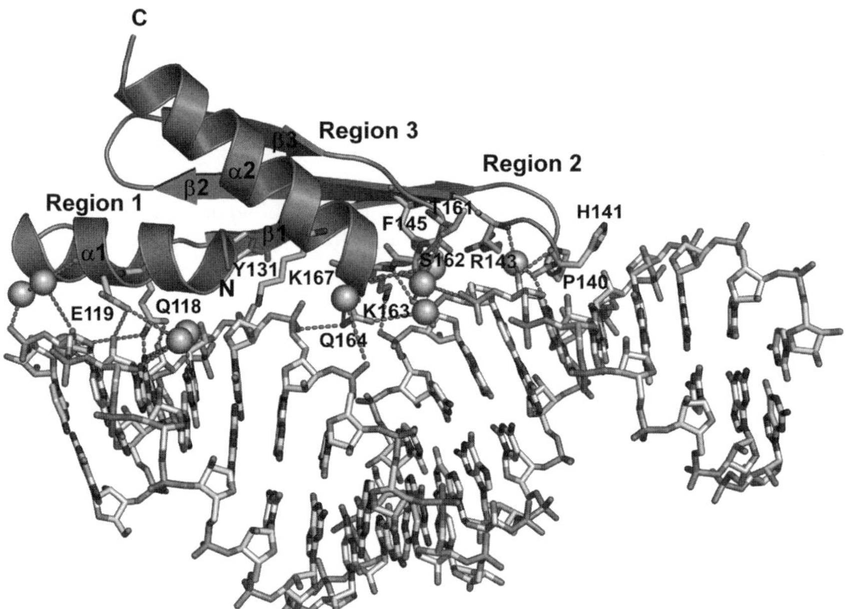

Figure 15.3 Crystal structure of the second *Xenopus laevis* XIrpba double-stranded RNA-binding domain at 1.9 Å. The protein interacts with the RNA minor grooves and across the intervening major groove on one face of the dsRNA helix. Protein is shown as ribbon in blue. Side-chain residues interacting with the RNA minor groove in regions 1and 2 are colored in green, and side chains interacting with the RNA major groove in region 3 are colored in orange. The amino acid side chains Y131 and F145 position K167 and K163, respectively, and are colored in orange. Oxygen atoms are shown in red, and nitrogen atoms are in blue. Hydrogen bonds are indicated by dashed lines. Green spheres represent water molecules. (*See color insert.*)

common sequence motif identified in many dsRBDs is a cluster of lysines (KKXXK) that are important for function. Deletions or mutations to this motif reduce dsRNA binding and indicate that it is important for recognition (33–37).

The dsRBD structure has an approximately ellipsoidal shape. All structurally characterized dsRBDs have similar structures, especially regarding the position of α-helix 1. The C-terminal α-helix (α2) packs against all three strands, while the N-terminal α-helix (α1) only contacts to β1 and α2. Thus, for most of its length, α2 is the more deeply buried of the two helices. This is reflected in greater hydrophobic residue conservation in α2 than in α1 in all the dsRBDs (32). The similarity between the dsRBD and N-terminal domain of the S5 family of ribosomal proteins is more extensive, since the two domains share not only the same topology, but also a number of conserved residues in their hydrophobic cores (31). This indicates that the dsRBD has an ancient origin, and it supports the idea that dsRBD may have evolved from a primitive ribosomal protein (see also discussion of cytidine deaminase origins in Chapter 16). Although the individual dsRBDs share many conserved residues, the overall length of the domain is somewhat variable.

15.2.2 *Xenopus laevis* Xlrbpa

Residues implicated in the RNA-binding cluster are on one face of the dsRBD near the beginning of the C-terminal α-helix (31, 32). Important amino acids for the interaction with the A-form helix of a dsRNA were revealed in both biochemical and spectroscopic studies (31, 32, 38–40). The crystal structure of the second dsRBD of *Xenopus laevis* XIrbpa (XIrbpa-2) complexed with dsRNA has been solved to a resolution of 1.9 Å, and a precise description of the dsRBD–dsRNA interaction has been reported (Figure 15.3) (38). The dsRBD contacts two neighboring minor grooves and spans across a central major groove on one face of the dsRNA helix (Figure 15.3). These interactions span 16 base pairs and collectively bury a surface area of 1680 $Å^2$. The interactions can be divided into three regions: Region 1 involves interaction of the N-terminal α-helix with the RNA minor groove, region 2 involves interaction of the loop between β-strands 1 and 2 with the RNA minor groove, and region 3 involves interaction of the C-terminal α-helix across the RNA major groove.

In region 1 of XIrbpa-2, the N-terminal three-turn α-helix interacts with the minor groove of the RNA (Figure 15.3). Two side chains (Gln 118 and Glu 119) and one peptide backbone group (carbonyl of Glu 119) are involved in direct and water-mediated interactions with four 2′-OH groups and five base functional groups in the minor groove of the RNA (Figure 15.3). These direct and water-mediated interactions satisfy all of the hydrogen-bonding potential in this region of the RNA minor groove. In region 2, the loop between β-strands 1 and 2 interacts with the adjacent minor groove of the RNA (Figure 15.3). One amino acid side chain (His 141) and three peptide backbone groups (the carbonyl of Pro 140, the carbonyl of His 141 and the amine of Arg 143) make direct and water-mediated interactions with two 2′-OH groups and one base functional group in the RNA minor groove (Figure 15.3). In region 3, the protein interacts across the major groove of the RNA. Six nonbridging oxygens of the phosphodiester backbone are contacted either directly or via water molecules (Figure 15.3). A surprisingly short peptide segment just preceding and including the

N-terminal portion of the C-terminal α-helix (residues 161–167) provides all of the major groove interactions. On the side of the major groove proximal to region 2, the backbone amine groups of residues 163 and 164, which are at the N-terminal end of the five-turn α-helix, interact with nonbridging oxygen of the phosphodiester backbone. The side chain of Lys 163 hydrogen-bonds directly to an adjacent phosphodiester group on this same side of the major groove, and amino acid side chains further along in the α-helix (Gln 164 and Lys 167) interact on the other side of the major groove (proximal to region 1) such that the α-helix bridges the major groove.

15.2.3 ADAR2 dsRBD1 and dsRBD2

Similarly, Stefl and colleagues studied the interaction of the two dsRBDs from rat ADAR2 with a 71 nucleotide RNA encoding the R/G site of the GluR-B (37). They found that both dsRBD1 (residues 74–147) and dsRBD2 (residues 231–301) interact with a defined location on the R/G stem loop. These findings indicate that the dsRBDs recognize the RNA substrate by themselves, without contributions from the deaminase domain. The binding preference of dsRBD1 for a stem loop containing a stable GCU (A/C)A pentaloop is similar to Rnt1P dsRBD's recognition of a AGNN tetraloop and to Staufen's dsRBD-3 interaction with a stem loop capped by a UUCG tetraloop (40, 41). In contrast, dsRBD2 favors RNA duplex RNA containing two A–C mismatches separated by 10 base pairs adjacent to the R/G site. Functional analyses revealed specific contributions in substrate recognition for the individual ADAR2 dsRBDs. Deletion of dsRBD1 decreased editing of the R/G site by 3-fold, whereas deletion of dsRBD2 decreased activity by 10-fold (37). Collectively, both dsRBDs appear to play a role in editing site selection.

Recent results show that ADARs act as dimers in mammals and *Drosophila* (35, 42, 43). ADAR1 and ADAR2 can form homodimers that are required for RNA editing *in vitro*. These interactions are proposed to align the active site in the correct position over the target adenosine. It is not known whether dimer interactions are cooperative for substrate binding. The N-terminal dsRBDs have been implicated in forming the dimerization interface, and RNA interactions may be important (35, 44). There are discrepancies in this model, though. Two studies have shown that ADAR dimerization does not involve RNA interactions (43, 45). Interestingly, ADAR1 and ADAR2 with mutations in dsRBDs that prevented RNA binding were still capable of dimerization thus indicating that protein–protein interactions are important. Moreover, ADAR dimers appear to require two subunits with functional dsRBDs to bind dsRNA substates and for editing activity. The precise mechanism of how RNA binding facilitates deamination is still uncertain, but studies suggest that formation of a ternary complex with two ADAR proteins per substrate may be required for activity (42). Collectively, editing activity may be regulated by ADAR dimerization, but further studies are required to verify this hypothesis.

15.2.4 Deaminase Domain

ADARS from all organisms have a conserved C-terminal catalytic deaminase domain (24, 25) (see also Chapter 16 for structural considerations of cytidine deaminase domain). Purified human ADAR2 deaminates substrates *in vitro* in the absence of any added cofactors (46), and deletion of dsRBD1 results in an active protein that accurately edits an RNA substrate (47). It has also been shown that a protein consisting of only the deaminase domain is active *in vitro*, although it deaminates RNA less efficiently than the full-length protein.

The crystal structure of the catalytic domain of human ADAR2 has been solved to a resolution of 1.7 Å (48). The structure of the domain is spherical in shape with a diameter of 40 Å. The active site is indicated by a zinc ion, which coordinates a water molecule that presumably displaces ammonia during the deamination reaction. The coordination of the zinc ion by His 394, Cys 451, Cys 516, and a hydrogen bond from a water molecule mediated by Glu 396 (Figure 15.4) is essentially identical to the geometry seen at the catalytic centers of cytidine deaminase (CDA) and tRNA adenosine deaminase (TadA) (49, 50). About 77 residues form a structurally conserved "deamination motif" comprising two helices (α2 and α5), four strands (β1, β2, β5, and β8), and connecting loops (Figure 15.4). Other residues of the hADAR2 deaminase domain do not have structural equivalents in CDA and TadA (48). The site of nucleophilic attack during the ADAR deamination reaction (C6 of adenine) lies deep in the major groove of the dsRNA substrate. Because this site is probably inaccessible by the enzyme, ADARs may use a base flipping mechanism to orient the target adenosine inward toward the active site, similar to other enzymes that modify double-stranded polynucleotides (51–53). The catalytic zinc center is found in a deep positively charged pocket that probably interacts with dsRNA. In contrast, TadA uses an alternative mechanism of substrate selection that probably involves recognition of the anticodon stem/loop of tRNA (50). The catalytic zinc of hADAR2 is shallower than that observed for TadA enzymes. Another surprising, but important, feature of the hADAR2 catalytic domain is a cavity that contained an inositol hexakisphosphate (IP_6) molecule. This cavity is extremely basic and is lined with arginine and lysine residues that are highly conserved in other ADAR proteins. In the crystal structure, the center of the inositol ring is more than 10 Å from the protein surface. This suggests that human ADAR2 requires IP_6 for activity and biochemical experiments have confirmed this assertion (48). The requirement of IP_6 for ADAR and ADAT activity suggests many interesting links between RNA editing and diverse processes such as metabolism, cell signaling, and transportation, thus setting the stage for future studies.

15.2.5 Zα and Zβ Domains

The RNA editing enzyme ADAR1 contains two homologous domains, Zα and Zβ. The Zα domain has been shown to bind both Z-DNA and Z-RNA (18, 54, 55). ADAR1 is present in cells in two forms: **(a)** a constitutively expressed form containing only Zβ and **(b)** a longer interferon-induced form, where both Zα and Zβ are present. The ADAR1 Zα domain is the prototype for a family of related peptides, which all share common amino acid sequence motif (54).

By carrying out controlled proteolysis of the Zab domain of ADAR1, it was possible to isolate a stable Zα domain that could be overexpressed and purified in sufficient quantities for crystallographic studies (56). Schwartz and co-workers were able to co-crystallize Zα with a segment of Z-DNA and the structure of this complex

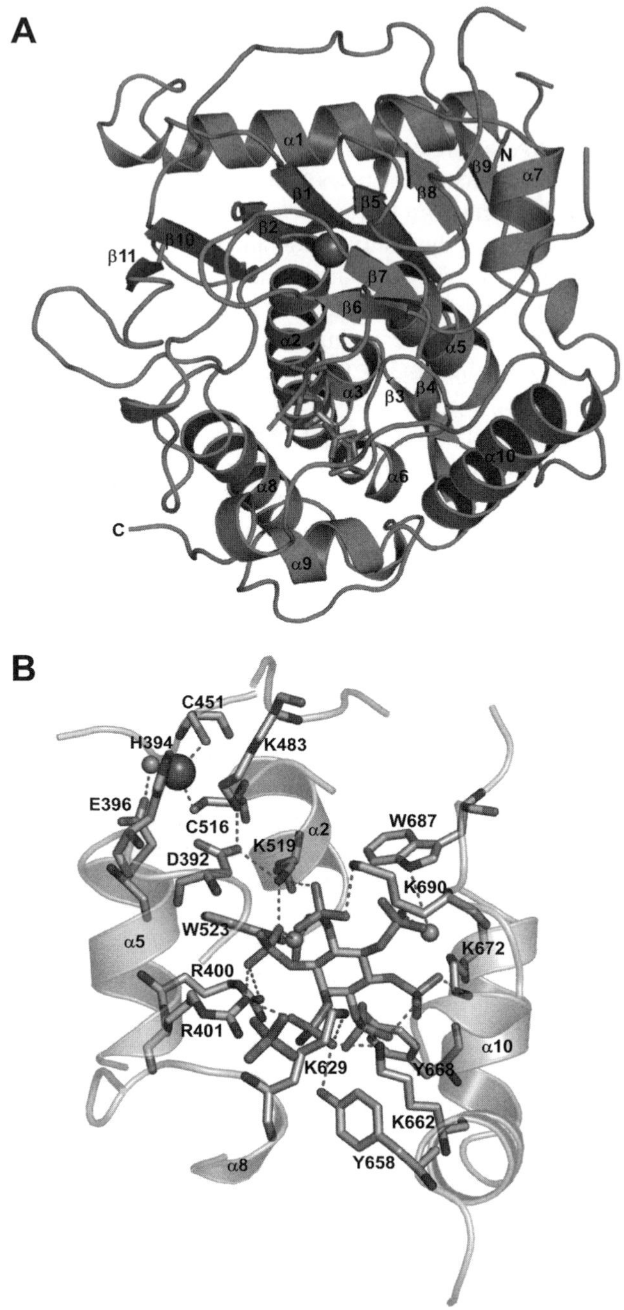

Figure 15.4 Crystal structure of the ADAR2 catalytic domain at a resolution of 1.9 Å. (A)The deaminase domain has a spherical shape constructed by 10 α-helices and 11 β-strands. (**A**) blue-gray sphere represents the active-site zinc ion. (**B**) Residue interactions at the active site. Zinc and inositol hexakisphosphate (IP6) are shown. The zinc ion and E396 coordinate the nucleophilic water (green sphere). The hydrogen-bond relay between zinc and IP6 is shown as dashes, as are hydrogen bonds between conserved residues and IP6 (gray sticks). IP6 interactions with W523 and W687 are mediated by waters (green spheres). (*See color insert.*)

has been solved at 2.1-Å resolution (57). The structure that emerged was quiet surprising and revealed the mechanism Zα uses for recognizing Z-DNA. One Zα monomer binds to one strand of the left-handed DNA, while a second monomer binds to the opposite strand of DNA; however, the two proteins do not interact with each other (Figure 15.5A). It is unlikely that Zα in the native ADAR1 protein interacts with Z-DNA in such a fashion do to steric constraints.

15.2.6 Zα Structure

The Zα domain has a compact α/β architecture containing a three-helix bundle (α1 to α3) flanked on one side by a twisted antiparallel sheet (β1 to β3). The N-terminal helix α1 is followed by the short-strand β1, which positions the C-terminal β hairpin with two hydrogen bonds. This arrangement of β hairpin is a common feature of helix–turn–helix (HTH) proteins with α/β topology (58). Helices α2 and α3 reside between β1 and β2 and form the HTH motif. Aliphatic residues from the three helices, together with Trp 195 in strand β3, interdigitate and form a hydrophobic core. The protein domain is folded in the form of a helix–turn–helix (HTH), a motif that is widely used in transcription factors for the recognition of specific B-DNA sequences. In the case of B-DNA-binding HTH proteins, generally the recognition helix fits into the broad, deep major groove of B-DNA and contacts specific base pairs. Zα utilizes a recognition helix, but it does not bind in the major groove. Instead, it runs along the side of the Z-DNA double helix together with an adjacent β-sheet and recognizes five adjacent phosphate residues in a zigzag backbone using a complex of 11 different hydrogen bonds. This recognition is shown in Figure 15.5B. The contact between Zα and the DNA strand consists of a single continuous recognition surface. This surface, which is complementary to the DNA in terms of shape and electrostatic nature, consists of α3 and the C-terminal β-hairpin. All of the polar interactions in the complex are with the sugar–phosphate backbone. Direct contacts are made between Lys 169, Lys 170, Asn 173, Arg 174, and Tyr 177 in helix α3, as well as Thr 191 and five consecutive phosphates on the nucleic acid backbone. In addition, several water-mediated phosphate interactions were observed. Lys 169, Asn 173, and Trp 195 coordinate two ordered water molecules embedded within the protein–DNA interface. These waters are completely shielded from the surface. Furthermore, Thr 191 and Arg 174 also bind to the furanose oxygens of G2 and G6, respectively. This ensemble of residues forms an extensive network of hydrogen bonds. The crystal structure reveals a striking complementary in the van der Waals surfaces of the protein and the DNA. Peptide backbone atoms of helix α3 (Lys 170 and Arg 174) contact the ribose of G4. As shown in Figure 15.5B, the face of the aromatic ring of Tyr 177 makes a close van der Waals contact with the exposed C8 of the purine base G4 of the Z-helix. This interaction is a stabilizing edge-to-face contact. Immediately on the other side of the tyrosine residue, a tryptophan is in van der Waals contact with the tyrosine in a second stabilizing edge-to-face interaction. The tyrosine residue in this conformation can interact with guanine in the *syn* conformation. Thus, the Zα domain recognizes Z-DNA by its two most distinct features, which distinguish it from B-DNA: the zigzag phosphate backbone and the *syn* conformation of a purine nucleotide. Trp195 plays important role in both protein stability and DNA binding. Pro 192 and Pro 193 in the

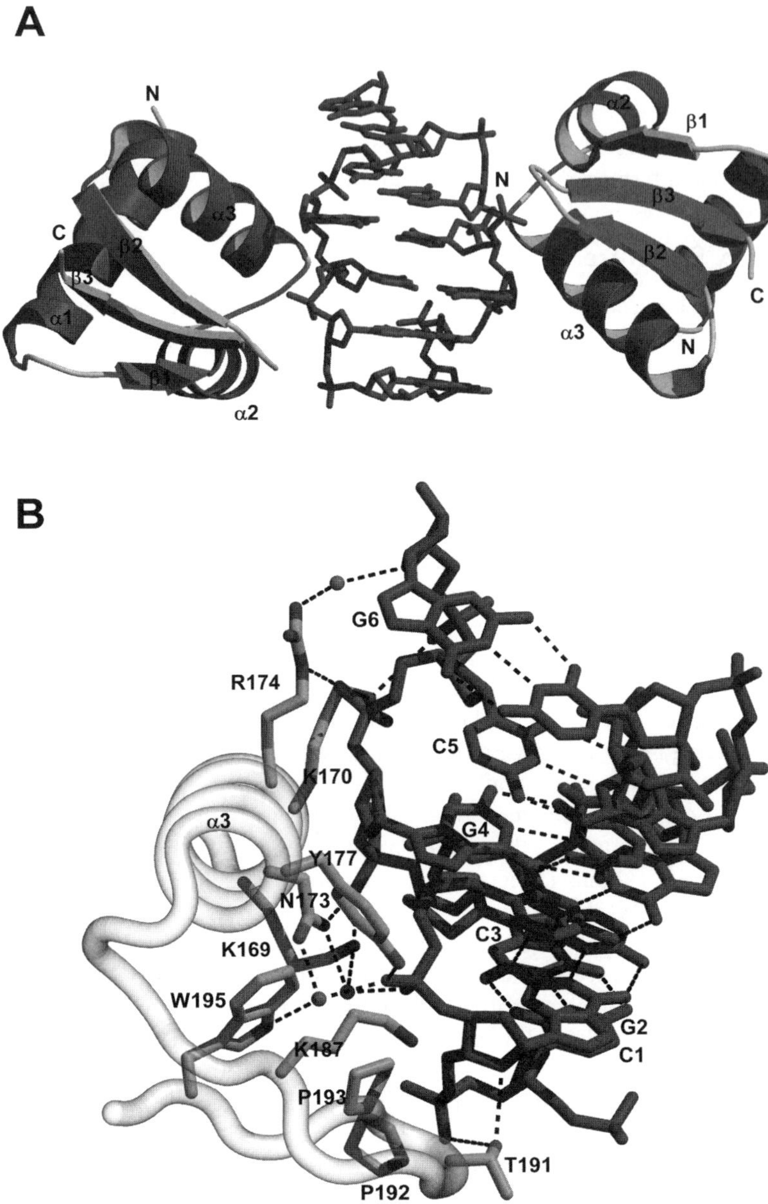

Figure 15.5 Overview of the Zα domain bound to left-handed Z-DNA. (**A**) Residues 134–198 of two symmetry-related Zα monomers and the 6-base-pair left-handed DNA duplex d (CGCGCG)2 are shown. Labels indicate N- and C-termini of the proteins as well as helices (α1–α3) and strands (β1–β3). (**B**) Protein residues involved in DNA contact. Hydrogen bonds are indicated by dashed lines. Water molecules in key positions are indicated by green spheres. (*See color insert.*)

C-terminal β-hairpin form another important van der Waals interaction with the DNA. The absolute conservation of Pro 192 within the known homologs of Zα (54) underlines the importance of this local structure. There is extensive electrostatic and van der Waals complementarity that explains the high affinity of Zα for Z-DNA. The major difference between the interaction of Zα and Z-DNA and the structurally similar helix–turn–helix domain recognizing B-DNA is that the recognition helix has a different "angle of attack." In the B-DNA interaction, the recognition helix is almost perpendicular to the axis of B-DNA, while in the Z-DNA interaction, the recognition helix is rotated so that it is more in line with the helix axis.

It is likely that Z-DNA binding is evolutionarily maintained, because the residues that contact Z-DNA in Zα are well-conserved within the family of Zα homologs, including the second Z-DNA-binding domain of ADAR1, Zβ (57). The unique zigzag pattern of sugar–phosphate backbone and the *anti–syn* alteration of the bases are characteristic of Z-DNA and distinguish it from other nucleic acid structures. The Zα domain has a tailored fit, which recognizes both features specifically.

15.2.7 Zβ Structure

Cellular proteins containing Zα domains bear a second homologous domain termed Zβ that appears to lack the ability to bind left-handed nucleic acids. In ADAR1, the Zβ domain is located ~80 residues downstream of Zα and coincides exactly with the site of translational initiation for the constitutively expressed shorter form of this editing enzyme. The Zβ domain is highly conserved, more so than Zα, among ADAR1 sequences and maintains all hydrophobic residues important for the winged-HTH fold. This suggests that the two domains Zα and Zβ, although apparently related in structure and evolution, probably serve different functions. Comparing the domain organization of constitutive ADAR1 and ADAR2, the Zβ domain represents a unique feature of ADAR1. Thus, it appears likely that the Zβ domain is a determinant of ADAR1 function by affecting its targeting, localization, or regulation. The Zβ domain of ADAR1 was overexpressed and purified for crystallographic studies. The structure of the Zβ domain has been solved to a resolution of 0.97 Å and has shown how it differs from other members of the Z-domain family (55). A Zβ dimer with an extensive dimerization surface was observed in the crystals. The Zβ monomer consists of four α-helices and three β-strands arranged in the scheme: α1-β1-α2-α3-β2-β3-α4 (Figure 15.6A), with most elements contributing side-chains to the extensive hydrophobic core of the domain.

Based on its topology, Zβ belongs to the winged HTH domains with helices 2 and 3 forming the HTH motif and a short "wing" connecting β2 and β3. An additional helix, α4, is attached to the winged-HTH domain, replacing the second wing, which is found only occasionally among winged-HTH domains. Structure-based alignment of the Zβ domain of ADAR1 homologs from human, mouse, rat, *Xenopus*, *Danio rerio*, *Takifugu rubripes*, and *Tetraodon fluviatilis* reveals four clusters of residues are highly conserved in Zβ (55). Cluster I corresponds to helix α1, where conserved residues include Glu 301, and Cys 304 are involved metal binding. Clusters II (residues 316–324) and III (residues 337–345) span the end of helices α2, α3, and the corresponding turns following them. Conserved residues of

A

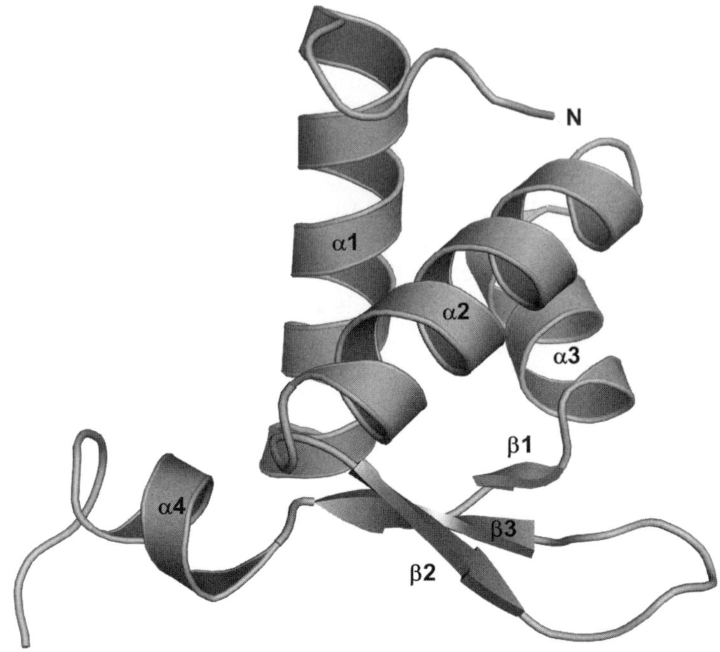

N
α1
α2
α3
β1
α4
β3
β2
C

B

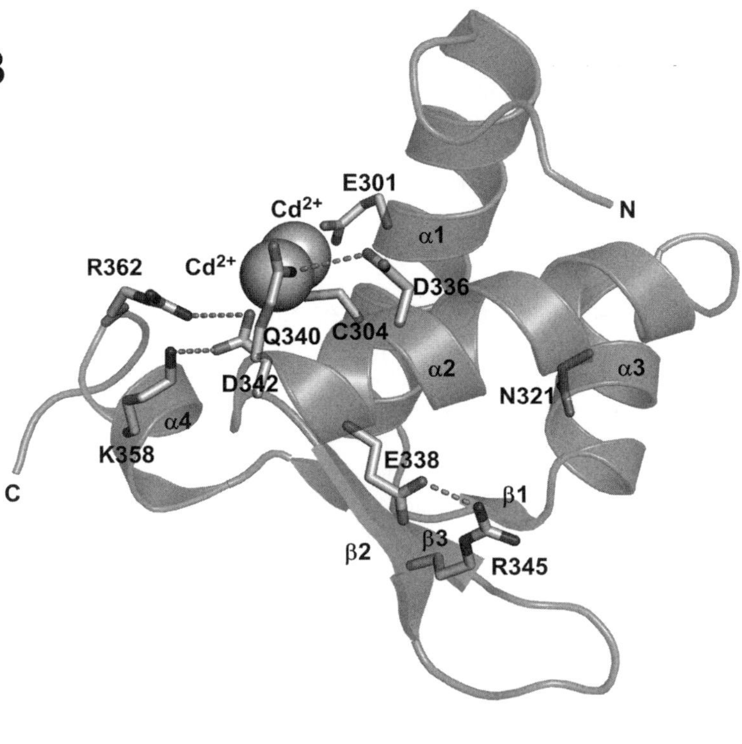

E301
N
Cd2+
α1
Cd2+
R362
D336
Q340
C304
α2
α3
D342
N321
α4
K358
E338
C
β1
β3
β2
R345

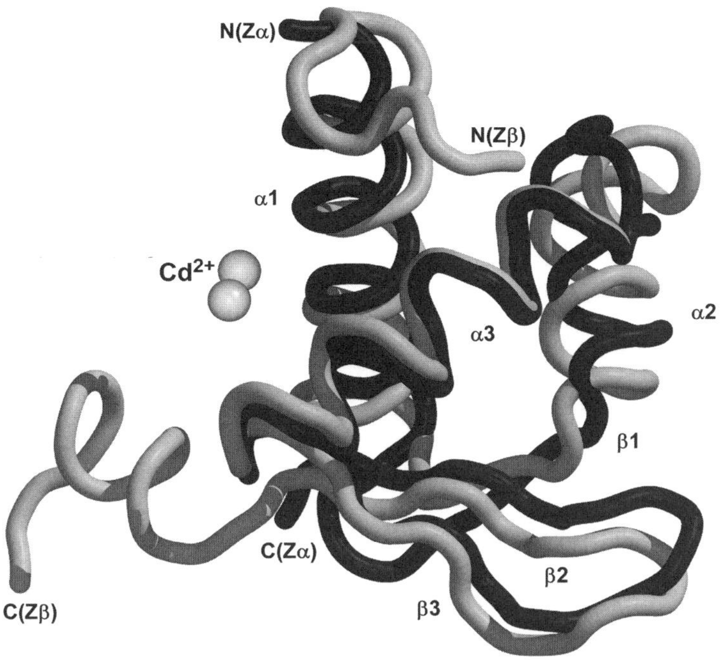

Figure 15.7 Superposition of the Zα and Zβ domains. The ADAR1 Zα and Zβ domains have the same topology and overall structure. Metal ions identified in the Zβ structure are shown as spheres. (*See color insert.*)

clusters II and III have a structural role in the formation of these tight turns. Helix 4 (residues 356–363) at the C-terminus of the winged-HTH motif represents the last highly conserved motif (motif IV) and forms an integral part of the domain. It is positioned diametrically opposite to helix α2 and approximately antiparallel to it, with helix α1 lying in between in a perpendicular orientation (Figure 15.7). The conserved Asp 342 of β2 has a central role in positioning α4 by forming salt bridges with Arg 362 and Lys 358, while Met 363 at the end of helix α4 makes hydrophobic contacts with Phe 308 and Leu 355. Overall, the Zβ domain shows a compact structure and very short loops connecting secondary structure elements. It is interesting that the Zβ domains of DLM-1 and PKZ proteins do not show conservation in the corresponding region of helix α4 (55), suggesting a unique role for helix α4 within the ADAR1 Zβ domain.

Figure 15.6 The ADAR1 Zβ domain. (**A**) Ribbon diagram of the ADAR1 Zβ domain. (**B**) Conserved residues and those involved in metal-binding residues are shown. Cadmium ions are depicted as spheres. Residues E301 and C304 are involved in metal binding. Residue Q340 forms a hydrogen bond with D336. The conserved D342 forms salt bridges with R362 and K358. E338 forms two hydrogen bonds with R345. A hydrogen bond is formed between N321 and the backbone carbonyl moiety of N317. (*See color insert.*)

Structural comparison shows that the Zβ domain has an additional secondary structure elements, the C-terminal H4, which is characterized by the presence of a highly conserved and charged KKRER sequence. The turns following helixes α1 and α2 are three residues shorter (α1) and one residue longer (α2), relative to both Zα/DNA complex structures. This represents a signature of the Zβ folding pattern. The shorter α1–β1 turn may be the cause of an ~1-Å dislocation of helix α1 relative to the rest of the domain (Figure 15.7). Zα domains interact with Z-DNA through a surface formed by the "recognition helix" α3 and the wing involving five direct side-chain contacts with the DNA backbone, all of which are conserved in all Zα/DNA structures. Superposition of Zα with Zβ shows that only Arg 328 and Asn 331 form similar contacts in Zβ, while residues Ala 327, Ala 332, and Ile 335 do not have potential hydrogen bonding to the backbone of Z-DNA. Mutagenesis studies of these residues in the context of Zα have shown that all three residues are important for DNA binding by Zα (59). This lack of hydrogen bonding potential may be the reason for the inability of Zβ to interact with Z-DNA. The residue differences seen in the potential Zβ DNA binding site results in a significant reduction of the positive electrostatic potential when compared to that seen in the Zα/DNA interface. In the Zβ structures, a conserved surface forms an unexpected double binding site for CD^{2+} and mediates the formation of a Zβ dimmer. Metal-mediated dimerization of Zβ may have potential implications for the function of ADAR1, given the prevalence of ion channels among ADAR substrates (55).

15.2.8 Structural Comparison of Zα and Zβ

Zα and Zβ share a similar hydrophobic core. A central role in the Z-domain fold is played by Trp 195 (Zα numbering), which is absolutely conserved in all known Z-binding domain sequences. Nevertheless, the two domains display different behavior with respect to their interaction with nucleic acids. The conservation among surface residues of the Zα and Zβ subfamilies reveals that distinct surfaces of the domains are crucial for their function. Since Zβ is the only domain that does not participate in the enzymatic activity of ADAR1 *in vitro*, it would be interesting to see if it has any role for complex formation. Zβ maintains a winged-HTH fold with the addition of a C-terminal helix. Mapping of the Zβ conservation profile on the Zβ surface reveals a new conserved surface formed partly by the terminal helix 4, involved in metal binding and dimerization and absent from Zα domains.

15.3 CONCLUSIONS

Although the structures of several ADAR subdomains have been solved, a structure of a full-length ADAR protein has not. The complete functions and roles of the ADAR proteins are also not well understood. Future studies on ADAR structure and function will likely elucidate mechanisms of how ADAR proteins alter genetic information as well as protein function. Ideally, structures of more functionally complete ADAR proteins will provide insight into these areas of interest.

15.4 SUBSTRATES

15.4.1 Overview and General Features

The ADAR enzymes perform two types of A-to-I editing activities, a site-selective form, and a promiscuous "hyper-editing." Site-selective editing of single adenosines typically occurs on short stretches of dsRNA that are punctuated with extrusions such as mismatches, bulges, and loops (24, 60–62). In contrast, hyper-editing (also termed hypermutation) occurs on long stretches of wholly paired double-stranded RNA and many adenosines are deaminated in a non-selective manner (see also Chapters 1, 6, and 9). Thus, two distinct types of double-stranded RNA substrates exist for the ADARs: (a) those with bulges and mismatches and (b) those that are virtually based-paired entirely. The ADAR3 enzyme is capable of interacting with single-stranded RNA, in addition to double-stranded RNA, but a *bona fide* editing substrate has not yet been found. ADAR1 contains two unique domains, Zα and Zβ, which have been shown to interact with left-handed Z-DNA or Z-RNA. The role of these Z-binding domains in substrate recognition is not clear.

Most of the genes edited by ADAR enzymes in mammals, *Drosophila melanogaster*, *Caenorhabditis elegans*, and squid are expressed in the nervous system (see also Chapter 1). Prominent examples exist where deamination of single adenosines leads to single amino acid changes in the resulting proteins, often with profound functional consequences. RNA editing can create or destroy splice sites, altering the protein product made from a gene. Editing has also been implicated in microRNA processing and maturation. In the case of hyper-editing, deamination is completely random and is thought to have a role in host antiviral defense.

15.5 DOUBLE-STRANDED RNA TARGETS AND STRUCTURAL FEATURES

15.5.1 Site-Selective A-to-I Editing

Editing in coding regions by ADAR can have important physiological effects, as evidenced by editing of the pre-mRNAs encoding subunits of the α-amino-3-hydroxy-5-methyl-4-isoxazolepropionic acid-sensitive glutamate receptor (GluR-B), the serotonin-2C receptor, the GABA type A receptor subunit, and several ion channels (63–67).

The best-described example of mammalian editing is that of the glutamate receptor, the most abundant excitatory neurotransmitter. GluR-B pre-mRNAs are edited at multiple sites. The most important physiological editing site in GluR-B is the Q/R site, which is edited to ~100% (Figure 15.8A) (68). This glutamine-to-arginine amino acid change causes a significant reduction in calcium permeability (69). ADAR2$^{/}$ mice die young after repeated epileptic seizures due to reduced editing at the Q/R site (70). The genetic inactivation of ADAR in *Drosophila* resulted in flies with extreme behavioral defects and neurological symptoms such as paralysis, locomotor uncoordination, and tremors, which increased in severity with age (71).

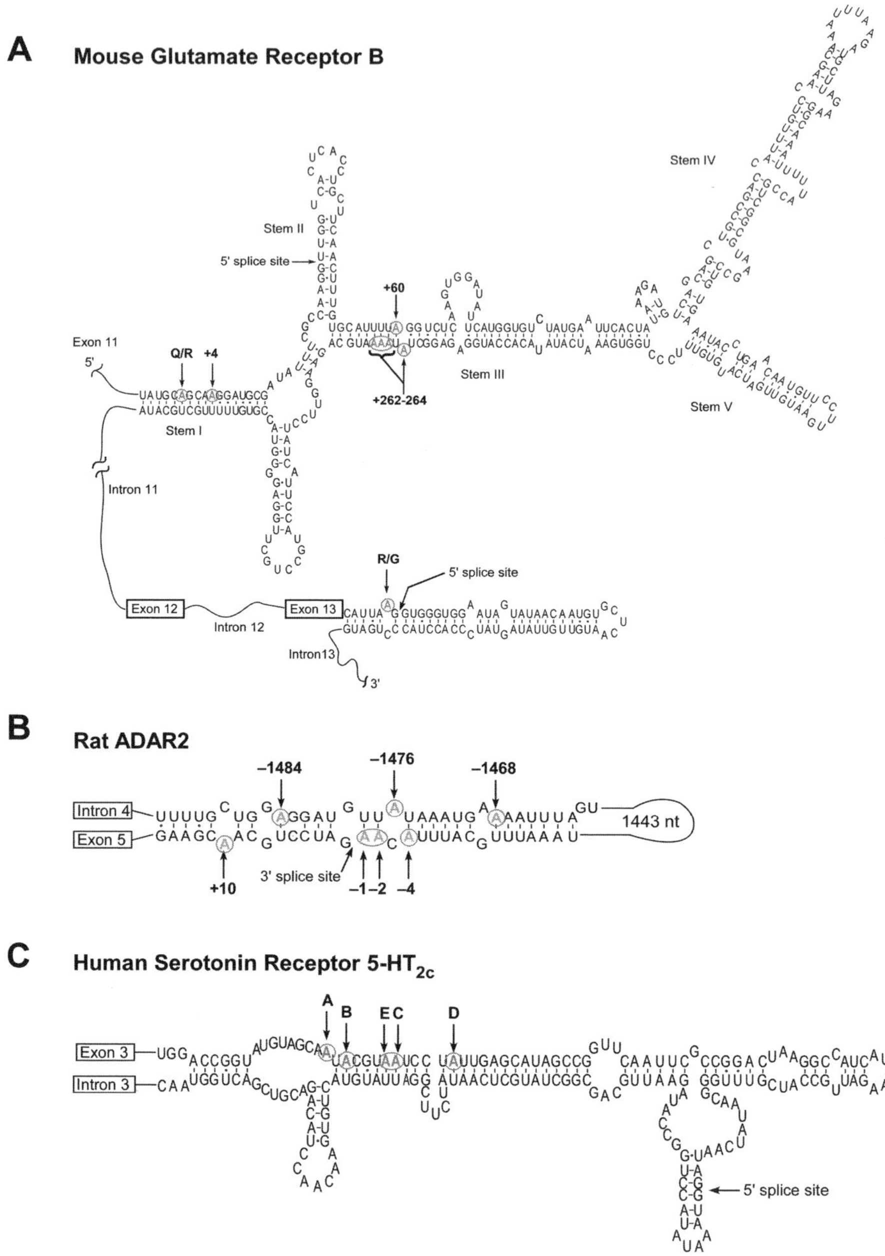

Figure 15.8 Representative ADAR family site-selective editing substrates. (**A**) The mouse glutamate receptor pre-mRNA. Editing sites and relative positioning are indicated (122–124). (**B**) Rat ADR2 pre-mRNA showing the most evolutionary conserved vertebrate editing sites. The positions of seven editing sites are indicated. Deaminated adenosines are highlighted. Numbering of editing sites is relative to the proximal 3′- splice junction in intron 4 (74). (**C**) Serotonin receptor 5-HT_{2c} showing the five editing sites 125. (**D**) $GABA_A$ receptor subunit α3 (Gabra-3) pre-mRNA. The I/M site and target adenosine are indicated (65). (**E**) Hepatitis delta virus genomic RNA showing the amber/W editing site 24, 76.

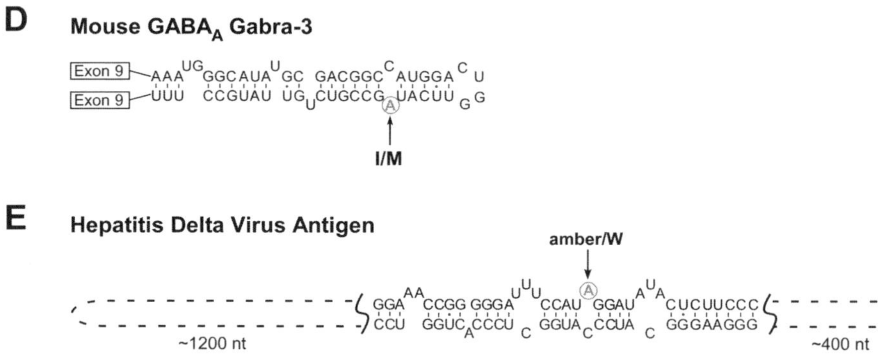

Figure 15.8 (*Continued*)

Another well-defined site is the R/G site (Figure 15.8A), where a genomically encoded AGA is modified to IGA, resulting in an arginine-to-glycine amino acid change in the protein. This change affects the biophysical properties of the ion channel, allowing the edited isoform to recover faster from desensitization (72).

ADAR-mediated editing was also found to occur in noncoding regions of pre-mRNAs (63), most notably in the transcripts of the editing enzyme ADAR2 (73). By editing its own transcripts, an alternative splice acceptor site is created in intron 4, leading to alternative splicing resulting in a nonfunctional protein. Analysis of rat ADAR2 pre-mRNA transcripts identified 16 editing sites within the base-paring interaction of intron 4 and exon 5 (Figure 15.8B) (74). Thus, auto-editing may be a negative feedback loop where ADAR2 regulates its own expression. Interestingly, the only ADAR-like gene in *Drosophila*, DADAR, which is most closely related to the mammalian ADAR2 protein, is also self-edited within its coding region (6).

Different isoforms of the serotonin receptor (5-HT$_{2c}$) are generated by RNA editing. The A-to-I conversion at five specific sites reduces G-protein coupling interactions (Figure 15.8C). Editing affects three amino acids located within the second intracellular loop region of 5-HT2CR and changes the genetically encoded Ile, Asn, and Ile at positions 156, 158, and 160, respectively. The fully edited receptor has a 20-fold reduction in G-protein coupling activity compared to the unedited receptor.

A recently identified RNA editing target is the α3 subunit of the gama-aminobutyric acid type A receptor (GABA$_A$), Gabra-3 (65). The target site can be edited by ADAR1 and ADAR2, and the A-to-I change recodes an isoleucine for a methionine (Figure 15.8D). This editing change is low at birth but reaches approximately 100% in adult brains. The functional consequences are unknown, but are speculated to be important for normal brain development.

Some non-neuronal substrates for A-to-I editing have also been described—for instance, hepatitis delta virus (HDV) RNA (75, 76). HDV is often found associated with the hepatitis B virus in patients with severe liver disease and encodes a single protein, the hepatitis delta antigen. This protein occurs in two forms due to an ADAR-mediated A-to-I modification in the HDV antigenomic RNAs which converts a stop codon to tryptophan codon (UAG to UGG) and extends the open reading frame by 19 amino acids (Figure 15.8E) (75, 76). The shorter form is essential for genomic replication, whereas the longer one inhibits replication and is required for viral packaging (77, 78). Thus, A-to-I editing enables the virus to switch from genome replication to viral packaging.

Any RNA that is moderately double-stranded represents a potential substrate for A-to-I editing (see also Chapters 5, 12, and 13). This raises the question of whether other cellular processes that rely on double-stranded RNA molecules as functional entities, such as microRNAs, will be influenced by A-to-I RNA editing (see also Chapter 9). For the RNAi mechanism operating in the cytoplasm, the most significant member of the ADAR gene family is the 150-kD form of ADAR1. The potency of the dsRNAs mediating RNAi was significantly reduced following A-to-I editing *in vitro* by ADAR proteins (79). Thus, ADAR1 may interact with precursors of the recently identified class of short RNAs such as micro-RNAs and small transient RNAs and affect their synthesis.

A-to-I editing has also recently been shown to occur in microRNAs and their precursors (80–84), although it has been suspected for some time (85). It has been estimated that A-to-I editing sites are in 6% of microRNA precursors (82), and perhaps up to 50% of pri-miRNAs may have specific editing sites (84). Editing of these precursors could have drastic effects on miRNA processing and expression of functionally mature miRNAs.

15.5.2 Structural Features of Site-Selective A-to-I Editing

The structural features that ADARs use to recognize their substrates are not fully understood. Obviously, dsRBD interactions with dsRNA targets by dsRBDs are an important facet of substrate recognition. It is thought that typical dsRBDs only provide general binding with minor sequence specificity, but the distance and number of dsRBDs may provide some selectivity. Structural and biochemical studies have shown that different ADAR dsRDBs have distinct binding specificities and that they individually contribute to the selection of the editing site (37, 86). The dsRBD-1 domain of ADAR2 prefers to interact with irregular RNA features like stem loops or non-Watson–Crick base pairs at the exclusion of regular dsRNA. Alternatively, dsRBD-2 preferentially binds to mismatches and bulges (37). Moreover, some editing sites are preferentially edited by an individual ADAR family member. For example, the Q/R site of GluR-B and the D site of the serotonin receptor are edited exclusively by ADAR2, whereas ADAR1 selectively deaminates the A and B sites of the serotonin receptor (Figure 15.8). These ADAR-specific selectivities may be due to the individual binding preferences of dsRBDs; the number of, and spacing between, dsRBDs; and ADAR–ADAR dimer interactions mediated through the dsRBDs.

Typically, short double-stranded RNAs (~20–30 base pairs) or longer partially double-stranded RNAs in which the duplex structure is interrupted by mismatches, bulges, and internal loops are edited selectively (Figure 15.8) (60, 74, 87–89). In these substrates, less than 10% of the adenosine residues are deaminated. In contrast, long, completely base-paired duplexes (>100 bp) are edited nonselectively, and up to 50% of adenosines are edited. The model for site-specific recognition is that mismatches, bulges, and loops in a substrate restrict ADAR, binding, and the target adenosine is then flipped out for deamination (90). It appears that internal loops act as helix termini for ADARs. Internal loops could destabilize adjacent helixes and convert long, promiscuously deaminated substrates into series of shorter, selectively edited targets (60). Typically, loops with greater than six nucleotides acted as helix ends, whereas loops with less than four nucleotides did not. Thus, the secondary structure features of the dsRNA and the length of the double-stranded regions are important elements in site selectivity.

The structure of a target RNA seems to determine the selectivity of editing rather than the sequence. Thus, there is no common sequence motif found in all substrates. However, distinct 5′-nearest-neigbor preferences for ADAR1 and ADAR2 have been observed: A = U > C > G. A 3′-nearest-neigbor preference for ADAR2 has also been identified: U = G > C = A. Recent bioinformatic studies have found that A-to-I editing is more frequent at adenosines involved in A–C mismatches than any other base pair (91–93). These sites may promote easier base flipping of the target adenosine during deamination.

Interestingly, many ADAR substrates all share a general feature: Editing is directed by the interaction of exonic sequences with partner sequences that lie in adjacent introns (see also Chapters 1 and 6). Editing sites are formed by intermolecular base-pairing between the sequence encompassing the editing site and a downstream editing site complementary sequence (ECS), which is often located in a neighboring intron (63). The evolutionary conservation of these intronic ECSs emphasizes their important role (94, 95).

Overall, selective substrate recognition by ADARs is a complex event that is just begining to be understood. It depends on both the overall secondary structure and local structure (such as mismatches, bulges, and loops) of the target RNA; the binding preference and number of dsRBDs; and the dimerization state of the enzyme. Collectively, these aspects permit differentiation of short site-specific adenosine targets from those in long, perfectly base-paired dsRNA.

15.5.3 Promiscuous Editing

Promiscuous editing occurs in extended, perfectly double-stranded RNA in which up to 50% of A's can be converted to I. This type of editing is sensitive to the length of the dsRNA duplex. Perfect RNA duplexes <15 base pairs are modified only inefficiently *in vitro*, and perhaps not at all *in vivo* (96). Optimal deamination requires dsRNAs to be at least 25–30 base pairs, and preferably >100 base pairs, in length (14, 96). In long, perfectly base-paired duplexes, approximately 50% of the adenosines on each strand are edited in a promiscuous fashion, except for a clear 5′-neighbor preference for A or U (51). The hyper-edited RNAs contain I–U base pairs that make the RNA duplex unstable (14). Considering that the target A's in these dsRNA are buried deep within the major groove, a transient "flipping-out" of the bases before deamination appears plausible.

ADAR1 has the ability to deaminate adenosine nonspecifically in long regions of double-stranded RNA. This apparently random conversion of adenosine to inosine is thought to play a role in the defense mechanism of host cells against viruses that form dsRNA during their life cycle where double-stranded RNA intermediates are formed in the cytoplasm as the result of complementary transcripts from opposing promoters. Hyper-editing caused by adenosine-to-inosine conversions disrupts viral open reading frames (11, 97, 98). In addition, expression of the longer isoform of ADAR1 is inducible by interferon, a cytokine typically produced during viral infections. Not only is gene activity increased, but the protein distribution also shifts from being concentrated primarily in the nucleus to both nuclear and cytoplasmic distribution.

Hyper-editing has been observed in viral RNAs isolated from host cells infected with human parainfluenza virus (99), vesicular stomatitis virus (100), avian leukosis virus (101), and respiratory syncytial virus (102). Nonviral examples of hyper-editing include pre-mRNAs encoding a voltage-gated potassium channel (sqKv2) from squid (103) and an extended 3′-UTR hairpin structure of poly A^+ RNA from *Caenorhabditis elegans* (104). Extensive A-to-I conversions were found in RNAs of measles virus from patients suffering from subacute sclerosing panencephalitis (SSPE) and measles inclusion body encephalitis (MIBE) (105). Particularly, hyper-editing occurs on mRNAs encoding the matrix protein, which is thought to be essential for viral budding. Instead of normal lytic infection cycle, viral persistence most likely takes place because matrix protein synthesis is prevented (106). The fate of promiscuously edited RNAs has been most clearly studied in the mouse polyoma virus model system. It has been suggested that during the late phase of infection by polyoma virus, early-strand viral RNAs hybridize with intronic sequences of multimeric late-strand transcripts in the host cell nucleus. These double-stranded regions are targets for ADAR-mediated hyper-editing, resulting in the deamination of approximately half of the adenosines in early strand RNAs (107). The hypermodified RNAs are retained in the nucleus and only the late-strand mRNAs can be translated, allowing the transition from the early phase to the late phase of viral infection (107).

The retention of the hyper-edited dsRNAs in the nucleus has been shown to involve a complex containing the RNA-binding protein p54nrb, the splicing factor PSF, and the inner nuclear matrix protein matrin 3 (mat3) (108). Nuclear retention caused by hyper-editing of viral RNAs could be a defense mechanism against viral infection. Moreover, cellular mRNAs hyper-edited by mistake could also be retained in the nucleus in order to prevent the synthesis of aberrant proteins. The accumulation of hyper-edited RNA in any compartment could be deleterious for the cell, suggesting the existence of a mechanism for their disposal. Recently, an endonuclease activity specific for inosine-containing dsRNAs has been described (109). Cleavage occurs at specific sites that consist of alternating I·U and U·I base pairs. Interestingly, this nuclease is only found in the cytoplasm, suggesting a possible function in the elimination of hyper-edited RNAs that have escaped from the nucleus or of RNAs that became hyper-edited in the cytoplasm by the interferon-inducible form of ADAR1.

15.6 SINGLE-STRANDED RNA TARGETS

ADAR-mediated editing generally requires a double-stranded RNA substrate; however, ADAR3 can bind to both single-stranded RNA (ssRNA) and dsRNA. It interacts with ssRNA via an arginine and lysine-rich domain (R-domain) located at the N-terminus region (10). This ssRNA-binding domain makes ADAR3 distinct from the other two ADAR enzymes. Six consecutive arginines located in the center of the R-domain are predominantly responsible for its affinity to ssRNA. ADAR3 interacts with GluR-B and 5-HT_{2C}R RNAs, both of which contain dsRNA and ssRNA regions; however, ADAR3 has not been shown to edit these substrates, nor synthetic long dsRNA substrates. The affinity of ADAR3 for dsRNA binding is lower than that of ADAR1 or ADAR2, but ADAR3 seems to interact uniquely with RNA substrates because of its dual capacity for binding dsRNA and ssRNA. An ADAR3 mutant lacking the entire dsRNA-binding region was able to bind with GluR-B and 5-HT_{2C}R RNAs through the R-domain with comparable affinity to the wild-type protein. An ADAR3 mutant lacking the six consecutive arginine residues of the R-domain did not bind ssRNA at all.

15.7 Z-DNA AND Z-RNA TARGETS

Since the discovery of Z-DNA in 1979 and Z-RNA in 1984, many groups have striven to understand whether specific biological functions are associated with these unusual nucleic acid conformations. The discovery that the N-terminus of ADAR1 bound Z-DNA with high affinity and a detailed structural view of the interaction has shed much light into this interesting system. Less is known about Z-RNA. The Zα domain was also shown to interact with Z-RNA, and a co-crystal structure of Zα bound to an RNA duplex has recently been determined (18, 19). Thus, Z-RNA is also a potential target for the ADAR1 enzyme.

The crucial step in the editing process is the formation of a hairpin or fold-back structure in the pre-mRNA molecule, resulting in the formation of an RNA duplex (63). The duplex RNA substrate is frequently formed by the pairing of an intron with an exon, and the exon is edited to change the amino acid codon. This has a number of interesting consequences. The control of the editing system rests in particular intronic sequences that are complementary to exonic sequences. In addition, it raises the question of how the enzyme manages to carry out all of its editing activity before the introns are removed by the splicing apparatus, which is known to be attached to the end of the nascent mRNA chain. This is where the postulated role of the Z-DNA-binding domain becomes more important. The problem that the editing enzyme has is that of finding an actively transcribing gene in contrast to a gene that is not transcribing. Actively transcribing genes with their moving RNA polymerases generate the negative torsional strain upstream of the polymerase that transiently stabilizes Z-DNA while the polymerase is moving (110). Hence, transcribing genes have Z-DNA in them, while nontranscribing genes do not. It is likely that the high-affinity Z-DNA binding domain at the N-terminus of ADAR1 localizes itself on the Z-DNA as a way of targeting a transcribing gene, as distinct from a nontranscribing one. In effect, it may increase the local concentration of the ADAR1 editing enzyme near areas undergoing active transcription.

It is also reasonable to assume that the targets of the Z-domains are Z-RNA rather than Z-DNA. Higher levels of ADAR1 editing have been observed in dsRNA substrates containing Z-forming purine–pyrimidine repeats (111). Interferon-induced forms of ADAR1 have also been associated with multiple viral genomes, including the measles virus and hepatitis C and D viruses (76, 112–114). Viruses that have dsRNA genomes are obvious targets, but various mechanisms are employed by these viruses to prevent the exposure of naked dsRNA. This suggests an effect of dsRNA on cell physiology (115). Double-stranded RNA may also exist in ssRNA viruses as a replicative intermediate in infected cells. Isolation of both positive-and negative-sense RNA bound to the dsRNA-binding protein 2′,5′-oligoadenylate synthetase from EMCV-infected cells seem to support this, although it is unclear how much true dsRNA between these exists in infected cells (116). In DNA viruses, overlapping convergent transcription could result in the appearance of dsRNA. If viral transcripts fail to terminate at discrete sites at the end of genes, complementary mRNAs are produced from transcription occurring in the opposite direction (117–121).

15.8 FUTURE DIRECTIONS

The discovery of RNA editing challenged the dogma that genetic sequences predict protein sequences. RNA editing is now recognized as an important mechanism by which to generate protein diversity in addition to alternative splicing, polyadenylation, differential promoter usage, or translational frame shifting. Several questions remain to be answered. Many more RNA substrates are expected to exist, and it will be important to identify them. Bioinformatics efforts have begun to elucidate many novel targets, but with these findings come the daunting task of sorting out which targets are edited by what ADAR and determining how specific recognition is achieved. In addition, functional studies with human, mouse, rat, *Drosophila*, and *C. elgans* ADAR mutants will help to detect new *in vivo* substrates and to uncover the physiological implications of editing. No complete structures have been determined for RNA-dependent deaminases, and we are just beginning to understand how these enzymes recognize their RNA targets. Thus, it will be important to obtain structural information of these proteins in complexes with their RNA substrates. A third goal will be to understand how RNA editing is linked to other important cellular processes, such as RNAi, RNA modification, and pre-mRNA splicing.

REFERENCES

1. Bass, B. L., Nishikura, K., Keller, W., Seeburg, P. H., Emeson, R. B., O'Connell, M. A., Samuel, C. E. and Herbert, A. (1997) A standardized nomenclature for adenosine deaminases that act on RNA. *RNA* **3**, 947–949.
2. Stapleton, M., Carlson, J. W., and Celniker, S. E. (2006) RNA editing in *Drosophila melanogaster: New targets and functional consequences. RNA* **12**, 1922–1932.

3. Gerber, A. P., and Keller, W. (2001) RNA editing by base deamination: More enzymes, more targets, new mysteries. *Trends. Biochem. Sci.* **26**, 376–384.
4. Keegan, L. P., Gallo, A., and O' Connell, M. A. (2001) The many roles of RNA editor. *Nat Rev Genet* **2**, 869–878.
5. Hough, R. F., and Bass, B. L. (2000) *Adenosine Deaminases that Act on RNA in RNA Editing*, Oxford University Press, Oxford.
6. Palladino, M. J., Keegan, L. P., O'Connell, M. A., and Reenan, R. A. (2000) dADAR, a *Drosophila* double-stranded RNA-specific adenosine deaminase is highly developmentally regulated and is itself a target for RNA editing. *RNA* **6**, 1004–1018.
7. Hough, R. F., Lingam, A. T., and Bass, B. L. (1999) Caenorhabditis elegans mRNAs that encode a protein similar to ADARs derive from an operon containing six genes. *Nucleic Acids Res* **27**, 3424–3432.
8. Wagner, R. W., Yoo, C., Wrabetz, L., Kamholz, J., Buchhalter, J., Hassan, N. F., Khalili, K., Kim, S. U., Perussia, B., and McMorris, F. A. *et al.* (1990) Double-stranded RNA unwinding and modifying activity is detected ubiquitously in primary tissues and cell lines. *Mol Cell Biol* **10**, 5586–5590.
9. Melcher, T., Maas, S., Herb, A., Sprengel, R., Higuchi, M., and Seeburg, P. H. (1996) RED2, a brain-specific member of the RNA-specific adenosine deaminase family. *J Biol Chem* **271**, 31795–31798.
10. Chen, C. X., Cho, D. S., Wang, Q., Lai, F., Carter, K. C., and Nishikura, K. (2000) A third member of the RNA-specific adenosine deaminase gene family, ADAR3, contains both single- and double-stranded RNA binding domains. *RNA* **6**, 755–767.
11. Patterson, J. B., Thomis, D. C., Hans, S. L., and Samuel, C. E. (1995) Mechanism of interferon action: double-stranded RNA-specific adenosine deaminase from human cells is inducible by alpha and gamma interferons. *Virology* **210**, 508–511.
12. Herbert, A., Lowenhaupt, K., Spitzner, J., and Rich, A. (1995) Chicken double-stranded RNA adenosine deaminase has apparent specificity for Z-DNA. *Proc Natl Acad Sci USA* **92**, 7550–7554.
13. Eckmann, C. R., Neunteufl, A., Pfaffstetter, L., and Jantsch, M. F. (2001) The human but not the *Xenopus* RNA-editing enzyme ADAR1 has an atypical nuclear localization signal and displays the characteristics of a shuttling protein. *Mol Biol Cell* **12**, 1911–1924.
14. Bass, B. L., and Weintraub, H. (1988) An unwinding activity that covalently modifies its double-stranded RNA substrate. *Cell* **55**, 1089–1098.
15. Kim, U., Garner, T. L., Sanford, T., Speicher, D., Murray, J. M., and Nishikura, K. (1994) Purification and characterization of double-stranded RNA adenosine deaminase from bovine nuclear extracts. *J Biol Chem* **269**, 13480–13489.
16. Lai, F., Chen, C. X., Carter, K. C., and Nishikura, K. (1997) Editing of glutamate receptor B subunit ion channel RNAs by four alternatively spliced DRADA2 double-stranded RNA adenosine deaminases. *Mol Cell Biol* **17**, 2413–2424.
17. Gerber, A., O'Connell, M. A., and Keller, W. (1997) Two forms of human double-stranded RNA-specific editase 1 (hRED1) generated by the insertion of an Alu cassette. *RNA* **3**, 453–463.
18. Brown, B. A., 2nd, Lowenhaupt, K., Wilbert, C. M., Hanlon, E. B., and Rich, A. (2000) The Zα domain of the editing enzyme dsRNA adenosine deaminase binds left-handed Z-RNA as well as Z-DNA. *Proc Natl Acad Sci USA* **97**, 13532–13536.
19. Placido, D., Brown, B. A., J2nd., Lowenhaupt, K., Rich, A., and Athanasiadis, A. 2007 A left-handed RNA double helix bound by the Zα domain of the RNA-editing enzyme ADAR1. *Structure* **15**, 395–404.
20. Ha, S. C., Lokanath, N. K., Van Quyen, D., Wu, C. A., Lowenhaupt, K., Rich, A., Kim, Y. G., and Kim, K. K. (2004) A poxvirus protein forms a complex with left-handed Z-DNA: Crystal structure of a Yatapoxvirus Zalpha bound to DNA. *Proc Natl Acad Sci USA* **101**, 14367–14372.
21. Kim, Y. G., Lowenhaupt, K., Oh, D. B., Kim, K. K., and Rich, A. (2004) Evidence that vaccinia virulence factor E3L binds to Z-DNA in vivo: Implications for development of a therapy for poxvirus infection. *Proc Natl Acad Sci USA* **101**, 1514–1518.
22. Kwon, J. A., and Rich, A. (2005) Biological function of the vaccinia virus Z-DNA-binding protein E3L: Gene transactivation and antiapoptotic activity in HeLa cells. *Proc Natl Acad Sci USA* **102**, 12759–12764.

23. Rothenburg, S., Deigendesch, N., Dittmar, K., Koch-Nolte, F., Haag, F., Lowenhaupt, K., and Rich, A. (2005) A PKR-like eukaryotic initiation factor 2alpha kinase from zebrafish contains Z-DNA binding domains instead of dsRNA binding domains. *Proc Natl Acad Sci USA* **102**, 1602–1607.
24. Bass, B. L. (2002) RNA editing by adenosine deaminases that act on RNA. *Annu Rev Biochem* **71**, 817–846.
25. Maas, S., Rich, A., and Nishikura, K. (2003) A-to-I RNA editing: recent news and residual mysteries. *J Biol Chem* **278**, 1391–1394.
26. St. Johnston, D., Brown, N. H., Gall, J. G., and Jantsch, M. (1992) A conserved double-stranded RNA-binding domain. *Proc Natl Acad Sci USA* **89**, 10979–10983.
27. Green, S. R., and Mathews, M. B. (1992) Two RNA-binding motifs in the double-stranded RNA-activated protein kinase. *DAI Genes Dev* **6**, 2478–2490.
28. Bass, B. L., Hurst, S. R., and Singer, J. D. (1994) Binding properties of newly identified Xenopus proteins containing dsRNA-binding motifs. *Curr Biol* **4**, 301–314.
29. Bevilacqua, P. C., and Cech, T. R. (1996) Minor-groove recognition of double-stranded RNA by the double-stranded RNA-binding domain from the RNA-activated protein kinase PKR. *Biochemistry* **35**, 9983–9994.
30. Hitti, E., Neunteufl, A., and Jantsch, M. F. (1998) The double-stranded RNA-binding protein X1rbpa promotes RNA strand annealing. *Nucleic Acids Res* **26**, 4382–4388.
31. Bycroft, M., Grunert, S., Murzin, A. G., Proctor, M., and St. Johnston, D. (1995) NMR solution structure of a dsRNA binding domain from *Drosophila* staufen protein reveals homology to the N-terminal domain of ribosomal protein S5. *EMBO J* **14**, 3563–3571.
32. Kharrat, A., Macias, M. J., Gibson, T. J., Nilges, M., and Pastore, A. (1995) Structure of the dsRNA binding domain of *E. coli* RNase III. *EMBO J.* **14**, 3572–3584.
33. Lai, F., Drakas, R., and Nishikura, K. (1995) Mutagenic analysis of double-stranded RNA adenosine deaminase, a candidate enzyme for RNA editing of glutamate-gated ion channel transcripts. *J Biol Chem* **270**, 17098–17105.
34. Liu, Y., and Samuel, C. E. (1996) Mechanism of interferon action: Functionally distinct RNA-binding and catalytic domains in the interferon-inducible, double-stranded RNA-specific adenosine deaminase. *J Virol* **70**, 1961–1968.
35. Gallo, A., Keegan, L. P., Ring, G. M., and O'Connell, M. A. (2003) An ADAR that edits transcripts encoding ion channel subunits functions as a dimer. *EMBO J* **22**, 3421–3430.
36. Sansam, C. L., Wells, K. S., and Emeson, R. B. (2003) Modulation of RNA editing by functional nucleolar sequestration of ADAR2. *Proc Natl Acad Sci USA* **100**, 14018–14023.
37. Stefl, R., Xu, M., Skrisovska, L., Emeson, R. B., and Allain, F. H. (2006) Structure and specific RNA binding of ADAR2 double-stranded RNA binding motifs. *Structure* **14**, 345–355.
38. Ryter, J. M., and Schultz, S. C. (1998) Molecular basis of double-stranded RNA–protein interactions: Structure of a dsRNA-binding domain complexed with dsRNA. *EMBO J* **17**, 7505–7513.
39. Ramos, A., Grunert, S., Adams, J., Micklem, D. R., Proctor, M. R., Freund, S., Bycroft, M., St. Johnston, D., and Varani, G. (2000) RNA recognition by a Staufen double-stranded RNA-binding domain. *EMBO J.* **19**, 997–1009.
40. Wu, H., Henras, A., Chanfreau, G., and Feigon, J. (2004) Structural basis for recognition of the AGNN tetraloop RNA fold by the double-stranded RNA-binding domain of Rnt1p RNase III. *Proc Natl Acad Sci USA* **101**, 8307–8312.
41. Stefl, R., and Allain, F. H. (2005) A novel RNA pentaloop fold involved in targeting ADAR2. *RNA* **11**, 592–597.
42. Jaikaran, D. C., Collins, C. H., and MacMillan, A. M. (2002) Adenosine to inosine editing by ADAR2 requires formation of a ternary complex on the GluR-B R/G site. *J Biol Chem* **277**, 37624–37629.
43. Cho, D. S., Yang, W., Lee, J. T., Shiekhattar, R., Murray, J. M., and Nishikura, K. (2003) Requirement of dimerization for RNA editing activity of adenosine deaminases acting on RNA. *J Biol Chem* **278**, 17093–17102.
44. Poulsen, H., Jorgensen, R., Heding, A., Nielsen, F. C., Bonven, B., and Egebjerg, J. (2006) Dimerization of ADAR2 is mediated by the double-stranded RNA binding domain. *RNA* **12**, 1350–1360.

45. Chilibeck, K. A., Wu, T., Liang, C., Schellenberg, M. J., Gesner, E. M., Lynch, J. M., and MacMillan, A. M. (2006) FRET analysis of in vivo dimerization by RNA-editing enzymes. *J Biol Chem* **281**, 16530–16535.
46. O'Connell, M. A., Gerber, A., and Keller, W. (1997) Purification of human double-stranded RNA-specific editase 1 (hRED1) involved in editing of brain glutamate receptor B pre-mRNA. *J Biol Chem* **272**, 473–478.
47. Macbeth, M. R., Lingam, A. T., and Bass, B. L. (2004) Evidence for auto-inhibition by the N terminus of hADAR2 and activation by dsRNA binding. *RNA* **10**, 1563–1571.
48. Macbeth, M. R., Schubert, H. L., Vandemark, A. P., Lingam, A. T., Hill, C. P., and Bass, B. L. (2005) Inositol hexakisphosphate is bound in the ADAR2 core and required for RNA editing. *Science* **309**, 1534–1539.
49. Betts, L., Xiang, S., Short, S. A., Wolfenden, R., and Carter, C. W., Jr. (1994) Cytidine deaminase. The 2. 3 A crystal structure of an enzyme: Transition-state analog complex. *J Mol Biol* **235**, 635–656.
50. Kuratani, M., Ishii, R., Bessho, Y., Fukunaga, R., Sengoku, T., Shirouzu, M., Sekine, S., and Yokoyama, S. (2005) Crystal structure of tRNA adenosine deaminase (TadA) from *Aquifex aeolicus*. *J Biol Chem* **280**, 16002–16008.
51. Polson, A. G., and Bass, B. L. (1994) Preferential selection of adenosines for modification by double-stranded RNA adenosine deaminase. *EMBO J* **13**, 5701–5711.
52. Stephens, O. M., Yi-Brunozzi, H. Y., and Beal, P. A. (2000) Analysis of the RNA-editing reaction of ADAR2 with structural and fluorescent analogues of the GluR-B R/G editing site. *Biochemistry* **39**, 12243–12251.
53. Cheng, X., and Roberts, R. J. (2001) AdoMet-dependent methylation, DNA methyltransferases and base flipping. *Nucleic Acids Res* **29**, 3784–3795.
54. Herbert, A., Alfken, J., Kim, Y. G., Mian, I. S., Nishikura, K., and Rich, A. (1997) A Z-DNA binding domain present in the human editing enzyme, double-stranded RNA adenosine deaminase. *Proc Natl Acad Sci USA* **94**, 8421–8426.
55. Athanasiadis, A., Placido, D., Maas, S., Brown, B. A., 2nd, Lowenhaupt, K., and Rich, A. (2005) The crystal structure of the Zbeta domain of the RNA-editing enzyme ADAR1 reveals distinct conserved surfaces among Z-domains. *J Mol Biol* **351**, 496–507.
56. Schwartz, T., Lowenhaupt, K., Kim, Y. G., Li, L., Brown, B. A., 2nd. Herbert, A. and Rich, A. (1999) Proteolytic dissection of Zab, the Z-DNA-binding domain of human ADAR1. *J Biol Chem* **274**, 2899–2906.
57. Schwartz, T., Rould, M. A., Lowenhaupt, K., Herbert, A., and Rich, A. (1999) Crystal structure of the Zalpha domain of the human editing enzyme ADAR1 bound to left-handed Z-DNA. *Science* **284**, 1841–1845.
58. Wintjens, R., and Rooman, M. (1996) Structural classification of HTH DNA-binding domains and protein–DNA interaction modes. *J Mol Biol* **262**, 294–313.
59. Schade, M., Turner, C. J., Lowenhaupt, K., Rich, A., and Herbert, A. (1999) Structure-function analysis of the Z-DNA-binding domain Zalpha of dsRNA adenosine deaminase type I reveals similarity to the (alpha + beta) family of helix-turn-helix proteins. *EMBO J* **18**, 470–479.
60. Lehmann, K. A., and Bass, B. L. (1999) The importance of internal loops within RNA substrates of ADAR1. *J Mol Biol* **291**, 1–13.
61. Valente, L., and Nishikura, K. (2007) RNA binding-independent dimerization of adenosine deaminases acting on RNA and dominant negative effects of nonfunctional subunits on dimer functions. *J Biol Chem* **282**, 16054–16061.
62. Ohman, M., (2007) A-to-I editing challenger or ally to the microRNA process. *Biochimie* **89**, 1171–1176.
63. Higuchi, M., Single, F. N., Kohler, M., Sommer, B., Sprengel, R., and Seeburg, P. H. (1993) RNA editing of AMPA receptor subunit GluR-B: A base-paired intron–exon structure determines position and efficiency. *Cell* **75**, 1361–1370.
64. Burns, C. M., Chu, H., Rueter, S. M., Hutchinson, L. K., Canton, H., Sanders-Bush, E., and Emeson, R. B. (1997) Regulation of serotonin-2C receptor G-protein coupling by RNA editing. *Nature* **387**, 303–308.
65. Ohlson, J., Pedersen, J. S., Haussler, D., and Ohman, M. (2007) Editing modifies the GABA(A) receptor subunit alpha3. *RNA* **13**, 698–703.

66. Reenan, R. A., Hanrahan, C. J., and Barry, G. (2000) The mle(napts) RNA helicase mutation in drosophila results in a splicing catastrophe of the para Na^+ channel transcript in a region of RNA editing. *Neuron* **25**, 139–149.
67. Hoopengardner, B., Bhalla, T., Staber, C., and Reenan, R. (2003) Nervous system targets of RNA editing identified by comparative genomics. *Science* **301**, 832–836.
68. Sommer, B., Kohler, M., Sprengel, R., and Seeburg, P. H. (1991) RNA editing in brain controls a determinant of ion flow in glutamate-gated channels. *Cell* **67**, 11–19.
69. Burnashev, N., Khodorova, A., Jonas, P., Helm, P. J., Wisden, W., Monyer, H., Seeburg, P. H., and Sakmann, B. (1992) Calcium-permeable AMPA-kainate receptors in fusiform cerebellar glial cells. *Science* **256**, 1566–1570.
70. Higuchi, M., Maas, S., Single, F. N., Hartner, J., Rozov, A., Burnashev, N., Feldmeyer, D., Sprengel, R., and Seeburg, P. H. (2000) Point mutation in an AMPA receptor gene rescues lethality in mice deficient in the RNA-editing enzyme ADAR2. *Nature* **406**, 78–81.
71. Palladino, M. J., Keegan, L. P., O'Connell, M. A., and Reenan, R. A. (2000) A-to-I pre-mRNA editing in *Drosophila* is primarily involved in adult nervous system function and integrity. *Cell* **102**, 437–449.
72. Seeburg, P. H., and Hartner, J. (2003) Regulation of ion channel/neurotransmitter receptor function by RNA editing. *Curr Opin Neurobiol* **13**, 279–283.
73. Rueter, S. M., Dawson, T. R., and Emeson, R. B. (1999) Regulation of alternative splicing by RNA editing. *Nature* **399**, 75–80.
74. Dawson, T. R., Sansam, C. L., and Emeson, R. B. (2004) Structure and sequence determinants required for the RNA editing of ADAR2 substrates. *J Biol Chem* **279**, 4941–4951.
75. Casey, J. L., and Gerin, J. L. (1995) Hepatitis D virus RNA editing: Specific modification of adenosine in the antigenomic RNA. *J Virol* **69**, 7593–7600.
76. Polson, A. G., Bass, B. L., and Casey, J. L. (1996) RNA editing of hepatitis delta virus antigenome by dsRNA-adenosine deaminase. *Nature* **380**, 454–456.
77. Kuo, M. Y., Chao, M., and Taylor, J. (1989) Initiation of replication of the human hepatitis delta virus genome from cloned DNA: Role of delta antigen. *J Virol* **63**, 1945–1950.
78. Ryu, W. S., Bayer, M., and Taylor, J. (1992) Assembly of hepatitis delta virus particles. *J Virol* **66**, 2310–2315.
79. Scadden, A. D., and Smith, C. W. (2001) RNAi is antagonized by A $\rightarrow$ I hyper-editing. *EMBO Rep* **2**, 1107–1111.
80. Luciano, D. J., Mirsky, H., Vendetti, N. J., and Maas, S. (2004) RNA editing of a miRNA precursor. *RNA* **10**, 1174–1177.
81. Pfeffer, S., Sewer, A., Lagos-Quintana, M., Sheridan, R., Sander, C., Grasser, F. A., van Dyk, L. F., Ho, C. K., Shuman, S., Chien, M., Russo, J. J., Ju, J., Randall, G., Lindenbach, B. D., Rice, C. M., Simon, V., Ho, D. D., Zavolan, M., and Tuschl, T.(2005) Identification of microRNAs of the herpesvirus family. *Nat Methods* **2**, 269–276.
82. Blow, M. J., Grocock, R. J., van Dongen, S., Enright, A. J., Dicks, E., Futreal, P. A., Wooster, R., and Stratton, M. R. (2006) RNA editing of human microRNAs. *Genome Biol* **7**, R27.
83. Nishikura, K. (2006) Editor meets silencer: Crosstalk between RNA editing and RNA interference. *Nat Rev Mol Cell Biol* **7**, 919–931.
84. Yang, W., Chendrimada, T. P., Wang, Q., Higuchi, M., Seeburg, P. H., Shiekhattar, R., and Nishikura, K. (2006) Modulation of microRNA processing and expression through RNA editing by ADAR deaminases. *Nat Struct Mol Biol* **13**, 13–21.
85. Bass, B. L. (2000) Double-stranded RNA as a template for gene silencing. *Cell* **101**, 235–238.
86. Stephens, O. M., Haudenschild, B. L., and Beal, P. A. (2004) The binding selectivity of ADAR2's dsRBMs contributes to RNA-editing selectivity. *Chem Biol* **11**, 1239–1250.
87. Ohman, M., Kallman, A. M., and Bass, B. L. (2000) In vitro analysis of the binding of ADAR2 to the pre-mRNA encoding the GluR-B R/G site. *RNA* **6**, 687–697.
88. Wong, S. K., Sato, S., and Lazinski, D. W. (2001) Substrate recognition by ADAR1 and ADAR2. *RNA* **7**, 846–858.
89. Kallman, A. M., Sahlin, M., and Ohman, M. (2003) ADAR2 A $\rightarrow$ I editing: Site selectivity and editing efficiency are separate events. *Nucleic Acids Res* **31**, 4874–4881.

90. Hough, R. F., and Bass, B. L. (1997) Analysis of *Xenopus* dsRNA adenosine deaminase cDNAs reveals similarities to DNA methyltransferases. *RNA* **3**, 356–370.
91. Athanasiadis, A., Rich, A., and Maas, S. (2004) Widespread A-to-I RNA editing of Alu-containing mRNAs in the human transcriptome. *PLoS Biol* **2**, e391.
92. Blow, M., Futreal, P. A., Wooster, R., and Stratton, M. R. (2004) A survey of RNA editing in human brain. *Genome Res* **14**, 2379–2387.
93. Levanon, E. Y., Eisenberg, E., Yelin, R., Nemzer, S., Hallegger, M., Shemesh, R., Fligelman, Z. Y., Shoshan, A., Pollock, S. R., Sztybel, D., Olshansky, M., Rechavi, G., and Jantsch, M. F. (2004) Systematic identification of abundant A-to-I editing sites in the human transcriptome. *Nat Biotechnol* **22**, 1001–1005.
94. Hanrahan, C. J., Palladino, M. J., Ganetzky, B., and Reenan, R. A. (2000) RNA editing of the *Drosophila para* Na(+) channel transcript. Evolutionary conservation and developmental regulation. *Genetics* **155**, 1149–1160.
95. Aruscavage, P. J., and Bass, B. L. (2000) A phylogenetic analysis reveals an unusual sequence conservation within introns involved in RNA editing. *RNA* **6**, 257–269.
96. Nishikura, K. (1992) Modulation of double-stranded RNAs in vivo by RNA duplex unwindase. *Ann N Y Acad Sci* **660**, 240–250.
97. Liu, Y., George, C. X., Patterson, J. B., and Samuel, C. E. (1997) Functionally distinct double-stranded RNA-binding domains associated with alternative splice site variants of the interferon-inducible double-stranded RNA-specific adenosine deaminase. *J Biol Chem* **272**, 4419–4428.
98. Lei, M., Liu, Y., and Samuel, C. E. (1998) Adenovirus VAI RNA antagonizes the RNA-editing activity of the ADAR adenosine deaminase. *Virology* **245**, 188–196.
99. Murphy, D. G., Dimock, K., and Kang, C. Y. (1991) Numerous transitions in human parainfluenza virus 3 RNA recovered from persistently infected cells. *Virology* **181**, 760–763.
100. O'Hara, P. J., Nichol, S. T., Horodyski, F. M., and Holland, J. J. (1984) Vesicular stomatitis virus defective interfering particles can contain extensive genomic sequence rearrangements and base substitutions. *Cell* **36**, 915–924.
101. Hajjar, A. M., and Linial, M. L. (1995) Modification of retroviral RNA by double-stranded RNA adenosine deaminase. *J Virol* **69**, 5878–5882.
102. Rueda, P., Garcia-Barreno, B., and Melero, J. A. (1994) Loss of conserved cysteine residues in the attachment (G) glycoprotein of two human respiratory syncytial virus escape mutants that contain multiple A-G substitutions (hypermutations). *Virology* **198**, 653–662.
103. Patton, D. E., Silva, T., and Bezanilla, F. (1997) RNA editing generates a diverse array of transcripts encoding squid Kv2 K^+ channels with altered functional properties. *Neuron* **19**, 711–722.
104. Morse, D. P., and Bass, B. L. (1999) Long RNA hairpins that contain inosine are present in *Caenorhabditis elegans* poly$(A)^+$ RNA. *Proc Natl Acad Sci USA* **96**, 6048–6053.
105. Cattaneo, R., Schmid, A., Eschle, D., Baczko, K., ter Meulen, V., and Billeter, M. A. (1988) Biased hypermutation and other genetic changes in defective measles viruses in human brain infections. *Cell* **55**, 255–265.
106. Sheppard, R. D., Raine, C. S., Bornstein, M. B., and Udem, S. A. (1985) Measles virus matrix protein synthesized in a subacute sclerosing panencephalitis cell line. *Science* **228**, 1219–1221.
107. Kumar, M., and Carmichael, G. G. (1997) Nuclear antisense RNA induces extensive adenosine modifications and nuclear retention of target transcripts. *Proc Natl Acad Sci USA* **94**, 3542–3547.
108. Zhang, Z., and Carmichael, G. G. (2001) The fate of dsRNA in the nucleus: A p54(nrb)-containing complex mediates the nuclear retention of promiscuously A-to-I edited RNAs. *Cell* **106**, 465–475.
109. Scadden, A. D., and Smith, C. W. (2001) Specific cleavage of hyper-edited dsRNAs. *EMBO J* **20**, 4243–4252.
110. Liu, L. F., and Wang, J. C. (1987) Supercoiling of the DNA template during transcription. *Proc Natl Acad Sci USA* **84**, 7024–7027.
111. Koeris, M., Funke, L., Shrestha, J., Rich, A., and Maas, S. (2005) Modulation of ADAR1 editing activity by Z-RNA in vitro. *Nucleic Acids Res* **33** 5362–5370.
112. Cattaneo, R. (1994) Biased (A $\rightarrow$ I) hypermutation of animal RNA virus genomes. *Curr Opin Genet Dev* **4**, 895–900.

113. Horikami, S. M., and Moyer, S. A. (1995) Double-stranded RNA adenosine deaminase activity during measles virus infection. *Virus Res* **36**, 87–96.
114. Taylor, D. R., Puig, M., Darnell, M. E., Mihalik, K., and Feinstone, S. M. (2005) New antiviral pathway that mediates hepatitis C virus replicon interferon sensitivity through ADAR1. *J Virol* **79**, 6291–6298.
115. Jacobs, B. L., and Langland, J. O. (1996) When two strands are better than one: The mediators and modulators of the cellular responses to double-stranded RNA. *Virology* **219**, 339–349.
116. Gribaudo, G., Lembo, D., Cavallo, G., Landolfo, S., and Lengyel, P. (1991) Interferon action: Binding of viral RNA to the 40-kilodalton 2′-5′-oligoadenylate synthetase in interferon-treated HeLa cells infected with encephalomyocarditis virus. *J Virol* **65**, 1748–1757.
117. Boone, R. F., Parr, R. P., and Moss, B. (1979) Intermolecular duplexes formed from polyadenylylated vaccinia virus RNA. *J Virol* **30**, 365–374.
118. Colby, C., and Duesberg, P. H. (1969) Double-stranded RNA in vaccinia virus infected cells. *Nature* **222**, 940–944.
119. Colby, C., Jurale, C., and Kates, J. R. (1971) Mechanism of synthesis of vaccinia virus double-stranded ribonucleic acid in vivo and in vitro. *J Virol* **7**, 71–76.
120. Duesberg, P. H., and Colby, C. (1969) On the biosynthesis and structure of double-stranded RNA in vaccinia virus-infected cells. *Proc Natl Acad Sci USA* **64**, 396–403.
121. Moss, B. (1990) Poxviridae and their replication. In: Fields, B. N., and Knipe, D. M. (eds.), *Fundamental Virology*. 2nd ed., Raven Press, New York 2079–2111.
122. Kohler, M., Kornau, H. C., and Seeburg, P. H. (1994) The organization of the gene for the functionally dominant alpha-amino-3-hydroxy-5-methylisoxazole-4-propionic acid receptor subunit GluR-B. *J Biol Chem* **269**, 17367–17370.
123. Yang, J. H., Sklar, P., Axel, R., and Maniatis, T. (1995) Editing of glutamate receptor subunit B pre-mRNA in vitro by site-specific deamination of adenosine. *Nature* **374**, 77–81.
124. Bass, B. L. (1997) RNA editing and hypermutation by adenosine deamination. *Trends Biochem Sci* **22**, 157–162.
125. Price, R. D., Weiner, D. M., Chang, M. S., and Sanders-Bush, E. (2001) RNA editing of the human serotonin 5-HT2C receptor alters receptor-mediated activation of G13 protein. *J Biol Chem* **276**, 44663–44668.

CHAPTER **16**

CHEMISTRY, PHYLOGENY, AND THREE-DIMENSIONAL STRUCTURE OF THE APOBEC PROTEIN FAMILY

Celeste MacElrevey
Joseph E. Wedekind

"If you want to understand function, study structure" – Francis Crick (in What Mad Pursuit)

THE ABILITY of the mammalian enzyme APOBEC1 to catalyze C-to-U deamination or "editing" of the apolipoprotein B mRNA, thereby generating a truncated protein of altered function, has been known for nearly two decades. The recent discovery of a family of human APOBEC1-related proteins has stirred excitement in the field, especially since several family members edit single-stranded (ss) DNA and have proven essential for immunoglobulin gene diversification as well as for innate antiretrotransposon and antiretroviral activities. Although crystal structures of APOBEC family members would do much to elucidate structure–function relationships at the molecular level, such analyses are in a nascent stage at present. Nonetheless, in this chapter we provide a demonstration of the available tools and approaches that can be employed to glean useful three-dimensional information about APOBEC1-related proteins, which should prove useful to gain molecular insight and guide experimentation until high-resolution structures become available.

16.1 INTRODUCTION TO NUCLEIC ACID DEAMINATION WITH IMPLICATIONS FOR BIOLOGICAL ACTIVITY

Nucleobase oxidation in the context of DNA or RNA is generally considered an undesirable biological phenomenon due to its potential to threaten genomic integrity.

RNA and DNA Editing: Molecular Mechanisms and Their Integration into Biological Systems,
Edited by Harold C. Smith

However, there are a growing number of circumstances in higher organisms in which enzymes of the APOBEC family perform site-specific deamination of DNA or RNA substrates; refer to Chapters 2, 10, and 11 for a review of biological functions. APOBEC enzymes are highly specialized and are required for such activities as proteomic diversification, codon modification, adaptation of the humoral immune response, and innate cellular immunity against invading retroviruses or retrotransposable elements. As one can imagine, these "editing" activities must be highly regulated and compartmentalized since indiscriminate DNA deamination can lead to genome instability. Indeed, the overexpression or misregulation of APOBEC deaminases has been associated with the development of cancers, lymphomas, and other neoplastic diseases (1–10). In this chapter, we will be concerned especially with the chemistry of deamination mediated by the action of APOBEC family members. Using a hierarchical approach, it will become apparent (i) how a defining amino acid signature sequence (the Zn^{2+}-dependent deaminase motif) folds into a conserved three-dimensional structure, (ii) how this fold is part of a conserved subunit that is shared by a diverse group of deaminase enzymes, and (iii) how variations to the core fold impart specialization in terms of biological function. With regard to the latter point, structural precedents drawn from well-characterized enzymes will be used to gain insight into the APOBEC family, for which there is limited structural data. A review of current three-dimensional data for the APOBEC family will be followed by a phylogenetic, sequence-driven analysis with the aim of relating family members to each other. Finally, a discussion of the influences of selective pressure and chromosome remodeling highlights the evolutionary forces that have sculpted the form and function of the APOBEC family.

16.2 THE CHEMISTRY OF THE ZINC-DEPENDENT DEAMINASE AMINO ACID SIGNATURE MOTIF

The enzymes that catalyze the deamination of cytidine to uridine are known as cytidine deaminases or CDAs (enzyme classification or EC 3.5.4.5). Much of the chemistry and structural details of the CDA reaction were established by Wolfenden and Carter, who demonstrated that the Zn^{2+} ion is bound tightly by the His and Cys residues of the bacterial enzyme from *E. coli* that acts on the free nucleosides (11). The Zn^{2+} ion is an essential catalytic component that serves as a Lewis acid to activate a water molecule for nucleophilic aromatic substitution at the C4 position of the cytosine ring (12–14) (Figure 16.1). As such, maintenance of the Zn^{2+} ion coordination and stereochemical positioning of water are essential aspects of the CDA mechanism of action. In function studies, site-directed mutations of the respective Cys or His residues to Ala resulted in activity losses on the order of 10^3-fold in k_{cat} and 3- to 10-fold in K_M relative to wild type (11). These changes, as well as metal composition analysis of the mutant enzymes, suggested that loss of function is mostly due to the absence of Zn^{2+} participation in catalysis, rather than its participation in substrate binding. A conserved Glu serves as a proton shuttle that facilitates transfer of hydrogen from the activated water to the imino N3 position of the pyrimidine ring (14). Site-directed mutagenesis of this Glu to Ala resulted in a

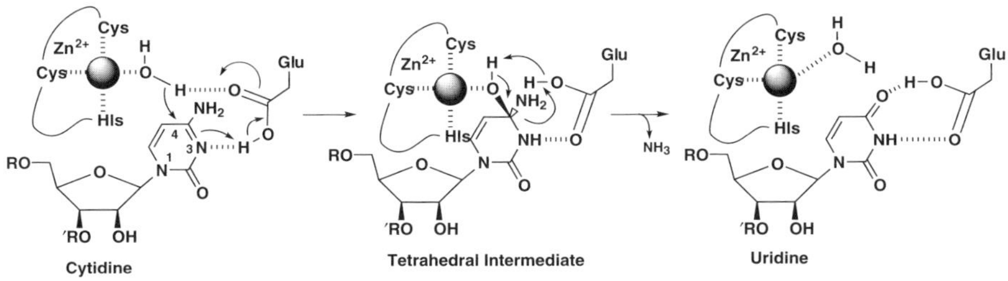

Figure 16.1 Schematic diagram for the cytidine deaminase mechanism of action. (Left) The cytidine-containing substrate binds in the CDA active site and proceeds to undergo nucleophilic attack by an activated water coordinated to Zn^{2+}. A conserved Glu donates a proton to the N3 position. (Middle) The tetrahedral intermediate at C4 breaks down through loss of ammonia, which gains a proton from the conserved Glu. (Right) The uridine product is free to diffuse away, leaving a protonated Glu that is poised for the next cycle of catalysis. A new water molecule enters the Zn^{2+} coordination sphere. (*See color insert.*)

10^6-fold loss in k_{cat}, but a 30-fold gain in affinity for the cytidine substrate as reflected in K_M. A comparable gain in affinity was attained for the product uridine, as measured by the equilibrium K_D (15). These results suggest that Glu contributes significantly to the chemistry of catalysis, but not to substrate binding. Collectively, these amino acids form part of a conserved amino acid signature sequence known as the zinc-dependent deaminase (ZDD) motif, which implies a common chemistry and three-dimensional fold (16). Detection of this active site constellation within the amino acid sequence of a protein can be distilled into the pattern (H/C) $xEx_{25-30}PCxxC$, where x is any residue (17). As such, the ZDD motif is a defining characteristic of the CDA superfamily that includes a number of pyrimidine metabolism deaminases, as well as APOBEC family members. Moreover, the ZDD sequence is also a common feature of adenosine deaminases that act on tRNA (ADAT Chapter 5) and adenosine deaminases that act on RNA (ADARs, Chapters 1, 6, 9, and 15). Although the latter enzymes deaminate adenosine to inosine, conservation of the ZDD motif is evidence of a common ancestor, which supports classification of these molecules into distinct clades based on amino acid sequence identity, but ultimately establishes their grouping into a single CDA superfamily.

16.3 THE ZDD SIGNATURE MOTIF IMPLIES A SPECIFIC THREE-DIMENSIONAL ARRANGEMENT OF AMINO ACIDS

To the structural biologist, the ZDD motif invokes not only an image of the catalytic residues within the linear amino acid sequence, but also their arrangement in three-dimensional space leading to stereospecific substrate binding and catalysis (16). Residues of the ZDD interact in the context of a helix–strand–helix super-secondary structure element (Figure 16.2A). This motif is integral to the formation of the CDA active site and is representative of all known structures of this superfamily. In this motif, the three Zn^{2+} coordinating residues reside at the N-terminal ends of two

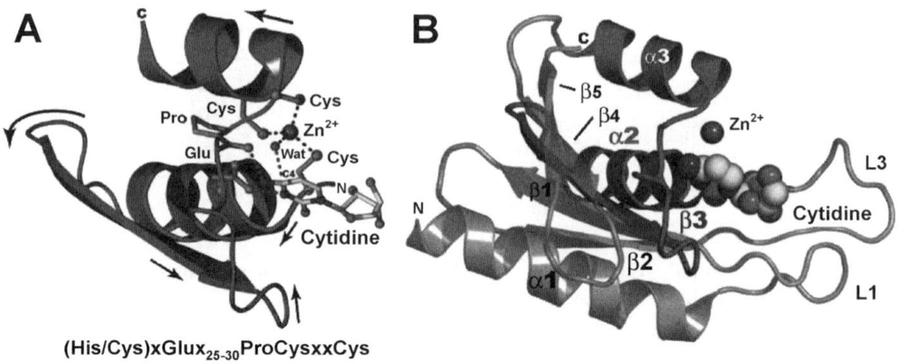

Figure 16.2 Ribbon diagrams of the representative zinc-dependent deaminase motif bound to cytidine. The coordinates were derived from the mouse cytidine deaminase crystal structure (43). (**A**) The helix–strand–helix structure of the ZDD motif indicating the spatial orientation of key residues. Dashed lines (blue) indicate ionic coordination to Zn^{2+}; pink lines indicate hydrogen bonds. The activated water makes a close contact to atom C4 of the cytidine ring (gray line). Black arrows indicate the progression of the polypeptide chain from N- to C-terminus. (**B**) The ZDD signature motif in the context of the CDA domain (gray, semitransparent). The cytidine substrate is depicted as a space-filling model. (*See color insert.*)

adjacent α-helices that are connected topologically by a single, intervening β-strand that imparts a right-handed crossover element (Figure 16.2A). The coordination geometry of the Zn^{2+} ion is tetrahedral and is fixed by the three ZDD ligands such that a water molecule completes the coordination sphere. This activated water is located ~2.8 Å from atom C4 on the *si* (i.e., left-handed) face of the cytosine ring in the crystal structure, as represented by the CDA enzyme structure from mouse (Figure 16.2A). In contrast, the leaving group, NH_3, departs from the cytosine ring from the opposite *re* or right-handed face. The departure of ammonia is guided via formation of a hydrogen bond between NH_3 and the carbonyl oxygen of the residue preceding a conserved Pro of the ZDD. The role of this Pro is likely to be conformational restriction of the backbone while simultaneously assuring the presence of a pocket to receive the leaving group (14, 18). In the crystal structure of the *E. coli* CDA in complex with the transition-state mimic 5-fluorozebularine, the attacking water hydrogen bonds to the amide backbone of Cys immediately C-terminal to Pro in the PCxxC motif (14). Therefore, the presence of Pro has a direct influence on the orientation of catalytic groups through positioning of the local backbone. Moreover, the evolutionary selection of an imino acid at this position (i.e., one that is missing a backbone amide to donate a hydrogen bond) may serve to eliminate nonproductive hydrogen-bond interactions between the attacking water and the main chain on both *si* and *re* faces of the cytosine ring. Finally, the conserved Glu residue of the ZDD motif resides at the N-terminal end of the first α-helix where it is poised to serve as a proton shuttle as depicted in Figure 16.1. In the crystal structure, this activity appears plausible due to the apparent hydrogen bonds to the Watson–Crick face of cytidine at the N3 (imino) position (Figure 16.2A, broken pink lines), along with the leaving ammonia (not shown).

Despite the high level of detail known about the mode of CDA interactions with individual nucleoside substrates, no structural information exists for any member of the APOBEC family on the mode of single-stranded DNA or RNA recognition. As such, crystal structures of individual APOBEC family members will provide a valuable resource for understanding substrate affinity and specificity. One aspect that is certain about all CDA family members is that the ZDD motif is part of a larger core structure (Figure 16.2B). The need to embed the ZDD within this large, folded structure is sensible since the hydrophobic core of the protein imparts a great number of peripheral interactions that stabilize the helix–strand–helix motif while simultaneously providing elements that bind and orient the substrate, such as Loop 3 (Figure 16.2B). In this chapter, the larger α/β fold is referred to as the "CDA fold" or "CDA domain," which implies independent folding of the polypeptide chain. Based upon current knowledge, it is very likely that APOBEC family members will conform to the core CDA fold as well.

16.4 RATIONALE FOR A COMBINED STRUCTURAL AND PHYLOGENETIC APPROACH TO UNDERSTAND APOBEC EVOLUTION

Although not implied in Crick's adage, it has been proven that three-dimensional structure is a better prognosticator of function than amino acid sequence identity. This is because tertiary (three-dimensional) structure is more evolutionarily conserved than primary structure (19). In this section, we examine the relationship between comparative primary sequence analysis and three-dimensional structure analysis. In doing so, we wish to convey that combining both methods is a powerful tool to infer function and evolutionary relatedness, even in cases of low amino acid homology.

In Section 16.3, the ZDD sequence motif was revealed as a useful consensus predictor of deaminase function in amino acid sequence searches seeking to unveil candidate members of the APOBEC family. {*Note*: For those in doubt, try the string [HC]xEx(25,30)PCxxC in psi-blast (http://www.ncbi.nlm.nih.gov/blast/) in combination with one known CDA sequence}. However, beyond the apparent amino acid sequence homology within the ZDD, overall sequence similarities among members of the CDA superfamily are often quite poor (17, 20). Nonetheless, the secondary structure elements of the core CDA fold are spatially well-conserved (Figure 16.2B), which provides a powerful restraint in our efforts to relate APOBEC structure to function. The relationship between structural conservation and amino acid sequence identity was formalized more than two decades ago by Chothia and Lesk, who showed that sequence similarity between structures provides a predictable metric for the variation in overall core folding (21). This relationship established a useful limit regarding what three-dimensional information one may accurately infer about an unknown structure based solely on protein sequence. As a heuristic, the ability to confidently model a core three-dimensional fold requires sequence identity between the empirically defined "template" structure and unknown target sequence of $\geq$35%

(22); of course, the greater the identity the more reliable the result of modeling. Under circumstances of low identity (e.g., $\leq$30%) but conserved function, the core architecture and functional residues can be expected to remain largely unaltered. Changes to the peripheral loops and elements flanking the core fold are likely to be divergent, thus giving rise to specialized functions such as substrate binding, subunit oligomerization, or allosteric regulation (see Sections 16.5 and 16.6). The latter two points provide some of the most powerful tenets for structure-based alignments and will be the premise of our later analysis. Yet, despite the utility of structure as a tool to relate the amino acids of proteins in a pairwise manner, phylogeny is often inferred solely through sequence alignments unguided by structural restraints. This realization is sobering, especially in light of the tens of thousands of experimentally derived, high-resolution three-dimensional structures available in the Protein Data Bank (23) (www.pdb.org).

To illustrate the power of structure-based alignment, let us consider a comparison of the amino acid sequences from the α-chain of human hemoglobin, human myoglobin, and leghemoglobin from lupine, a flowering plant. (*Note*: see reference 19 for a more complete discussion.) Despite the fact that the primary sequences of these molecules do not exhibit more than 16% identity, the pairwise root mean square (rms) deviations between lupine hemoglobin or human myoglobin and human hemoglobin are 1.9 Å and 1.4 Å, respectively. Although not necessarily conveyed in the rms numbers to the average reader, this degree of structural similarity is considered high and is interpreted by the structural biologist to be the result of selective pressure that favored the evolution of respective proteins to coordinate heme for O_2 binding and transport (24). Significantly, the relatedness of these proteins would likely be missed with a simple primary sequence comparison (25), whereas knowledge of their common function implies structural similarity that favors the use of pairwise approaches.

Of course, there are also circumstances in which proteins possess the same fold even though they exhibit diverse biological activities. One example is the remarkable structural similarity between the ATP-binding domain of the mammalian heat shock protein 70 (HSP70), solved by McKay and colleagues, and the globular actin subunit determined by Holmes (26). Again, these proteins exhibit ~16% sequence identity, but their pairwise rms deviation is 1.6 Å between 193 shared atoms. The high degree of structural similarity suggests that they evolved from a common ancestor and subsequently diverged to fulfill special biological functions (i.e., protein folding and formation of the cytoskeleton). Such proteins are designated paralogs to indicate that they arose from a gene duplication event; this term also describes the evolutionary relationship between APOBEC1 and AID of the APOBEC family (20). The take-home message here is that structural conservation does not automatically equal functional equivalence, but it is an excellent means to establish evolutionary commonality. [*Note*: The reader is referred to the DALI Server as a resource to search for structural homologs based on an input three-dimensional structure (27). See http://www.ebi.ac.uk/dali/.]

As pointed out by Chothia and Lesk, the use of simple sequence alignments can be especially informative under circumstances with moderate to high identity (>40%) (21). Under such conditions, these comparisons may establish a conserved core fold, as well as evolutionary "embellishments" (deletions and insertions) that diversify an ancient fold for a variety of uses (e.g., the ATP binding domain of actin versus HSP70). Such a scenario appears to be the case for APOBEC family members. However, inferring evolutionary relatedness to more distant ZDD-containing proteins quickly reveals the tenuousness of such relationships, especially when one considers that three-dimensional folding topology, subunit oligomerization, and substrate binding may have been altered significantly through evolution to fulfill specialized biological niches. This knowledge can be crucial in the choice of a three-dimensional structural template for homology modeling (20). In the end, there is a balancing act between (a) inferring common structure based solely on sequence relatedness and (b) assigning common function based on known three-dimensional structure (e.g., the case of paralogs). In this review, we will walk this "tightrope" to relate sequence and structure, insofar as the experimental details are available. This strategy allows us to relate members of the APOBEC family to more ancient CDA folds to identify a general set of principles that provide insight into the maintenance of conserved chemical function versus core diversification elements among CDA family members.

16.4.1 The Starting Point for Structural and Phylogenetic Analyses of APOBEC Family Members

When considering how to proceed in terms of a phylogenetic comparison between APOBEC family members and enzymes of related function, it is appropriate to ask what these proteins have in common. The clear answer is that both the conserved ZDD amino acid signature motif (H/C)xEx$_{25-30}$PCxxC and the surrounding residues (17) form the basis for much of the phylogenetic classification of individual APOBEC family members (Figure 16.3). Because it resides within the core fold and imparts deaminase activity, the ZDD signature sequence is slow to change, making it the most reliable region for sequence comparisons among family members (28–30). To illustrate this point, we cite a comprehensive phylogenetic analysis that was generated from protein alignments of a $\sim$150-residue segment encompassing the ZDD (31). As seen in Section 16.4 (above) regarding the relationship between form and function, the observation that the ZDD motif is conserved across the ADAR, ADAT, and AID/APOBEC1/APOBEC3 genes implies that these families share a common ancestor with the CDA, dCMP DA, and riboflavin DA families (Figure 16.3). The ZDD commonality exists despite the diverse array of substrates targeted by each branch of the superfamily, and it is significant because the ZDD motif itself is not shared by all deaminases known in biology. As such, deaminase enzymes as a whole are considered an example of convergent evolution. That is, the adenosine deaminases of purine metabolism and bacterial cytosine deaminases developed through selective pressure to achieve the same chemical reaction as members of the CDA superfamily. However, each enzyme class achieves catalysis by its own unique three-dimensional fold (32, 33). Such themes are common in biology. For example, carbonic anhydrases come in multiple forms. Whereas the α-form of the human enzyme adopts an α/β fold, the γ-enzyme from the archaeon *Methanosarcina thermophila* is a left-handed β-helix

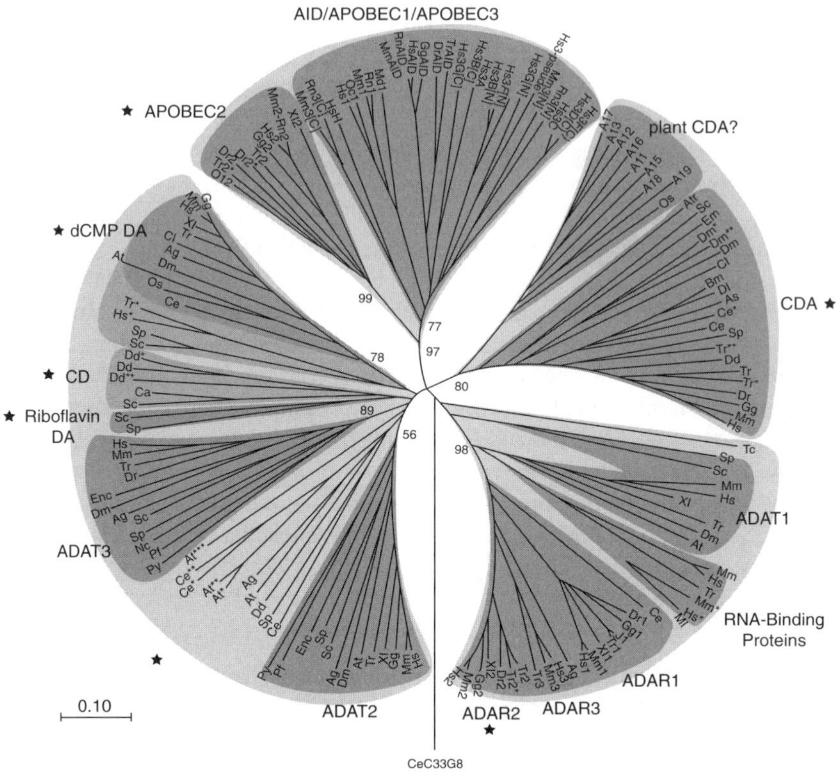

Figure 16.3 Relationships of the deaminase sequences in the vicinity of their ZDD motif. The neighbor-joining phylogenetic tree was generated from a protein alignment of the region encompassing the ZDD motif. Stars indicate families for which a representative crystal structure is discussed. Cytosine deaminases are labeled CD, whereas cytidine deaminases are identified as CDA. DA is an abbreviation of "deaminase." The scale bar represents the number of substitutions per amino acid. Clusters highlighted in dark gray are supported by a bootstrap value greater than 50, those grouped in a medium shading of gray are clustered by function, and the largest groupings in the lightest shading reflect gene families based on sequence similarity. Bootstrap values for selected nodes are indicated. The labels at the end of each branch indicate the organism from which the sequence is derived. (*Anopheles gambiae*, Ag; *Ascaris suum*, As; *Aspergillus terreus*, Atr; *Arabidopsis thaliana*, At; *Brugia malayi*, Bm; *Candida albicans*, Ca; *Caenorhabditis elegans*, Ce; *Ciona intestinalis*, Ci; *Dictyostelium discoideum*, Dd; *Dirofilaria immitis*, Di; *Drosophila melanogaster*, Dm; *Danio rerio*, Dr; *Encephalitozoon cuniculi*, Enc; *Echinococcus multilocularis*, Em; *Gallus gallus*, Gg; *Homo sapiens*, Hs; *Monodelphis domestica*, Md; *Macaca fascicularis*, Mf; *Mus musculus*, Mm; *Neurospora crassa*, Nc; *Oryctolagus cuniculus*, Oc; *Oryzias latipes*, Ol; *Oryza sativa*, Os; *Plasmodium falciparum*, Pf; *Plasmodium yoelii yoelii*, Py; *Rattus norvegicus*, Rn; *Saccharomyces cerevisiae*, Sc; *Schizosaccharomyces pombe*, Sp; *Trypanosoma cruzi*, Tc; *Tetraodon fluviatilis*, Tf; *Takifugu rubripes*, Tr; and *Xenopus laevis*, Xl). Within each darker shaded area, multiple sequences from the same organism are distinguished by use of asterisks (where their distinct functions are unknown) or by use of digits, where they can be assigned to separate groupings (e.g. 1, 2, 3 for ADAR1, ADAR2, or ADAR3 family members; 1, 2, 3 for APOBEC1, APOBEC2, APOBEC3 or AID family members; and 1, 2,3, . . ., 9 for hypothetical plant CDAs). In the case of double-domain-containing APOBEC3 proteins, an [N] or [C] indicates whether the sequence is derived from the N-terminal or C-terminal ZDD motif. CeC33G8 is a gene of unknown function. (This figure was adapted and reproduced with permission from Oxford University Press and originally appeared in reference 31.)

(31). The common feature of both enzymes is that they each coordinate to Zn^{2+} to catalyze conversion of CO_2 into carbonic acid or HCO_3^-. However, these enzymes have such disparate three-dimensional folds that they could not possibly have arisen from a common ancestor. Such is the case for ADAR enzymes relative to free nucleoside adenosine deaminases, suggesting that it is highly improbable that the former evolved from the latter or vice versa.

The starting point for exploring the evolutionary relationship between APOBEC and CDA family members was the structure determination of the CDA from *E. coli* (14). This crystal structure was published nearly a decade before any three-dimensional information became available for the RNA (or DNA) editing enzymes listed in Figure 16.3. As such, the *E. coli* enzyme was employed as a template in early homology models for members of the APOBEC family (20, 35, 36). In retrospect, one fascinating but cautionary aspect of the *E. coli* CDA structure was the observation that the enzyme folded as a dimer comprised of two identical polypeptide subunits related by twofold symmetry (14). Moreover, each polypeptide exhibited two highly conserved CDA folds itself, despite the fact that only a single ZDD motif was detectable per amino acid chain (Figure 16.4). This observation led to the suggestion that the bacterial enzyme evolved by gene duplication of a single CDA domain that gave rise to a second CDA-like domain that lost its ability to bind Zn^{2+} through divergent evolution (36). Such divergence is reminiscent of the relationship between actin and HSP70 except that it occurred within the same polypeptide chain. Hence, although N- and C-terminal halves of the amino acid sequence exhibit ~16% identity, each polypeptide segment adopts a highly similar

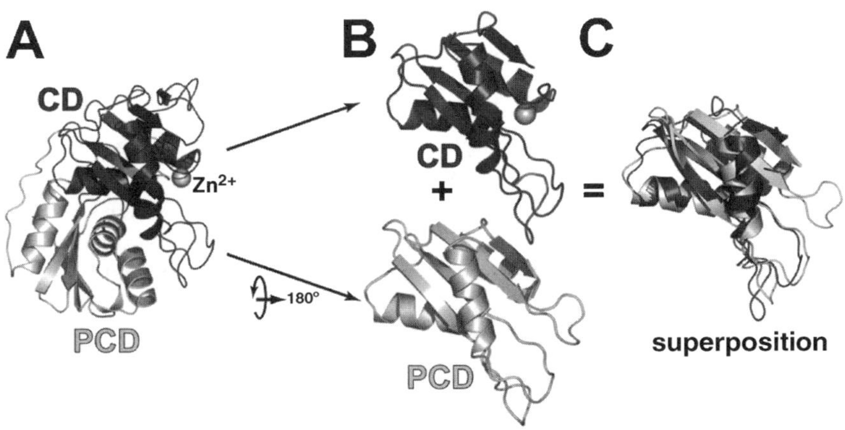

Figure 16.4 Superposition of N- and C-terminal domains from the *E. coli* CDA. (**A**) The CDA subunit in monomeric form. The respective catalytic domain (CD) harboring Zn^{2+} is dark-colored. The C-terminal "pseudo" catalytic domain (PCD) is light-colored. (**B**) Separation of the domains by removal of the long linker observed in part A at the far left. The PCD can be oriented similarly to the CD by 180° rotation about the *x*-axis, which is indicative of the internal symmetry of CDA subunits. (**C**) Translation of the CD (downwards) and the PCD (upwards) results in the superposition of these core domains.

CDA fold (i.e., their Cα backbones superpose with an rms deviation of 1.5 Å). The inability of the C-terminal CDA to bind Zn^{2+} or substrate led to its classification as a catalytically inert "pseudo-catalytic" domain or PCD, as compared to the bona fide catalytic domain (CD) at the N-terminus (14, 17). Retention of the PCD as part of the *E. coli* subunit fold appears to stabilize the global structure by burying a large hydrophobic area in the dimer interface. Maintenance of this interface is necessary to form the active site. The internal symmetry of the CDA subunit from *E. coli*, coupled with the intermolecular twofold relationship between subunits, gives rise to pseudo-222 or D2 symmetry of the dimer(14).

The example of symmetry in the *E. coli* CDA is important to illustrate a possible mode of folding and subunit intermolecular organization by other CDA superfamily members. Specifically, several members of the APOBEC3 family clearly contain tandem ZDD motifs, implying two CDA domains within a single polypeptide (37). These double-deaminase enzymes appear to coordinate zinc in both active sites, although frequently only a single CDA domain demonstrates deaminase activity (38, 39). In this case, the inactive ZDD has been implicated in alternative functions including RNA binding (40).

Following the structure determination of the cytidine deaminase from *E. coli*, the enzymes from other species were solved, including *B. subtilis, S. cerevisiae*, human, and mouse. These enzymes also exhibited an overall globular structure, but without the CD–PCD domain architecture (20, 41–43). Instead, the equivalent of a single CD domain was the fundamental catalytic subunit. Rather than dimerization with pseudo-222 symmetry, these enzymes oligomerized as tetramers and exhibited proper 222 symmetry (20, 41). As such, each tetrameric enzyme coordinated four Zn^{2+} ions rather than two as observed for the *E. coli* CDA. The yeast enzyme (Cdd1) was solved in the Wedekind lab and was among a handful of CDA structures used as a template for APOBEC1 and AID comparative modeling (20). Although classified as a pyrimidine metabolism enzyme (44), Cdd1 is distinguished by its ability to edit apolipoprotein B mRNA in reporter assays (45). This discovery provided a tenable connection between the structure and function of metabolic nucleoside deaminases and those that act on DNA or RNA (17, 20). The resulting comparative model of APOBEC1 was constrained to be a dimer, as documented experimentally. With evidence that both proteins were dimeric and of comparable size (APOBEC1: 27 kD versus *Ec*CDA: 31.5 kD), the N- and C-terminal domains were assumed to adopt similar CD–PCD architectures whereby formation of the active site would depend on dimerization requiring loops that originate from each of the two identical polypeptide chains (20). Overall, the APOBEC1 model satisfactorily explained dominant negative effects of the dimer (46) and how its active site accommodated single-stranded RNA (17) or DNA substrates (47, 48). However, the model lacked sufficient molecular detail to account for substrate specificity (20, 49), thereby underscoring a need for experimental structural approaches. As tRNA editing TadA enzymes were solved, it was proposed that editing enzymes more closely mimicked the deoxycytidylate and cytosine deaminases, which typically exist as dimers rather than tetramers (50, 51). This negated the requirement for a PCD to exist to complete the anticipated dimeric architecture of APOBEC1- or AID-like family members. Most recently, the crystal structure of APOBEC2 solidified the finding that APOBEC family members do not exhibit the CD–PCD architecture within a single ZDD, but instead fold as a single domain (52).

16.4.2 The CDA Superfamily: Overview of Conserved Fold Topology in the Core and Common Variations

In Section 16.3, we alluded to the fact that the ZDD signature motif is embedded in a larger fold that maintains the integrity of the active site and Zn^{2+} coordination by contributing core structural elements that flank the ZDD βαβ super-secondary structure (Figure 16.2B). A comparison of available structures within the CDA superfamily demonstrates that the core of the CDA domain is a conserved fold that extends beyond the boundaries of the embedded ZDD signature motif (Figure 16.2). This fold is characterized by a five-stranded mixed (parallel and antiparallel) β-sheet flanked by α-helices on either broad face of the pleated surface. Typically, the CDA sheet is triangular in shape when looking at its broadest face with the C-terminal strands being shorter than those of the N-terminus. Two helices contribute the necessary residues for both Zn^{2+} coordination and catalysis (Figure 16.2). All of the conserved ZDD residues reside on a single face of the β-sheet (Figure 16.2B); a long N-terminal helix usually juxtaposes this motif on the opposing side of the triangular surface (Figure 16.2B, helix α1). The arrangement of these secondary structure elements, as one progresses along the polypeptide chain from N- to C-terminus, is α1β1β2α2β3α3β4β5 with the Zn^{2+} coordinating helices assigned as α2 and α3. However, the β-strand order in the three-dimensional structure is 2-1-3-4-5 when viewing the top of the sheet from the N- to C-terminal direction. This 'topological' description is useful since it defines the unique arrangement of β-strands as they are arranged spatially, which is often sufficient to define the uniqueness of a particular three-dimensional fold in relation to other known structures. In this chapter, we define a "standard" orientation of the CDA that places the zinc ligands at the top of the diagrams, but behind the β-sheet, which gives an apparent topology of 5-4-3-1-2 from left to right (Figure 16.5, top left panels). Strands β2, β3, and β4 are directed toward the active site, whereas β1 is directed downwards away from the active site (Figures 16.2 and 16.5). Variability is observed in the topological orientation of strand β5, providing the basis for dividing this superfamily into two smaller groups (53).

The free nucleotide cytidine deaminases (fnCDAs), including the *E. coli, B. subtilis,* and mammalian CDAs, are distinguished by the downward directionality of strand β5, which points away from the catalytic zinc ion and is antiparallel to strand β4 (Figure 16.5) (14, 41–43). This orientation allows for a rather short loop between the fourth and fifth strands. In all other CDA superfamily members, including the editing enzymes ADAR2, TadA and APOBEC2, the orientation of strand β5 is upward toward the active site, parallel to strand β4 (Figures 16.6–16.8) (52, 54, 55). In the latter case, the loop from strand β4 to β5 is often longer because the change in direction of β5 necessitates the presence of an extended topological cross-over element. This connector region commonly exhibits helical character (Figures 16.6 and 16.7). The connectivity of these family members is therefore α1β1β2α2β3α3β4<u>α4</u>β5. Further complicating matters, even in those deaminases where β5 runs antiparallel to strand β4, the α4

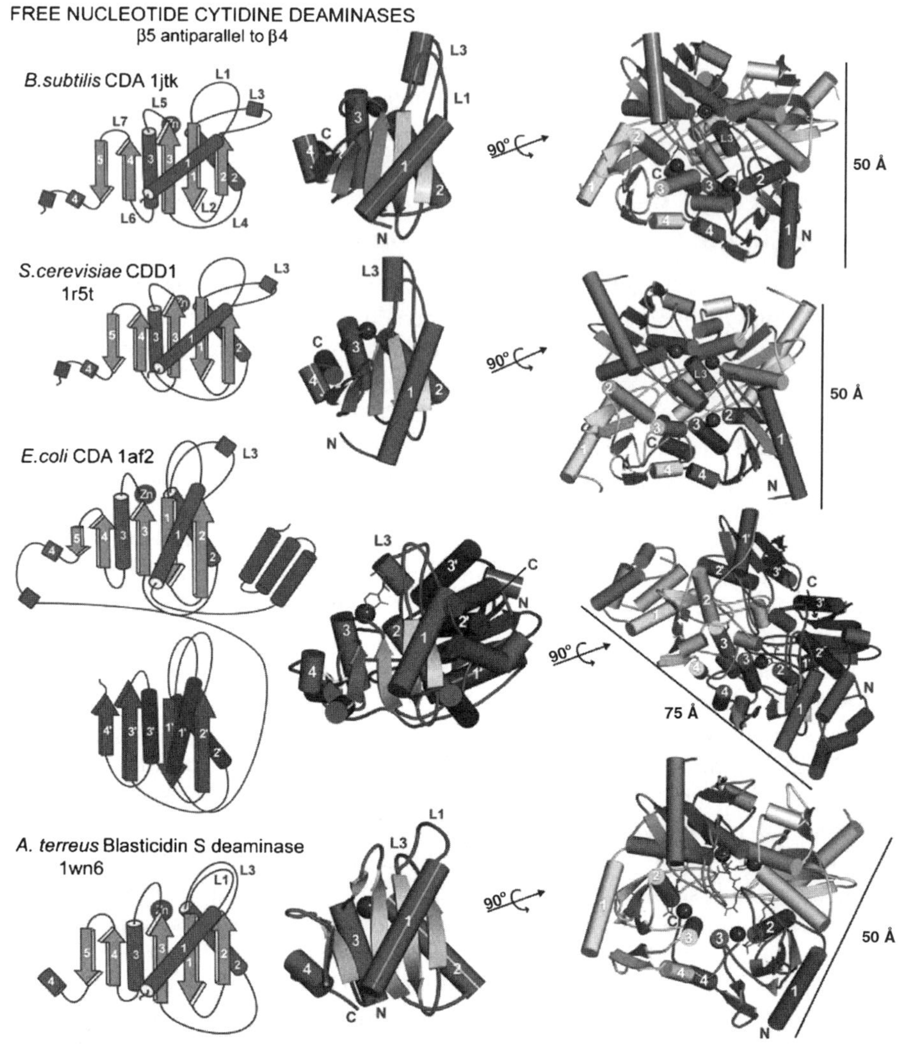

Figure 16.5 Free nucleotide cytidine deaminases (fnCDAs): β-strand 5 antiparallel to β-strand 4. Representative members of the fnCDA family are labeled with species, enzyme name, and PDB ID and are shown in three forms. (Left) A simplified topology diagram displays the connectivity of secondary structure elements. Secondary structure elements that comprise the conserved core fold characteristic of the CDA superfamily are shown in red and blue and numbered sequentially. Embellishments to the core fold are colored gray and are excluded from the numbering scheme. All loops are labeled in the first example, and only as pertinent to the text throughout the rest of the figures. The zinc ion is shown in green. (Center) Ribbon diagram displays the three-dimensional structure of a subunit. Secondary structure elements are colored as in the topology diagram. Numbering is limited to α-helices and selected loops in the interest of clarity. N- and C-termini are indicated. The second *E. coli* CDA domain, believed to have resulted from a domain duplication event, is colored in darker shades of red and blue. (Right) Ribbon diagram displays the three-dimensional structure of the oligomer. Scale bars indicate the length of the protein. A single subunit is colored as in subunit panel, but rotated 90° along the axis of the arrow. Remaining oligomers are colored green (and dark green for the duplicated domain of *E. coli* CDA), orange, and yellow, with nonconserved embellishments shown in tan. The final perspective looks down from the top to view the substrate binding region and catalytic zinc ion, shown in green. Substrates, if available, are depicted in gray stick format. (*See color insert.*)

helix is spatially, although not topologically, conserved to yield α1β1β2α2β3α3β4β5<u>α4</u>. This is illustrated by comparing the *B. subtilis* CDA (Figure 16.5) to the cytosine deaminase of *S. cerevisiae* (Figure 16.6). In the bacterial enzyme, the smaller C-terminal helix follows rather than precedes β5 in the polypeptide sequence. However, structural superpositions place both helices, labeled α4, in the same location. Defining the orientation of such structural elements among APOBEC3 family members is especially important since this region has been shown to bind the HIV virion infectivity factor (Vif) (56), which leads to APOBEC3G and APOBEC3F degradation through polyubiquitination and subsequent targeting to the proteasome (57, 58) (Chapter 11).

C-terminal helices are also observed in those enzymes with strand β5 oriented upward (i.e., parallel) relative to strand β4. In these cases (exemplified in Figure 16.7), the C-terminal sequences are directed upward toward the active site, and the significant diversification observed in this region is situated to provide substrate specificity with minimal disturbance to the core fold. From a practical perspective, the significance of the alternate orientation of strand β5 becomes clear when designing experiments for family members that have not yet been structurally characterized. For example, the application of domain boundaries derived from a bacterial fnCDA onto an APOBEC protein would undoubtedly result in truncation to the core fold of the expressed sequence.

As observed for actin and HSP70, selective pressures yielding specialized function manifest as embellishments at the periphery of the core CDA fold. For example, the N- and C-termini are diversified commonly in the CDA superfamily, as well as in the loops that connect the core β-strand and α-helical secondary elements. Loop 3 (L3) is an especially common site of diversification, because it occurs between strand β2 and helix α2 in the ZDD motif and is therefore situated near the active site (Figure 16.2B); L3 commonly exhibits helical character. However in T4 dCMP deaminase, a 60-residue insertion in this region coordinates a second, non-catalytic Zn^{2+} ion (59) (Figure 16.7, middle panel). In comparison, APOBEC2 exhibits a rather shortened L3 (Figure 16.7, lower panel) as a result of an elongated strand β2 (52). This compensatory alteration leaves the putative active site more exposed, as might be expected for a protein fold designed to edit ssDNA or ssRNA (Figure 16.3).

16.4.3 Comparison of the Common CDA Superfamily Core Reveals Broad Peripheral Diversification

Members of the CDA superfamily share a core fold comprising a series of topologically conserved secondary structure elements that encompass the ZDD motif and facilitate deamination chemistry via Zn^{2+} coordination and cytidine binding. Diversifications to the core fold have imparted differences in substrate specificity and functional regulation among family members (49). As such, members of the CDA superfamily target a remarkably diverse array of substrates including RNA, DNA, deoxycytidylate, cytidine, cytosine, guanine, and Blasticidin S, as well as intermediates of riboflavin and purine biosynthesis. Most surprisingly, the adaptations that have enabled such functional diversity have evolved in a manner that does not alter the core

CDA-like polypeptide fold. Consequently, the locations in primary sequence that demonstrate divergence are restricted to insertions or deletions on the periphery of the core fold, and they are identifiable by structural comparisons among family members. For example, the gray elements shown in Figures 16.6–16.8 demonstrate how non-core elements fold in a manner that flanks the conserved Zn^{2+} ion held in place by the ZDD motif. Differentiating between functional adaptations and the conserved core fold continues to be of special interest for researchers aiming to target regions that impart specialized biological activity. For example, CDA superfamily members have long been targets in the development of chemotherapeutic agents. However, the cross-reactivity and ensuing toxicity resulting from such therapies (60–63) underscores the importance of discerning one family member from the next at the molecular level to fine-tune drug specificity. Greater discussion of the modes of substrate interaction will be presented in Section 16.6.

16.5 MODES OF OLIGOMERIZATION

16.5.1 Free Nucleotide Cytidine Deaminases (CDA): Strand β5 Antiparallel to Strand β4

A dominant theme for the diversification of the CDA superfamily is variation in the mode of subunit oligomerization. The structurally well-characterized fnCDAs, in which strand β5 runs antiparallel to strand β4, exist as tetramers (Figure 16.5), with the exception of the enzyme from *E. coli*. Although even the latter case, with its CD-PCD configuration (Figure 16.4), generates a comparable pseudo-tetrameric assembly comprised of four CDA-like domains (Figure 16.5, third panel). The subunits of the tetramer are related by 222, or *D*2 symmetry, as described in Section 16.4.1. The oligomerization interface of each tetramer is extensive and, in the case of the yeast enzyme Cdd1, is maintained by contributions from core helices 2, 3, and 4, a C-terminal 3_{10} helix (5), loops 1, 3, and 4, and a few residues from strand β5 (Figure 16.5 and reference 20). The take-home message is that the Cdd1 subunit interface is representative of enzymes of this class and that multiple polypeptide chains (three for the strict tetramers) are essential to complete the formation of a single active site (i.e., *trans* complementation).

16.5.2 Cytosine Deaminase, Guanine Deaminase, and TadA: Strand β5 Parallel to β4

Cytosine deaminase (CD), guanine deaminase (GD), and TadA exist as dimers, and they share a common mode of oligomerization (Figure 16.6) (64–66). Subunits are related by a proper twofold rotation axis in which each active site forms near the interface of the two subunits. The interface is maintained by contributions from the core helices 2, 3, and 4 with additional contributions from residues in loops 3 and 4. Again, the formation of a single active site appears dependent upon contributions from two polypeptide chains. Interestingly, a putative case of 3-D domain swapping was identified in the guanine deaminase structure whereby helix α4 and strand β5 are exchanged between the two subunits, thus increasing the buried surface area while maintaining all the elements of the core fold; for more information on 3-D domain swapping, see references 67 and 68. In yeast cytosine deaminase the C-terminal helices fold over the active site and contact, among other things, a nonconserved helical element present in L4′ of the opposing subunit. This bulky configuration has been noted previously to restrict access of the substrate to the active site, possibly discriminating free cytosine from larger nucleosides (49). In contrast, the C-terminal helix of TadA (a tRNA editing deaminase that converts adenosine to inosine) extends straight upward and has little opportunity to engage in oligomerization. The same would be true for guanine deaminase, but for the observed 3-D domain swapping effect (Figure 16.6, two bottom-most panels).

The TadA structures are more similar to the GD enzyme as compared to the CD enzyme, as might be expected considering their preference for purines. TadA does not engage in domain swapping. However, two unique features of GD are present in TadA that are not present in CD. Helix α2 is elongated on the lower (C-terminal) end, such that when inspecting the dimer, this helix protrudes from the opposing subunit and is available to aid in substrate binding. The dimerization interface of both TadA and GD exhibit a more staggered packing of helices 2 and 3 as compared to CD, which further enhances the protrusion of helix α2.

16.5.3 Deoxcytidylate Deaminases (T4, *N. e*) and APOBEC2: Strand β5 Parallel to β4

The deoxycytidylate deaminases (dCDs) exhibit a comparable helical oligomerization interface when compared to the CD and GD enzymes (Figure 16.6 versus Figure 16.7). A dimeric form of this enzyme derived from the nitrogen-fixing bacteria *N. europaea* is diversified to include an additional N-terminal helix and β-strand (Figure 16.7, upper panel). The extra β-strand interdigitates with the equivalent region from the opposing monomer and contributes to subunit dimerization. This structure was solved by the Midwest Structural Genomics Consortium, and the endogenous substrate for this protein has not been identified. As such, the classification of this enzyme is tenuous at present.

The T4 bacteriophage dCMP deaminase structure tells a unique story. The active form of the enzyme is a hexamer and therefore each subunit has two oligomeric interfaces (59). This molecule is allosterically regulated by dCTP and dTTP in a region of the protein at the C-terminal (bottom) end of α-helix 2 (69) (Figure 16.7, middle panel). A point mutation at this site, R115E, resulted in a dimeric enzyme (as characterized by gel filtration) that was largely devoid of allosteric regulation (70). Nonetheless, this mutant form of the enzyme crystallized as a hexamer, perhaps as a result of ionic shielding due to the high salt conditions of crystallization (59). In the crystal structure the R115E mutation localized to the helical oligomerization interface, thereby providing a rationale for its contribution to oligomerization. Notably, an R115Q mutation was dimeric in the absence of dCTP, but hexameric in the presence of dCTP, whereas the wild-type enzyme is hexameric regardless of dCTP concentration (70). Together, these findings implicate allosteric regulation through modulation of an oligomerization interface, which

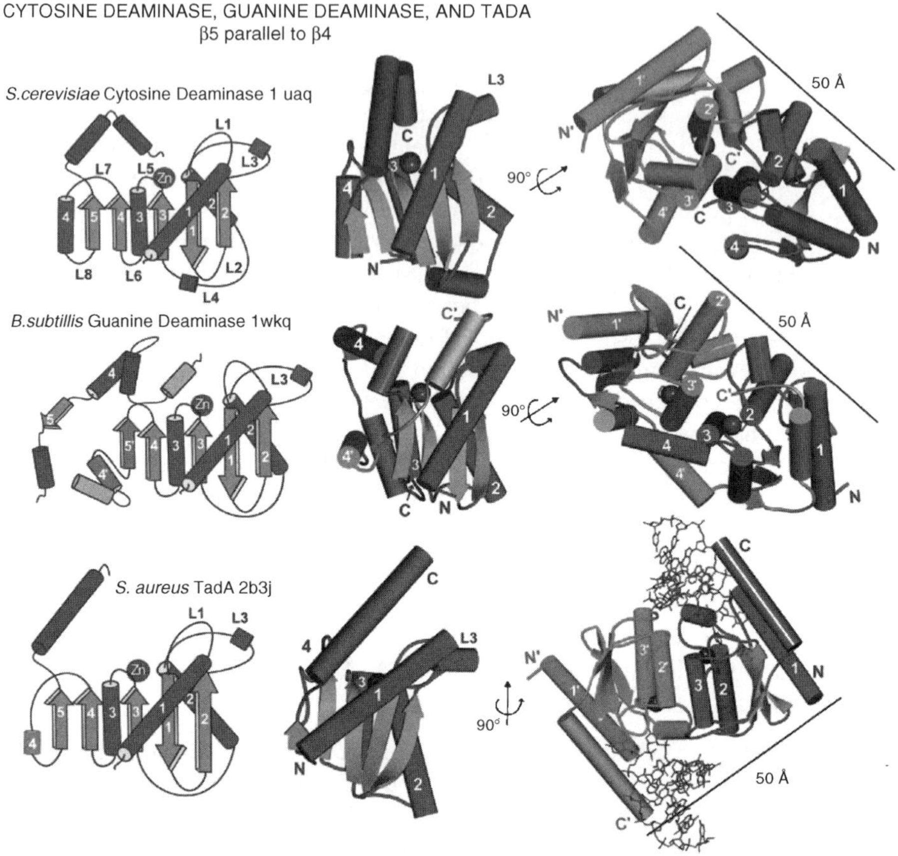

Figure 16.6 Cytosine deaminase, guanine deaminase, and TadA: β-strand 5 parallel to β-strand 4. Figure layout is as described for Figure 16.5. Three-dimensional domain swapping in guanine deaminase is indicated in the following manner: The subunit is colored in red, blue, and gray as described previously. However, α-helix 4 and β-strand 5 do not contribute to the core fold of the subunit, but rather to the core of the opposing monomer. α-Helix 4′ and β-strand 5′ of the opposing monomer, which contribute to the core fold of the displayed subunit, are labeled with primes (′) and colored in green; nonconserved embellishments of the opposing subunit are shown in tan. (*See color insert.*)

is a common mode of regulation (71, 72). The elements maintaining this interface are extensive and familiar with respect to other CDA superfamily members, including core helices 2, 3 and 4 and loops 3 and 4. This information is particularly relevant in light of current hypotheses that suggest that both AID and APOBEC3 family members, which act on DNA, may be allosterically inhibited through RNA binding (73, 74). We will see in Section 16.8 that RNA has a definite influence on the molecular mass of APOBEC3G.

Interestingly, the dimeric interface of T4 dCD is created through the antiparallel association of two β2 strands that form a continuous β-sheet surface (Figure 16.7, middle panel, yellow inset). This subunit interaction is stabilized

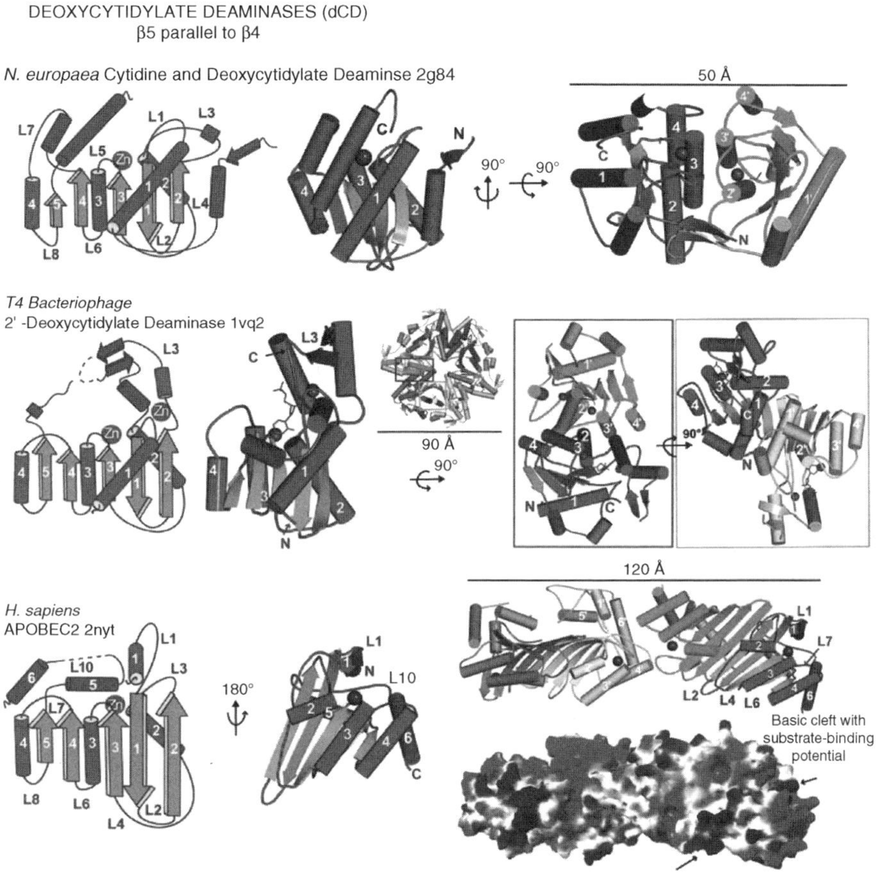

Figure 16.7 Deoxycytidylate Deaminases and APOBEC2: β-strand 5 parallel to β-strand 4. Figure layout is as described for Figure 16.5. (Upper) For *N. europea,* multiple rotation axes indicate the relationship of the subunit to its position in the oligomer. The view of oligomer is still from the top down, displaying the substrate binding pocket. (Middle) The ~20 residues missing from L3 of the T4 dCD crystal structure are represented by a dashed line. T4 dCD exists as a hexamer shown above rotation axis. For clarity, the two subunit interfaces are boxed and expanded; the green box denotes the helical interface, whereas the yellow box denotes the β-sheet interface. The β-sheet interface boxed in yellow is orientated similarly to the subunit depiction. (Lower) In APOBEC2, the subunit represented in the topology diagram and in the subunit ribbon diagram is that of the outermost subunit of the tetramer. A depiction of the electrostatic potentials is mapped onto the APOBEC2 surface to demonstrate the relative charge of the protein surface. Red indicates a negative charge; blue indicates a positive charge, as predicted at pH 7.0. The orientation of the tetramer is as shown in the ribbon diagram. Arrows indicate a basic cleft with the potential to bind nucleic acid substrate. [Figure generated by Swiss-PDBViewer (198).] (*See color insert.*)

further through interactions between the N- and C-terminal helices, the former of which pack side by side lengthwise with additional contributions from helix α2 and loop L1 (59). Although α-helical packing is the most common mode of subunit oligomerization for members of the CDA superfamily, edge-to-edge β-sheet-mediated oligomerization is an emergent theme, as observed recently for APOBEC2.

APOBEC2 has the most extensive edge-to-edge β-sheet subunit interface of any structurally characterized deaminase (Figure 16.7, lower panel). This interface is maintained through the canonical carbonyl and amide backbone interactions in a largely sequence-independent interaction, with additional contacts made by two residues at the bottom of helix 2 (52). The N-terminal 40 residues of APOBEC2 were omitted from this construct, presumably to obtain crystals. Hence, the possibility exists that additional supportive contacts might originate from this unique N-terminal region and further stabilize the dimeric interface, akin to the interdigitating strands of *N. europaea* dCD or the extra four-helix bundle of *Ec*CDA, although there is no apparent sequence homology among the respective N-termini of these structures to support this assertion.

APOBEC2 includes all the canonical elements of the CDA superfamily topology, although these elements have been rearranged somewhat. Helix 1 is shortened and juts out from the β-sheet surface rather than laying across it. Thus, tertiary contacts from α-helix 1 are made primarily to the C-terminal features of APOBEC2, rather than to the β-sheet surface typically contacted by the α-helix 1 of other deaminases (Figure 16.7; compare upper and lower panels). APOBEC2 encodes two C-terminal helices beyond strand β5; the first of these (α5) traverses the broad face of the β-sheet, akin to α-helix 1 of other deaminases (52). A long intervening loop (L10) enables the second C-terminal helix (α6) to cap the side of the β-sheet, contacting β-strand 5 and α-helix 4. In this manner, the C-terminus is directed away from the active site, which is in contrast to the comparable, but upwardly directed, helix observed in TadA (55) (Figure 16.6, lower panel).

Like the T4 dCMP enzyme, APOBEC2 migrated as a dimer in gel filtration studies, but crystallized as a higher-order tetramer (52). The amount of buried surface area in the tetramer interface was 1745 $Å^2$, which is near the average value observed in a survey of biologically relevant protein–protein interactions (75). The tetramer interface exhibits contributions from loops L1 and L7, helix α4, and the last C-terminal helix, α6. The rearrangement of the C-terminal helices places the tetramer interface on the opposite end from the dimer interface, such that the tetramer observed in the crystal lattice is formed through a tail-to-tail interaction (Figure 16.7, lower panel—right). Gel filtration results not withstanding, if the tetrameric form is relevant, there is nothing to prevent APOBEC2 from further oligomerizing to form an extended polymer. In contrast, the helical interface of the T4 dCD enzyme is located on the back surface of the protein rather than its edge, thus resulting in formation of a toroidal hexamer (Figure 16.7, middle panel). Further investigation of APOBEC2 will be necessary to establish how its oligomeric state relates to biological function.

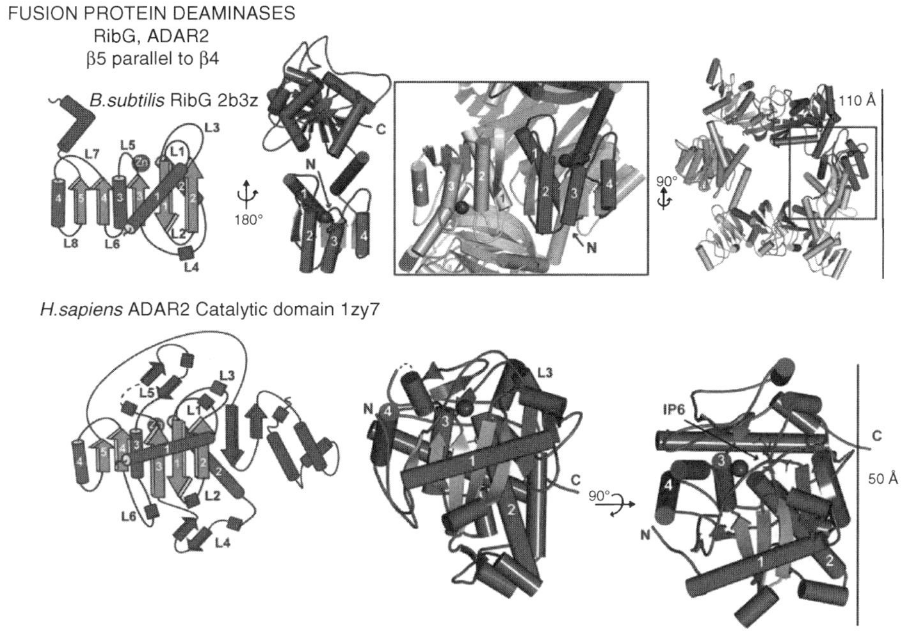

Figure 16.8 Fusion protein deaminases RibG and ADAR2: β-strand 5 parallel to β-strand 4. The figure layout is as described for Figure 16.5. (Upper) For RibG, the monomer shown in the ribbon diagram is rotated 180° from the topology diagram, such that the "backside" helices face forward. The fusion domain, comprised of a Rossmann fold, is shown in gray. The oligomeric region of interest is the β-sheet subunit interface; this is expanded and boxed in red. The full tetramer is shown at far right for perspective. (Lower) ADAR2 is depicted as a single subunit and does not appear to oligomerize in the crystal structure. The topological orientation is similar to that of the three-dimensional subunit orientation. A second subunit orientation rotated 90° is depicted at the far right. (*See color insert.*)

16.5.4 Multidomain Enzymes RibG and ADAR2 of the CDA Superfamily: Strand β5 Parallel to β4

The CDA superfamily members RibG and ADAR2 demonstrate that the CDA-like fold can be fused to additional, discrete domains within a single polypeptide chain (Figure 16.8) (51, 76, 77). In RibG, the additional domain performs a discrete chemical reaction and is classified as a Rossmann fold, which is a classical nucleotide binding structure (78–80). RibG is a bifunctional deaminase and reductase involved in purine metabolism (51, 81). Tetramer formation is mediated by one surface of each domain. The deaminase domain interface is mediated by an antiparallel side-by-side arrangement of the β2 strands (Figure 16.8, upper panel insets) similar to that of T4 dCD and APOBEC2. In RibG, the β2 strands are twisted and not positioned to interact as a sheet, but engage in a few side-chain-mediated contacts further supported by contributions from α-helices 1 and 2 and loops 1, 2, and 3.

ADAR proteins incorporate both a deaminase domain and one or more copies of a double-stranded RNA binding domain (dsRBD). The latter domain(s) contributes to both RNA-independent dimer formation and substrate recognition (82–84). Significantly, the crystallization construct of Bass and colleagues contained only the deaminase domain, which was demonstrated to be a moderately active monomer (54). The monomeric state of this construct has been disputed in gel filtration studies but validated by FRET analyses (84, 85). Assuming that the A-to-I editing activity of the monomeric form is genuine, this property represents a unique characteristic of this CDA family member. The active site of most other family members arises through *trans* complementation of subunits engaged in oligomerization.

ADAR2 is the most highly diversified fold yet observed for members within the CDA superfamily, and the self-contained fold supports the case for monomeric activity. The interfaces observed to engage in oligomerization in other family members are unavailable in the ADAR2 structure, such that if this construct is dimeric, the mode of oligomerization is likely to be novel among family members. Both ends of the central ADAR2 β-sheet exhibit auxiliary β-strands, beyond the canonical five strands of the CDA fold, that would interfere with a strand β2-mediated interface, such as that observed in APOBEC2. Additionally, the "back side" of helices 2 and 3 are covered by a number of additional helices, thereby preventing formation of a dimer like that observed in TadA enzymes. These additional helices in ADAR2 coordinate an inositol hexakisphosphate (IP6) molecule that is essential for activity (54). Significantly, ADAT1 proteins have been shown to exhibit a similar IP6 dependence, but ADAT2/3 proteins, the human homologs of the more primitive TadA proteins, do not (54). Phylogenetic analyses further support this division of relatedness between tRNA editing enzymes (Figure 16.3) (31). Structural details about ADARs are provided in Chapter 15.

16.6 MODES OF SUBSTRATE INTERACTION

Although there is currently no structural information that describes how members of the immediate APOBEC family bind and deaminate ssDNA or RNA substrates, a comparison of members of the CDA superfamily reveals a central theme in substrate recognition. Namely, these deaminase enzymes share a common active site fold that originates from restraints imposed by the ZDD motif (Figure 16.9). Like many enzymes that utilize β-scaffolds, the substrate binds at the periphery of the sheet, most likely because this region is easiest to diversify through alteration of loops. Similar observations have been made for triose phosphate isomerase (TIM) α/β barrels, which typically bind substrate via loops connecting α-to-β elements at the carboxylic end of the barrel (86, 87). Likewise, the ZDD motif places the essential catalytic Zn^{2+} ion at the periphery of the beta sheet while using loops, particularly L3, to bind and orient the substrate for nucleophilic attack by activated water (Figure 16.2).

Substrate specificity is achieved through variations on this theme, including alterations in the size of the binding pocket, the extent of solvent exposure, contributions from neighboring subunits, and supplemental contributions from additional elements within the structure, such as C-terminal helices (for a review see reference 49). APOBEC2 is a prime example of the variance that can be achieved with

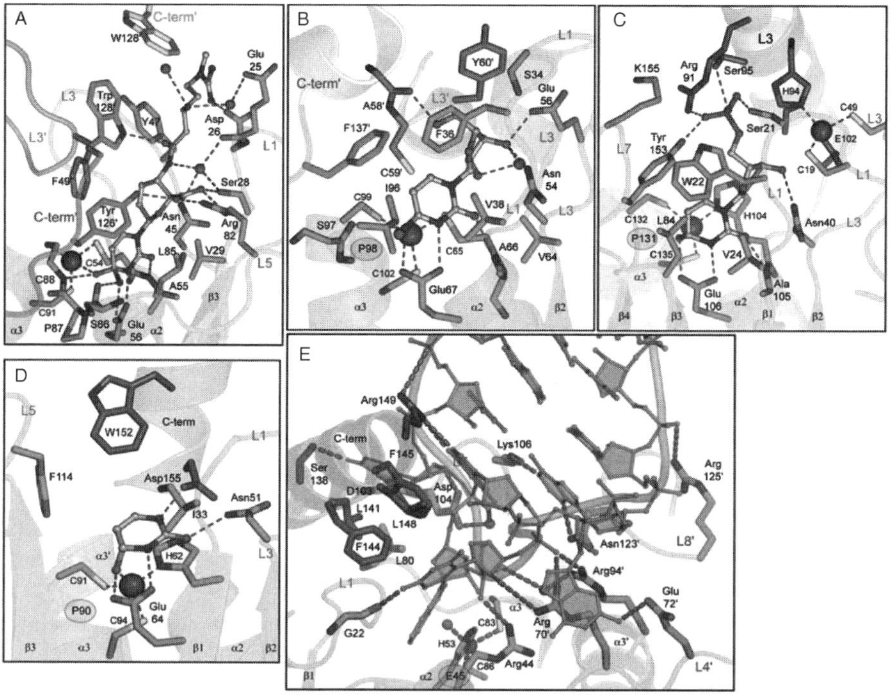

Figure 16.9 Schematic diagrams of substrate binding in the active sites of representative CDA superfamily members. (**A**) blasticidin S deaminase in complex with blasticidin S (**B**) mouse cytidine deaminase in complex with activated cytidine, (**C**) T4 deoxycytidylate deaminase in complex with deoxycytidine monophosphate, (**D**) yeast cytosine deaminase in complex with the substrate analog 4-(*R*)-hydroxyl-3,4-dihydropyrmidine, and (**E**) TadA deaminase from *S. aureus* in complex with the anticodon stem-loop RNA substrate (orange). Subunits are colored as in Figures 16.5–16.8, substrates are colored white. Oxygen atoms are red, and nitrogen atoms are blue. Residues omitted for clarity are represented as shaded ovals. Those residues implicated in hydrogen bonds to substrate are represented with three-letter codes; hydrogen bonds are indicated by dashed gray lines. Ionic interactions with Zn^{2+} are indicated by dashed green lines. Selected waters that mediate substrate–protein interactions are shown as red spheres. Residues are represented by side chain only, except for those where the backbone atoms are implicated in hydrogen bonding. Secondary structure features are selectively displayed and labeled according to Figures 16.5–16.8. A cartoon secondary structure is provided for the RNA backbone. The perspective for each panel is not identical throughout the panels. (*See color insert.*)

regard to solvent exposure (Figure 16.7, lower panel). Here, the putative active site is strikingly open relative to all other family members, including those editing enzymes that also bind polynucleotide substrate such as ADARs and TADs (52). The fnCDAs offer another "variation on a theme". As described in Section 16.5.1, all of the fnCDAs require multiple subunits to form their active sites, yet the size of the active site varies (Figure 16.5). For example, the active site of Blasticidin S deaminase is significantly larger than that of mouse cytidine deaminase (Figure 16.9A versus Figure 16.9B). Several factors, including a rearrangement of loops 1 and 3, contribute to the more open active site of the former enzyme. These observations demonstrate the remarkable variability of CDA superfamily members despite a conserved core fold and the presence of the ZDD motif.

16.6.1 Tetrameric fnCDAs Favor Flexible Flaps: RNA Editing and the Case of Cdd1 from Yeast

Difficulties in preparing crystal structures of bona fide APOBEC family members led investigators to seek three-dimensional structural information from more operable sources, such as Cdd1 from yeast. This fnCDA site-specifically edited reporter apolipoprotein B mRNA constructs during late log phase growth in yeast (45), albeit at levels considerably lower than APOBEC1. Furthermore, both APOBEC1 and Cdd1 have been shown to target ssDNA in a bacterial mutator assay, although again, Cdd1 activity was inferior to that of APOBEC1 (47, 88). These similarities in RNA and DNA editing function suggested that Cdd1 might provide structural insights into APOBEC1. In light of the recent APOBEC2 structure, this now seems less likely (20, 52). Nonetheless, the Cdd1 structure suggests that the auxiliary nucleic acid editing activity of Cdd1 may be attributable to the accessible nature of fnCDA active site, as well as the consequence of protein overexpression.

In addition to Cdd1, several ZDD containing yeast proteins were examined for their ability to target the reporter apolipoprotein B substrate, but only Cdd1 exhibited significant activity (45). Notably, Cdd1 was the only fnCDA family member tested; other family members included a deoxycytidylate deaminase, a cytosine deaminase, and a deaminase active in riboflavin biosynthesis. Compared to these other enzymes, the Cdd1 active site appears to be the most accessible in terms of the degree to which multiple structural elements from neighboring subunits fold over the active site (e.g., compare the Zn^{2+} accessibility of Figure 16.5, *Cdd1* versus Figure 16.6, *Cytosine deaminase*, Figure 16.7, *deoxycytidylate deaminases* and Figure 16.8, *RibG*). Upon analyzing the x-ray structure, Wedekind and colleagues tested the requirement for C-terminal flexibility by making a series of chimeric constructs predicted to possess flexible C-termini versus highly ordered termini, such as the *E. coli* CDA. Editing analysis revealed that constructs with flexible C-termini were capable of editing reporter apolipoprotein B RNA, whereas constructs with rigid C-termini or the intact *E. coli* CDA, whose active site is occluded by a rigid linker connecting the CD to the PCD, were incapable of C-to-U deaminase activity (20). As such, the working hypothesis for the Cdd1 editing of RNA (or DNA) is that the C-terminus is flexible and can open to accommodate a polynucleotide substrate (Figure 16.9B, flap harboring F137′).

Notably, Cdd1 required late log phase growth in the presence of galactose to elicit RNA editing (45, 89). This observation suggests that the synthesis of a new protein may be required for Cdd1's RNA editing activity, which would make such activity more of an exception than a rule. Overexpression of APOBEC enzymes also resulted in general promiscuous editing activity on noncanonical substrates (6, 90–93). Thus, both of these enzymes exhibit auxiliary activities in the appropriate context. It is not known if the explanation behind these activities is the same. Given the precedent of the APOBEC2 structure, it seems unlikely. Although Cdd1 editing of *apoB* mRNA may be the result of structural flexibility in the C-terminus, "promiscuous" APOBEC1 activity may rely more upon cellular localization and the availability of a cofactor necessary for specific substrate recognition.

16.6.2 A Topological Transformation Obstructs Active Site Accessibility in Dimeric Deaminases that Bind Bases

A parallel orientation of strands β4 and β5 facilitates the engagement of C-terminal helices in substrate binding. This is accomplished to an extreme extent in yeast CD, where two well-developed C-terminal helices fold over the active site and sequester substrate to such an extent that only the monomer contacts cytosine (Figures 16.6 and 16.9D) (64). Oligomerization remains important for sequestering substrate from solvent and aids in defining the binding pocket in each of the examples presented here. In guanine deaminase, the dimer contributes to pocket formation and substrate binding to a larger extent than yeast CD (Figure 16.6). Substrate modeling suggested that two aromatic residues from the opposing subunits are likely to interact directly with substrate, one from the opposing helix $\alpha3'$ and another from the domain swapped C-terminal helix (65). The T4 dCD enzyme has a relatively underdeveloped C-terminal helix and contacts substrate directly with only one subunit (Figure 16.7, middle panel). Nucleotide recognition is dominated by a 60-residue insertion in loop 3 that contains two helices, two small β-strands, and 20 residues of unknown structure that were not visible in the crystal structure (59). Although absent in the human ortholog, this region coordinates a second zinc ion that acts to stabilize the binding pocket and ameliorate the localized negative charge of the substrate's phosphate group (Figure 16.9C).

The take-home message is that nature evolves enzymes to accommodate a specific substrate, but there appear to be a finite number of approaches employed to achieve this goal to preserve the fundamental CDA architecture. Knowledge of the common variations is significant as a means to direct experiments on structurally uncharacterized enzymes. As such, the suggested *modus operandi* is to align sequences of interest to known structures to identify segments that have been evolutionarily embellished (i.e., C-terminal helices or L3 expansions). These areas can then be targeted experimentally to define functional relevance. We propose that it may be necessary to compare the sequence in question to a representative from each of the major CDA classes presented (Figures 16.5–16.8).

16.6.3 Substrate Selection by Polynucleotide Editing Enzymes Remains Elusive

Family members that edit polynucleotide substrates must accommodate relatively large dsRNA, ssRNA, or ssDNA substrates. Structures representative of each editing family (TADs, ADARs, and APOBECs) suggest that they achieve such targeting uniquely, although only the TadA enzyme from bacteria has been solved in the presence of a substrate analog (Figures 16.6 and 16.9E) (94). The TadA enzyme looks strikingly similar to GD, but binds a much larger substrate (Figure 16.6, middle versus lower panels). The absence of 3-D domain swapping in TadA allows for a more open active site. In particular, a single, long C-terminal helix extends upward from strand β5 and engages in several contacts to the tRNA substrate. Contacts on the opposite side of the active site are provided by the protrusion of the opposing subunit's long helix α2′. Together, these elements extend the binding pocket up from the active site, such that no fewer than five nucleotides are accommodated (Figure 16.9E) (94). This arrangement is in stark contrast to yeast CD, which cannot accommodate a single nucleoside (Figure 16.9B).

The active site of ADAR2 represents the opposite scenario in terms of accessibility (Chapter 15) . Each subunit harboring the fundamental CDA domain appears to fold autonomously as a monomer, with helices α2 and α3 of the ZDD poised such that the active site zinc resides slightly above the plane of the β-sheet and is thus far more solvent exposed by comparison to the TadA enzymes (Figure 16.6, lower panel versus Figure 16.8, lower panel). This shallowness seems sensible in light of the requirement for ADAR2 to bind an adenosine in the context of dsRNA. To gain access to the edited position, the base is hypothesized to flip out of the duplex into the superficial opening revealed in ADAR2. This flipping mechanism is supported by fluorescence studies from Beal's lab (95). Consequently, the substrate would not be able to extend as far into the core of ADAR2 as the anticodon loop of tRNA does with TadA without significant melting of the duplex structure.

The APOBEC family recognizes single-stranded nucleic acid substrates, and as with TadA and ADAR2, the crystal structure is reflective of the functional data. The putative active site of the APOBEC2 structure is strikingly open (52) (Figure 16.7, bottom panel). This effect is achieved by (i) the placement of the C-terminal helices on the opposite face of the β-sheet rather than folded over the active site, and (ii) the unique head-to-head and tail-to-tail modes of oligomerization that avoid the more common interaction between helices α2, α3, and α4 (Figure 16.7, bottom panel). Notably, these helices would be exposed in ADAR2 as well, if not for the considerable adaptations necessary to bind the IP6 molecule. The β-strands in APOBEC2 are considerably elongated relative to other CDA domains, and loop 1 is also expanded. Collectively, these elements limit substrate entry from the "top" as compared to ADAR2 and TadA.

16.7 THE APOBEC FAMILY: INSIGHTS INTO A STRUCTURALLY UNDERREPRESENTED FAMILY

Phylogenetic analysis places a protein in the context of evolution, between its ancestors and descendents. The preceding sections used structure to provide a phylogenetic perspective of the APOBEC family by placing this family in the context of the CDA superfamily. This served to highlight common themes and unique features for the investigator interested in pinpointing regions of functional significance. However, the absence of structural information for the bulk of APOBEC enzymes prevents such lucid comparisons between APOBEC family members. Consequently, alternative methods must be considered in the analysis of these enzymes. Comparative sequence analysis, the tool most commonly used to predict evolutionary relationships, can be combined with information about gene structure to glean insight into how other APOBEC enzymes might compare to the structure of APOBEC2. For example, sequence comparisons of APOBEC family members demonstrate a number of conserved regions (Figure 16.10). Based on the APOBEC2 structure, some of these sequences appear to stabilize the core CDA fold (^{70}LCY^{72},^{122}YVSS^{125}) or C-terminal

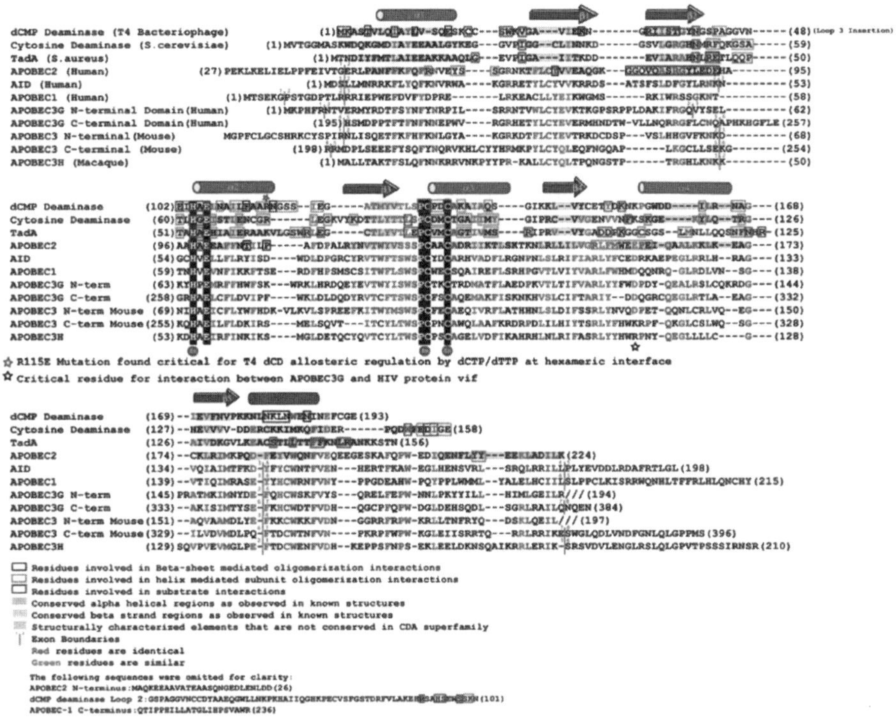

Figure 16.10 Structure-based sequence alignment of representative CDA superfamily members and a subset of APOBEC sequences. Four known structures corresponding to T4 dCD, *Sc*CD, *Sa*TadA, and APOBEC2 are aligned with APOBEC family members whose structures are unknown. Red letters denote identical residues, and green letters indicate similarity. Residues that coordinate Zn^{2+} are indicated by green spheres. Residues comprising the ZDD signature motif are shown as white letters boxed in black. Secondary structure elements are indicated approximately by cartoons above the alignment; specific boundaries for known structures are indicated by shaded backgrounds such that blue represents β-strand, orange indicates α-helix, and gray indicates nonconserved C-terminal helical "embellishments." Residues that contact substrate, if known, are boxed in pink. Residues that contribute to oligomerization via the helical interface are boxed in blue; those that contribute to a β-sheet interface are boxed in black. Exon junction boundaries are indicated by a vertical line (golden). Representative members of the various types of APOBEC3 sequence are included: human APOBEC3G is a Z1a-Z1b combination, mouse APOBEC3 is comprised of a Z1a-Z2 combination and macaque APOBEC3H is an active example of the only primate Z2. (*See color insert.*)

helical structures (^{187}VWxxFV^{192},^{219}LxxIL^{223}), whereas others are positioned to contribute to either the tetramer interface or substrate recognition (^{153}RLF^{155}). Clearly, there is an obvious benefit in distinguishing between such regions when designing experiments.

Sequence-based analyses can also suggest evolutionary timelines including the order and pace with which family members evolved. The pace at which a protein evolves is representative of the selective pressure it has experienced. It has been asserted by Gillespie and others that the evolution of proteins is consistent with the principles of natural selection to some extent (96, 97). As such, external influences can cause a protein to remain relatively the same over time (purifying selection), or select for a new functionality (positive selection). The hallmark of positive selection is an excess of nonsynonymous substitutions (i.e., those that alter the amino acid being encoded) relative to synonymous substitutions (that retain the encoded amino acid). These changes can be benchmarked by pairwise sequence comparisons of orthologous proteins from different species. The ratio between these two values (nonsynonymous versus synonymous) is referred to as the d_N/d_S ratio. Values greater than one are suggestive of positive selection (98). For example, proteins involved in viral defense commonly exhibit rapid rates of evolution, as indicated by high d_N/d_S ratios, as a means to counteract rapidly evolving viruses (99).

After considerable discussion of the similarities of the APOBEC family members to other deaminases in which strands β4 and β5 are parallel (Sections 16.5.2 to 16.5.4), it is interesting to observe that the APOBEC proteins form a distinct phylogenetic cluster (Figure 16.3). This branching is indicative of the age of the APOBEC family and supports the rationale for initial modeling efforts derived from the CDA enzyme structures rather that the CD enzymes; that is, APOBEC proteins appeared equally related to either class, but the CDA structures were elucidated first (20, 35). This branching also supports the prediction that the structures of other APOBEC family members will more closely resemble the APOBEC2 structure rather than other members of the CDA superfamily.

Estimation of the evolutionary relationship of APOBEC family members requires identification of a progenitor member. The APOBEC proteins are divided into five groups according to function and order of identification. These groups include AID, APOBEC1, APOBEC2, APOBEC3, and APOBEC4. Each class is represented by a single member with the exception of the APOBEC3 clade, which has undergone rapid expansion in primates (31). Members of the AID, APOBEC1, and APOBEC3 clades exhibit fairly nonspecific deamination activity on ssDNA without any cofactor or ATP requirement (100–102). No editing activity has yet been demonstrated for the APOBEC2 or APOBEC4 proteins (28). APOBEC1 is unique among family members in that its endogenous substrate is RNA rather than DNA, albeit an RNA-binding cofactor is required to achieve specificity (see Chapter 10). This precedent suggests that other family members may have the capacity to target specific RNA, although the substrate sequences and cofactors have not been forthcoming. In the next sections, we review available phylogenetic and structural data for the APOBEC family to highlight the relationships between individual enzymes and thereby guide future research efforts.

16.7.1 Activation-Induced Deaminase (AID)—an Ancient Enzyme with Essential Roles in Adaptive Immunity

AID was identified in 1999 independent of its relationship to other APOBEC family members, thus giving rise to its unique nomenclature (103). The gene encoding AID is located on human chromosome 12 and contains five exons (104). The protein is expressed in germinal center B cells, and is essential for class switch recombination and somatic hypermutation during antibody gene diversification, which is found in all extant jawed vertebrates (105–108; see Chapter 2). As expected for a protein performing a highly conserved function, the d_N/d_S ratio generated by comparing human and chimp orthologs indicates strong purifying selection (109). Two variants of AID were identified in lamprey (*P. marinus*), a jawless vertebrate that emerged ~480 million years ago (110, 111). One variant, PmCDA1, represents a 208-residue protein from a single exon, whereas PmCDA2 is a 331-residue protein from four exons. Each of these genes exhibits a single characteristic ZDD signature motif. The longer PmCDA2 also contains an AT-hook motif in the C-terminus, which is a small DNA-binding motif comprised of Gly, Arg, and Pro residues that recognizes the minor groove of AT-rich DNA sequences (112, 113). Expression of these ancient proteins was detectable in lymphocytes from blood and hematopoietic tissues (110). While AID has now been identified in early vertebrates, it is absent in the protochordate, *Ciona intestinalis* (114). Thus, AID is currently believed to be the oldest APOBEC family member and emerged in early vertebrate evolution. As such, it is the most likely progenitor for the more recently evolved APOBEC1 and APOBEC3 families, as evidenced by the conservation of multiple exon boundaries within each of these genes (Figure 16.10) (31, 37).

The APOBEC2 structure represents the strongest model available to provide insight into AID. Prochnow et al. functionally examined a number of AID mutations using this assertion (52). AID is hypothesized to dimerize utilizing the essential region ^{47}G to ^{54}G (115). This corresponds with the β-sheet interface of APOBEC2 (Figure 16.10). Interestingly, mutants R112C and Y114A/F115A (analogous to the conserved ^{153}RL^{155}F patch described above) exhibited no deaminase activity (52). This patch has been hypothesized as important for tetramer formation in APOBEC2, but may also play a role in substrate recognition or regulation. Both AID and APOBEC3G have exhibited unique properties in the presence of RNA. AID has been shown to require RNAse treatment before deamination can occur on DNA (73). Based on the precedent of allosteric regulation in T4 dCD via nucleotide binding within the hexameric interface (Section 16.5.3) and given that both APOBEC2 and T4 dCD enzymes were characterized as dimers in gel filtration but crystallized as higher-order oligomers (52, 70), investigation of this interface as a site of regulation is warranted. If a tetramer interface of AID were a site of allosteric regulation, analogous to that of T4 dCD, the regulatory molecule would be in a prime location to influence active site architecture and substrate recognition.

R24E is another inactive AID mutant formulated from the APOBEC2 model (52). In APOBEC2, the equivalent residue (R65) forms a cation-π interaction with Y61 to stabilize the β-hairpin conformation observed in Loop1 of the outer monomers (described in Section 16.7.2). It is hypothesized that the R24E mutation inactivates the

AID enzyme by collapse of loop1 and subsequent inactivation of the active site, although further investigation will be required to definitively prove this assertion.

16.7.2 APOBEC2—A Divergent Ancestral Protein of Unknown Function

If AID is the oldest family member, then APOBEC2 is the penultimate member (116, 117). Discovered in 1999 during a search for APOBEC1 homologs, this protein has been identified in the teleostei, including pufferfish and zebrafish, but not in the chondrichthyes (cartilaginous fish) (31). Interestingly, the bony fish contain two paralogous copies of the APOBEC2 gene, presumably a product of the genome duplication that took place in the ray-finned fish (118, 119). It is not understood why there is only one copy of AID in these animals, but the AID and APOBEC2 genes are located on different chromosomes and gene deletion events are common in evolution. The APOBEC2 gene is located on human chromosome 6 and contains three exons (116). The core APOBEC fold, including all C-terminal helices, is encoded within the second exon; the first exon contributes the N-terminal 40-amino-acid residues that are absent from all other APOBEC family members (and were omitted from the crystallization construct). Together, the unique exon boundaries and sequence-driven phylogenetic analyses suggest that APOBEC2 and AID diverged prior to the evolution of APOBEC1 and APOBEC3 family members (Figures 16.3 and 16.10) (31).

Akin to AID, the d_N/d_S ratio calculated between human and orangutan orthologs of APOBEC2 indicates strong purifying selection (109). Robust protein expression was observed in cardiac and skeletal muscle tissues and, to a lesser extent, in a number of other tissues including kidney, liver, and small intestine (116, 117, 120). Expression in human liver cells is regulated by pro-inflammatory cytokines via NFκB activation, and two NFκB regulatory elements (NREs) were identified in the 5′ flanking region of the APOBEC2 gene (120). These results suggest that APOBEC2 has a function, although it has not been identified. Davidson and colleagues demonstrated that APOBEC2 could inhibit the RNA editing activity of APOBEC1 (117). However, deletion of the mouse ortholog exhibited no identifiable phenotype (121). Lastly, although multiple labs have reported low levels of deamination activity on free nucleotides in vitro, it has been suggested that this activity may be due to assay contamination (116, 117, 121).

Importantly, analysis of the APOBEC2 structure provides some insight into the apparent absence of detectable activity. Within the tetrameric form of APOBEC2, two distinct loop conformations were observed at the active site (52). The external subunits at the ends of the tetramer are characterized by a β-hairpin turn between helix α1 and strand β1 that allows for a more open active site. The Zn^{2+} ion of each outer subunit is coordinated to the His, Cys, and Cys residues of the ZDD motif, although the distance between Zn^{2+} and the second Cys residue is slightly longer than expected for an inner-sphere coordinated Zn^{2+}. Additionally, the fourth ligand for an active deaminase is water, which is present in only one of the two outer monomers and is also characterized by an unusually long Zn^{2+} coordination distance of 3.5 Å. These discrepancies are suggestive of local disorder. The arrangement of these outer subunits contrasts with the two internal subunits, in which Glu60 serves as the fourth Zn^{2+} ligand, thereby displacing the nucleophilic water and "inactivating" this subunit. Glu60 localizes to L1, which is more curved in the two internal subunits as a result of the encroaching tetramer interface (52). To clarify, it is Glu100, not Glu60, that is believed to be the conserved proton shuttle depicted in Figure 16.1. Interestingly, these inactivated subunits exhibit the expected Zn^{2+} coordination distances, suggesting greater definition of the atoms within the experimental electron density maps.

One possible reason for the suboptimal conformation of the outer, "active" subunits of APOBEC2 involves crystal packing contacts. The formation of the APOBEC2 crystal lattice involves a series of four hydrogen bonds to residues in the L1 regions of the outer subunits. The possibility that these interactions are the driving force behind the more structured versions of L1 cannot be dismissed, and it warrants further investigation using experimental approaches. Hence, it is plausible that there are only two active sites in the APOBEC2 tetramer, but it is also conceivable that neither is active in the solution state, which would account for the preponderance of biochemical evidence to date.

Predictions of the mode of substrate binding for APOBEC2 are purely speculative at present since the protein has no known substrate or biological editing activity (116, 117, 121). However, comparative structural analysis supports the hypothesis that C-terminal helices will contribute to substrate binding (Section 16.6.2). Additionally, the presence of a phosphate backbone in the substrate suggests that whatever the trajectory of the RNA, it is likely to be complemented by a trail of positively charged residues on APOBEC2. Indeed, such a trail is observed in maps of electrostatic surface potentials (Figure 16.7, bottom panel), prompting the hypothesis that an RNA strand might wrap around the outermost end of APOBEC2, nestled between helices $\alpha1$ and $\alpha6$ to contact basic residues in L1 and L7, then extending down the backside of helices $\alpha2$ and $\alpha3$ toward a patch of basic residues contributed by loops 2, 4, and 6.

Basic residues will occur on the surface of a protein for a number of reasons, not the least of which is to create the hydrophilic exterior necessary to maintain solubility. Likewise, aromatic residues readily engage in base-stacking with single-stranded nucleic acid (Figure 16.9E) and can be indicators of substrate binding when located on a protein surface; however, such residues may also contribute to protein–protein interactions. Such disparate possibilities demand a synergistic approach that implements structural and functional methods. Moreover, without a known function or assayable substrate, there is no way to directly test the molecular details of substrate recognition or to define them crystallographically. However, all is not lost. The shrewd investigator recognizes that amino acid sequence similarity equals structural similarity. As such, the next rational step is to extend knowledge of the APOBEC2 coordinates onto APOBEC family members better suited for experimental validation.

16.7.3 APOBEC1—The Historical Archetype of C-to-U Editing Enzymes

APOBEC1 was the first C-to-U editing enzyme to be characterized and therefore is the namesake of the family—that is, the **APO**lipoprotein **B** **E**diting **C**atalytic subunit 1 (122–125). In isolation, this enzyme exhibits strong mutator activity on ssDNA in a bacterial assay, as observed for several other family members (100). However, the endogenous substrate of APOBEC1 is a single cytidine within the ~14,000-nucleotide mRNA transcript that encodes the lipid transporter, apolipoprotein B. Deamination activity influences lipid transport, and it is achieved in the context of a 27S editosome ribonucleoprotein complex comprised minimally of APOBEC1, an obligate RNA binding auxiliary factor named APOBEC complementation factor (ACF), and the apolipoprotein B mRNA (Chapter 10) (126). The specificity of APOBEC1 for substrate is accomplished through ACF, which contains 3 RRMs (RNA recognition motifs) (127). Several groups have suggested that the apolipoprotein B mRNA substrate exhibits secondary structure that enhances recognition (128, 129); however, such structures appear to be melted away by the ACF cofactor such that APOBEC1 acts on a ssRNA substrate (129). Thus far, APOBEC1 is the only family member with an identified essential cofactor.

APOBEC1 editing is limited to the small intestine in humans (130). In mice and rats (and to a lesser extent, dogs and horses), editing also occurs in the liver due to the presence of three additional 5′-exon regions that encode a hepatic-specific promoter (131). Notably, the APOBEC1 variant present in these species is seven residues shorter on the C-terminus and the mRNA is more widely expressed, having been identified in spleen, kidney, lung, muscle, and heart. The APOBEC1 gene is located ~1 Mb from AID on chromosome 12 in humans (132, 133). The location of the rodent versions are syntenic with human chromosome 12, but are located only 30 kb apart and the transcriptional orientation of APOBEC1 is reversed. Genome comparisons suggest that a 1-Mb inversion containing the APOBEC1 locus occurred after the rodent/primate divergence (31), although it has not been indicated whether this inversion could be the underlying cause for loss of APOBEC1 expression in the human liver.

Current estimates of the age of APOBEC1 hinge on the observation of apoB mRNA editing in the American opossum (134), but not in amphibians or birds (130). This indicates that APOBEC1 is younger than either AID or APOBEC2, since marsupials diverged ~170 million years ago (135). Together, the conservation of sequence, exon junction boundaries, and chromosome assignment support the hypothesis that APOBEC1 evolved from AID (Figure 16.10). The more recent evolution of the APOBEC1 protein presents the possibility that its activity on RNA may be a uniquely evolved adaptation. In contrast, if the ancient AID were found to deaminate RNA, it would be a strong argument that RNA deamination is a conserved trait of the APOBEC family, which appears to be the case for ADARs. The d_N/d_S ratios for APOBEC1 vary considerably according to species; however, it is generally agreed that this protein has experienced a moderate level of positive selection, similar to that observed for the apoB protein itself (29, 131), and greater than that of either AID or APOBEC2 (109). Moreover, a comparison of human and orangutan orthologs suggests that current pressures are focused on the N-terminal 100 residues (109).

Investigations into the cellular trafficking patterns of APOBEC1 demonstrated that nuclear localization requires residues within the segments 1–56 and 97–172 and may be ACF-dependent, whereas a C-terminal leucine-rich region (173L–187L) is sufficient for cytoplasmic retention (136). This enzyme is believed to function as a dimer, and truncation studies suggest that the C-terminus is a key component of this interaction (137), although the N-terminus has also been implicated (138). If the C-terminal data are accurate, the tetrameric, helical interface of the APOBEC2 structure may be more relevant to APOBEC1 than the β-sheet interface. This highlights how variations on the themes of oligomerization and substrate recognition may contribute to the diverse activities of APOBEC enzymes.

A number of APOBEC1 truncation and point mutations generated by Teng and Navaratnam can now be mapped onto the APOBEC2 model for further consideration (137, 138). R16A and R17A mutants retain dimerization capability but are unable to crosslink or edit RNA (138). In APOBEC2, the equivalent residues localize before the start of helix 1 on the front face of the β-sheet. This patch is not likely to contribute to either oligomerization or catalysis, but may be involved in substrate recognition or stabilization of the long flexible Loop 10. The single-point mutations R33A and K34A do not significantly reduce editing; however, the double mutant edits RNA at levels only 3% of wild-type (137). Interestingly, R33 in APOBEC1 is analogous to R24 in AID (Section 16.7.1). It is not understood why R24E yields an inactive AID enzyme, whereas R33A APOBEC1 is fully functional. Regardless, this mutation is likely to contribute to substrate recognition or active site stabilization.

Other residues implicated in substrate recognition include F66 and F87, which abolish editing when mutated to Leu (139). Both of these residues are conserved across the APOBEC family; position 66 is always Phe, whereas position 87 is observed to be Phe or Tyr. In APOBEC2, the residue equivalent to F66 (F103) extends from the bottom of helix α2 into the cleft between helix 2 and helix 3. In this manner, it could contribute to either substrate recognition or the stability of the core fold, although the inability of F66L APOBEC1 to edit free nucleotide cytidine supports the latter of these two possibilities (139). The residue equivalent with F87 (Y122) extends from the front face of the β-sheet just below helix 5 in APOBEC2, where it most likely stabilizes the protein fold. I185 is another highly conserved residue that yields an inactive enzyme upon mutation to Ala in APOBEC1 (137). In APOBEC2, this residue appears to stabilize the orientation of C-terminal helix α6 with respect to helix α4. A number of additional mutations have been made to the C-terminus of APOBEC1; however, poor sequence conservation between family members in this region precludes rational prediction. The observation of numerous Pro and Leu residues in the C-terminus of APOBEC1 has generated hypotheses about protein–protein interactions, but it remains to be seen if these residues are important for recognition of ACF. As stated for APOBEC2, the identification of hydrophobic residues on a protein surface may be suggestive of protein–protein interactions. As such, the APOBEC2 coordinates provide a tenable starting point to initiate such an analysis.

16.7.4 APOBEC4—Pushing the Envelope of APOBEC Boundaries

APOBEC4 is the least well-studied member of the APOBEC family and was identified through an iterative database search (28). The protein has no known function, although EST analysis indicated expression in testes. As reported in the Ensembl database, this protein is found on chromosome 1 and contains two exons. The ORF is confined to exon 2 and is predicted to encode a protein of 367 amino acids. The length of this sequence is suggestive of a double deaminase, but there is a single copy of the ZDD consensus motif; interestingly, the HxE sequence exists as HPE in both APOBEC4 and the noncatalytic N-terminal deaminase domain of APOBEC3G. Additionally, the PCxxC motif involved in Zn^{2+} coordination has been altered to PCx_6C in APOBEC4, which calls into question the assignment of this protein as a "canonical" APOBEC family member. Sequence motifs conserved within the APOBEC family are also present in APOBEC4, although most are distinguished by conservative mutations. For example, the $^{153}RL^{155}F$ motif found in APOBEC2 is present in human APOBEC4 as $^{156}QL^{158}Y$. Analysis of the d_N/d_S ratio between mouse and human orthologs indicate modest selective pressure, similar to that of APOBEC1 (28). The gene for APOBEC4 has been identified in mammals, frogs, and chickens, but not in fish, suggesting that it emerged after AID and APOBEC2 but prior to APOBEC1 and APOBEC3. Further research will be necessary to clarify the role of this protein and its relationship to other family members.

16.8 APOBEC3—RADIATIVE EXPANSION OF PROTEINS INVOLVED IN VIRAL DEFENSE

The APOBEC3 genes have received extensive investigative interest due to the exciting discovery that many proteins function as defenders against retroviruses and retrotransposons (Chapter 11). In light of this significance, along with the multiple members that comprise the APOBEC3 family, the remainder of this chapter will focus on these proteins. The emphasis will be upon understanding the origin and sequence relationships between family members, which is paramount to undertaking a comparative modeling analysis. Although their exact mechanisms of action are unknown, it appears that the efficacy of some APOBEC3 family members in restricting viral infectivity is due to C-to-U deaminase activity that targets viral genome intermediates resulting from reverse transcription (140–142). A deaminase-independent activity has been invoked as well (143–145). The expression profile further supports a protective role. APOBEC3 proteins are most commonly and abundantly expressed in lymphoid tissues, which are relevant to retroviral replication, and in the germline where genome restricted mobile genetic elements need to transpose to ensure generational survival (37, 102, 146–150).

Among the APOBEC family members, the APOBEC3 genes are the most recently evolved and appear to be confined to mammals. To date, a single APOBEC3 gene has been identified in rodents, cats, pigs, and sheep, two in cows, three in dogs and horses, and eight in humans (149). There is, as yet, no standard nomenclature for the APOBEC3 genes identified in lower mammals. However, in humans the genes are labeled alphabetically in order of occurrence. A–G (including D/E, which are now considered one gene) are arranged tandemly on chromosome 22, followed 14 kb downstream by APOBEC3H (17, 31, 37). Lastly, a pseudo-gene devoid of introns [the likely product of a retrotransposition event (151)] is hypothesized to be the most recent duplication of APOBEC3G and was identified on chromosome 12 (31). Given current estimates of mammalian evolution (152) and the absence of APOBEC3 in avian lineages (153), the window of emergence for the APOBEC3 gene is between 100 and

300 million years ago. Based on sequence homology, exon boundaries, and chromosomal location, AID is the suggested parent of this family (31). Furthermore, the respective sites on human chromosomes 12 and 22 where the AID and APOBEC3 genes reside have been hypothesized to originate from the same ancestral chromosome (154).

Sequence-based phylogenetic trees reveal three distinct groupings of the APOBEC3 CDA domain, referred to as Z1a, Z1b, and Z2. The amino acid sequence characteristics driving this classification have not been systematically documented, although one example is the four-residue insertion exclusive to loop 1 of Z2 domains (Figure 16.10, C term of mouse A3 and macaque A3H located between $\alpha1$ and $\beta1$). Notably, all three APOBEC3 domain forms are present in dogs, horses, cows, and humans (149), indicating that three forms of APOBEC3 emerged early on in the evolution of this gene. Furthermore, this division would seem to support unique functionalities for each subtype, such as modes of oligomerization or substrate recognition, based on the general principles observed for the CDA superfamily. Of unknown significance is the observation that in all species to date, the Z2 domain is present in a single copy, either as the C-terminal portion of a fusion gene (rodents, cows, and pigs) or as a single domain (dogs, horses and humans) (Figure 16.11) (149). In hominids, including the gorilla, chimpanzee, and human, the sole Z2 gene representative (i.e., APOBEC3H) exhibits poor ssDNA deaminase activity, resulting from a truncation mutation that occurred sometime after the separation of gibbons from other hominids ~18 million years ago (149, 155, 156). Notably, a fully active -3H that efficiently hypermutates retroviral genomes is present in Old World monkeys including the macaque and sooty mangabey (149).

The APOBEC3 proteins are the only members of the APOBEC family that are found in both single-CDA and double-CDA forms. Akin to the *E. coli* CDA enzyme (Section 16.4.1), a gene duplication event within the APOBEC3 cluster is hypothesized to have produced a single polypeptide chain that encodes the equivalent of an AID dimer within a single subunit (Figure 16.10). Such double-CDA domain APOBEC3 variants are present in rodents, sheep, cows, and pigs, as well as primate -B, -D/E, -F, and -G forms (37, 147, 157, 158). Single-domain forms of APOBEC3 genes are also present, as in the examples of cat, dog, and horse (Figure 16.11), as well as the primate forms -A, -C, and -H (37, 149, 159). There is no correlation between the type (Z1a, Z1b, Z2) of APOBEC3 CDA domain and its occurrence as either (a) a single-domain CDA protein or (b) the N-terminal or C-terminal portion of a double-CDA protein. Several of the double-CDA forms of APOBEC3 exhibit deaminase activity in only one CDA domain. In mice (Z1a, Z2), the N-terminal domain is catalytically active (160), whereas in human APOBEC3G (Z1a, Z1b) the C-terminal domain is catalytically active (38–40). In human APOBEC3B (Z1a, Z1b), the only double-CDA domain protein that traffics to the nucleus, both domains exhibit deaminase activity (145, 158, 161, 162). Thus, there is no apparent correlation between domain type and catalytic activity. Oligomerization may contribute to the differential activity of these domains, given the apparent inactivation of the APOBEC2 active site at the tetramer interface (52). Notably, lack of deaminase activity is not necessarily an indicator of functional relevance, as deaminase-independent viral inhibition has been reported (143, 163).

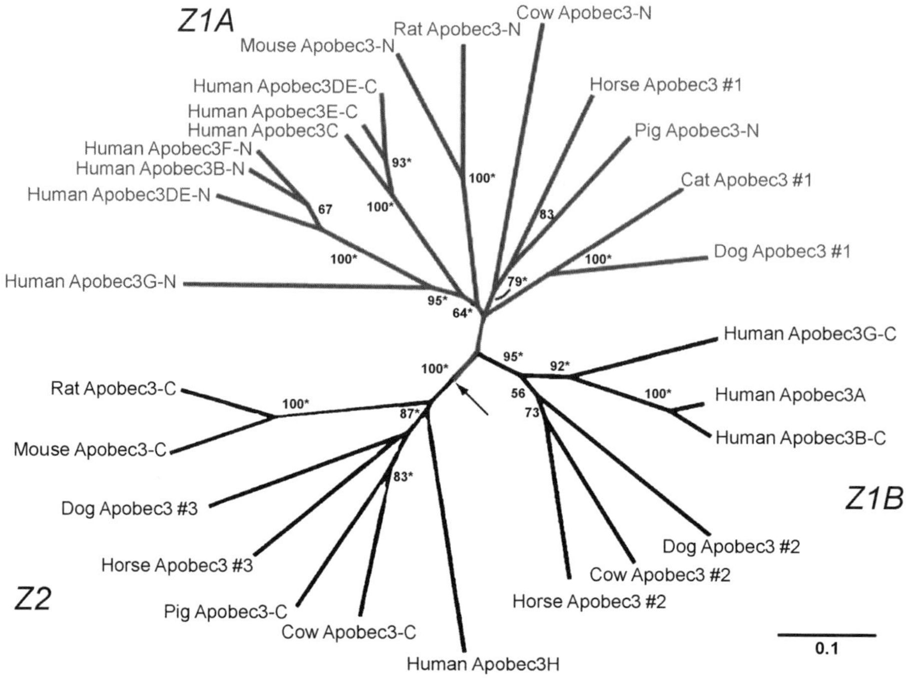

Figure 16.11 Phylogenetic tree of APOBEC3 family members illustrating the division into three main 'Z' groups. A neighbor-joining tree was constructed in CLUSTAL_X based on a protein alignment of APOBEC3 sequences from humans and nonprimate eutherian (placental) mammals. The N- and C-terminal domains of double-domain APOBEC3 proteins have been split and are designated -N and -C, respectively. APOBEC3 proteins from mammals in which more than one APOBEC3 was identified are numbered. An APOBEC3H-like domain has been conserved in a number of mammalian species, including mice, rats, dogs, horses, pigs, cows, and primates. This phylogeny is rooted with the most closely related enzyme, AID, as an outgroup lineage; placement of this root is shown by an arrow. The scale bar represents a mutational frequency of 0.1 substitutions per residue. Bootstrap support for the groupings is indicated by numbers next to the relevant branches; those nodes supported by a Bayesian maximum-likelihood analysis are indicated with an asterisk. This figure was reproduced with permission from the American Society for Microbiology and originally appeared in reference 149.

16.8.1 Mechanisms of Primate-Specific Expansion of the APOBEC3 Proteins

The CDA structural domains of the tandemly arranged APOBEC3A-G are either Z1a or Z1b (31, 149) and occur in the same transcriptional orientation. Based on conservation of both noncoding nucleotide sequence and protein sequence, these genes are paralogs—that is, the result of local duplication events (31, 37). For example, APOBEC3C is highly related to the C-terminal domain of APOBEC3F. Similarly, the single-CDA domain APOBEC3A is the result of a partial gene duplication of the double-CDA protein APOBEC3B. APOBEC3A has a unique exon 1 derived from the adjacent chromosome 22 sequence, but the next four exons are derived from the last three exons of APOBEC3B (37, 164). In fact, a 29.5-kB gene deletion polymorphism described by Eichler and colleagues (165) eliminates the C-terminal portion of APOBEC3A and most of APOBEC3B. However, due to the homologous nature of these regions, the resultant mRNA encodes an intact APOBEC3A followed by the 3′-UTR of APOBEC3B. This deletion is rare in Africans and Europeans, more common in Asians, and nearly fixed in the Oceanic populations of the Far East. The implications of this polymorphism with respect to human disease are unknown at present.

Based on the observation of unique Z-type architecture, it seems likely that the initial duplication event that gave rise to the double-domain proteins in primates is independent of that which occurred in lower mammals. All double-CDA APOBEC3 proteins in lower mammals exhibit a Z1a, Z2 architecture; the only Z2 present in primates is APOBEC3H. In contrast, it has been hypothesized that the first double-CDA APOBEC3 protein generated in the primate locus was duplicated in its entirety to generate additional double-CDA APOBEC3 genes (37). This is supported by the conservation of several repeated sequences present in the 5′ region as well as within intron 4 (i.e., the intron that separates the two CDA domains) (37). Repeats and duplications of any kind, once established, promote further rearrangement through their own misalignment and subsequent recombination (166–168). These misalignments can result in further duplications or deletions. Thus, a chromosome with two or three duplicated APOBEC3 genes is more likely to experience further remodeling events.

The extensive expansion of APOBEC3 genes is believed to be exclusive to primates, but is not necessarily identical in all primates (149). One indicator of how this process may have conferred a selective advantage is the dramatic decline in retrotransposon activity observed in primates over the past 35–50 million years (169). This trend is not seen in the mouse genome, presently believed to encode only one APOBEC3 enzyme, where a number of transposable elements are still active. In fact, it is estimated that about 1 in 600 mutations in the human genome are caused by retrotransposons, whereas in mice the odds are 1 in 10 (170). The discussion of retrotransposons brings up a rather ironic scenario. APOBEC3 proteins may have evolved to curb the very mechanism that facilitated their initial expansion. Primate genomes, as a whole, experience a greater incidence of segmental duplications compared to other mammals (171), a finding that has been partially attributed to Alu repeat elements (166). Significantly, both Alu and LINE repeat elements were identified in the BAC from which APOBEC3 genes were originally identified (37). In fact, chromosome 22 exhibits the highest incidence of segmental duplications among the autosomal chromosomes (171), as well as the third highest percentage of sequence encoding Alu elements (18%) (172). Not surprisingly, this chromosome contains a number of gene families, in addition to APOBEC3, that have diversified via duplication events (164, 173). Lastly, the current address of the APOBEC3 locus at 22q13.1 is believed to have resulted from a double translocation event that occurred between human chromosomes 12 and 22 subsequent to the divergence of anthropoids from prosimians (~50 million years ago) (174). Consideration of such relationships is common among structural biologists since they are useful in the generation of homology models. As such, it is apparent that the duplicated nature of APOBEC3

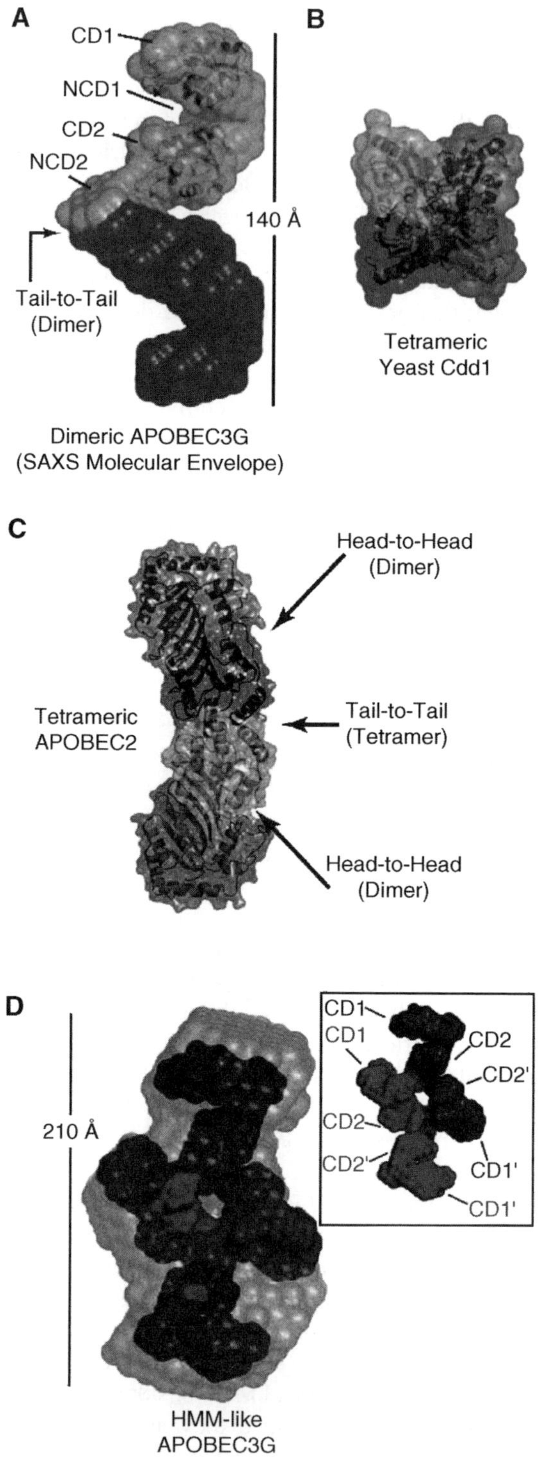
A
CD1
NCD1
CD2
NCD2
Tail-to-Tail
(Dimer)
140 Å
Dimeric APOBEC3G
(SAXS Molecular Envelope)
B
Tetrameric
Yeast Cdd1
C
Head-to-Head
(Dimer)
Tetrameric
APOBEC2
Tail-to-Tail
(Tetramer)
Head-to-Head
(Dimer)
D
210 Å
CD1
CD1
CD2
CD2'
CD2
CD2'
CD1'
CD1'
HMM-like
APOBEC3G

family members precludes a facile description of their three-dimensional structures without a more sophisticated experimentally defined starting model than APOBEC2. This observation is especially true for a tandem CDA protein such as APOBEC3G.

16.8.2 Alternative Methods to Obtain Structure: The Molecular Envelope of APOBEC3G by Small-Angle X-Ray Scattering

Although the use of X-ray crystallography to define three-dimensional structures of APOBEC3 family members represents the "gold standard" in the field, other biophysical approaches can provide low-resolution structural information that can contribute to understanding the shape and oligomeric state of a protein in solution. As such, Wedekind and Smith and colleagues restored the molecular envelope of human APOBEC3G by use of small-angle X-ray scattering (SAXS) methods (175). This technique provided the global shape of APOBEC3G in solution, at ~3-nanometer resolution, thus making it possible to detect where specific CDA structural domains are located based on the volumetric distribution of the particle relative to known structures (e.g., Figures 16.5–16.8). Because RNA binding is an inherent property of APOBEC3G, it was necessary to digest samples with ribonuclease A during the purification process (similar to findings with AID). The SAXS results from RNA depleted samples revealed that the molecule adopts a 140-Å-long extended shape in solution (175) (Figure 16.12A). The SAXS data, as well as gel filtration and dynamic light scattering, suggested that APOBEC3G forms a homodimer in solution of ~95 kD (175). These observations implied that previously determined fnCDA structures (Figure 16.5) were unsuitable as templates for APOBEC3G comparative modeling, since their oligomeric subunit arrangements are distinctly "square" in shape (Figure 16.12A versus Figure 16.12B) with dimensions of 55 Å by 55 Å by 40 Å. By comparison, the elongated large–small–large–small volume distribution of APOBEC3G is evident based upon visual inspection of the SAXS envelope (Figure 16.12A). Therefore, although an fnCDA tetramer did not fit into the electron density for the SAXS dimer, simple volumetric arguments suggested that an isolated

Figure 16.12 The global molecular envelope for APOBEC3G based on shape restorations from SAXS, and subunit interactions of crystallographically defined cytidine deaminases. (**A**) APOBEC3G (RNase treated) at 3.4-nm resolution. The putative subunit boundaries are indicated in green and magenta. Possible domains and subdomains are labeled CD1, NCD1, CD2, and NCD2. Individual CDA domains from part B are docked into the envelope at CD1 and CD2. (**B**) Transparent surface model and underlying ribbon representation of fnCDA Cdd1 from yeast defined crystallographically (PDB entry 1R5T). Individual CDA domains are colored green, red, blue, and purple. (**C**) Transparent surface model and underlying ribbon diagram for the APOBEC2 crystal structure (PDB entry 2NYT). The scale bar in part A applies to parts A–C. (**D**) The molecular envelope restoration for a minimal, HMM-like variant of APOBEC3G. Computational searches led to a "dimer-of-dimers" configuration of LMM-like subunits within the HMM molecular boundary (cyan). The remaining molecular envelope may comprise RNA, which is hypothesized to bridge neighboring subunits via an intermolecular CD1 (red) to CD1 (purple) interaction (*inset*), thereby sequestering the DNA deaminase sites at CD2. (*See color insert.*)

CDA domain (e.g., Figure 16.12B, green subunit) could be fit into the SAXS envelope at either of two distinct locations within a single APOBEC3G subunit (Figure 16.12A, ribbon diagrams at CD1 and CD2). Of course, the exact orientation of the CDA domains is unknown due to the resolution limits of the SAXS method. However, this model is appealing because it agrees with the tandem CDA domain composition apparent in the linear amino acid sequence (Figure 16.10). As such, the current working model for APOBEC3G is that the individual APOBEC3G subunits engage in a tail-to-tail interface with a domain/subdomain distribution comprising: (i) an N-terminal, CDA-like catalytic domain (CD1), (ii) an N-terminal noncatalytic domain (NCD1) with no known structural homology, (iii) a second, C-terminal CDA (CD2), and (iv) a C-terminal noncatalytic domain (NCD2) comparable to NCD1 (Figure 16.12A).

Following the SAXS analysis of APOBEC3G, Goodman and Chen reported the crystal structure of APOBEC2 (52). Importantly, this molecule forms an elongated 127-Å tetramer featuring a tail-to-tail subunit interaction analogous to that proposed for the APOBEC3G SAXS envelope (Figure 16.12C and Section 16.7.2). At 184 amino acids, the APOBEC2 crystal structure is <50% of the length of APOBEC3G (Figure 16.10). However, the fact that the tandem ZDD motifs of APOBEC3G arose by gene duplication of an ancestral AID-like gene (Section 16.8) makes it possible to align each of its N- and C-terminal CDA domains with an APOBEC2 subunit (Figure 16.10). Despite this primary structure similarity, the APOBEC2 tetramer, which appears cylindrical, does not fit readily into the boundaries of the dimeric APOBEC3G molecular envelope (Figure 16.12A versus Figure 16.12C). This difference may suggest a fundamental difference in the orientation of hA3G domains and subunits compared to APOBEC2, or could be a consequence of the SAXS method itself, which is representative of a solution structural ensemble (175). Hence, although the level of sequence identity suggests that the core ZDD fold of each APOBEC3G domain is represented by the APOBEC2 subunit, modes of substrate binding, dimerization, and Vif interactions cannot be deduced for APOBEC3G at present and will require high-resolution crystallographic studies.

16.8.3 Characterization of Structural Changes in APOBEC3G Morphology in the Presence of RNA

A somewhat controversial aspect of the APOBEC3G story is the observation that the enzyme has been isolated from activated CD4+T cells as an HMM (high molecular mass) ribonucleoprotein complex (176–178). The dC-to-dU deaminase activity of such species is nominal, but can be restored in vitro by digestion with ribonuclease, which coincides with formation of enzymatically active, LMM (low molecular mass) variants whose mass is consistent with a monomer or dimer (~47–95 kD) (176). As such, it has been hypothesized that APOBEC3G binding to RNA, and subsequent formation of HMM complexes from LMM variants, serves to switch off its deaminase-dependent antiviral activity (177, 179). Although this phenomenon appears to contradict the antiretroviral activity of this enzyme (Chapter 11), it has been demonstrated that HMM complexes incorporate HIV RNA, as well as nonautologous Alu and retrotransposable hY RNAs (177). This RNA-binding activity may have evolved to inhibit retro-element reverse transcription and retrotransposition into the primate genome (180), rather than as a means to combat lentiviruses. The observation of an HMM complex also supports evidence for the observed deaminase-independent antiviral activity of APOBEC3G (143), which may simply be a consequence of the enzyme's innate ability to bind nonspecifically to ssRNA (181).

Genuine HMM complexes are localized within cellular P-bodies and stress granules that comprise multiple proteins, many of which are associated casually with APOBEC3G via RNA binding as ribonucleoprotein complexes (177, 181). Nonetheless, HMM variants of reduced complexity, ranging in size from 5 to 15 MD, have been isolated from CD4 + T cells using gel filtration chromatography (176). HMM-like complexes from recombinant sources display similar size distributions and deaminase activity in vitro, and these complexes can be reduced to deaminase-active LMM-like species via RNase A treatment (175, 176, 182, 183), making them tractable targets for structural and functional investigations.

To address the question of how RNA affects the morphology and deaminase activity of APOBEC3G, Wedekind and Smith collaborated to examine a minimal, HMM-like form by SAXS (175). The results revealed a significantly larger particle than the LMM-like sample with a maximum dimension of 210 Å (Figure 16.12A versus Figure 16.12D). The results of a circular dichroism analysis of LMM- versus HMM-like variants indicated no significant changes in the overall protein secondary structure composition in the presence of RNA, implying that RNA did not significantly alter the global protein fold (175). Efforts to fit envelopes of LMM-like dimers into the HMM-like electron density utilized six-dimensional computational searches. This effort yielded two possible packing configurations, one of which is depicted in Figure 16.12D. In both instances, optimal packing solutions suggested the possibility that APOBEC3G associates symmetrically with itself as a dimer of dimers (Figure 16.12D, inset). Work by Landau and colleagues identified previously that the N-terminal CDA domain (CD1) of APOBEC3G binds RNA, whereas the C-terminal CDA (CD2) deaminates ssDNA (40). If the proposed mode of packing in the minimal HMM particle is correct, it implies that the N-terminal CDA domains of closely packed LMM-like particles (i.e., the dimer of dimers) could be bridged via intermolecular RNA interactions. Coalescence of multiple HMM-like particles would result in sequestration of CD2 domains responsible for dC-to-dU deaminase activity. Thus, inactivation of APOBEC3G deaminase function by RNA and formation of HMM complexes may simply be a property of locking the enzyme into an inactive conformation through molecular crowding and steric overlap at the active site. RNase degradation would necessarily break the RNA crosslinks and free each active site to engage in substrate deamination. Importantly, the HMM-like model also predicts the presence of discrete protein–protein interaction sites between APOBEC3G molecules (Figure 16.12D, inset), but these interactions alone may be insufficient to form a core (stable) structure without RNA binding.

Whether the HMM-like molecular envelope from SAXS is representative of a bona fide cellular complex has yet to be determined. However, recent studies provide evidence for the existence of minimal HMM-like ribonucleprotein complexes inside the viral capsid. Pulse labeling suggested that newly translated APOBEC3G exists in the cell in the LMM form for only a short period ($t_{1/2} = 20$ min) before interacting with cellular RNA to become HMM (74). This investigation also implied that only the LMM form of APOBEC3G interacts with viral components that become integrated into viral particles. Interestingly, LMM APOBEC3G in viral particles is HMM-like in that it is associated with the RNA genome and exhibits reduced deaminase activity. Moreover, the deaminase activity of this Intra-Virion APOBEC3G Complex (IVAC) is recovered only after post-entry viral replication. The authors suggested that restoration of deaminase activity is the result of the RNase H activity of HIV reverse transcriptase, which removes viral genomic RNA from RNA–DNA duplexes. This investigation further contends that prior work in which APOBEC3G was isolated with viral particles, but did exhibit deaminase activity on ssDNA in vitro, was caused by enzyme trapped within the viral particle, but outside the core assembly (i.e., not bound to viral RNA). This condition was attributed to overexpression of APOBEC3G in producer cells. Collectively, these data suggest that conversion of minimal HMM to LMM variants in vivo would promote APOBEC3G antiviral deaminase activity and deaminase-independent activity through binding to reverse transcription initiation and elongation complexes (184, 185). Again, the relevance of the IVAC with respect to the HMM-like particles from SAXS (Figure 16.12D) is uncertain, but the results of the biophysical characterization are consistent with a protein that undergoes morphological changes that are modulated by RNA, but do not impact the global fold of the fundamental protein structure (175).

It is not yet known if this RNA-mediated HMM/LMM modulation is common to other APOBEC3 family members. Nor is it known if APOBEC3 family members will share a single mode of oligomerization. APOBEC3G is the first APOBEC3 enzyme to be analyzed from a biophysical perspective. Prolific duplication events, such as that observed for APOBEC3 enzymes, have been hypothesized to play an important role in primate adaptive evolution, particularly when a diversity of responses was advantageous (171). Thus, we might expect each enzyme to exhibit unique features with regard to both its structure and function. With so many versions present, the forces of selection can tailor each APOBEC3 enzyme to fulfill a specific niche.

16.8.4 Positive Selection Exerted on the APOBEC3 Family

Studies of the selective pressures on the APOBEC3 family have also focused on APOBEC3G, as this was the enzyme first shown to diminish HIV infectivity (186). Protein alignments of the APOBEC3G sequence among primate species demonstrated high sequence variation, thereby prompting an analysis of the evolutionary forces that influenced this protein. Positive selection is often observed for host proteins that are directly involved in pathogen defense (187), as suggested for a number of the APOBEC3 family members. This type of genetic conflict between host and pathogen drives rapid change in interacting host and pathogen proteins as they attempt to increase or decrease interactions with one another in the battle for host cell dominance. As assessed across several primate species by the d_N/d_S ratio, the APOBEC3G enzyme has been under strong positive selection for at least 33 million years (109, 149, 169). Actually, with the sole exception of APOBEC3A, all human APOBEC3 enzymes exhibit some degree of positive selection as compared to chimpanzee. APOBEC3A, akin to AID, is under purifying selection (109). The d_N/d_S ratio for APOBEC3H is on par with that of A3G over a wide range of primate species, implying that these proteins have experienced similar evolutionary influences (149). APOBEC3D/E, as compared between humans and chimpanzees, is the only example that appears to experience greater selective pressure than APOBEC3G. This pressure is specific to the "E" domain. When considered independently, the "E" domain exhibits the highest d_N/d_S ratio of any APOBEC3 protein, whereas the "D" domain experiences mild positive selection comparable to that of APOBEC1 (109).

The take-home message of these results is that most APOBEC3 family members show evidence of positive selection, with the possible exception of APOBEC3A. This would suggest that each protein has evolved on a species-specific basis and, moreover, that the needs of each species are unique. In other words, there are two aspects to the selection observed in APOBEC3 family members. Upon a gene duplication event, there is the evolution of a specific activity for each family member. That is not observed in an analysis of d_N/d_S ratios, because it likely occurred prior to the speciation events compared here. Rather, the positive selection described here is more supportive of the species-specific adaptations (e.g., human APOBEC3B versus chimpanzee APOBEC3B).

Analysis of d_N/d_S ratios are often more informative when generated for small regions of a sequence because, as established previously, not all regions of the CDA fold are equally amenable to mutation. Such sliding-window analyses of APOBEC3G suggest that pressures on APOBEC3G are widely dispersed throughout its sequence (109, 169). Specifically, it has been estimated that 30% of the APOBEC3G protein residues have undergone strong positive selection (109). The proportion of nonsynonymous substitutions exceeds that of synonymous substitutions in all but one region of the sequence (169). Reports of the strongest site of selection differ according to methods of analysis and species selected for comparison (109, 169, 188). Together, they indicate that episodes of genetic conflict have applied pressure throughout the APOBEC3G sequence. This pressure is observed in a variety of primate species and is not exclusive to those susceptible to HIV/SIV infection (169).

Such dispersed pressure supports Zhang's hypothesis that APOBEC3G is experiencing diversifying positive selection rather than directional selection (169). Directional selection occurs when natural selection favors a single allele, whereas diversifying selection may favor multiple alleles. This hypothesis was based on the observation that the HIV Vif infectivity factor that targets APOBEC3G for proteasomal degradation must evolve specific structures and sequences to bind APOBEC3G (directional selection), but APOBEC3G need only acquire changes that abolish viral recognition while maintaining deaminase activity (diversifying selection). Such escape behavior by APOBEC3G is evidenced by the preference for charge reversal observed in nonsynonymous substitutions (169). A specific example is the observation that Asp128 in human APOBEC3G co-varies with the first basic amino acid in the HIV Vif sequence ^{14}D $\underline{R}$MR^{17}. Asp128 maps to Loop7 of the APOBEC2 structure (Figure 16.10); the equivalent APOBEC2 residue (E159) engages in a salt bridge across the tetramer interface (52). The comparable position in APOBEC3G from several nonhuman primates such as African green monkey is Lys(128), which interacts with SIV Vif through the equivalent sequence $^{(14)}$SERQ$^{(17)}$ (194). As an analogy for escape behavior, when a fox (Vif) chases a rabbit (APOBEC3G), the rabbit can choose from a number of escape routes; the fox has only one route, that of the rabbit. This observation is a reversal of the typical roles played by pathogen and host and perhaps is the reason why APOBEC3G continues to maintain a tandem CDA domain structure (i.e., one is a decoy for Vif at the N-terminus, whereas the other C-terminal CDA deaminates ssDNA).

In light of these correlations between HIV and APOBEC3G, it is important to note that the evidence of APOBEC3G selection far precedes the emergence of modern lentiviruses in the primate population (109), which can be traced back only as far as ~7 million years (195). Thus, although APOBEC3G may currently be "running" from the HIV Vif protein, this is only a recent phenomenon on the evolutionary timescale. Indeed, APOBEC3G exhibits antiviral activity against other, non-HIV viruses, such as the hepatitis B virus (140). The very nature of APOBEC3G's antiviral dC-to-dU deaminase activity supports the hypothesis of a broad target range. While each APOBEC3 enzyme exhibits a preference for the nucleotide neighbors nearest the editing site (e.g., APOBEC3G prefers the context 5′-<u>C</u>CA-3′) (196, 197), no specific consensus sequence has been identified that is equivalent to the APOBEC1 mooring sequence (Section 16.7.3).

16.9 CONCLUSIONS AND FUTURE PROSPECTS

It is appropriate to reflect on the generalities that are now apparent regarding the chemistry, phylogeny, and three-dimensional structure of the APOBEC protein family. A hierarchical examination revealed that the ZDD signature motif, (H/C)xEx$_{25-30}$PCxxC, is highly conserved throughout evolution in terms of its primary and tertiary structures. Phylogenetic analysis revealed that the perpetuation of the ZDD motif in numerous species is attributable to the successful spatial presentation of key amino acids, as well as a Lewis acid Zn^{2+} ion, that form the basis of a robust hydrolytic deamination module. The ZDD motif itself is incapable of adopting the requisite three-dimensional structure for function, and therefore it must be supported by a larger self-organizing CDA domain of the form α1β1β2α2β3α3β4β5 with the conserved β-strand topology 2-1-3-4-5, in which helices α2 and α3 coordinate the essential Zn^{2+}. Changes in the direction of the C-terminal β-strand, as well as deletions and insertions to loop elements that join core secondary structure elements, appear to influence active site accessibility and substrate binding. Adaptations have culminated in specialized enzymes that act on free nucleobases and nucleosides, as well as on nucleotides embedded within single-stranded DNA or RNA substrates. The mode of subunit oligomerization plays a major role in substrate binding, activity, and regulation. However, despite a rather divergent evolution, the core CDA domain remains intact, thus providing consistency with respect to the stereochemistry and regiospecificity of the overall reaction. One exciting result from the human genome project was the discovery of the APOBEC3 subfamily of CDAs, which exhibit antiretroviral activity that most likely evolved to quell the explosion in retrotransposable elements in primates ~40 million years ago. At present, some family members, such as APOBEC3F and APOBEC3G, exhibit an innate anti-HIV activity that manifests as a dC-to-dU hypermutation in viral DNA. The challenge of the next generation of scientists will be to develop 'shields' for these natural defense proteins that protect them from proteasomal degradation by masking their binding to viral proteins, such as the HIV Vif factor. To accomplish this task, it will be necessary to understand the molecular basis of host–viral protein interactions in order to target the viral binding site while leaving the host protein function intact. With this challenge, we return to Crick's adage, "*If you want to understand function, study structure.*"

REFERENCES

1. Kotani, A., et al. (2007)Activation-induced cytidine deaminase (AID) promotes B cell lymphomagenesis in Emu-cmyc transgenic mice. *Proc Natl Acad Sci USA* **104** (5),1616–1620.
2. Kotani, A., et al. (2005)A target selection of somatic hypermutations is regulated similarly between T and B cells upon activation-induced cytidine deaminase expression. *Proc Natl Acad Sci USA***102** (12),4506–4511.
3. Okazaki, I. M., et al. (2003) Constitutive expression of AID leads to tumorigenesis. *J Exp Med* **197** (9), 1173–1181.
4. Ramiro, A. R., et al. (2004) AID is required for c-myc/IgH chromosome translocations in vivo. *Cell* **118** (4), 431–438.
5. Yamanaka, S., et al. (1995)Apolipoprotein B mRNA-editing protein induces hepatocellular carcinoma and dysplasia in transgenic animals. *Proc Natl Acad Sci USA* **92** (18),8483–8487.
6. Yamanaka, S., et al. (1997) A novel translational repressor mRNA is edited extensively in livers containing tumors caused by the transgene expression of the apoB mRNA-editing enzyme. *Genes Dev* **11** (3), 321–333.
7. Skuse, G. R., et al. (1996) The neurofibromatosis type I messenger RNA undergoes base-modification RNA editing. *Nucleic Acids Res* **24** (3), 478–485.
8. Durandy, A. and Honjo, T. (2001) Human genetic defects in class-switch recombination (hyper-IgM syndromes). *Curr Opin Immunol* **13** (5), 543–548.
9. Revy, P., et al. (2000) Activation-induced cytidine deaminase (AID) deficiency causes the autosomal recessive form of the Hyper-IgM syndrome (HIGM2). *Cell* **102** (5), 565–575.
10. Zhu, Y., et al. (2003) Type two hyper-IgM syndrome caused by mutation in activation-induced cytidine deaminase. *J Med Dent Sci* **50** (1), 41–46.
11. Smith, A. A., et al. (1994) Mutations affecting transition-state stabilization by residues coordinating zinc at the active site of cytidine deaminase. *Biochemistry* **33** (21), 6468–6474.
12. Frick, L., et al. (1989) Binding of pyrimidin-2-one ribonucleoside by cytidine deaminase as the transition-state analogue 3, 4-dihydrouridine and the contribution of the 4-hydroxyl group to its binding affinity. *Biochemistry* **28** (24), 9423–9430.
13. Xiang, S., et al. (1995) Transition-state selectivity for a single hydroxyl group during catalysis by cytidine deaminase. *Biochemistry* **34** (14), 4516–4523.
14. Betts, L., et al. (1994) Cytidine deaminase. The 2. 3 Å crystal structure of an enzyme: Transition-state analog complex. *J Mol Biol* **235** (2), 635–656.
15. Carlow, D. C., et al. (1995) Major contribution of a carboxymethyl group to transition-state stabilization by cytidine deaminase: Mutation and rescue. *Biochemistry* **34** (13), 4220–4224.
16. Mian, I. S., et al. (1998) Statistical modelling and phylogenetic analysis of a deaminase domain. *J Comput Biol* **5** (1), 57–72.
17. Wedekind, J. E., et al. (2003) Messenger RNA editing in mammals: New members of the APOBEC family seeking roles in the family business. *Trends Genet* **19** (4), 207–216.
18. Xiang, S., et al. (1996) Cytidine deaminase complexed to 3-deazacytidine: A "valence buffer" in zinc enzyme catalysis. *Biochemistry* **35** (5), 1335–1341.
19. Berg, J. M., Tymoczko, J. L., and Stryer, L. (2002) Biochemistry, 5th ed., W.H. Freeman and Company, New York, p. 894.
20. Xie, K., et al. (2004) The structure of a yeast RNA-editing deaminase provide insight into the fold and function of activation-induced deaminase and APOBEC-1. *Proc Natl Acad Sci USA* **101** (21), 8114–8119.

21. Chothia, C., and Lesk, A. M. (1986) The relation between the divergence of sequence and structure in proteins. *EMBO J* **5** (4), 823–826.
22. Marti-Renom, M. A., et al. (2000) Comparative protein structure modeling of genes and genomes. *Annu Rev Biophys Biomol Struct* **29**, 291–325.
23. Berman, H. M., et al. (2000) The Protein Data Bank. *Nucleic Acids Res* **28** (1), 235–242.
24. Harutyunyan, E.H., Safonova, T.N., Kuranova, I.P., Popov, A.N., Teplyakov, A.V., Obmolova, G.V., Rusakov, A.A., Vainshtein, B.K., Dodson, G.G., Wilson, J.C., and Perutz, M.F. (1995) The structure of deoxy– and oxy–leghaemoglobin from lupin. *J Mol Biol* **251**(1), 104–115.
25. Bashford, D., Chothia, C., and Lesk, A. M. (1987) Determinants of a protein fold: Unique features of the globin amino acid sequences. *J Mol Biol* **196**, 199–216.
26. Flaherty, K. M., et al. (1991) Similarity of the three-dimensional structures of actin and the ATPase fragment of a 70-kDa heat shock cognate protein. *Proc Natl Acad Sci USA* **88**, 5041–5045.
27. Holm, L. and Sander, C. (1993) Protein structure comparison by alignment of distance matrices. *J Mol Biol* **233** (1), 123–138.
28. Rogozin, I. B., et al. (2005) APOBEC4, a new member of the AID/APOBEC family of polynucleotide (deoxy)cytidine deaminases predicted by computational analysis. *Cell Cycle* **4** (9), 1281–1285.
29. Chang, B. H. and Chan, L. (1998) Evolutionary analysis of RNA editing enzymes. *Methods* **15** (1), 41–50.
30. Philippe, H., et al. (2005) Heterotachy and long-branch attraction in phylogenetics. *BMC Evol Biol* **5**, 50.
31. Conticello, S. G., et al. (2005) Evolution of the AID/APOBEC family of polynucleotide (deoxy) cytidine deaminases. *Mol Biol Evol* **22** (2), 367–377.
32. Wang, Z. and Quiocho, F. A. (1998) Complexes of adenosine deaminase with two potent inhibitors: X-ray structures in four independent molecules at pH of maximum activity. *Biochemistry* **37** (23), 8314–8324.
33. Ireton, G. C., et al. (2002) The structure of *Escherichia coli* cytosine deaminase. *J Mol Biol* **315** (4), 687–697.
34. Kisker, C., et al. (1996) A left-hand beta-helix revealed by the crystal structure of a carbonic anhydrase from the archeon *Methanosarcina thermophila*. *EMBO J* **15**, 2323–2330.
35. Scott, J., Navaratnam, N., and Carter, C. (1999) Molecular modelling of the biosynthesis of the RNA-editing enzyme APOBEC-1, responsible for generating the alternative forms of apolipoprotein B. *Exp Physiol* **84** (4), 791–800.
36. Carter, C. W. Jr. (1998) In: Grosjean, H. and Benne, R. (eds), *Nucleoside Deaminase for Cytidine and Adenosine: Comparison with Deaminases Acting on RNA. In Modification and Editing of RNA*, ASM Press,Washington, D.C., pp. 591–596.
37. Jarmuz, A., et al. (2002) An anthropoid-specific locus of orphan C to U RNA-editing enzymes on chromosome 22. *Genomics* **79** (3), 285–296.
38. Hache, G., Liddament, M. T., and Harris, R. S. (2005) The retroviral hypermutation specificity of APOBEC3F and APOBEC3G is governed by the C-terminal DNA cytosine deaminase domain. *J Biol Chem* **280** (12), 10920–10924.
39. Newman, E. N., et al. (2005) Antiviral function of APOBEC3G can be dissociated from cytidine deaminase activity. *Curr Biol* **15** (2), 166–170.
40. Navarro, F., et al. (2005) Complementary function of the two catalytic domains of APOBEC3G. *Virology* **333** (2), 374–386.
41. Johansson, E., et al. (2002) Crystal structure of the tetrameric cytidine deaminase *from Bacillus subtilis* at 2.0 Å resolution. *Biochemistry* **41** (8), 2563–2570.
42. Chung, S. J., Fromme, J. C., and Verdine, G. L. (2005) Structure of human cytidine deaminase bound to a potent inhibitor. *J Med Chem* **48** (3), 658–660.
43. Teh, A. H., et al. (2006) The 1.48 Å resolution crystal structure of the homotetrameric cytidine deaminase from mouse. *Biochemistry* **45** (25), 7825–7833.
44. Kurtz, J. E., et al. (1999) New insights into the pyrimidine salvage pathway of *Saccharomyces cerevisiae*: Requirement of six genes for cytidine metabolism. *Curr Genet* **36** (3), 130–136.
45. Dance, G. S., et al. (2001) Identification of the yeast cytidine deaminase CDD1 as an orphan C → U RNA editase. *Nucleic Acids Res* **29** (8), 1772–1780.
46. Oka, K., et al. (1997) Tissue-specific inhibition of apolipoprotein B mRNA editing in the liver by adenovirus-mediated transfer of a dominant negative mutant APOBEC-1 leads to increased low density lipoprotein in mice. *J Biol Chem* **272** (3), 1456–1460.

47. Harris, R. S., Petersen-Mahrt, S. K., and Neuberger, M. S. (2002) RNA editing enzyme APOBEC1 and some of its homologs can act as DNA mutators. *Mol Cell* **10** (5), 1247–1253.
48. Petersen-Mahrt, S. K., Harris, R. S., and Neuberger, M. S. (2002) AID mutates *E. coli* suggesting a DNA deamination mechanism for antibody diversification. *Nature* **418** (6893), 99–103.
49. Smith, H. C., Wedekind, J. E., Xie, K., and Sowden, M. P. (2005) Fine-tuning of RNA functions by modification and editing in mammalian C to U editing. In: Grosjean H. (ed.), Topics in Current Genetics, Springer-Verlag, Berlin, pp. 1610–2096.
50. Zaim, J. and Kierzek, A. M. (2003) Domain organization of activation-induced cytidine deaminase. *Nat Immunol* **4** (12), author reply on p. 1153–1154.
51. Chen, S. C., et al. (2006) Crystal structure of a bifunctional deaminase and reductase from Bacillus subtilis involved in riboflavin biosynthesis. *J Biol Chem* **281** (11), 7605–7613.
52. Prochnow, C., et al. (2007) The APOBEC-2 crystal structure and functional implications for the deaminase AID. *Nature* **445** (7126), 447–451.
53. Murzin, A. G., et al. (1995) SCOP: A structural classification of proteins database for the investigation of sequences and structures. *J Mol Biol* **247** (4), 536–540.
54. Macbeth, M. R., et al. (2005) Inositol hexakisphosphate is bound in the ADAR2 core and required for RNA editing. *Science* **309** (5740), 1534–1539.
55. Kim, J., et al. (2006) Structural and kinetic characterization of *Escherichia coli* TadA, the wobble-specific tRNA deaminase. *Biochemistry* **45** (20), 6407–6416.
56. Huthoff, H., and Malim, M. H. (2007) Identification of amino acid residues in APOBEC3G required for regulation by human immunodeficiency virus type 1 Vif and Virion encapsidation. *J Virol* **81** (8), 3807–3815.
57. Sheehy, A. M., Gaddis, N. C., and Malim, M. H. (2003) The antiretroviral enzyme APOBEC3G is degraded by the proteasome in response to HIV-1 Vif. *Nat Med* **9** (11), 1404–1407.
58. Liu, B., et al. (2005) Regulation of Apobec3F and human immunodeficiency virus type 1 Vif by Vif-Cul5-ElonB/C E3 ubiquitin ligase. *J Virol* **79** (15), p. 9579–9587.
59. Almog, R., et al. (2004) Three-dimensional structure of the R115E mutant of T4-bacteriophage 2′-deoxycytidylate deaminase. *Biochemistry* **43** (43), 13715–13723.
60. Eliopoulos, N., Cournoyer, D., and Momparler, R. L. (1998) Drug resistance to 5-aza-2′-deoxycytidine, 2′,2′-difluorodeoxycytidine, and cytosine arabinoside conferred by retroviral-mediated transfer of human cytidine deaminase cDNA into murine cells. *Cancer Chemother Pharmacol* **42** (5), 373–378.
61. Greco, O. and Dachs, G. U. (2001) Gene directed enzyme/prodrug therapy of cancer: Historical appraisal and future prospectives. *J Cell Physiol* **187** (1), 22–36.
62. Liou, J. Y., et al. (2003) Assessment of the effect of phosphorylated metabolites of anti-human immunodeficiency virus and anti-hepatitis B virus pyrimidine analogs on the behavior of human deoxycytidylate deaminase. *Mol Pharmacol* **63** (1), 105–110.
63. Vauthey, J. N., et al. (1996) Arterial therapy of hepatic colorectal metastases. *Br J Surg* **83** (4), 447–455.
64. Ko, T. P., et al. (2003) Crystal structure of yeast cytosine deaminase. Insights into enzyme mechanism and evolution. *J Biol Chem* **278** (21), 19111–19117.
65. Liaw, S. H., et al. (2004) Crystal structure of Bacillus subtilis guanine deaminase: The first domain-swapped structure in the cytidine deaminase superfamily. *J Biol Chem* **279** (34), 35479–35485.
66. Kuratani, M., et al. (2005) Crystal structure of tRNA adenosine deaminase (TadA) from Aquifex aeolicus. *J Biol Chem* **280** (16), 16002–16008.
67. Bennett, M. J., Choe, S., and Eisenberg, D. (1994) Refined structure of dimeric diphtheria toxin at 2. 0 A resolution. *Protein Sci* **3** (9), 1444–1463.
68. Liu, Y. and Eisenberg, D. (2002) 3D domain swapping: As domains continue to swap. *Protein Sci* **11** (6), 1285–1299.
69. Moore, J. T., et al. (1994) Identification of a site necessary for allosteric regulation in T4-phage deoxycytidylate deaminase. *Biochemistry* **33** (8), 2104–2112.
70. Keefe, R. G., et al. (2000) A T4-phage deoxycytidylate deaminase mutant that no longer requires deoxycytidine 5′-triphosphate for activation. *J Biol Chem* **275** (17), 12598–12602.
71. Mora, P., Rubio, V., and Cervera, J. (2002) Mechanism of oligomerization of *Escherichia coli* carbamoyl phosphate synthetase and modulation by the allosteric effectors. A site-directed mutagenesis study. *FEBS Lett* **511** (1–3), 6–10.

72. Kashlan, O. B., et al. (2002) A comprehensive model for the allosteric regulation of mammalian ribonucleotide reductase. Functional consequences of ATP- and dATP-induced oligomerization of the large subunit. *Biochemistry* **41** (2), 462–474.
73. Bransteitter, R., et al. (2003) Activation-induced cytidine deaminase deaminates deoxycytidine on single-stranded DNA but requires the action of RNase. *Proc Natl Acad Sci USA* **100** (7), 4102–4107.
74. Soros, V. B., Yonemoto, W., and Greene, W. C., (2007) Newly synthesized APOBEC3G is incorporated into HIV virions, inhibited by HIV RNA, and subsequently activated by RNase H. *PLoS Pathog* **3** (2), e15.
75. Lo Conte, L., Chothia, C., and Janin, J. (1999) The atomic structure of protein–protein recognition sites. *J Mol Biol* **285** (5), 2177–2198.
76. Cheong, C. G., et al. (2004) Crystal structures of human bifunctional enzyme aminoimidazole-4-carboxamide ribonucleotide transformylase/IMP cyclohydrolase in complex with potent sulfonyl-containing antifolates. *J Biol Chem* **279** (17), 18034–18045.
77. Keegan, L. P., et al. (2004) Adenosine deaminases acting on RNA (ADARs): RNA-editing enzymes. *Genome Biol* **5** (2), 209.
78. Adams, M. J., et al. (1970) Structure of lactate dehydrogenase at 2–8 Å resolution. *Nature* **227** (5263), 1098–1103.
79. Adams, M. J., et al. (1970) The structure of the nicotinamide-adenine dinucleotide coenzyme when bound to lactate dehydrogenase. *J Mol Biol* **51** (1), 31–38.
80. Stockwell, G. R. and Thornton, J. M. (2006) Conformational diversity of ligands bound to proteins. *J Mol Biol* **356** (4), 928–944.
81. Richter, G., et al. (1997) Biosynthesis of riboflavin: Characterization of the bifunctional deaminase-reductase of *Escherichia coli* and *Bacillus subtilis*. *J Bacteriol* **179** (6), 2022–2028.
82. Cho, D. S., et al. (2003) Requirement of dimerization for RNA editing activity of adenosine deaminases acting on RNA. *J Biol Chem* **278** (19), 17093–17102.
83. Gallo, A., et al. (2003) An ADAR that edits transcripts encoding ion channel subunits functions as a dimer. *EMBO J* **22** (13), 3421–3430.
84. Chilibeck, K. A., et al. (2006) FRET analysis of *in vivo* dimerization by RNA-editing enzymes. *J Biol Chem* **281** (24), 16530–16535.
85. Valente, L. and Nishikura, K. (2007) RNA-Binding independent dimerization of ADAR and dominant negative effects of nonfunctional subunits on dimer functions. *J Biol Chem* **282** (22), 16054–16061.
86. Babbitt, P. C., et al. (1996) The enolase superfamily: A general strategy for enzyme-catalyzed abstraction of the alpha-protons of carboxylic acids. *Biochemistry* **35** (51), 16489–16501.
87. Farber, G. K. and Petsko, G. A. (1990) The evolution of alpha/beta barrel enzymes. *Trends Biochem Sci* **15** (6), 228–234.
88. Smith, H. C. (2007) Measuring editing activity and identifying C to U mRNA editing factors in cells and biochemical isolates. In: Gott J. M. (ed.), In: Methods in Enzymology, RNA Editing & Modification, Academic Press, New York, pp. 390–419.
89. Dance, G. S., et al. (2000) APOBEC-1 dependent cytidine to uridine editing of apolipoprotein B RNA in yeast. *Nucleic Acids Res* **28** (2), 424–429.
90. Yamanaka, S., et al. (1996) Hyperediting of multiple cytidines of apolipoprotein B mRNA by APOBEC-1 requires auxiliary protein(s) but not a mooring sequence motif. *J Biol Chem* **271** (19), 11506–11510.
91. Siddiqui, J. F., et al. (1999) Disproportionate relationship between APOBEC-1 expression and apolipoprotein B mRNA editing activity. *Exp Cell Res* **252** (1), 154–164.
92. Sowden, M., Hamm, J. K., and Smith, H. C. (1996) Overexpression of APOBEC-1 results in mooring sequence-dependent promiscuous RNA editing. *J Biol Chem* **271** (6), 3011–3017.
93. Sowden, M. P., Eagleton, M. J., and Smith, H. C. (1998) Apolipoprotein B RNA sequence 3′ of the mooring sequence and cellular sources of auxiliary factors determine the location and extent of promiscuous editing. *Nucleic Acids Res* **26** (7), 1644–1652.
94. Losey, H. C., Ruthenburg, A. J., and Verdine, G. L. (2006) Crystal structure of *Staphylococcus aureus* tRNA adenosine deaminase TadA in complex with RNA. *Nat Struct Mol Biol* **13** (2), 153–159.

95. Stephens, O. M., Yi-Brunozzi, H. Y., and Beal, P. A. (2000) Analysis of the RNA-editing reaction of ADAR2 with structural and fluorescent analogues of the GluR-B R/G editing site. *Biochemistry* **39** (40), 12243–12251.
96. Gillespie, J. H. (1984) The molecular clock may be an episodic clock. *Proc Natl Acad Sci USA* **81** (24), 8009–8013.
97. Gillespie, J. H. (1986) Natural selection and the molecular clock. *Mol Biol Evol* **3** (2), 138–155.
98. Hurst, L. D. (2002) The Ka/Ks ratio: Diagnosing the form of sequence evolution. *Trends Genet* **18** (9), 486.
99. Van Valen, L. (1973) A new evolutionary law. *Evol Theory* **1**, 1–30.
100. Petersen-Mahrt, S. K. and Neuberger, M. S. (2003) In vitro deamination of cytosine to uracil in single-stranded DNA by apolipoprotein B editing complex catalytic subunit 1 (APOBEC1). *J Biol Chem* **278** (22), 19583–19586.
101. Beale, R. C., et al. (2004) Comparison of the differential context-dependence of DNA deamination by APOBEC enzymes: Correlation with mutation spectra in vivo. *J Mol Biol* **337** (3), 585–596.
102. Yu, Q., et al. (2004) APOBEC3B and APOBEC3C are potent inhibitors of simian immunodeficiency virus replication. *J Biol Chem* **279** (51), 53379–53386.
103. Muramatsu, M., et al. (1999) Specific expression of activation-induced cytidine deaminase (AID), a novel member of the RNA-editing deaminase family in germinal center B cells. *J Biol Chem* **274** (26), 18470–18476.
104. Muto, T., et al. (2000) Isolation, tissue distribution, and chromosomal localization of the human activation-induced cytidine deaminase (AID) gene. *Genomics* **68** (1), 85–88.
105. Muramatsu, M., et al. (2000) Class switch recombination and hypermutation require activation-induced cytidine deaminase (AID), a potential RNA editing enzyme. *Cell* **102** (5), 553–563.
106. Flajnik, M. F. (2002) Comparative analyses of immunoglobulin genes: Surprises and portents. *Nat Rev Immunol* **2** (9), 688–698.
107. Saunders, H. L. and Magor, B. G. (2004) Cloning expression of the AID gene in the channel catfish. *Dev Comp Immunol* **28** (7–8), 657–663.
108. Zhao, Y., et al. (2005) Identification of the activation-induced cytidine deaminase gene from zebrafish: An evolutionary analysis. *Dev Comp Immunol* **29** (1), 61–71.
109. Sawyer, S. L., Emerman, M., and Malik, H. S. (2004) Ancient adaptive evolution of the primate antiviral DNA-editing enzyme APOBEC3G. *PLoS Biol* **2** (9), E275.
110. Rogozin, I. B., et al. (2007) Evolution and diversification of lamprey antigen receptors: evidence for involvement of an AID-APOBEC family cytosine deaminase. *Nat Immunol.* **8** (6), 647–656.
111. Hedges, S. B. (2001) In: Ahlberg, P. E. (ed.), Major Events in Early Vertebrate Evolution, Taylor & Francis, London, pp. 119–134.
112. Reeves, R. and Nissen, M. S. (1990) The A. T-DNA-binding domain of mammalian high mobility group I chromosomal proteins. A novel peptide motif for recognizing DNA structure. *J Biol Chem* **265** (15), 8573–8582.
113. Aravind, L. and Landsman, D. (1998) AT-hook motifs identified in a wide variety of DNA-binding proteins. *Nucleic Acids Res* **26** (19), 4413–4421.
114. Azumi, K., et al. (2003) Genomic analysis of immunity in a Urochordate and the emergence of the vertebrate immune system: "Waiting for Godot." *Immunogenetics* **55** (8), 570–581.
115. Wang, J., et al. (2006) Identification of a specific domain required for dimerization of activation-induced cytidine deaminase. *J Biol Chem* **281** (28), 19115–19123.
116. Liao, W., et al. (1999) APOBEC-2, a cardiac- and skeletal muscle-specific member of the cytidine deaminase supergene family. *Biochem Biophys Res Commun* **260** (2), 398–404.
117. Anant, S., et al. (2001) ARCD-1, an apobec-1-related cytidine deaminase, exerts a dominant negative effect on C to U RNA editing. *Am J Physiol Cell Physiol* **281** (6), C1904–C1916.
118. Taylor, J. S., et al. (2003) Genome duplication, a trait shared by 22000 species of ray-finned fish. *Genome Res* **13** (3), 382–390.
119. Christoffels, A., et al. (2004) Fugu genome analysis provides evidence for a whole-genome duplication early during the evolution of ray-finned fishes. *Mol Biol Evol* **21** (6), 1146–1151.
120. Matsumoto, T., et al. (2006) Expression of APOBEC2 is transcriptionally regulated by NF-kappaB in human hepatocytes. *FEBS Lett* **580** (3), 731–735.

121. Mikl, M. C., et al. (2005) Mice deficient in APOBEC2 and APOBEC3. *Mol Cell Biol* **25** (16), 7270–7277.
122. Teng, B., Burant, C. F., and Davidson, N. O. (1993) Molecular cloning of an apolipoprotein B messenger RNA editing protein. *Science* **260** (5115), 1816–1819.
123. Navaratnam, N., et al. (1993) The p27 catalytic subunit of the apolipoprotein B mRNA editing enzyme is a cytidine deaminase. *J Biol Chem* **268** (28), 20709–20712.
124. Chen, S. H., et al. (1987) Apolipoprotein B-48 is the product of a messenger RNA with an organ-specific in-frame stop codon. *Science* **238** (4825), 363–366.
125. Powell, L. M., et al. (1987) A novel form of tissue-specific RNA processing produces apolipoprotein-B48 in intestine. *Cell* **50** (6), 831–840.
126. Sowden, M. P., et al. (2002) The editosome for cytidine to uridine mRNA editing has a native complexity of 27S:identification of intracellular domains containing active and inactive editing factors. *J Cell Sci* **115** (Pt 5), 1027–1039.
127. Mehta, A. and Driscoll, D. M. (1998) A sequence-specific RNA-binding protein complements apobec-1 To edit apolipoprotein B mRNA. *Mol Cell Biol* **18** (8), 4426–4432.
128. Hersberger, M., Patarroyo-White, S., Arnold, K. S. and., and Innerarity, T. L. (1999) Phylogenentic Analysis of the Apolipoprotein B mRNA-editing Region. *J Biol Chem* **274** (49), 34590–34597.
129. Maris, C., et al. (2005) NMR structure of the apoB mRNA stem-loop and its interaction with the C to U editing APOBEC1 complementary factor. *RNA* **11** (2), 173–186.
130. Greeve, J., et al. (1993) Apolipoprotein B mRNA editing in 12 different mammalian species: Hepatic expression is reflected in low concentrations of apoB-containing plasma lipoproteins. *J Lipid Res* **34** (8), 1367–1383.
131. Nakamuta, M., et al. (1995) Alternative mRNA splicing and differential promoter utilization determine tissue-specific expression of the apolipoprotein B mRNA-editing protein (Apobec1) gene in mice. Structure and evolution of Apobec1 and related nucleoside/nucleotide deaminases. *J Biol Chem* **270** (22), 13042–13056.
132. Espinosa, R. 3rd., et al. (1994) Assignment of the gene encoding the human apolipoprotein B mRNA editing enzyme (APOBEC1) to chromosome 12p13.1. *Genomics* **24** (2), 414–415.
133. Lau, P. P., et al. (1994) Dimeric structure of a human apolipoprotein B mRNA editing protein and cloning and chromosomal localization of its gene. *Proc Natl Acad Sci USA* **91** (18), 8522–8526.
134. Fujino, T., et al. (1999) C → U editing of apolipoprotein B mRNA in marsupials: identification and characterisation of APOBEC-1 from the American opossum *Monodelphus domestica. Nucleic Acids Res* **27** (13), 2662–2671.
135. Kumar, S. and Hedges, S. B. (1998) A molecular timescale for vertebrate evolution. *Nature* **392** (6679), 917–920.
136. Yang, Y., Sowden, M. P., and Smith, H. C. (2001) Intracellular trafficking determinants in APOBEC-1, the catalytic subunit for cytidine to uridine editing of apolipoprotein B mRNA. *Exp Cell Res* **267** (2), 153–164.
137. Teng, B. B., et al. (1999) Mutational analysis of apolipoprotein B mRNA editing enzyme (APOBEC1). structure-function relationships of RNA editing and dimerization. *J Lipid Res* **40** (4), 623–635.
138. Navaratnam, N., et al. (1998) Escherichia coli cytidine deaminase provides a molecular model for ApoB RNA editing and a mechanism for RNA substrate recognition. *J Mol Biol* **275** (4), 695–714.
139. Navaratnam, N., et al. (1995) Evolutionary origins of apoB mRNA editing: Catalysis by a cytidine deaminase that has acquired a novel RNA-binding motif at its active site. *Cell* **81** (2), 187–195.
140. Turelli, P., et al. (2004) Inhibition of hepatitis B virus replication by APOBEC3G. *Science* **303** (5665), 1829.
141. Suspene, R., et al. (2005) Extensive editing of both hepatitis B virus DNA strands by APOBEC3 cytidine deaminases *in vitro* and *in vivo*. *Proc Natl Acad Sci USA* **102** (23), 8321–8326.
142. Mangeat, B., et al. (2003) Broad antiretroviral defence by human APOBEC3G through lethal editing of nascent reverse transcripts. *Nature* **424** (6944), 99–103.
143. Hulme, A. E., et al. (2007) Selective inhibition of Alu retrotransposition by APOBEC3G. *Gene* **390** (1–2), 199–205.
144. Nguyen, D. H., Gummuluru, S., and Hu, J. (2007) Deamination-independent inhibition of hepatitis B virus reverse transcription by APOBEC3G. *J Virol* **81** (9), 4465–4472.

145. Bogerd, H. P., et al. (2007) The intrinsic antiretroviral factor APOBEC3B contains two enzymatically active cytidine deaminase domains. *Virology* **364** (2), 486–493.
146. Bogerd, H. P., et al. (2006) Cellular inhibitors of long interspersed element 1 and Alu retrotransposition. *Proc Natl Acad Sci USA* **103** (23), 8780–8785.
147. Dang, Y., et al. (2006) Identification of APOBEC3DE as another antiretroviral factor from the human APOBEC family. *J Virol* **80** (21), 10522–10533.
148. Liddament, M. T., et al. (2004) APOBEC3F properties and hypermutation preferences indicate activity against HIV-1 *in vivo*. *Curr Biol* **14** (15), 1385–1391.
149. OhAinle, M., et al. (2006) Adaptive evolution and antiviral activity of the conserved mammalian cytidine deaminase APOBEC3H. *J Virol* **80** (8), 3853–3862.
150. Chen, H., et al. (2006) APOBEC3A is a potent inhibitor of adeno-associated virus and retrotransposons. *Curr Biol* **16** (5), 480–485.
151. Hurles, M. (2004) Gene duplication: The genomic trade in spare parts. *PLoS Biol* **2** (7), E206.
152. Murphy, W. J., Pevzner, P. A., and O'Brien, S. J. (2004) Mammalian phylogenomics comes of age. *Trends Genet* **20** (12), 631–639.
153. Glazko, G. V., Rogozin, I. B., and Glazkov, M. V. (2001) Comparative study and prediction of DNA fragments associated with various elements of the nuclear matrix. *Biochim Biophys Acta* **1517** (3), 351–364.
154. Jaillon, O., et al. (2004) Genome duplication in the teleost fish Tetraodon nigroviridis reveals the early vertebrate proto-karyotype. *Nature* **431** (7011), 946–957.
155. OhAinle, M., Emerman, M., and Malik, H. (2007) Evolution of antiretroviral Activity of APOBEC genes in primates. In: *Conference on Retroviruses and Opportunistic Infections*, Los Angeles, CA.
156. Sibley, C. G. and Ahlquist, J. E. (1984) The phylogeny of the hominoid primates, as indicated by DNA-DNA hybridization. *J Mol Evol* **20** (1), 2–15.
157. Mariani, R., et al. (2003) Species-specific exclusion of APOBEC3G from HIV-1 virions by Vif. *Cell* **114** (1), 21–31.
158. Jonsson, S. R., et al. (2006) Evolutionarily conserved and non-conserved retrovirus restriction activities of artiodactyl APOBEC3F proteins. *Nucleic Acids Res* **34** (19), 5683–5694.
159. Lochelt, M., et al. (2005) The antiretroviral activity of APOBEC3 is inhibited by the foamy virus accessory Bet protein. *Proc Natl Acad Sci USA* **102** (22), 7982–7987.
160. Hakata, Y. and Landau, N. R. (2006) Reversed functional organization of mouse and human APOBEC3 cytidine deaminase domains. *J Biol Chem* **281** (48), 36624–36631.
161. Bonvin, M., et al. (2006) Interferon-inducible expression of APOBEC3 editing enzymes in human hepatocytes and inhibition of hepatitis B virus replication. *Hepatology* **43** (6), 1364–1374.
162. Kinomoto, M., et al. (2007) All APOBEC3 family proteins differentially inhibit LINE-1 retrotransposition. *Nucleic Acids Res* **35** (9), 2955–2964.
163. Noguchi, C., et al. Dual effect of APOBEC3G on hepatitis B virus. *J Gen Virol* (2007) **88** (Pt 2), 432–432.
164. Bailey, J. A., et al. (2002) Human-specific duplication and mosaic transcripts: The recent paralogous structure of chromosome 22. *Am J Hum Genet* **70** (1), 83–100.
165. Kidd, J. M., et al. (2007) Population Stratification of a Common APOBEC Gene Deletion Polymorphism. *PLoS Genet* **3** (4), e63.
166. Bailey, J. A., Liu, G., and Eichler, E. E. (2003) An Alu transposition model for the origin and expansion of human segmental duplications. *Am J Hum Genet* **73** (4), 823–834.
167. Eichler, E. E. (2001) Recent duplication, domain accretion and the dynamic mutation of the human genome. *Trends Genet* **17** (11), 661–669.
168. Stankiewicz, P. and Lupski, J. R. (2002) Genome architecture, rearrangements and genomic disorders. *Trends Genet* **18** (2), 74–82.
169. Zhang, J. and Webb, D. M. (2004) Rapid evolution of primate antiviral enzyme APOBEC3G. *Hum Mol Genet* **13** (16), 1785–1791.
170. Lander, E. S., et al. (2001) Initial sequencing and analysis of the human genome. *Nature* **409** (6822), 860–921.
171. Bailey, J. A. and Eichler, E. E. (2006) Primate segmental duplications: Crucibles of evolution, diversity and disease. *Nat Rev Genet* **7** (7), 552–564.

172. Grover, D., et al. (2004) Alu repeat analysis in the complete human genome: Trends and variations with respect to genomic composition. *Bioinformatics* **20** (6), 813–817.
173. Scherer, S. E., et al. (2006) The finished DNA sequence of human chromosome 12. *Nature* **440** (7082), 346–351.
174. Wienberg, J. (2005) Fluorescence in situ hybridization to chromosomes as a tool to understand human and primate genome evolution. *Cytogenet Genome Res* **108** (1–3), 139–160.
175. Wedekind, J. E., et al. (2006) Nanostructures of APOBEC3G support a hierarchical assembly model of high molecular mass ribonucleoprotein particles from dimeric subunits. *J Biol Chem* **281** (50), 38122–38126.
176. Chiu, Y. L., et al. (2005) Cellular APOBEC3G restricts HIV-1 infection in resting CD4+ T cells. *Nature* **435** (7038), 108–114.
177. Chiu, Y. L., et al. (2006) High-molecular-mass APOBEC3G complexes restrict Alu retrotransposition. *Proc Natl Acad Sci USA* **103** (42), 15588–15593.
178. Kreisberg, J. F., Yonemoto, W., and Greene, W. C. (2006) Endogenous factors enhance HIV infection of tissue naive CD4 T cells by stimulating high molecular mass APOBEC3G complex formation. *J Exp Med* **203** (4), 865–870.
179. Cullen, B. R. (2006) Role mechanism of action of the APOBEC3 family of antiretroviral resistance factors. *J Virol* **80** (3), 1067–1076.
180. Esnault, C., et al. (2005) APOBEC3G cytidine deaminase inhibits retrotransposition of endogenous retroviruses. *Nature* **433** (7024), 430–433.
181. Kozak, S. L., et al. (2006) The anti-HIV-1 editing enzyme APOBEC3G binds HIV-1 RNA and messenger RNAs that shuttle between polysomes and stress granules. *J Biol Chem* **281** (39), 29105–29115.
182. Chelico, L., et al. (2006) APOBEC3G DNA deaminase acts processively $3' \rightarrow 5'$ on single-stranded DNA. *Nat Struct Mol Biol* **13** (5), 392–399.
183. Opi, S., et al. (2006) Monomeric APOBEC3G is catalytically active and has antiviral activity. *J Virol* **80** (10), 4673–4682.
184. Bishop, K. N., Holmes, R. K., and Malim, M. H. (2006) Antiviral potency of APOBEC proteins does not correlate with cytidine deamination. *J Virol* **80** (17), 8450–8458.
185. Guo, F., et al. (2006) Inhibition of formula-primed reverse transcription by human APOBEC3G during human immunodeficiency virus type 1 replication. *J Virol* **80** (23), 11710–11722.
186. Sheehy, A. M., et al. (2002) Isolation of a human gene that inhibits HIV-1 infection and is suppressed by the viral Vif protein. *Nature* **418** (6898), 646–650.
187. Yang, Z. (2005) The power of phylogenetic comparison in revealing protein function. *Proc Natl Acad Sci USA* **102** (9), 3179–3180.
188. Ortiz, M., et al. (2006) Patterns of evolution of host proteins involved in retroviral pathogenesis. *Retrovirology* **3**, 11.
189. Bogerd, H. P., et al. (2004) A single amino acid difference in the host APOBEC3G protein controls the primate species specificity of HIV type 1 virion infectivity factor. *Proc Natl Acad Sci USA* **101** (11), 3770–3774.
190. Mangeat, B., et al. (2004) A single amino acid determinant governs the species-specific sensitivity of APOBEC3G to Vif action. *J Biol Chem* **279** (15), 14481–14483.
191. Marin, M., et al. (2003) HIV-1 Vif protein binds the editing enzyme APOBEC3G and induces its degradation. *Nat Med* **9** (11), 1398–1403.
192. Schrofelbauer, B., Chen, D., and Landau, N. R. (2004) A single amino acid of APOBEC3G controls its species-specific interaction with virion infectivity factor (Vif). *Proc Natl Acad Sci USA* **101** (11), 3927–3932.
193. Xu, H., et al. (2004) A single amino acid substitution in human APOBEC3G antiretroviral enzyme confers resistance to HIV-1 virion infectivity factor-induced depletion. *Proc Natl Acad Sci USA* **101** (15), 5652–5657.
194. Schrofelbauer, B., et al. (2006) Mutational alteration of human immunodeficiency virus type 1 Vif allows for functional interaction with nonhuman primate APOBEC3G. *J Virol* **80** (12), 5984–5991.
195. Katzourakis, A., et al. (2007) From the cover: discovery and analysis of the first endogenous lentivirus. *Proc Natl Acad Sci USA* **104** (15), 6261–6265.

196. Wiegand, H. L., et al. (2004) A second human antiretroviral factor, APOBEC3F, is suppressed by the HIV-1 and HIV-2 Vif proteins. *EMBO J* **23** (12), 2451–2558.
197. Schumacher, A. J., Nissley, D. V., and Harris, R. S. (2005) APOBEC3G hypermutates genomic DNA and inhibits Ty1 retrotransposition in yeast. *Proc Natl Acad Sci USA* **102** (28), 9854–9859.
198. Guex, N. and Peitsch, M. C. (1997) SWISS-MODEL and the Swiss-PdbViewer: An environment for comparative protein modeling. *Electrophoresis* **18** (15), 2714–2723.

INDEX

RNA and DNA Editing: Molecular Mechanisms and Their Integration into Biological Systems, Edited by Harold C. Smith